A

61.00

68I

D0024597

61.00

68I

Paleontology:
The Record of Life

Paleontology:
The Record Of Life

COLIN W. STEARN
Logan Professor of Geology
McGill University

ROBERT L. CARROLL
Strathcona Professor of Biology
McGill University

with illustrations by
Linda Angeloff Sapienza

WILEY

John Wiley & Sons, Inc.
New York • Chichester • Brisbane • Toronto • Singapore

Cover Photo Credit
Neg./Trans. No. 277069
Courtesy Department of Library Services
American Museum of Natural History
Fossil Crinoids from the Chalk Beds of Kansas

Library of Congress Cataloging-in-Publication Data:

Stearn, Colin William, 1928-
  Paleontology: the record of life / Colin W. Stearn, Robert L. Carroll.
    p. cm.
  Bibliography:
  Includes index.
  ISBN 0-471-84528-0
  1. Paleontology. I. Carroll, Robert L., 1938- . II. Title
QE711.2.S74 1989
560—dc19   88-37526

Printed in the United States of America

10 9 8 7 6 5 4 3

To a pioneer in the study of the
record of life in Canada.

SIR WILLIAM DAWSON
(1820–1899)

Preface

This book is designed as a textbook for a first course in paleontology. Many students in such classes are geology majors, but majors in biology and other sciences also attend the courses we teach. We have written the book for a typical student of an elementary paleontology course who has a general science education up to the college level but little knowledge of geology and a high school preparation in biology. A short chapter on the geological time scale is included for students with no preparation in earth sciences. To make the text attractive to a wide range of students, we have not stressed classification and used as few technical terms as possible in describing the parts of organisms. We believe it is more important for the student to know how the organism lived than the names of its parts.

This book differs from most textbooks at this level in several important respects. First, it includes both descriptions of the organisms and their distribution in earth history and also extensive discussions of the principles of paleontology deduced from the fossil record. The descriptions of fossil animals and plants are presented first so that knowledge of the organisms can be applied to an understanding of the principles, which are given in the third part of the book, Lessons from the Record.

Second, the book integrates studies of invertebrates, vertebrates, and plants. The practice of dividing these aspects of paleontology into separate courses and teaching only invertebrate paleontology to geology majors, followed in many curricula, is unfortunate because the student needs to appreciate the interdependence and interaction of the various types of life in earth history. For example, no discussion of reefs can be complete without considering the role of algae, nor can any investigation of vertebrate evolution be complete without a knowledge of the plants that shaped the environment in which the animals evolved.

Third, we have emphasized the history of life by arranging the chapters in historical order rather than strictly taxonomically. This allows us to place chapters on Archean and Proterozoic life and the origin of the metazoans in a logical position. It also allows for discussions in the appropriate order of major events in the history of life, such as its origin, the evolution of skeletons, and times of adaptive radiations and extinctions.

Fourth, we have emphasized a biological approach to paleontology, for it is in this direction that we believe the science is moving. Information from fossil and living species is integrated to provide a unified view of the history of life. The interaction of extinct organisms with their environments, as reflected in the sediments in which they are preserved, is emphasized not only in the chapter on paleoecology but also throughout the book. Paleontology has long been studied within the geological sciences, but biologists too need to study paleontology, for species that happen to be extinct may be of as much importance to a biologist as species that happen to be alive.

In the first part of the book we introduce the methodology of phylogenetic systematics as a means of determining relationships between modern and extinct groups. We discuss current controversies in evolution in three chapters in the third section. The influence of continental motions on paleobiogeography is discussed in Chapter 18. Because many geologists see more fossils in thin sections of limestones than in the field, a chapter on fossils as builders of rocks is included. The rapidly developing field of trace fossil study is surveyed in the last chapter.

The complete text can be used for a full-year course in paleontology, or chapters can be selected for semester or quarter courses. Although the teacher may choose to use only certain chapters as a direct supplement to lecture material, other chapters covering the larger spectrum of paleontology can be used for additional readings or reference. Teachers who do not wish to follow the emphasis on earth history in their lectures can select chapters in the book that are appropriate to a more taxonomically based approach.

The fossil drawings have been specially and carefully prepared for this book by Linda Sapienza, under our immediate supervision. We are grateful not only for her artistic talents but also for the enthusiasm with which she worked with both of us. Other illustrations have been supplied by our colleagues, museums, and other institutes, without whose contributions we could not have produced this book. The contributions of all these are acknowledged in the captions.

During the preparation of the book we have been helped by the advice and critical reading of the chapters by many fellow paleontologists. Without in any way holding them responsible for the final result, we would like to thank specifically the following: Carlton Brett, Thomas Bolton, Murray Copeland, Brian Chatterton, Owen Dixon, Felix Gradstein, Roger Hewitt, Hans Hofmann, Andrew Knoll, Alfred Lenz, Pierre Lesperance, Kenneth McNamara, Alfonso Mucci, Guy Narbonne, Godfrey Nowlan, John Pojeta, Henry Reiswig, June Ross, and Gerd Westermann. In the preparation of a wide-ranging work such as this one, we have been impressed by the willingness of colleagues to share their time and knowledge. Without such sharing by the scientific community, textbooks could not be written.

Montreal   Colin W. Stearn
December 1988   Robert L. Carroll

Contents

Preface   *vii*

PART ONE: INTRODUCTION   *1*

1. Fossils   3
2. Time   14
3. Names, Relationships, and Classifications   21

PART TWO: THE FOSSIL RECORD   *37*

4. The Time of Microscopic Life   39
5. Origin and Diversification of the Metazoa   54
6. The Invertebrates Dominate the Seas: Acoelomates and Lophophorates   75
7. The Invertebrates Dominate the Seas II: Schizocoelomates and Enterocoelomates   116
8. Land Plants and Their Ancestors   169
9. Paleozoic Vertebrates   184
10. The Second Wave: Invertebrates of the Mesozoic and Cenozoic Eras   213
11. Evolution of Vertebrates During the Mesozoic   243
12. Cenozoic Vertebrates   277

PART THREE: LESSONS FROM THE RECORD   *315*

13. Biostratigraphy   317
14. Adaptation and Functional Morphology   327
15. The Mechanisms of Evolution   340

16.   Evolution in Earth History   362
17.   Paleoecology   374
18.   Paleobiogeography   402
19.   Fossils as Builders of Sedimentary Rocks   410
20.   Paleoichnology: The Study of Trace Fossils   422

Index   433

PART ONE

INTRODUCTION

Chapter 1

Fossils

INTRODUCTION

Scientists recognize more than 3 million different kinds of plants, animals, and microorganisms in the modern biota, but these organisms are only the most recent products of almost 4 billion years of evolution. An immensely greater variety of plants and animals lived in the past but are now extinct and recorded only by their fossil remains. The study of modern plants and animals convinced Darwin that all shared a common ancestry and were the products of a long period of evolutionary history. The history of life can be established only through a study of the fossil record. Fossils document the emergence of humans from apelike predecessors and the changes in plants and animals from their origin in the seas to their emergence into terrestrial environments.

The scientific study of fossil remains is paleontology. This book is an introduction to that study. It provides an outline of the history of life and discusses methods used in the study of fossils and the interpretation of fossils as the remains of formerly living organisms.

WHAT IS A FOSSIL?

Definition:

In the Middle Ages the word *fossil* applied to any specimen dug from the ground and included minerals and rocks. By the eighteenth century the word was used only for shells, bones, and other evidence of dead organisms found in rocks. The objects we now recognize as fossils are all remains of organisms that lived in the distant past and have been buried and preserved in sediments and sedimentary rocks. Fossils may be the whole organism or parts of it, impressions of the organism made as trails or footprints, and artifacts made by animals, such as excreta, burrows, and nests.

Although fossils are usually considered to be of prehistoric age, no strict lower limit to their age is recognized. However, studies of prehistoric human artifacts, such as arrowheads, are made by archeologists and those of the bones of ancient humans, by anthopologists.

Occurrence:

Geologists recognize three classes of rocks:

(1) Igneous rocks have cooled from a melted, or molten, condition. An example is lava.

(2) Sedimentary rocks consist of the products of the mechanical breakup and dissolution of other rocks when they are attacked by agents of weathering, such as rain, frost, wind, and ice. These products—mud, sand, gravel, and materials in solution—are carried by streams to lakes and the sea where they are deposited as shale, sandstone, and conglomerate or precipitated as limestone and evaporites.

(3) Metamorphic rocks are igneous and sedimentary rocks that have been changed in structure, texture, and composition by heat, pressure, chemically active fluids and gases, and stresses deep in the earth's crust.

Most sedimentary rocks contain fossils; those deposited in ancient shallow seas are rich in them (Fig. 1–1). Igneous rocks almost never contain fossils. Only rarely was an animal caught in a lava flow or overwhelmed by a fall of volcanic ash as were the inhabitants of Pompei in A.D. 79. Fossils may survive the drastic changes in texture that take place in the formation of metamorphic rocks, but these changes are often so intense that the fine structures are

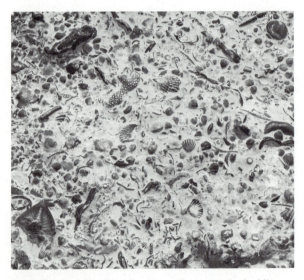

Figure 1-1   Fossils of marine invertebrates on the bedding surface of a limestone of Middle Silurian age. (Courtesy of Redpath Museum, McGill University.)

obscured and the fossils yield little information to the paleontologist.

WHY STUDY FOSSILS?

People are curious about the history of the human race, the environment in which they live, and the animals and plants that share it. Fossils are an element of the natural world that require explanation. To understand ourselves we must know where we have come from and how we are related to other living things.

Evolution:

The chain that connects organisms to one another through the ages of earth history is referred to as evolution. Some of the processes of evolution have been discovered through the comparative study of living organisms, and some can be tested in the laboratory. Scientists disagree on the extent to which mechanisms of evolution are revealed by the fossil record, but all agree that the course of evolution can be followed only through a study of fossils. Without a knowledge of fossils we would be unaware of the many species of extinct animals and plants that formed the links in the evolutionary chain.

Geochemistry:

Without knowledge of the history of life we cannot understand many processes that have taken place on the surface of the earth. Since almost the beginning of geological time, life processes such as respiration, photosynthesis, and secretion of skeletons have influenced the conditions on this planet. Life has controlled the chemistry of the oceans and atmosphere for the last 3.5 billion years. For example, calcium and carbon have been removed by many different organisms and oxygen has been added by the photosynthesis of plants. Changes in ocean chemistry, probably caused by microorganisms, resulted in the precipitation of uranium and iron ores during one interval of geological time about 2 billion years ago.

Dating Rocks:

Because life has changed with time, and the major features and many of the details of this change are now known, paleontologists can determine the relative age of most fossils with great precision. Many paleontologists, particularly those hired by oil companies and government services, are involved in this branch of paleontology, which is called biostratigraphy.

Environmental Reconstruction:

One of the principal objectives of geologists studying sedimentary rocks is the reconstruction of the environment in which the rocks were deposited. Correct reconstruction may result in the discovery of oilfields, deposits of salts, or base metals. As evidence of ancient environments the geologist has the structures, textures, and composition of the grains that make up sediments, and the fossils preserved in them. Organisms are far more sensitive to, and therefore diagnostic of, the environment in which they live than are the grains of sediment to the environment in which they were deposited. A study of the way of life and environmental requirements of organisms that became fossils therefore yields the most accurate information about the environment in which sediments were formed.

Paleogeography:

Reconstruction of the position of continents and oceans in the geological past is aided by a knowledge of fossil distribution. At present the distribution of animals and plants, both terrestrial and marine, is influenced by the pathways available for them to spread, and these pathways are dependant on the positions of land, sea, mountains, and lowlands—in other words, on the geography of the earth. In the past the earth's geography was different. Not only were the outlines of the continents unlike those of today, but the continents themselves were in different places on the globe. Many disciplines of the earth sciences contribute to the reconstruction of the geography of the past (paleogeography). Discontinuities in the distribution of modern organisms and fossils were among the first evidence that led geologists to suspect that continents in the past were in different positions from that of the present day.

Paleontology is the study of animals and plants that have lived in the past and their distribution in time and space. The information supplied by paleontological studies is essential to the understanding of earth history, evolution, paleogeography, and

sedimentation. All these topics are examined in greater detail in subsequent chapters.

HISTORY OF PALEONTOLOGY

Neanderthals placed fossils around their dead in burials, perhaps attributing to them a magic that would conquer death. These primitive people could hardly have been able to appreciate the symbolism inherent in fossils which, though dead, have achieved the immortality of preservation in stone. The true significance of the strangely shaped stones that fascinated cave dwellers was not generally appreciated for tens of thousands of years after these burial ceremonies.

Fossils are sufficiently abundant that they must have caused our distant ancestors to wonder how rocks could take regular forms resembling living animals and plants. The first to record their speculations were Greek philosophers. The oldest recorded statement that fossils are remains of once-living animals that have been entombed in rocks was made by Xanthos of Sardis in about 500 B.C. Aristotle (born 384 B.C.), who passed down the most comprehensive account of Greek science, suggested that fossils of fish were remnants of sea animals that swam into cracks in rocks and were stranded there. These early speculations were neither added to, nor improved upon, for the next 2000 years, during which Roman civilization succeeded the Greek and itself passed into the intellectual night of the Dark Ages.

During the Middle Ages and the Renaissance, fossils were illustrated and described as figured stones by learned doctors and clerics (1). The origin of these stones was a matter of dispute. Some ascribed them to the working of an arcane force within the earth (*vis plastica*), others to the works of the devil, and still others thought they had fallen from the sky as thunderbolts during storms or as falling stars. The starlike form of the plates of fossil crinoids (see Chapter 7) and the radial patterns on fossil corals suggested the sky as their source, and these objects were called "star stones" (*Lapis stellaris* or *Astroites stellis*) (Fig. 1–2). By the middle of the eighteenth century, the similarity of fossils to living organisms had convinced many naturalists that figured stones were the remnants of animals and plants that had somehow been preserved in rocks. The more perceptive understood that the sea must have once covered large areas of the continents in order for

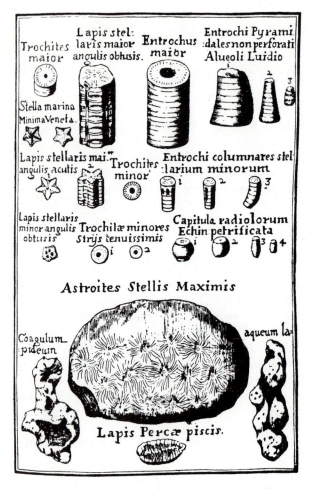

Figure 1-2   "Figured Stones" from the "Historia Lapidum Figuratorium Helvetiae" published in Venice, 1708. (*Source:* F. D. Adams, *Birth and Development of the Geological Sciences.*)

HOW ARE FOSSILS FORMED?

Some organisms are composed entirely of soft, unstable, organic compounds—largely of carbon, hydrogen and oxygen—but many have bodies reinforced with stable, hard compounds—commonly of calcium carbonate or calcium phosphate—that constitute a shell or skeleton. Paleontologists speak of unstable organic compounds as "soft parts" and shells and skeletons as "hard parts".

Life may be considered as the sum of the processes by which organisms keep inherently unstable carbon compounds together. When life ends, bacteria from within and without break down these complex compounds into simpler molecules, and eventually into stable compounds such as water, carbon dioxide, and methane. When these processes have been completed, no trace of the original soft parts of the organism remains. A visit to any marine or terrestrial environment in which organisms are abundant will illustrate the efficiency of decaying agents. Although organisms in such places are dying by the hundreds each day, no accumulation of their soft parts remains for more than a few days and usually no permanent accumulation of organic matter takes place. Living organisms build organic molecules from soils, gases, and water, and these molecules find their way back into the environment when the organisms that synthesized them decompose directly, or when the organisms feed a predator or scavenger, which in turn dies and decays. An organism composed entirely of soft parts can be preserved to enter the fossil record only if it is removed rapidly from predators, scavengers, and decayers.

Fortunately for the paleontologist, many marine organisms secrete shells of calcium carbonate, and many marine and terrestrial animals secrete skeletons of calcium phosphate. These are relatively resistant to postmortem destruction and are capable of survival while the soft parts of the organism decay. Almost the whole of the paleontological record consists of such hard parts that have persisted through millions of years, either in the state that they were secreted, or altered in various ways to be described.

Despite the relative resistance of these shells and bones to bacterial decay and their lack of appeal to predators and scavengers, only a small proportion of such hard parts formed by organisms are entombed in sediments to become part of the fossil

marine organisms to have been deposited as far inland as some fossils were found. Most attributed them to animals destroyed in the Biblical flood—a view still held by some fundamentalist Christians. The flood theory would not stand up to the evidence brought forward by paleontologists of the late eighteenth and early nineteenth centuries that the succession of sedimentary rocks reveals a progression of life forms and could not be the result of a single catastrophe. By the beginning of the nineteenth century the figured stones controversy was over, and knowledgeable naturalists agreed that fossils were remnants of organisms that lived in the distant past. By the middle of the century the groundwork had been laid for the recognition of fossils as evidence of evolution.

record. On the seashore where wave motion moves shells against each other and against mineral grains in the sand, abrasion reduces hard parts first to sandsized particles and then to mud. Other shells disintegrate into a muddy residue when the organic matrix that binds their crystals decays. Many marine organisms, including sponges, barnacles, gastropods, sea urchins, and a variety of "worms," bore into or abrade the surfaces of hard parts, reducing them eventually to unrecognizable particles. This process is called bioerosion and it may be so effective in tropical environments that it prevents any permanent accumulation of hard parts, even where animals that secrete shells and form skeletons thrive. In terrestrial environments many animals attack bones and horns left by dying animals. Although several groups of mammals, such as deer, moose, elk, and caribou, grow and discard large sets of antlers each year, complete sets can only rarely be found in forests where these animals live. The antlers are rapidly eaten by rodents and other small animals and bacteria infect them so that there is no permanent accumulation.

In order for hard parts to enter the permanent record, they must be protected from these destructive processes by being covered rapidly with sediment. Recycling of both soft and hard parts is the normal condition in most environments; removal of the remains from the recycling agents, enabling them to be preserved in the fossil record, is an exceptional event.

METHODS OF PRESERVATION

Unaltered Remains:

Few organisms are entombed and persist to the present day essentially complete with little change of form or composition. Ice Age mammoths found frozen in the tundra of North America and Siberia are so little changed that their flesh has been reported still palatable to wild animals, and their last meals remain undigested in their stomachs. A specimen of an Ice Age rhinoceros was preserved intact in a Polish oil seep because the petroleum penetrating the flesh had retarded bacterial decay. These are exceptionally rare specimens which are not representative of normal conditions of preservation.

After the soft parts decay, the hard parts of many organisms can be preserved essentially unaltered in sedimentary rocks. The shells of many marine organisms, once they have been entombed in sediments, are relatively stable. Some fossil shells formed millions of years ago cannot be distinguished from modern shells.

Permineralization:

Most shells and skeletons are not solid but contain canals and pores. The bones of land animals are generally highly porous, particularly in the central regions where marrow fills the voids when the animal is alive. When the soft parts decay and these porous materials are buried in sediments, the pores are filled with water that contains dissolved minerals. These waters precipitate calcium carbonate and silica in the pores of shells or bones, reinforcing and solidifying them. Almost all older fossils are subject to this process, which is called permineralization. Coniferous logs in the Petrified Forest of Arizona have been preserved through the infiltration of silica into the cellular structure of the wood, with so little disturbance that the microstructure appears to be that of a living tree. Yet these trees are millions of years old (Fig. 1–3).

Recrystallization:

After a shell has been covered with sediment and possibly also infiltrated with mineral matter, the crystals that were secreted to form the shell may be changed in form and size without changing in composition. In calcium carbonate shells, the original minute crystals often increase in size until the texture of the shell is a coarse calcite mosaic. The forms of the shell may remain faithfully defined, but the microstructure secreted by the organism has been destroyed. Such fossils are said to be recrystallized.

Replacement:

The water seeping through sedimentary rocks that contain fossil bones and shells may dissolve some of the hard parts and at the same time replace them with minerals that it carries in solution. The effect is to substitute another material for the original hard parts without changing the form of the shell or bone. Paleontologists cannot duplicate these processes, nor do they fully understand them, but the fossil record leaves no doubt that they occur. Commonly the replacing mineral is silica in its fine,

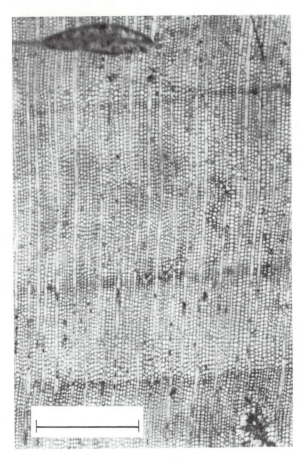

Figure 1-3 A thin section of silicified wood, showing the preservation of the cellular structure and the growth rings. Scale bar is 1 mm long. Locality unknown. (Courtesy of C. W. Stearn and Redpath Museum, McGill University.)

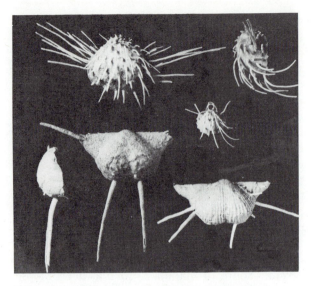

Figure 1-4 Silicified spiny brachiopods dissolved from Permian limestone. Permian of Texas. (Courtesy of G. A. Cooper and the Smithsonian Institution.)

crystalline form of the mineral chalcedony. Where silica has replaced fossils in limestones, the silicified fossils can be separated by dissolving the surrounding carbonate rock in dilute hydrochloric or acetic acid (Fig. 1–4). Delicate spines and fine structures on fossil shells, which would be broken off were the shells cracked out of the rock with the geologist's hammer, have been revealed by dissolving silicified fossils from the rock around them.

Many different minerals can replace fossils. In shales, shells are commonly replaced by the brass-colored mineral pyrite, to produce attractive specimens (Fig. 1–5). Pyrite is an iron sulfide mineral. Delicate structures of organisms replaced by pyrite have been preserved in the Hunsrück Slate in western Germany (Fig. 1–6) and may be revealed by passing X-rays through thin slabs of the slate. Iron

oxide in the form of the mineral hematite may also replace fossils. Under rare conditions, both the soft parts and hard parts of animals may be replaced by calcium phosphates (Fig. 1–7). About twenty different minerals are known to have replaced fossils.

Figure 1-5 Polished cross section of a Jurassic ammonite whose outer whorl is replaced by pyrite and whose inner whorls are not preserved. (Courtesy of C. W. Stearn and Redpath Museum, McGill University.)

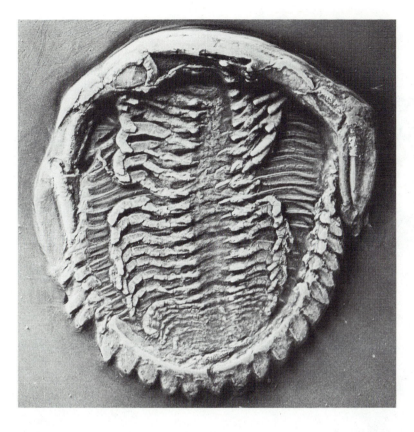

Figure 1-6   The ventral side of the Early Devonian trilobite *Rhenops* from the Hunsrück Slate, Germany. The skeleton has been replaced by pyrite, which has preserved the appendages, that are usually not preserved in unpyritized fossils. The specimen is about 3 cm across. (Courtesy of Jan Bergstrom and *Lethaia*, v. 17, p. 68.)

Molds and Casts:

Instead of replacing a fossil shell, water passing through the sediment may dissolve it completely. If the sediment packed around the shell has consolidated into rock before the shell is dissolved, a void will be formed. The walls of the void may preserve an impression of the face of the shell, which can be examined when the void is broken open (Fig. 1–8). These impressions are called molds. Their relief is opposite to that of the shell itself; knobs on the shell are represented by depressions and vice versa. If conditions change and the water deposits minerals into the void where a shell has been dissolved, the mold may be filled. The mineral deposit will take from the mold the external form of the fossil it has inherited but will have none of its internal structure. It is not a replacement of the shell but a cast that duplicates the shell after the original has been destroyed. The cast is a replica of the dissolved fossil. The mold is a negative impression of its surfaces.

Carbonization:

Plants are commonly preserved as thin films of carbon pressed on to bedding planes of sandstones and shales (Fig. 1–9). When a plant dies and is covered with sediment, the volatile constituents of the various carbohydrates of which it is composed disperse leaving a residue of coallike carbon—a black film which preserves the form of the fruit, leaves, or stem. The soft parts of marine animals have also been preserved in this way. Such preservation of soft parts rarely takes place, but when it does it is usually in environments where bacterial decay is inhibited by lack of oxygen. These environments are represented by black shales in the stratigraphic record.

Tracks and Trails:

Animals crossing the surface of sediment may leave behind footprints or impressions of parts of their body that are preserved when the sediment hardens

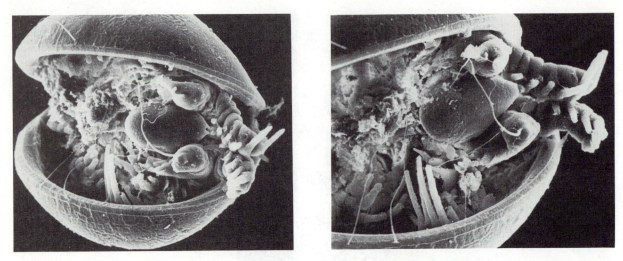

Figure 1-7   Late Cambrian ostracode, *Falites*, in which the soft parts have been preserved by replacement with phosphate. Four pairs of appendages and remains of the egg cases can be seen within the bivalved shell. Scanning electron micrograph. The specimen is about 0.16 mm across. (Courtesy of K. J. Muller and *Lethaia*, v. 12, p. 25.)

Figure 1-8   Preservation of fossils as molds. A. Gastropods in a Cenozoic sandstone. The long specimen is a mold of the exterior. In some specimens the interior of the shell has been filled with sand to produce a mold of the interior. B. Mold of the interior of a bivalve. Sediment has filled the space between the two valves producing a mold of their inner surfaces, and the shell has then been dissolved, leaving only the dark space around the internal mold. (Courtesy of Redpath Museum, McGill University.)

Figure 1-9   Preservation by carbonization. A. Dendroid graptolite *Rhabdinopora* from Silurian dolomite. B. Fernlike leaves of the Carboniferous plant *Pecopteris*. Both shown about natural size. (Courtesy of Redpath Museum, McGill University.)

to rock (Fig. 1–10). Animals living or feeding in soils or sediment leave behind burrows and tunnel systems that may become part of the fossil record. These types of fossils are called ichnofossils or trace fossils.

These seven ways in which fossils are preserved—unchanged, permineralized, recrystallized, replaced, carbonized, as molds and casts, and as ichnofossils—are those that the paleontologist is likely to encounter. Other mechanisms for preservation are confined to particular times and places and are considered in appropriate places in the following chapters.

BIASES OF THE FOSSIL RECORD

Soft-Bodied Organisms:

The variety of barriers to entry into the fossil record discussed in this chapter suggest that the

chances of preservation for an individual are extremely small and that all organisms do not have an equal chance. The soft parts of animals and plants, and whole organisms composed completely of soft organic matter, cannot be preserved unless some unusual condition inhibits bacterial decay. As a result, large groups of living invertebrate animals that do not secrete hard parts are unrepresented, or scarcely represented, in the fossil record. Jellyfish are examples of marine organisms that are abundant at present but have a poor fossil record. We know from impressions of their surfaces that they existed half a billion years ago, and we suspect that they have been abundant since then without leaving a significant record. Table 1–1 illustrates the direct relation between the paleontological representation of marine invertebrates and the possession of hard parts. All animals with hard parts have a good representation, and those without are poorly represented.

Figure 1-10   Trackway of the Permian amphibian *Laoporus* from the Grand Canyon. (Courtesy of the Smithsonian Institution.)

TABLE 1-1   Estimates of the total number of living marine invertebrate species described and of the number of these that may be fossilizable because they have hard skeletons.

	Total living species described	Total described species "fossilizable"
Foraminiferida and		
Radiolaria	20,000	20,000
Porifera (sponges)	2,250	1,750
Cnidaria (largely		
corals)	9,500	6,200
"Worms"	20,000	1,000
Bryozoa	3,000	3,000
Brachiopoda	225	225
Mollusca	54,300	53,100
Arthropoda	23,000	10,500
Echinodermata	6,000	5,000
Others	2,000	
TOTALS	140,275	100,775

Source: Modified from Valentine.

Sediment Burial:

Animals and plants that do not live in environments where sediments can bury them when they die (out of the reach of scavengers, bioeroders, and agents of decay) are unlikely to become fossils. Land animals and plants living in highlands where erosion rather than deposition is taking place are unlikely to be preserved unless they are washed into lowland swamps and lakes. The fossil record therefore entombs a far larger proportion of land organisms that lived in lowlands than those that lived in hills, high plains, or mountains.

Collection Biases:

Deposits of the past are not equally available to paleontologists for collection of fossils. Most of the marine sediments whose fossils have been studied were laid down in shallow seas that flooded the interiors of continents. Although organisms are assumed to have existed in the abyssal depths of oceans, where sediment accumulates to entomb them, most of these deposits remain at abyssal depths and are difficult to sample except by expensive techniques of deep-sea drilling. As a result our knowledge of shallow marine faunas far exceeds that of abyssal faunas.

Our knowledge of the paleontology of Europe and North America is more extensive than that of the southern continents because more scientists live in the northern hemisphere. Therefore, a fossil organism from the south is less likely to come to the attention of paleontolgists than one from the north. Our knowledge of the paleontology of large areas of the earth is limited because they have few outcroppings of sedimentary rocks. Tropical areas beneath rain forests and polar areas beneath ice are examples of such limitations.

Inadequacy or Bias:

Citing these deficiencies, some have claimed that our knowledge of the fossil record can never be adequate to supply a detailed history of life on our planet. The adequacy of the fossil record to supply answers depends on the nature of the questions we ask. The record is effective in solving some problems and is unable to solve others.

Paleontologists prefer to describe the record as biased rather than inadequate. A biased record of events has come to mean one that is slanted by the observer's prejudices. The fossil record is biased against the recovery of soft-bodied organisms, highland organisms, and abyssal organisms, but, if these prejudices are recognized, paleontologists can make corrections for them in order to arrive at a balanced view of the life of the past.

REFERENCE

1. Adams, F. D., 1936, The birth and development of the geological sciences: Philadelphia, Williams and Wilkins, 506 p.

A SELECTION OF RECENT TEXTBOOKS AND REFERENCE BOOKS ON PALEONTOLOGY

Boardman, R. S., Cheetham, A. H., and Rowell, A. J., 1987, Fossil Invertebrates: Palo Alto, Blackwell Scientific Publications, 713 p.

Carroll, R. L., 1987, Vertebrate paleontology and evolution: New York, Freeman, 698 p.

Clarkson, E. N. K., 1986, Invertebrate paleontology and evolution, 2nd ed: London, Allen & Unwin.

Fairbridge, R. W., and Jablonski, D., 1979, Encyclopedia of paleontology. Encyclopedia of earth sciences, v. 7: Stroudsburg, Dowden, Hutchinson and Ross, 886 p.

Gregory, W. K., 1957, Evolution emerging: New York, Macmillan Co.

Hildebrand, M., 1988, Analysis of vertebrate structure: New York, Wiley.

Lane, N. G., 1986, Life of the past. 2nd ed: Columbus, Ohio, Merrill, 326 p.

Murray, J. W. (ed.), 1985, Atlas of invertebrate macrofossils: New York, Wiley, 241 p.

Nield, E. W., and Tucker, V. C. T., 1985, Paleontology, an Introduction: Oxford, Pergamon Press, 178 p.

Piveteau, J., 1952–1959 (ed.). Traité de Paléontologie, 7 vols Masson, Paris.

Raup, D. M., and Stanley, S. W., 1978, Principles of paleobiology, 2nd ed.: San Francisco, Freeman, 481 p.

Romer, A. S., and Parsons, T. S., 1977, The vertebrate body: Philadelphia, Saunders.

Tasch, P., 1980, Paleobiology of the invertebrates, 2nd ed.: New York, Wiley, 975 p.

Young, J. Z., 1981, The life of vertebrates: Oxford, Oxford Univ. Press (Clarendon).

Ziegler, B., 1983, Introduction to paleobiology: general paleontology: New York, Wiley, 225 p.

Chapter 2

Time

Paleontology is an historical science. From fossils paleontologists reconstruct long-dead organisms and the world in which they lived. Unlike chemists, physicists, or biologists, they cannot experiment with the materials they study but, like all scientists, they examine the products of past events and try to reconstruct what happened. Chemists can combine solutions, whose properties are known, repeatedly under controlled conditions in order to understand the product of their reaction. The events that paleontologists study are unique in time. They do not recur under exactly the same conditions in earth history because the earth and its inhabitants have progressed along an irreversible course. Paleontologists observe in the fossil record the interaction of organisms with one another and with the physical environment. Their behavior can be deduced from the behavior of similar living organisms. Although paleontology does not deal with repeatable events, it shares with all sciences the gathering of data, its interpretation according to recurring patterns, and the formulation of general laws or relationships. These describe the manner in which life has changed through the millennia.

Events of the past must be placed in a sequence and calibrated with a time scale if their direction and rates of change are to be assessed. The succession of the beds that contain fossils must be determined, as well as their ages in years. Since the beginning of the twentieth century geologists have developed methods of dating many types of rocks in years.

DATING BY RADIOACTIVE ISOTOPES

Principles:

The regularity of decay of radioactive atoms makes dating of ancient rocks possible. The atoms change to daughter atoms by emission of alpha particles and beta rays, altering the number of protons and neutrons in the nucleus at rates that are unaffected by chemical or physical conditions. Once deposited in a rock, a mineral containing such radioactive atoms continues to accumulate the products of its change (daughter products), regardless of heat, pressure, or its chemical bonding to other atoms. If the rate of change can be measured in the laboratory and the amount of the parent and daughter atoms can be obtained by analysis, a relatively simple calculation yields the time at which the parent was emplaced and began to accumulate products. Since they came into existence, parent elements have been generating daughter products. The age measured is the time at which the parent, separated from its previously accumulated daughter products, began to accumulate them again in a new environment.

Materials Used in Dating:

The first minerals used to calculate ages were those containing uranium. The element uranium exists in two forms of different atomic weights that are called isotopes. Each of these decays through a series of unstable daughter products to a stable isotope of the element lead. Thorium, which is associated with uranium in most natural deposits, also decays to an isotope of lead different from those produced by uranium isotopes. Determination of the rates of decay, and analysis for parent and daughter products of the two uranium isotopes allow the geologist not only to determine the age of the mineral, but also to correct for loss of accumulated lead that may have taken place.

Elements other than uranium and thorium also have radioactive isotopes and have been used in dating. Potassium−40, an isotope that is widespread in small quantities in the earth's crust, changes into an isotope of calcium and one of argon. The amount of argon accumulated from this transformation is measured. The element rubidium, commonly associated with potassium in minerals, has a radioactive isotope which decays to an isotope of strontium. Several rare earth elements, such as samarium and neodymium, have also been used to measure the age of rocks (Table 2−1).

Rates of Radioactive Decay:

In order to be useful for dating rocks millions of years old, transitions must take place at a rate low enough that even very old rocks retain both parent and daughter isotopes, that is, the decay process must still be continuing. For short-lived isotopes the radioactive decay is completed and the radioactive "clock" "runs down" before a significant interval of geological time has passed. Rates of decay are measured by an interval called the half-life. During the first half-life interval, one-half of the parent atoms change into daughter products; in the

TABLE 2−1   RADIOACTIVE ISOTOPES USED IN DATING ROCKS AND THEIR HALF-LIVES

Parent Element	Atomic Wt. of radioactive isotope	Ultimate daughter element	Atomic Wt. daughter	Half life
Uranium	238	Lead	206	4.5×10^9
Uranium	235	Lead	207	0.7×10^9
Thorium	232	Lead	208	1.4×10^{10}
Rubidium	87	Strontium	87	5.0×10^{10}
Potassium	40	Argon	40	1.3×10^9
Samarium	147	Neodymium	143	1.0×10^{11}

next interval, one half of the remaining half, and so on (Fig. 2−1). Most of the half-lives of the commonly used decay series are of the order of billions of years (Table 2−1).

These measurements have shown that the earth is about 4600 million years old; that complex animals first arose about 700 million years ago; that animals possessing skeletons and shells first left their hard parts behind in the sea about 570 million years ago; that dinosaurs walked the earth between 200 and 66 million years ago; and that since then mammals have dominated the fauna of the land. These dates for salient events in life history are representative of the type of information available from

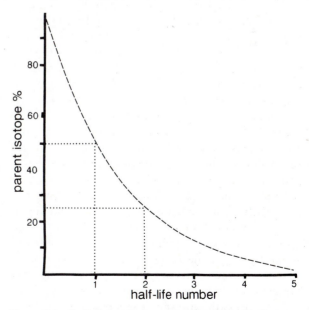

Figure 2-1   Radioactive decay curve to show the significance of the half-life. In one half-life interval, 50 percent of the parent changes into the daughter isotope; in another interval, a further 50 percent of the remainder changes, and so on.

the work of geochronologists, as those who date rocks are called. By dating many thousands of rocks annually, they provide paleontologists with a time scale for the history of life

ROCKS THAT CAN BE DATED

Sedimentary Rocks Are Difficult:

Not all rocks contain minerals that are radioactive. Unfortunately for paleontologists, the sedimentary rocks that contain the record of life rarely contain radioactive minerals suitable for dating. Most sedimentary rocks are the products of the chemical and physical destruction by the atmosphere and organic agencies (weathering) of other rocks. Sands and silts inherit, from the rocks that were weathered, particles containing radioactive minerals and daughter products that they have accumulated. The ages of such grains are the ages of the rocks weathered to produce the sediment, not of the time of deposition of the grains, because the processes producing the grains do not separate parent isotopes from their daughter products. Some sediments, like limestone, are deposited biochemically or chemically from seawater and are ultimately the product of weathering by dissolution. Unlike sedimentary rocks composed of grains weathered from other rocks, their mineral constituents are formed when they are deposited. However, the minerals are not, with one or two exceptions, composed of atoms that have radioactive isotopes. The exceptions, sediments that can be dated by their radioactive isotopes, are some highly organic shales that contain uranium, some sandstones composed of the potassium-rich mineral glauconite, shales in which the strontium isotopic composition has been homogenized during their consolidation, and some evaporites rich in potassium salts.

Most of the rocks that contain radioactive isotopes sufficient for dating are igneous and metamorphic rocks. Igneous rocks cool from a molten liquid in which the radioactive parent atoms are separated from the daughter products that accumulated before the melting. The processes that metamorphose sedimentary and igneous rocks, such as heat, pressure and shearing stress, do not affect the decay of radioactive isotopes, but the disruption of the rock may physically disperse accumulating daughter products. The dispersal of the gas argon resets the clock of the potassium–argon method so

that analyses yield the date of metamorphism, not the date of formation for the rock that was metamorphosed.

Using Igneous Rocks to Date Sedimentary Rocks:

The general lack of direct dating methods for sedimentary rocks means that their age must commonly be related to that of igneous rocks. If the extrusion of a submarine lava flow or the fall of ash from a volcanic eruption interrupted the accumulation of sediment in a marine basin, then the sediment below it must be slightly older than the date determined from the flow or ash bed, and the sediment above must be slightly younger. If sedimentation is essentially continuous, the beds above and below can be accurately dated by the flow or ash. In a sedimentary succession interrupted by repeated lava flows, the ages of most of the sedimentary beds can be accurately determined.

More commonly, the age of a sediment must be determined by the relation between igneous rocks that intrude a sedimentary succession and igneous rocks that are overlain unconformably by sedimentary rocks. The relationships are shown in Figure 2–2. Sedimentary rocks that are intruded by an igneous body are older than that body, and those that overlie the eroded surface of an igneous body are younger than that body.

RELATIVE TIME SCALE

Development of the Scale:

By the time geologists had developed methods of dating rocks by radioactivity, they had already outlined the physical and biological history of the earth. This was accomplished without knowledge of the age of the rocks in years.

The first classifications of sedimentary rocks were based on the principle of superposition, which simply states that the oldest of a succession of layers is at the base and the youngest is at the top. In central Europe the lowest rocks that can be seen are metamorphic and igneous rocks that were called Primary in the eighteenth century. Layered sedimentary rocks that were called Secondary overlie these rocks, and over them are poorly consolidated sands and gravels called Tertiary and Quaternary. When additional layers were found that overlie the Pri-

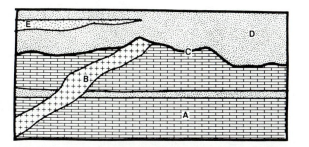

Figure 2-2   Use of igneous rocks that can be dated in years to date sedimentary rocks. The rocks labeled B and E are igneous rocks. Because B intrudes A, it must be younger than A. After the intrusion, A and B were eroded to a surface labeled C. Sediments of D were deposited over C to produce an unconformity. These sediments contain a lava flow, E, that may record the time of their deposition.

mary and are overlain by the Secondary, these were called the Transition. Fossils were found in many of these layers, and by 1825 the names Transition, Secondary, Tertiary and Quaternary were being replaced by terms that refer to life form: Paleozoic, Mesozoic, and Cenozoic (ancient, middle and recent life, respectively). Near the beginning of the nineteenth century William Smith (and, independently, Georges Cuvier and Alexandre Brongniart) discovered that the fossils found in sedimentary layers always occur in the same order so that certain fossils always characterize certain layers. They had discovered that life changed in time, and that these changes are recorded in successive sedimentary layers. Once the succession of fossils was known, isolated fossil-bearing layers, whose positions with respect to other layers were unknown on the basis of superposition, could be determined as either younger or older than other fossil-bearing layers.

As knowledge increased, the first divisions were broken up into smaller units and these units fitted into a continuous succession on the basis of their fossils. The smaller divisions of the Paleozoic, Mesozoic, and Cenozoic were called systems. In Britain the lower part of the Paleozoic was divided into the Cambrian, Ordovician, and Silurian systems, which were based on rocks in Wales and named after ancient tribes that lived there. The upper part of the Paleozoic was divided into the Devonian System, named from the English county; the Carboniferous System, which included and was named after the coal-bearing beds of Britain; and the Permian System, named from a province of Russia. The names of the divisions of the Mesozoic were based on the

threefold nature of the lower part in Germany (Triassic), the outcrops in the Jura Mountains between France and Switzerland (Jurassic), and the chalk (Latin, *creta*) beds of the English Channel (Cretaceous).

The divisions of the Cenozoic were first defined on the basis of the percentage of living (as opposed to extinct) molluscan species in the rocks. The Eocene was defined as the sedimentary rocks whose molluscan fauna included 3.5 percent molluscan species that are still alive today; the Miocene as the beds whose molluscs included 17 to 18 percent living forms; and the Pliocene, more than 35 percent. Further subdivisions were added to make the six listed in Figure 2–3, and recognizing divisions by percentage of surviving species was abandoned. These divisions of the Cenozoic are not as large as systems and are called series. The younger four (Miocene, Pliocene, Pleistocene, and Holocene) are recognized as comprising the Neogene System and the older three (Paleocene, Eocene, Oligocene), the Paleogene.

Periods Are Based on Systems:

As the succession of systems was being established in Europe, the names were also applied to sedimentary rocks in other parts of the world. For example, the term Silurian System was used in North America for rocks containing fossils similar to those of the Silurian rocks of Britain; the similarity of fossils demonstrated that the two had been deposited during the same time interval. The names of the systems that originally referred only to local successions of sedimentary layers came also to have time significance. The list of successive systems had become a time scale by which sedimentary successions all over the world could be related. The rocks deposited during the interval of the Silurian System in Wales, and identified as such by their fossils, were then said to be deposited during the Silurian Period. Specific systems are now recognized as the succession of rocks deposited anywhere in the world during a specific interval of geological time (a period), and the periods are defined by the boundaries of the system in the area where they were first described but the boundaries may be extended by information from other areas. Although this definition appears to be circular, it is not, because the areas where the systems were first described are unique standards called type sections.

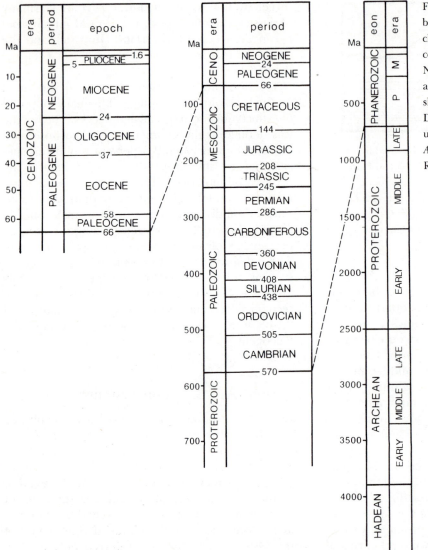

Figure 2-3   Time scale used in this book. The scale of the columns changes by a factor of 5 between each column. The last two epochs of the Neogene, the Holocene (last 0.1 Ma) and the Pleistocene (1.6 Ma), are too short to be shown clearly on this scale. Dates of boundaries between time units are from the *Decade of North American Geology Time Scale*. (*Source:* A. R. Palmer, *Geology*, v. 11, p. 503–504.)

This time scale was a relative one. That is, for all the nineteenth and part of the twentieth century the lengths of the various periods were unknown. Throughout this time most scientists thought that the total age of the earth was only one percent of what isotopic dating has now shown it to be. Despite this major misconception, the main outline of the earth's physical history and the history of life were established, but the time scale of events remained uncalibrated. No one appreciated how slowly the events had actually taken place or how long the history really was.

Although the succession of life-forms that allows the systems to be recognized worldwide is now ascribed to the processes we call evolution, the succession was recognized 60 years before the publication of Darwin's *The Origin of Species*. The establishment of the succession of life did not depend on Darwin's proposal of a mechanism to explain the change but on a simple observable fact; older beds contain fossils that are different from those in younger beds.

MAGNETIC REVERSALS

The earth's magnetic field is modeled as a bar magnet whose poles approximate the geographic poles of the planet. For over a hundred years geophysicists have known that this field has at times been reversed; that is, the end of the compass needle

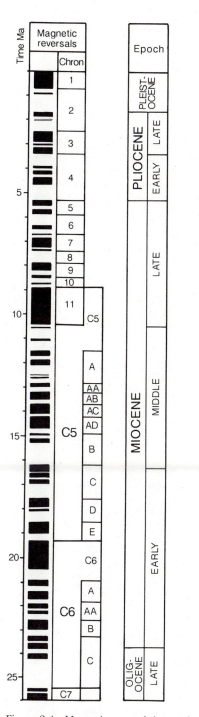

Figure 2-4 Magnetic reversal time scale for late Oligocene time to present. Intervals of normal polarity are represented by the black pattern. The intervals of magnetic polarity (chrons) are given in numbers and letters. (*Source:* W. Berggren and others, *Bulletin of the Geological Society of America*, v. 96, p. 1407–1418.)

that now points north would then have pointed south. Today's condition of the earth's field is said to be normal, the changed position reversed. Discovery of field reversal was possible because sedimentary and igneous rocks containing minerals susceptible to magnetic fields (mainly those containing large proportions of iron) preserve a record of the earth's field. When grains of these minerals in a sediment are deposited from a fluid, or crystals of them are frozen into place in a lava, they are aligned with the magnetic field around them and preserve a record of that field after it has changed. The phenomenon is called remanent magnetism. By dating the remanent magnetism of many samples, geologists have determined that the earth's field reversed itself frequently in some intervals of the past and at other times remained relatively stable. As more detailed work has been done, intervals that were once considered to represent a time of magnetic stability have been shown to be divided by several polarity changes. The larger intervals defined by magnetic polarity have been called chrons and the smaller ones events. Not many reversals of polarity have been dated isotopically. The relatively few chrons that have been dated in years are used through interpolation and extrapolation to calibrate the record of polarity reversals preserved in the crust of the ocean floor as it is formed at mid-ocean ridges and moved away. Polarity changes have been defined in this way back to Late Jurassic time. Figure 2–4 shows the polarity reversal record for Neogene time, which includes chrons 1 to 7. Forty eight polarity changes are recognized in this interval of 26 million years.

The polarity of a sample may help to place it accurately in the time scale if the approximate position is known from paleontological evidence. Unless a relatively precise age is available, knowledge of the normal or reversed polarity is useless, because polarity change has been so common in earth history.

STANDARD TIME SCALE

Events in earth history are now universally related to a standard time scale (Fig. 2–3). By dating igneous rocks that are associated with sedimentary rocks bearing fossils characteristic of the various systems, geologists are now able to determine the ages of the boundaries between periods of the relative time scale. The scale that once could be used only to place sedimentary layers in their relative position in earth history, that is, as older or younger than other layers, can now be used to estimate how many years ago a fossil-bearing layer was deposited. For example, if a sedimentary deposit contains a fossil that is known to occur only at the top of the Silurian System, the deposition of that bed can be dated as about 410 million years ago, even though no igneous rock near the layer can be isotopically dated. The end of Silurian time has been dated as 408 million years ago by the relationship between igneous rocks and Silurian layers bearing similar fossils at some other locality.

Subsequent discussion of life history in this book will be in terms of this standard time scale. The dates of the boundaries between the periods are constantly under review as new age determinations are made; the time scales in books may differ by a few million years. The methods of using fossils to divide sedimentary successions and determine their ages are considered in greater detail in Chapter 13.

SUGGESTED READINGS

Berggren, W. A., Kent, D. V., Flynn, J. J., and Van Couvering, J. A., 1985, Cenozoic geochronology. Geological Society of America Bulletin, v. 96, p. 1407–1418.

Berry, W. B. N., 1987, Growth of a prehistoric time scale, Revised ed.: Boston, Blackwell Scientific Publications, 202 p.

Harland, W. B. and others, 1982, A geologic time scale: Cambridge, England. Cambridge Univ. Press.

Faul, H. A., 1978, A history of geologic time: American Scientist, v. 66, p. 159–165.

Faure, G., 1977, Principles of isotope geology: New York, Wiley, 464 p.

Palmer, A. R., 1985, The decade of North American Geology 1983 time scale: Geology, v. 11, p. 503–504.

Chapter 3

Names, Relationships, and classification

NAMES

Botanists and zoologists give the various kinds of plants and animals Latin names. These names allow scientists of different countries to understand one another despite the fact that the common names of animals and plants are different in each language. Latin was the language of science and the church through the Middle Ages into the Renaissance and many organisms were given Latin names in the treatises of these times. Carl von Linné (1707–1778), a Swedish naturalist who published many editions of a book called *Systema Naturae* in the middle part of the eighteenth century, systematized new and old names into this catalogue of plants, animals, and fossils. Linné (or as he is sometimes called, Linneus, his Latinized name) used two Latin names for each organism. The first name indicated the type of animal or plant being named and was somewhat analoguous to our family names but, like Chinese family names, it was placed first. He gave all cats the first name *Felis*, the Latin for cat. Because this name identifies the type of animal, it is called the generic name (Latin *genus* = race or kind, plural is genera). The various cats were distinguished from one another by different second names.

Felis leo was the name given to the lion; *Felis domestica*, to the house cat; *Felis tigris*, to the tiger, and so on. Lions, tigers, and house cats were recognized as various kinds, or species, of cat and the second name was therefore referred to as the species name (plural = species), or sometimes as the trivial name. This name may be a Latin adjective (*domestica*) modifying the genus name which is a noun, or it may be another noun (*leo, tigris*), but in either case the species name does not start with a captial letter as does the generic name. In addition to hundreds of living animals and plants, Linné named in his *Systema Naturae* many fossils common in northern Europe. Such was the logic of Linné's scheme of uniform genus and species names that within a generation it was accepted by biologists throughout the world and has been systematically applied to both living and fossil organisms.

Describing and Naming Newly Discovered Organisms:

Since the time at Linné, biologists and paleontologists have been describing new species, genera, families, and so on, as they have recognized different forms of life. The description of newly recognized species is one of the most important activities of the research paleontologist. We are far from documenting all the forms of life that have been preserved, and new fossil species are discovered every day. Descriptions of new fossils are published in scientific journals. After a collection of fossils is made, the paleontologist determines whether the fossils present have been previously described by consulting these journals which together constitute what is called the paleontological literature. This literature, which has accumulated over the past 200 years, is so vast that no one can be familiar with it in its entirety, and most paleontologists therefore specialize in mastering small fractions of it. Most are experts in a single group of fossil organisms, and some are specialized to the extent that they are familiar only with those organisms for a particular interval of the geological past.

If the paleontologist finds that the fossil specimens belong to a species that has not yet been described, he or she may describe it for the first time and give it a name. Usually a new species can be assigned to a genus that has already been described, but if the paleontologist believes that no previously described genus can include the new species, he or she may decide that a new genus must also be established. The naming and describing of newly discovered forms of life are governed by Codes of Botanical and Zoological Nomenclature, which are administered by international commissions. These codes are complex documents but are based on the simple concept that the person who first recognizes a new form has the right to name it, and that the specific name once given, cannot be changed unless the proposal is proved incorrect. In order to help people search the literature, the name of the describer, the species author, is usually attached to the binomial name. For example, the house cat is *Felis domestica* Linné.

Biologists have found that the divisions originally proposed for classifying organisms, such as family, order and class, are insufficient. Since Linné's time subdivisions have been used so that we now have subspecies, subgenera, superfamilies, superorders, and so on. To illustrate how this set of terms is used to place an organism in relation to similar organisms, consider the classification of a familiar animal, man's best friend. All domestic dogs, despite the variety produced by manipulative breeding, belong to the species *familiaris* of the genus *Canis*. The genus also includes some species of wild dogs such as the jackal, *Canis aureus*. The genus *Canis* is united with other such genera as *Lupus* (the wolves) and *Vulpes* (the foxes) in the family Canidae. The Canidae are one family of the suborder Arctoidea which also includes the bear, panda, and racoon families. This suborder is within the order Carnivora, the meat-eating mammals. The carnivores are a member of the class Mammalia to which all mammals belong. With reptiles, amphibians, and the various fish groups, the mammals are included in the phylum Chordata whose members are characterized by a supporting structure, the notochord, along the back. Finally, the chordates share membership with a host of different animals without backbones in the kingdom Animalia. Although this may seem like a large number of divisions to use in classifying an organism, in fact, several subgroupings and supergroups have been left out of the preceding summary. Note in this account that the Latin name of a division is capitalized but its English equivalent is not, and that the Latin names above the level of genus are not placed in italics.

The term *taxon* (plural taxa) refers to any group in the classification, such as family, order, species, or genus. The science of naming and classifying organisms is taxonomy.

Family names can be recognized by the ending "idae" which is unique to this taxonomic rank. Subfamily names end in "inae". The ending "oidea" is used consistently for superfamilies, but it is used for some suborders as well. Names of vertebrate orders frequently end in "formes," but this is not a formal rule. There is no consistency in the endings for higher taxonomic groupings.

Biological classification is not static, but changes as new species are discovered and new knowledge makes the revision of old classifications and the subdivision of old groupings necessary. For example, since Linné, biologists have found new forms of cats, such as the North American puma (*Felis concolor*), and have agreed that is it no longer appropriate to consider all the large cats species of the genus *Felis*. At present some are placed in the genus *Panthera* and the lion is now known as *Panthera leo* (Linné). To indicate that the generic name is not the one assigned by the original describer, his name is placed in parentheses.

ESTABLISHING RELATIONSHIPS

Linné not only distinguished species and genera but grouped the genera into larger divisions called families to indicate relationships among them. The families were grouped into orders, the orders into classes, the classes into phyla, and the phyla into kingdoms. In this way a unified system that reflected similarities and differences between organisms was established.

In Linné's time, the observed similarities among species were attributed to some systematic aspect of their creation by God. With the acceptance of Darwin's evolutionary hypothesis in the mid-nineteenth century, the relative degree of similarity among species was attributed to the closeness of descent from a common ancestor. Animals and plants that are similar to one another were thought to have evolved from a common ancestor more recently than did those which differ from each other to a greater extent.

Evolutionary biologists agree that systems of classification should reflect the genetic relationships of organisms as closely as possible. Such systems of classification are referred to as natural, as opposed to an arbitrary system that might be based on similarities not resulting from descent. An unnatural classification might unite sharks and whales on the basis of the similarity of body form, but it would not reflect the fact that whales evolved from terrestrial ancestors only very distantly related to sharks.

A formal system of classification depends on establishing the relationships of organisms. Establishing relationships is consequently one of the most important activities of both paleontologists and biologists who study the diversity of living organisms.

From the time of the Greeks, systems of classification and the establishment of relationships were based primarily on overall similarities. The basic stability of Linné's system of classification shows the strength of this methodology. Today, two conceptually distinct methods of establishing relationships and expressing them in systems of classification can be recognized. These are *evolutionary systematics* and *phylogenetic systematics*. Both endeavor to establish classification on the basis of evolutionary relationships, and both depend on recognition of homologous structures.

With the realization that all organisms ultimately share a common ancestry, evolutionists reasoned that similar structures observed in closely related organisms had evolved from a similar structure in their immediate common ancestor. Structures that are so related are termed homologous. Within vertebrates, the upper bone of the forearm, the humerus, is homologous among bats, whales, and primates. It may differ greatly in form and function among these animals, but they share a common ancestry among early mammals in which this bone was present. Homology can usually be recognized among living organisms by the relative position of the structure and by similarities in the pattern of its embryological development.

However, very similar structures may evolve independently in different groups. This frequently occurs through the process of evolutionary convergence in which similar selective forces act on unrelated organisms. For example, sharks and whales show convergent patterns in the evolution of a streamlined body form necessary for rapid swimming. Structures that evolve convergently are termed analogous or homeoplastic rather than homologous.

Establishing relationships depends on the recognition of homologous structures, which reflect common ancestry, and their distinction from homeoplastic structures which do not reflect close relationship.

Another concept that is important for both phylogenetic and evolutionary systematists is that of the monophyletic group. If several species can be shown to share advanced characters that are not present in other groups, it is assumed that they share a common ancestry, distinct from all other species. Since they have a single ancestry these species are said to belong to a monophyletic group.

Evolutionary Systematics:

Like Linné, evolutionary systematists depend primarily on overall similarity to establish relationships. Where possible, they rely heavily on the fossil record to establish the homology of structures. The writings of evolutionary systematists such as Ernst Mayr and George Gaylord Simpson indicate that they do not consider classification a strictly objective task, but one which involves a degree of judgment and creativity, in addition to a detailed knowledge of the organisms being classified.

Numerical taxonomy, elaborated by Peter Sneath and Robert Sokal, may be considered as an effort to quantify and make more objective a system of classification based on overall similarity. They have used computers to analyze large data sets which give indexes of similarity. These indexes can be used as a basis for classification. Numerical classification (also referred to as phenetics) differs from evolutionary systematics in that little attention is given to establishing specific homologies. It is reasoned that overall similarity, measured by the summing of many different characteristics, is itself evidence of close genetic relationship.

To quote from Sneath and Sokal (1):

> A basic attitude of numerical taxonomists is the strict separation of phylogenetic speculation from taxonomic procedure. Taxonomic relationships are evaluated purely on the basis of the resemblances existing *now* in the material at hand. These phenetic relationships do not take into account the origin of the resemblance found nor the rate at which resemblances may have increased or decreased in the past.

Much has been written about numerical taxonomy but it has not been widely used among paleontologists.

Phylogenetic Systematics:

Within the past 25 years, a new methodology, phylogenetic systematics, has revolutionized concepts of the way in which relationships should be established and organisms classified. Phylogenetic systematics is based on the work of Willi Hennig (2). Much of his early research concentrated on the taxonomy of insects. Insects have a relatively incomplete fossil record, and most relationships must be established on the basis of the physiology and soft anatomy of living species. Hennig sought to develop an objective means of analyzing characters and placing organisms into a system of classification directly reflecting evolutionary relationships.

The most important concept stressed by Hennig was that relationships should be determined on the basis of advanced, or derived, characters rather than on the basis of overall similarity. For example, all fish groups share many characters, such as fins and gills. These characters are of no use in establishing specific relationships among the fish since they are possessed by all members of this assemblage. However, advanced, or derived, features shared by particular fish groups, such as the presence of a median rather than a paired opening for the external nostril in the lamprey and one group of Paleozoic fish (the Cephalaspidomorpha), suggests that they are more closely related to one another than are either to other groups of fish.

Hennig coined the term sister groups to describe two groups that uniquely share derived characters. For example, among mammals, placentals and marsupials are considered sister groups because both give birth to live young, as opposed to more primitive mammals, the monotremes, which lay eggs. Hennig also coined the terms plesiomorphic and apomorphic to describe, respectively, primitive and advanced (or derived) characters. A derived character that is possessed by a single group is termed an autapomorphy. An autapomorphy characterizes this single group, but is of no significance in establishing this group's relation to other groups, none of which have this character. A derived character that is shared by two groups is a synapomorphy. A shared primitive trait is a symplesiomorphy.

A particular character can be simultaneously an autapomorphy, a synapomorphy, and a symplesiomorphy, depending on the taxonomic level being discussed. The presence of fur was presumably an apomorphy or autapomorphy of a single lineage of early mammals, before this group began to differentiate. By itself, this character would have been of no value in establishing the relationships of this lineage. As mammals diverged into the lineages leading to marsupials and placentals, the presence of fur would be a synapomorphy reflecting the common ancestry of these two groups and differentiating them from all other groups of vertebrates. Within marsupials and placentals, fur is a plesiomorphic character which is of no value in establishing relationships among the many subgroups because it is common to all of them.

The use of these terms is becoming increasingly common in paleontology. Students wishing to understand the paleontological literature must be familiar with changes in the philosophy and practice of establishing relationships and systems of classification.

Because only characters that are derived at a particular taxonomic level can be used in determining relationships, a method must be developed to identify derived characters. Without some prior knowledge of the relationships of a species or its evolutionary history, a particular character cannot be identified as primitive or derived. For example, a reptile may have a particular number of teeth. Is this number a primitive or derived condition relative to other related reptiles? Did its ancestors have a smaller or a greater number of teeth? The direction of evolutionary change is referred to as the polarity of the character. A particular character may exist in either a primitive or a derived character state, or in several distinct derived conditions, all evolved from a single primitive character state. For example, the primitive character state for a group of clams may be a smooth shell. The derived state may be an ornamented shell, but this ornamentation might take many different forms, such as ridges, grooves, or spines, each of which would be an alternative derived state of the character.

Polarity:

Since Hennig's own research concentrated on a group with a poor fossil record, he applied methods of establishing the primitive or derived nature of traits, and hence polarity, among contemporary species. These include outgroup comparison, commonality, and embryological development.

Since outgroup comparison is considered the most useful method, it will be discussed first. An outgroup includes taxa that are not part of the particular group under consideration, but do share a distant common ancestry. For example, if a paleontologist is considering relationships among primates, the outgroup might include all other mammals. The polarity of a character may be judged by its distribution within the group being studied and its outgroup. A character (or character state) that is widely distributed in both the outgroup and the group under consideration is probably primitive. A character that is found only in the group under study is probably a derived feature of this group.

Another method is commonality. If a character is universally present in a group, it is almost certainly a primitive feature of this group, for example, hair in the modern mammal groups. Conversely, if a character is only rarely present in the group being studied, it is likely to have evolved within this group. There are obvious exceptions, however. Egg laying is very rare among mammals, being restricted to the platypus and the echidna living in the Australian area. This is certainly not a derived character of these genera, but a primitive feature, as established through consideration of their outgroup, other terrestrial vertebrates—reptiles, birds and amphibians—almost all of which lay eggs.

Another method of establishing the polarity of alternative character states is the study of the development of individuals from their embryos. Derived character states should be recognizable by the fact that they appear later in development. For example, primitive vertebrates have a large number of major blood vessels, the aortic arches, extending from the heart through the gill region. Adults of birds and mammals have only two major vessels, one that goes to the lungs, and the second to the rest of the body. Both develop from paired vessels, one of which is lost. In birds, the major vessel to the body develops from the right side, whereas in mammals it develops from the left side. Embryological evidence confirms that the condition in mammals and bird was derived, as was suggested by the distribution of these vessels in adults. It also confirms that the particular pattern in these two groups was independently derived, rather than evolving from a single ancestral group that had already reduced the paired condition to a single vessel.

Identifying ancestors:

Finding the ancestral species for a modern or fossil species should, on superficial consideration, be easy. We should need only to look for a fossil species that is similar but shows more primitive characteristics in sedimentary deposits slightly older than the ones in which the descendant species is found. Because change in organisms, particularly simple organisms, has been slow when averaged over the length of geological time, we would not expect great differences in form to separate ancestor from descendant. Because, in the long perspective, life has become more complex with time, we might expect the ancestor species to show its primitiveness by simpler construction. However, the chances of finding such an ancestor are small and the recognition of its primitiveness is difficult.

The transition from ancestor to descendant is unlikely to be captured in the fossil record because the chance of preservation of any particular species is small (Chapter 1), and the transition may have taken place in a remote area of the world or one where deposition was not taking place. Sedimentary series are the products of deposition of a succession of different environments. Each environment is characterized by a type of sediment and its own suite of species. The sequence of species collected layer by layer commonly reflects the coming and going of species adapted to the different environments, not the change of one species into another. Problems of locating fossil ancestors in the sedimentary rock record are similar to problems of people seeking their roots. The ancestry of most North Americans cannot be determined by examining local records of the last century because most of us are descended from immigrants from other countries, and in many of these countries records have been destroyed.

However, the earliest fossil record of a character may not accurately reflect its actual time of appearance. Evolution of particular characters occurs at different rates in different lineages, and some species (termed *living fossils*) retain many primitive features long after they have achieved the derived condition in other species within the same group. Use of the stratigraphic sequence to establish polarity is not infallible, but it is often effective in groups represented by a good fossil record.

Evolutionary systematists have long tried to discover ancestors of the major groups. Phylogenetic systematists argue that it is difficult, if not impossible, to demonstrate ancestor–descendent relationships, and unnecessary for phylogenetic analysis. Sister group relationships can be established on the basis of shared derived characters. Ancestors, conversely, can be identified only on the basis of the *absence* of one or more characters that are present in the descendant. It can be argued, however, that an ancestor can be recognized by shared derived characters that identify it as a member of the same monophyletic group as the putative descendant, along with the presence of one or more characters of a more primitive character state, and none of a more advanced state.

The basic philosophy of establishing relationships espoused by Willi Hennig, and more recently discussed by Peter Ax, (3) is not fundamentally different from that used by evolutionary systematists. Many systematists before Hennig did specify the use of derived, rather than primitive, characters in establishing relationships, but this was not elaborated as a distinct methodology. Phylogenetic systematics has provided a much more rigorous methodology in which the polarity of all variable characters should be established. The main problem with applying this methodology, as is the case with evolutionary systematics, is the difficulty of establishing polarity and homology.

As in the case of numerical taxonomy, phylogenetic systematics has attempted to apply computer technology in order to handle simultaneously large numbers of characters in many related groups. The results, however, still depend on the prior analysis of the polarity and homology of the characters chosen.

CLASSIFICATION

The great increase of interest in the methodology of establishing relationships has led to much more extensive and rigorous analysis of phylogenies of all taxonomic groups, fossil and living, in the past decade. Even greater changes are occurring in the overall methodology of classification. In retrospect, it seems strange that the basic system of classification proposed by Linné for a biological world without evolution could be accepted with almost no change by the followers of Darwin. Perhaps the slow and irregular growth in knowledge of the fossil record made this possible. Until relatively recently, many fossils were either classified directly with living groups, or recognized as members of extinct groups. Classification of extinct members of modern groups, such as molluscs, or lizards, could proceed much like classification of newly recognized species within the modern fauna. Extinct groups such as dinosaurs, however, could be classified independently of modern groups. In the past 20 years information linking fossil and modern groups has increased greatly, and many species that could be placed in true phylogenetic sequences have been discovered. With this new information, basic problems in the older methods of classification became evident.

One of the most important differences in classification between evolutionary systematics and phylogenetic systematics involves the concept of monophyletic groups. The term monophyletic was initially defined only in terms of the origin of a group. Evolutionary systematists defined a monophyletic group as one that had a single ancestor, rather than one that evolved from two or more ancestral groups. For example, the pinnipeds are a group of marine mammals including seals, sea lions, and walruses. If, as is frequently argued, seals evolved from a different family of terrestrial carnivores than did sea lions and walruses, the pinnipeds would be a polyphyletic group. Biologists frown on the recognition of polyphyletic groups as valid taxa.

Hennig and his followers argue that the term monophyletic should describe not only the ancestry of a group, but also its subsequent history. They contend that a monophyletic group should include both a single ancestral species and all its descendants. The term paraphyletic is applied to groups that have a common ancestry, but that do not include all their descendants. Hennig considered a monophyletic group as a logical and nonarbitrary assemblage of species. His definition, however, conflicts with many of the major taxonomic groups that are currently accepted. The most commonly discussed example of a paraphyletic group is the class Reptilia. This group includes the lizards, snakes, turtles, and crocodiles of the modern fauna, as well as the dinosaurs and many other extinct forms. Reptiles were defined on the basis of overall similarities in the structure and physiology of the modern species. As the fossil record became better known, both the birds and the mammals were recognized as ultimately evolved from animals whose skeletal anatomy resembled the pattern of living reptiles. We say that both birds and mammals evolved from reptiles. Clearly this concept of reptiles contradicts the definition of a monophyletic group proposed by Hennig; reptiles may have a single common ancestor, but they do not include all their descendants.

Both birds and mammals can be defined on the basis of unique derived characters, but reptiles must be defined on the basis of the absence of characters that distinguish their descendants. As the fossil record becomes better known, it is more and more difficult to formulate a point of distinction between reptiles (as currently defined) and birds, or reptiles and mammals, that is not arbitrary. This is evident if we look at the phylogentic relationships of the groups involved (Fig. 3–1). Among living groups, birds share a more recent common ancestry with crocodiles than that between crocodiles and lizards, or crocodiles and turtles. Birds also share a closer common ancestry with all these "reptile" groups than they do with the ancestors of mammals, which are also classified as reptiles.

In the case of reptiles, the generally accepted pattern of relationships is not reflected in the generally accepted classification. A natural classification, based on the relative recency of common descent, would group birds with crocodiles, and both of these groups with lizards and turtles, to the exclusion of mammals and their immediate ancestors.

The simplest solution would be to unite the mammals and their immediate (reptilelike) ancestors in one taxonomic group, and the birds and modern reptiles in a second group.

Both evolutionary and phylogenetic systematists recognize a single larger taxonomic group, the Amniota, which includes all the groups now termed

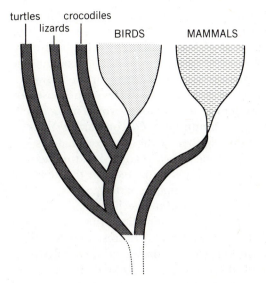

Figure 3-1 Simplified phylogeny of amniotes. The class Reptilia as customarily defined includes all the darkly stippled lineages. Reptiles have a common ancestry but are considered paraphyletic, because they do not include all their descendants. Birds and mammals can be defined on the basis of unique, shared derived characters. Reptiles can be defined as amniotes that lack the specialized characters of birds and mammals. From *Vertebrate Paleontology and Evolution* by Robert L. Carroll. Copyright © 1987 W. H. Freeman and Company. Reprinted with permission.

reptiles along with birds and mammals. All share derived features that distinguish them from more primitive terrestrial vertebrates, the amphibians.

However, the term reptile is familiar to most scientists, whereas no commonly used word is available that refers to a group including dinosaurs, crocodiles and birds, or these groups together with lizards and turtles. It is also difficult to accept a group that includes both warm-blooded, furry mammals and a host of markedly more primitive Paleozoic and Triassic genera, many of which have none of the attributes usually associated with mammals. Popular names have been supplanted in the past, and a classification more closely reflecting their relationships will probably eventually be accepted for the animals now termed reptiles. The living and fossil genera now grouped as amphibians are probably also a paraphyletic group, as are the bony fish. These groups too will eventually be reclassified. Major changes will also occur in the classification of invertebrates and plants as their evolutionary relationships become better known.

A host of practical problems makes application of the principles of phylogenetic classification to large taxonomic groups difficult. Among these are the proliferation of a great many new taxonomic ranks suggested by some workers, or the abandonment of all ranks above the species, as proposed by others. Either of these solutions would make it very difficult to communicate and compare systems of classification. Neither of these major changes has yet achieved general acceptance. For this reason, names and taxonomic ranks used in this text will follow the familiar Linnean pattern.

Another striking difference between evolutionary and phylogenetic systematics is the way in which relationships are graphically represented. Relationships have long been represented as family trees, with modern species on the terminal branches, and older and possibly ancestral forms on the lower branches and trunks (Fig. 3–2). This is the pattern followed in this text. It accords well with paleontological practices, since such family trees usually include a geological time scale. The relative distance between species corresponds roughly with their morphological differences, or presumed evolutionary distances. The exact shape of the tree is at the discretion of the author.

Hennig and his followers, in contrast, have consistently used a different pattern, termed a cladogram, to represent relationships (Fig. 3–3). It con-

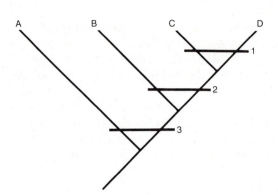

Figure 3-3   Cladogram showing the relationships between four species repesented by letters. The numbered bars represent shared derived characteristics (synapomorphies). Bar 1, joining species D and C, indicates that they share a derived characteristic that B does not. From this characteristic we deduce that D and C are more closely related than either is to B, because they share a more immediate common ancestor. The separation of the other three from A in the diagram is deduced from its lack of another derived characteristic (represented by bar 2) shared by B, C, and D. Its connection to a common ancestor must have been still more remote. All four will share many primitive characteristics (synplesiomorphies), but these are unimportant in the construction of the cladogram and in the determination of the relationships.

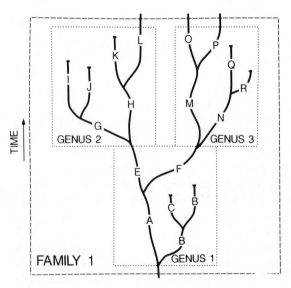

Figure 3-2   Hypothetical phylogeny showing the idealized relationship between species (A, B, C . . .), genera, and families.

sists of an angled axis, from which are drawn a number of parallel lines. Typically, taxa are represented only at the ends of the lines. Only sister groups, never ancestor–descendant relationships, are indicated. The sequence of origin of derived characters are listed along the main axis and the parallel lines. These show a nested hierarchy of derived charcters that can be used to form a nested, hierarchical system of classification (Fig. 3–4). Use of this system avoids the recognition of paraphyletic groups.

Cladograms are a very effective way to illustrate the distribution of derived characters. Typically, cladograms have lacked time scales, but there is no reason why this information cannot be incorporated. Phylogenetic systematics depends on determining the relative time of the divergence of sister groups and the appearance of derived characters. Establishment of either of these factors from the fossil record provides an important basis on which to test the accuracy of their determination by other means, such as outgroup comparison.

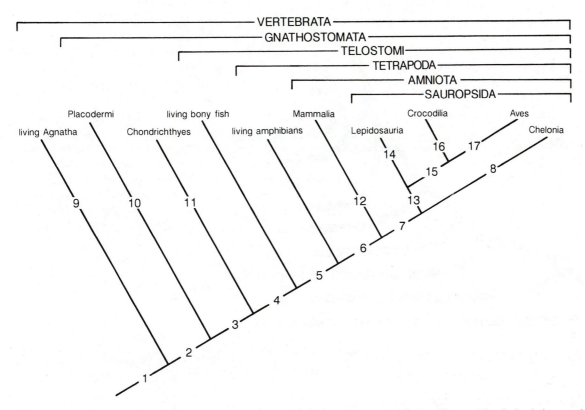

Figure 3-4   A cladogram showing a nested series of monophyletic groups among the vertebrates. The derived characteristics that define the nodes of the cladogram and characterize each group are as follows: 1, brain, specialized paired sense organs, ability to form bone; 2, jaws, gill filaments lateral to gill supports; 3, regular tooth replacement, palatoquadrate medial to adductor jaw musculature; 4, bone, a regular constituent of the endochondral skeleton. Swim bladder or lung; 5, paired limbs used for terrestrial locomotion; 6, extraembryonic membranes, allantois, chorion, and amnion, direct development without an aquatic larval stage; 7, loss of medial centrale of pes; 8, loss of ectopterygoid bone, closely integrated dorsal carapace, and ventral plastron; 9, gill filaments medial to gill supports; 10, palatoquadrate lateral to adductor jaw musculature; 11, prismatic cartilage; 12, fur, mammary glands; 13, dorsal and lateral temporal openings, suborbital fenestra; 14, large sternum on which the scapulocoracoid rotates; 15, hooked fifth metatarsal, foot directed forward for much of stride; 16, an akinetic skull with extensive pneumatization, elongate coracoid; 17, feathers. Neither osteichthyes nor modern amphibians can be defined on the basis of unique derived characteristics. From *Vertebrate Paleontology and Evolution* by Robert L. Carroll. Copyright © 1987 W. H. Freeman and Company. Reprinted with permission.

Molecular Phylogeny:

Comparisons of phylogenetic significance can be made at the molecular as well as at the morphologic level. The sequence of bases in chains of genetic material (DNA and RNA) and amino acid sequences of proteins are characteristic of each organism. Similar organisms have similar amino acid sequences, and the degree of genetic difference between organisms is reflected quantitatively in the differences of their proteins. The differences between amino acid sequences have been interpreted as a measure of the remoteness of a common ancestor. Most molecular comparisons have confirmed

conclusions about relationships reached by morphologic comparisons (Fig. 3–5).

Although proteins are subject to bacterial attack and may be decomposed rapidly after the death of an organism, collagen persists in bones up to a million years old. This allows comparisons to be made of the amino acid sequences of animals that became extinct in the Pleistocene Epoch with those of modern mammals. Direct comparison of amino acid sequences of fossils older than a million years is not yet possible. Proteins have been detected in fossil shells hundreds of millions of years old, but the molecules have been greatly changed and their original configuration is difficult to reconstruct.

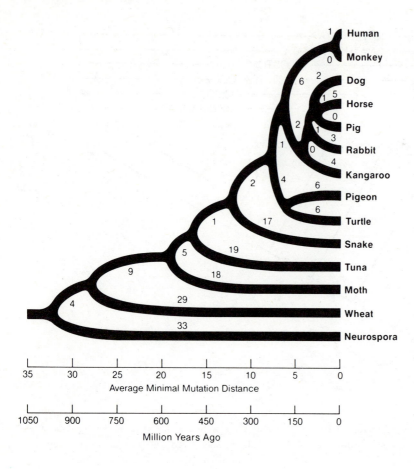

Figure 3-5   Relation between organisms based on the amino acid sequences in the protein cytochrome c in several mammals, a bird, two reptiles, a fish, an insect, a plant, and a yeast. The numbers on the figure are the mutational distances along the line of descent. If the divergence points are scaled against the fossil record, an average mutation rate of about one per 15 Ma for this protein of about 100 amino acids can be deduced. (Reproduced with permission, from *Annual Review of Earth and Planetary Sciences*, v. 14, © 1986 by Annual Reviews Inc.; courtesy of J. M. Lowenstein, modified from E. Margoliash.)

Although direct analysis of the amnio acid sequences in proteins is not possible in fossils older than one million years, phylogenetic analysis is possible by measuring the differences between the sequences in living groups, in the same way that relationships of living organisms were established on the basis of anatomical differences when little information was available from the fossil record.

Homologous proteins can be recognized, just as homologous anatomical structures can be recognized. For example, the oxygen-carrying globin proteins such as hemoglobin are common to all vertebrates. Particular globin proteins have similar numbers of amino acids in all vertebrate groups, but the specific amino acids differ from species to species. The total length of the molecule, its geometry, and some segments may be under stringent selection pressure to maintain an optimal pattern necessary for carrying oxygen, but some segments of the molecule can apparently function with any of a great variety of amino acid sequences. Through random mutations different amino acids are substituted in the ancestral sequence. The extent of substitution is thought to be more or less directly proportional to time. Although we can never directly establish the ancestral sequence, the amount of difference between any contemporary species is a good measure of the time that has passed since they diverged. This provides a measure of the relative time of divergence of a series of related species. Information from the fossil record helps to provide an absolute time scale. For example, we know that there has been a substitution of approximately 50 percent of the amino acids of particular globin molecules since the divergence of the ancestors of the modern reptile and mammal groups in the Late Carboniferous, 300 million years ago, and a 93 percent substitution since the divergence of sharks from the primitive vertebrate stock, 430 million years ago. This information suggests a substitution rate of 1 percent per 5 to 6 million years. The rate of change can then be applied to measure the time of divergence of groups for which the fossil record is less completely known. In human paleontology, this method has indicated humans diverged from apes only 5 or 6 million years ago. When this calculation was first published, it strongly contradicted estimates based on the fossil record which then suggested a time of divergence close to 20 million years ago. Fossils discovered more recently show that humans and living African apes might have diverged between 6 and 14 million years ago.

WHAT IS A SPECIES?

Whatever method is used to establish relationships and to classify organisms, the species is accepted as a fundamental biological unit. Biologists use two definitions or concepts to recognize species: one based primarily on morphology—form, size and proportions—and the other based on distribution and potential for interbreeding.

The Biogeographic–Genetic Approach to Species Definition:

Within the last 50 years biologists have formulated an approach to species definition based on the spatial distribution of organisms and their interbreeding. When the habitats of animals and plants are established, biologists find that groups of organisms that have been recognized as species on a morphologic basis occupy areas from which similar species of the same, or related, genera are commonly excluded. At the boundaries of their home range, species compete with related species for space and resources but do not interbreed with them or grade morphologically into them. The spatial boundaries of species are marked by morphologic and reproductive discontinuities. Within the boundaries, the individuals of the species interbreed and thus maintain a homogeneous population with a continuous range of variation. The boundaries of the species may be physical barriers to interbreeding, such as rivers, lakes, or mountains, or they may be more subtle. At the boundaries of their ranges two related species may be in contact but never interbreed because, for example, one mates in May and the other in June. A pair of insect species may never interbreed, even though their ranges overlap geographically, because one lives exclusively on one type of vegetation and the other on another type. The species in these examples are said to be genetically isolated from one another. The biogeographic–genetic definition of species is that they are populations of interbreeding individuals that are spatially segregated and genetically isolated from similar adjacent groups.

The morphology of organisms is controlled by the genetic material handed down from parents to offspring, that is, the genes. By interchanging genes in random interbreeding within a species population, the group maintains homogeneity. This does not mean that all individuals are alike, but that variation is continuous and within certain limits. The species can be considered either a collection of individuals or the totality of the genes that these individuals possess. This totality of genes is called the gene pool and individuals of the species can be thought of as determined by selections of sets of genes from this pool. The pool is kept homogeneous by the continual gene flow among the individuals of the species as they interbreed.

Within the geographic range of most species, the morphologic variation is continuous, but specific morphologies are clumped together rather than uniformly distributed. The environmental features of a species range are rarely uniform. Through natural selection, individuals adapting to local conditions separate into microgeographic races. This can be illustrated by a hypothetical example of mouse populations living in a range including highlands and plains. Suppose that a population of brown mice have genes giving them a temperature tolerance that favors their survival in the highlands, whereas black mice have genetically dictated features that favor survival in the plains. If black mice straying into the highlands did not mate as effectively in the cold nights there, and brown mice straying into the plains were debilitated by the hot noon-day sun, the resulting local populations would break the morphologic continuity of the species. Interbreeding between the brown and black mice would still take place on the boundaries of the highland and the variation between them would remain continuous, but the populations would have different proportions of the color phases as long as each type of mouse was at a disadvantage in the other's territory.

With further differentiation of interbreeding groups (typically on the basis of geographical distance) populations are recognized as subspecies, and the addition of a third Latin name to the standard genus and species names. The western bullfrog, *Bufo woodhousei woodhousei*, lives in the prairie states and is larger than the forest-dwelling eastern subspecies, *Bufo woodhousei fowleri*. However, they grade into and breed with one another over a belt that reaches a width of 300 km in Texas. Some populations recognized as subspecies are separated by physical barriers and cannot interbreed in nature.

For example, the salamander *Plethodon vandykei* is divided into a subspecies (*P. vandykei vandykei*) that lives in western Washington state and a subspecies (*P. vandykei idahoensis.*) that lives in the mountains of Idaho. The two are separated by the drylands of central Washington. They differ morphologically in skin markings.

The distinction between microgeographic races, subspecies, and species is sometimes unclear even after living populations have been intensely studied. Subspecies names have been used by some paleontologists for subdivisions of a species separated by morphologic differences smaller than those considered to be of specific rank.

The biogeographic–genetic approach to species definition is important in understanding the origin of species. Unfortunately, paleontologists cannot examine the breeding behavior of organisms that have been dead for millions of years, and they have no precise idea about the position or nature of the range boundaries for most fossil species. Fortunately, most modern species can be distinguished by morphological differences which reflect their genetic isolation. By studying the variation found in modern species, defined on the basis of interbreeding criteria, the paleontologist can estimate the range of morphological variation in fossil species.

Linnean Species:

Linné distinguished various species of organisms on the basis of their form. This species concept used by Linné, and still used by biologists and paleontologists for many groups of organisms, is therefore a morphologic definition. No two individuals of any species are exactly alike when examined critically, although to a casual observer the birds in a flock may appear to be identical. The various species could be distinguished by Linné because, despite the range of variation within each species, in nature one kind of animal or plant does not grade continuously in form into other kinds. That is, the Linnean system of identifying species is based on discontinuities in the range of variation in form of organisms. It is these discontinuities that are used to draw boundaries between species.

The morphologic discontinuities between species may be subtle and difficult to detect. The form of each individual of a species may be described by the proportions—such as length, height, width, and weight—of the whole organism and of each of its constituent parts. For species of simple organisms

consisting of a single cell, the number of measurements or characters completely describing its morphology would be small, perhaps less than 20. But for a complex animal like a dog, the number of characters would be in the millions. However, biologists, in defining such complex species, must demonstrate only that the diagnostic characters are sufficiently different from those of similar species to make objective recognition of the new species possible.

A paleontologist must often decide whether a collection of fossils contains only a single species of a particular genus, or two or more species. The paleontologist then looks for morphologic discontinuities that separate species in the collection. A similar problem, which must be approached in much the same way, is whether the characteristics of the

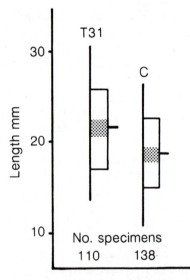

Figure 3-6 Dice diagram comparing the variation between two collections of the brachiopod *Atrypella phoca*. The vertical line represents the total range of length in the specimens collected. The empty box is two standard deviations long. The standard deviation can be calculated by taking the differences between each measurement and the average value (mean), squaring them to eliminate their positive or negative sign, summing them, taking their square root, and dividing by the number of measurements. This procedure yields a value that represents an averaging of the differences of values from the mean without regard to sign. The length of the filled box is ± 1 standard error of the mean and the horizontal line is the mean value. Note the considerable difference in the range and mean of two samples of the same species. (Courtesy of Brian Jones, The Paleontological Society, and *Journal of Paleontology*, v. 51, p. 466.)

group of fossils to be identified are within the range of previously described species variations. If they are not, the fossils may represent a new species. To solve these problems, the range of variation in species must be assessed.

Nature of Variation in Fossil Species:

In everyday life we are confronted with a wide range of variation in form of some species, such as *Homo sapiens* or *Canis familiaris*. The extent of variation can be expressed by measuring various dimensions of the species and quoting the range of values. For example, the variation in length of the brachiopod *Atrypella phoca* from a single locality is from 6 to 19 mm. Because the range in values is so large, some of the specimens in the collection are probably juveniles. A more accurate idea of the extent of variation can be obtained by calculating the mean (average) (14 mm) and standard deviation (1 mm) of the measurements that have been made, and plotting them on a Dice diagram (Fig. 3–6)(See an elementary statistics textbook and the figure caption for methods of calculation). The set of measurements can also be plotted as a histogram to illustrate

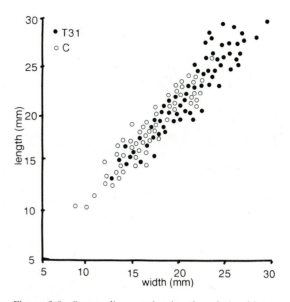

Figure 3-8 Scatter diagram showing the relationship between length and width in the specimens in two collections of the brachiopod *Atrypella phoca*. Although the diagram shows a large area of overlap, it also shows that some specimens from collection T31 are longer and wider than any from collection C. (Courtesy of Brian Jones, The Paleontological Society, and *Journal of Paleontology*, v. 51, p. 468.)

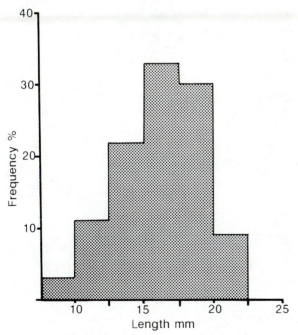

Figure 3-7 Histogram showing the frequency distribution of lengths of the brachiopod *Atrypella phoca*. (Courtesy of Brian Jones, The Paleontological Society, and *Journal of Paleontology*, v. 51, p. 476.)

graphically the extent of variation in this single character (Fig 3–7). The frequency distribution of the measurements shown in the diagram suggests that the few specimens represented by measurements of low value on the left are juveniles and that the many on the right are adults. Differentiation of juveniles also allows the paleontologist to estimate the average adult size.

Paleontologists illustrate the variation in a species and the discontinuities that separate it from similar species by means of diagrams such as Figure 3–8, in which the width and length of two collections of the brachiopod *Atrypella phoca* are plotted. Each dot on the diagram represents the width and length of an individual shell. Immature shells plot on the lower left; large adults on the upper right. In Figure 3–9 the proportions of a related species, *Atrypella foxi*, are compared with those of *Atrypella phoca*. The morphologic discontinuity between the two species is clearly illustrated. The proportional relation between three characters can be plotted on a two-dimensional graph, but the relation between more than three characters is almost impossible to illustrate graphically. However, mathematically, 10 or

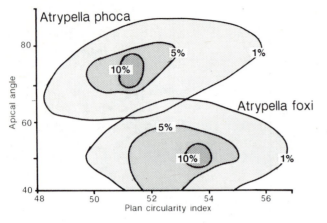

Figure 3-9   Bivariate plot of the dimensions of two species of the Silurian brachiopod *Atrypella*. The density of points of each specimen measured has been contoured. The plan circularity index is defined as 100 times the length divided by the sum of the length and width. Note that the two species can be separated clearly on the basis of these proportions. (Courtesy of Brian Jones, The Paleontological Society, and *Journal of Paleontology*, v. 51, p. 462.)

20 characters can be compared simultaneously in order to separate individuals of species that are morphologically close. Because each character is considered a dimension in the comparison, the ranges of variation of the groups are then multidimensional solids. Although multidimensional space and the solids that occupy it are difficult to imagine and impossible to illustrate, the comparison of many characters can be treated mathematically, with the aid of a computer, by various techniques.

In assessing the extent of variation within a species of living organisms, the biologist can relatively easily determine that the species sample consists only of adults. If the sample contained both juveniles and adults it would include a wide range of dimensions that change with growth, which would misrepresent the adult variation. In many animals, and some plants, species exist in male and female forms which have size differences greater than those between related species. Species which exist in two different forms are said to be dimorphic. Obviously, when collections are being compared to determine morphologic discontinuities, the biologist must be sure that adults are being compared with adults and males with males, or females with females.

In dealing with a collection of fossils, the paleontologist may have difficulty in defining the variation of a species because it is far more difficult to identify juveniles and dimorphism from skeletal parts than from living organisms. Specimens of ammonites (fossil shells described in Chapter 8) now believed to be male and female forms of the same species, were once described as belonging to different genera (Fig. 3–10). Many species of foraminiferans (single-celled organisms also described in Chapter 8) exist in two forms (microspheric and megalospheric) which are much different in overall size and size of the embryonic chamber. This is the result of an alternation in the animal's life history between sexual and asexual reproduction modes.

Despite such difficulties paleontologists can usually find characteristics to measure that are not age

Figure 3-10   Dimorphic forms (macroconch and microconch) of the Jurassic ammonite *Eliganticeras elegantulum*. (Courtesy of Ulrich Lehmann, *Ammonites and Their World*, Cambridge University Press.)

dependent, can differentiate adults from juvenile specimens, and can identify dimorphism.

Types:

When, through a search of the literature and possibly through statistical comparisons, a paleontologist or a biologist is convinced that an undescribed species has been found and prepares to describe it as a new species, he or she sets aside in a museum a single specimen that is a typical example of the morphology of the species. This specimen is called the holotype of the new species and is conserved by the museum so that future paleontologists who wish to compare their specimens with the new species have a sample of what was defined as typical by the original describer. As a single specimen cannot illustrate the range of variation of the new species, the describer may designate several other specimens as paratypes and include them in the museum's collection. It has been suggested that the use of holotypes is misleading, as no single specimen can represent the range of variation found in a species, and later workers may assume that their specimens must be identical with the holotype in order to be recognized as members of the species. This is a misrepresentation of the purpose of the holotype. The holotype merely assures that, if the description or illustration of the new species is inadequate, there is on record a concrete expression of what the author of that species had in mind.

Paleontological Species Concept:

The biogeographic–genetic species concept has had a marginal effect on the practice of species identification and discovery by paleontologists and biologists. An essentially Linnean species concept is still used; that is, species are defined as the lowest grouping of morphologically similar individuals. Although statistical analysis of measurements may help the paleontologist to reveal morphological discontinuities separating species, the identification of a group of specimens as a new species is largely a subjective one, based by the specialist on long experience with the range of variation commonly found in the group of fossils studied and the range found in modern species of related organisms. Nevertheless, the biogeographic–genetic species concept has

had a profound influence on our understanding of how species originate, evolve, and are distributed. It provides a theoretical foundation on which practical methods of species recognition can be based.

In this chapter, species have been treated largely as though they existed only in space, instead of in both time and space. The variation of species in time is the subject of evolution, and a species definition cannot be fully formulated without reference to the fact that organisms change from one species to another in time. Before we examine the problems of defining species in time, we will examine the history of life as preserved in the geological record.

REFERENCES

1. Sneath, P. H. A., and Sokal, R. R., 1973, Numerical taxonomy. The principles of numerical classification: San Francisco, Freeman, 573 p.
2. Hennig, W., 1966, Phylogenetic systematics: Urbana, Univ. of Illinois Press, 263 p.
3. Ax, P., 1987, The phylogenetic system. The systematization of organisms on the basis of their phylogenesis, New York, Wiley, 340 p.

SUGGESTED READINGS

Blackwelder, R. E., 1967, Taxonomy, a text and reference book: New York, Wiley, 698 p.

Lazarus, D. B., and Prothero, D. R., 1984, The role of stratigraphic and morphologic data in phylogeny: Journal of Paleontology, v. 58, p.163–172.

Lowenstein, J. M., 1986, Molecular phylogenetics: Annual Review of Earth and Planetary Sciences, v. 14, p. 71–83.

Mayr, E., 1963, Animal species and evolution: Cambridge, Mass., Belknap Press of Harvard Univ. Press, 797 p.

Patterson, C., 1982, Cladistics and Classification: New Scientist, v. 94, p. 303–306.

Runnegar, B., 1982, A molecular clock date for the origin of the animal phyla: Lethaia, v. 15, p. 199–205.

Runnegar, B., 1986, Molecular Palaeontology: Palaeontology, v. 29, p. 1–24.

Schaeffer, B., Hecht, M. K., and Eldredge., N., 1972, Paleontology and phylogeny: Evolutionary Biology, v. 6, p. 31–46.

Schoch, R. M., 1986, Phylogeny reconstruction in paleontology: New York, Van Nostrand Reinhold, 353 p.

Simpson, G. G., 1961, Principles of animal taxonomy: New York, Columbia Univ. Press, 247 p.

Wiley, E. O., 1981. Phylogenetics. The theory and practice of phylogenetic systematics: New York, Wiley, 439 p.

PART TWO

THE FOSSIL RECORD

Chapter 4

The Time of Microscopic Life

For most of geological time, organisms have lived in the earth's surface environments playing an essential part in the formation of the atmosphere and the seas, the erosion of the land, and the deposition of the sedimentary record. But living things were not always present on earth; there was a time when the earth's surface was lifeless. We now know that within the solar system, life is a unique feature of our planet and that even the nearest planets, Mars and Venus, probably never supported life.

Geologists and biologists have extended our knowledge of life history 3.5 billion years, almost back to the beginnings of the rock record. They have formulated hypotheses to explain the stages by which life could have formed from nonliving matter and evolved from the first living cells to the complex invertebrates of the beginning of the Paleozoic Era. At present, scientists are uncertain of the specific time when life began, and of the conditions on earth at that time. Nor have they have they been able to document in the fossil record the steps that led from complex carbon compounds to the first self-replicating, energy-regulating cells.

This chapter is concerned with the evolution of organisms that are capable of carrying on all of life's functions within a single cell. Some of them form colonies seen easily with the naked eye, but most noncolonial forms can be studied only under the magnification available in light and electron microscopes. These primitive organisms evolved slowly, and some that live today have almost identical ancestors in rocks 3500 million years old (Fig. 4–1). The record of their increasing complexity, adaptation to new environments and modification of environments by their life processes is contained in rocks of Archean and Proterozoic age. Three-quarters of the time span of life history is preserved in these rocks.

Pre-Paleozoic time can be divided into seven intervals (Table 4–1).

These divisions have not been standardized, nor have they been universally accepted. The eras are those recommended by the Decade of North American Geology Scale of 1983 (1), and the United States Geological Survey, who prefer the term pre-Archean for the Hadean Eon. A recent subcommission report recommends that the Early, Middle, and Late Proterozoic eras be called Proterozoic I, II, and III, respectively.

In the Hadean Eon, the crust of the earth was forming and the atmosphere and oceans appeared, setting the stage for the origin of life. The record of these events has not yet been found. They occurred prior to 3800 million years ago, the age of

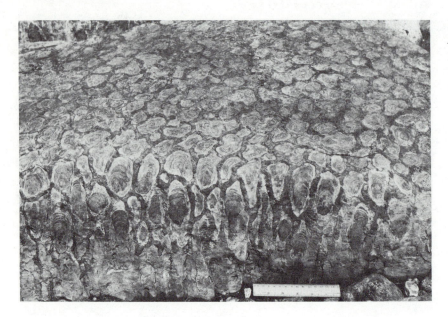

Figure 4-1   The columnar stromatolite *Mistassinia wabassiuon* from the Early Proterozoic Mistassini Group, Lake Mistassini, Quebec. (Courtesy of Hans Hofmann.)

the oldest earth rocks so far discovered. Our only evidence of this earlier period consists of mineral grains, dated as 4200 million years old, that have been incorporated as sedimentary particles in younger rocks. Our reconstruction of geological processes that occurred prior to 3800 million years ago must therefore be based solely on indirect evidence.

THE HADEAN AND EARLY ARCHEAN EARTH

Earth Origin and Hadean Environment:

The theory that the planets formed from a rotating disk of hydrogen and helium gases, water, ammonia, methane ices, and solid particles of iron and silicate minerals has been accepted by astronomers

TABLE 4–1   DIVISIONS OF PRE-PALEOZOIC TIME

EON	ERA	MILLION YEARS
PROTEROZOIC	Late Proterozoic	900 to 675
	Middle Proterozoic	1600 to 900
	Early Proterozoic	2500 to 1600
ARCHEAN	Late Archean	3000 to 2500
	Middle Archean	3400 to 3000
	Early Archean	3900 to 3400
HADEAN		4600 to 3900

for about 40 years. The earth formed from solids and gases of the dust cloud at low temperature, but early in its history it passed through a hot stage. The heating of the embryonic earth was probably caused partly by the accumulation of energy from impact of solid bodies during the accretion process, and partly by the energy released by radioactive isotopes that were abundant at this early stage but have since decayed to extinction. The whole earth may not have reached a liquid state during this hot phase, but it became hot enough for iron to be mobilized and drain toward the center of the planet to form the core. At the same time much of the gaseous material trapped in the accretion process escaped into space.

The only direct evidence of the nature of the Hadean environment in the vicinity of the earth comes from the moon (Fig. 4–2). The astronauts who visited the moon in the seventies brought back rocks from the highlands that have been dated as between 4200 and 4400 million years old. These rocks indicate that frequent collisions of solids from the dust cloud were still taking place in mid-Hadean time. The uppermost layers of the earth are believed to have been accreted in late Hadean time, following the hot phase.

The first atmosphere of light gases that the Hadean earth inherited from the dust cloud is believed to have been lost in early Hadean time. The second atmosphere, in late Hadean and Archean time, was probably formed of the gases that were trapped in the earth during the final phase of accretion and

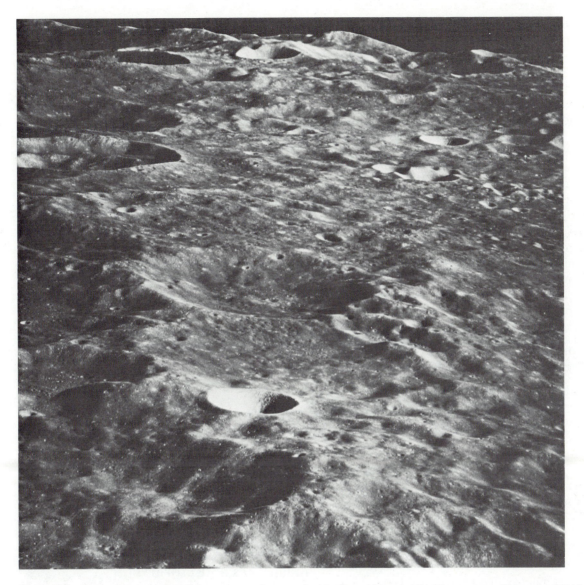

Figure 4-2   Cratered topography of the moon as photographed by the orbiting Apollo 10 astronauts. (Courtesy of NASA.)

subsequently leaked out. The escape of gases from the earth's interior, which is called degassing, must have been rapid then, but it still continues at lower rates as volcanoes and submarine fissures release water vapor, carbon dioxide, methane, hydrogen, and other gases, all of which are contributed to the atmosphere. Geologists do not agree on the specific composition of the second atmosphere.

The Early Archean Environment:

The first direct evidence of the environment of the earth's surface comes from the oldest rocks so far

dated, the 3800-million-year-old Isua Group of western Greenland. They are composed of metamorphosed sediments such as cherts, conglomerates, shales, and carbonate rocks. The sediments were deposited in water and some are composed of particles eroded from a continental source area. They indicate that by this time the earth was cool enough for water in liquid form to exist in a large body, and that continental masses rose above the surface of this primitive ocean. The presence of carbonates in the sediments shows that carbon dioxide was present in the atmosphere. Unfortunately, the rocks have been considerably changed by metamorphism and do not give additional definitive evidence of the nature of that atmosphere.

Scientists deduce the composition of the late Hadean atmosphere according to their assumptions about the timing of the hot phase, the formation of the core, and the availability of hydrogen at that time. One set of assumptions leads to the conclusion that the atmosphere was largely ammonia, methane, and hydrogen sulfide; another, that it was largely carbon dioxide, nitrogen, and water vapor. We know that by Early Archean time, water vapor derived from degassing had condensed into the original ocean. Space probes to both Mars and Venus have shown that their atmospheres are largely composed of carbon dioxide. This suggests that an atmosphere rich in carbon dioxide existed early in the history of all the inner planets and persists today in those planets where it was not subsequently modified by life processes.

Formation of Complex Carbon Compounds:

Compounds of carbon, hydrogen, nitrogen, and oxygen are the building blocks of life. The compounds that we recognize as necessary for life are complex polymers whose origin from simpler compounds must have taken place in Hadean time. Spectroscopic analysis of the light from interstellar gas tells us that some of these simple compounds, such as formic acid ($HCOOH$), methylacetylene (CH_3C_2H), and methylamine (CH_3NH_2), exist in space and were probably constituents of the primeval dust cloud trapped in the accreting earth. Simple organic compounds can be produced by the application of energy that would have been available from the sun, from lightning, or from thunder shock waves to the various gas mixtures that are possible models of Hadean and Early Archean atmospheres. Experiments in which such gas mixtures are subjected to electrical discharges and radiation show that such simple compounds as formic acid ($HCOOH$), formaldehyde ($HCHO$), alcohols (such as ethanol, CH_3CH_2OH), simple sugars, amino acids, and nitrogenous bases could have been formed under Hadean and Early Archean conditions. We conclude that the ocean of Early Archean time contained a variety of the simpler organic compounds left over from the dust cloud and produced by inorganic processes. However, biologists can only speculate on the path, or paths, by which simple carbon compounds passed to more complex poly-

mers and eventually formed a cell that, through its properties of replication (reproduction of itself) and energy intake and regulation, is defined as being alive. A. Graham Cairns-Smith (2) has speculated that clay crystals, on which inorganic ions were replaced by simple organic compounds, may have been intermediates between living and nonliving systems. However, fossils of intermediate forms have not been found, nor have intermediate compounds been identified in rocks of Early Archean age.

The accumulation of complex carbon compounds could only have taken place in an atmosphere in which oxygen was either absent, or in very short supply. Although scientists are in doubt about the composition of that atmosphere, they are sure that it lacked oxygen. Some idea of the nature of early life-forms may be obtained by examination of modern unicellular organisms living in oxygen-deficient environments. Deep in ocean sediments, in soil, or in marshy ground, removed from the present oxygen-rich atmosphere, environments still exist that duplicate the oxygen-deficient atmosphere of the early earth. In these areas unicellular organisms similar to those of the Archean persist.

TYPES OF UNICELLULAR LIFE

Organisms that live in environments devoid of oxygen are called anaerobes, or are said to be anaerobic. Organisms that live in oxygen-rich atmospheres are aerobes, and those that live in both are amphiaerobes.

Eukaryotes and Prokaryotes:

All organisms are either eukaryotes or prokaryotes. The differences between the two are so profound that the prokaryotes have been considered to constitute a kingdom separate from all other organisms, the kingdom Monera (Fig. 4–3).

Eukaryotic cells have a nucleus bounded by a well-defined membrane, and prokaryotic cells have no nuclear membrane. The eukaryotes enclose in the nucleus two copies of the genetic code which separate into sex cells in reproduction; the prokaryotes have a single copy of the code within the cell body which duplicates when the cell divides in asexual reproduction. Eukaryotic cells oxidize sugars as a source of energy, whereas many prokaryotic cells are anaerobes. Eukaryotic cells are usually considerably larger than prokaryotic cells, and this is one

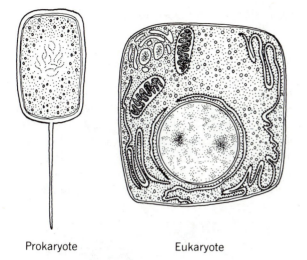

Prokaryote   Eukaryote

Figure 4-3   Major features of living prokaryotic and eukaryotic cells. Diagrammatic. Note the vague nucleoid in the prokaryotic cell and the membrane-bounded nucleus in the eukaryote. (*Source:* L. Margulis, *Early Life*.)

of the few criteria that can be used to distinguish them in the fossil record. Prokaryotes range in diameter from 0.3 to 20 micrometers and eukaryotes from 3 micrometers to several millimeters. The other differences either are behavioral or have not been preserved in the fossil record.

Unicellular Metabolism:

Modern unicellular organisms can be separated into six groups on the basis of their methods of obtaining energy. These methods probably evolved as life adapted to the earth's changing environments, but unfortunately direct evidence of metabolic pathways cannot be preserved.

Type 1—Anaerobic cells that obtain energy by assimilating complex carbon compounds. Examples are the bacteria that exist in the guts of most advanced animals and aid in the digestive process. The reaction used is that of fermentation.

Type 2—Anaerobic cells that use carbon dioxide to produce their own carbon compounds without the use of light. These bacteria are called methanogens because they produce methane from carbon dioxide.

Type 3—Anaerobic cells that use light to make carbon compounds from carbon dioxide, but use elemental hydrogen, or hydrogen sulfide, rather than water to supply the hydrogen for these carbohydrates. These bacteria use what is called photosystem-I in their photosynthesis, which is represented in highly simplified form by the equation: $CO_2 + 2H_2S + light = 2SO_2 + H_2O + CH_2O$ (organic compounds).

Type 4—Amphiaerobic bacteria that are capable of living in oxygen-rich or oxygen-deficient environments. These exist at present in sediments near the seafloor in which the oxygen level fluctuates.

Type 5—Aerobic cells containing chlorophyll which use light, carbon dioxide and water in what is called photosystem-II. This can be simplified in the equation: $CO_2 + 2H_2O + light = O_2 + H_2O + CH_2O$ (organic compounds).

Type 6—Aerobic cells that depend for energy on the oxidation of sugars. This process is called respiration.

The closeness of the relationship of one type of prokaryote to another is investigated by comparison of their amino acid sequences (Chapter 3). The biochemistry of these groups suggests that each did not give rise to the next most complex type in succession, but that the groups are stages in evolution reached in several lines of descent. Biochemistry confirms that the methanogens are a very primitive group that separated from the bacteria soon after the origin of life. To some, this justifies the formation of a separate kingdom, the Archaebacteria, which would also include sulfur- and sulfate-using bacteria. Photosynthesis apparently developed in many lines of bacterial descent and aerobic respiration did not develop once and spread but was a level of metabolism attained by different lineages (3,4).

DIRECT EVIDENCE OF PRE-PALEOZOIC LIFE

Some unicellular organisms secrete shells of mineral matter that are commonly preserved as fossils, but the primitive organisms that lived in Archean and Proterozoic time were completely soft-bodied and only rarely preserved. Until about 25 years ago they were almost unknown. Then the examination of black cherts in thin section showed minute cells preserved as organic residues in three dimensions. Since that time black cherts throughout the pre-

Figure 4-4   Tangle of algal filaments and spores from the Gunflint chert of Early Proterozoic age (approximate magnification ×800). (Courtesy of E. S. Barghoorn.)

Paleozoic succession have been examined (Fig. 4–4).

Black Chert Microfossils:

Many subspherical bodies the size of bacteria occur in pre-Paleozoic cherts but most of these were formed inorganically. Of the 22 reported occurrences of Archean microfossils, William Schopf and Malcolm Walter (5) recognize only two as including true microfossils. Two other collections include objects of doubtful significance and the rest contain only inorganic spheres and filaments. Because living bacteria are everywhere in the modern world, they can easily be included in samples collected for the purpose of finding pre-Paleozoic fossils and are easily mistaken for such fossils. Modern bacteria have undergone little evolutionary change since pre-Paleozoic times and still closely resemble their ancient ancestors in form. Therefore, great care must be taken to minimize contamination and to identify misleading bacteria that have entered cracks and pores of the samples after the rock was deposited. To qualify for acceptance as microfossils, the cells

must be shown to be embedded in the rock and to have been deposited at the same time as the rock was formed. Although authenticated fossil localities of Archean age are extremely rare, 24 localities are known to yield early Proterozoic microfossils.

The oldest fossils come from the Warrawoona Group in the region of western Australia called North Pole because of its remoteness (6,7). The rocks bearing the black cherts are dated as between 3300 and 3500 million years old by interbedded volcanic flows. The fossils include long sinuous threads with diameters ranging from a fraction of

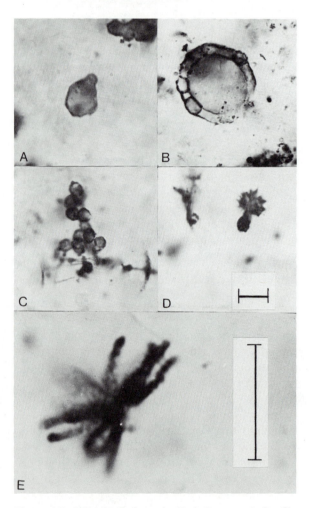

Figure 4-5   Microfossils from the Early Proterozoic Gunflint Formation, Lake Superior, Ontario. The scale bars are 10 micrometers long and that in Figure D applies to Figures A–D. A. *Huroniospora* showing budding. B. *Eosphaera tyleri*. C. *Huroniospora* in a cluster. D. *Kakabekia umbellata*, E. *Eoastrion simplex*. (Courtesy of Hans Hofmann and Geological Survey of Canada [neg. 140940] from *Iron Formations Facts and Problems*. Elsevier Science Publishers.)

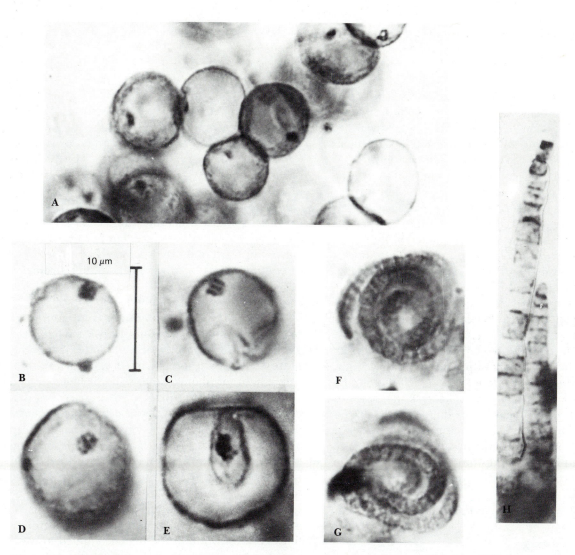

Figure 4-6   Microfossils from the Late Proterozoic Bitter Springs chert, Australia, magnified about 2500 times.
A. Green alga *Glenobotrydion aenigmatis* showing spots suggestive of intracellular structures. B–E. Single cells with internal granular areas suggestive of organelles and membranes. F, G. Coiled cyanobacterium *Contortothrix vermiformis*. H. Filamentous cyanobacterium *Cephalophytarion grande*. (Courtesy of J. W. Schopf, The Paleontological Society, *Journal of Paleontology*, and Kluiver Academic Publishers)

a micrometer to several micrometers, and globular groups of cells enclosed in a sheath. The largest filaments are divided by cross partitions and represent linear colonies of cells. These cellular filaments are similar in form to many modern prokaryotes and closest to cyanobacteria. The cyanobacteria (sometimes called blue-green algae) are photosynthetic bacteria of type 5. Isolated cells in the deposit have been identified as methanogens.

Black chert microfossils, from the Gunflint chert (8), were first described in detail by E. S. Barghoorn and S. A. Tyler in 1965. The fossils of Early Proterozoic age (1900 million years old) were found on the north shore of Lake Superior in western Ontario (Fig. 4–6). Most are long filaments consisting of cells similar to those that are today are characteristic of the cyanobacteria. Some of the filaments look like strung beads (Fig. 4–5). Other fossils in the Gunflint chert are more complex. Some large spherical cells have daughter cells arranged around them (*Eosphaera*); others are starlike in form (*Eoastrion*). The most complex fossil, which is called *Kakabekia umbellata*, consists of a bulbous cell joined by a shaft to a disk supported by radial arms, so that it looks like a tiny umbrella. The affinities of this form are problematic, but very similar anaerobic bacteria have been found in modern ammonia-rich soils. *Kakabekia* is unlikely to be indicative of an atmosphere in which ammonia was a principal constituent, for other geochemical indicators suggest that by early Proterozoic time oxygen was becoming abundant, and ammonia would be unstable unless separated in some way from the oxygen.

The microfossils of the black chert of the Bitter Springs Formation in Australia are typical of Late Proterozoic assemblages. They are about 900 million years old. A large variety of cyanobacterial filaments have been found accompanied by the fossils of advanced eukaryotic algae. Among the most interesting fossils are those showing cell division (Fig. 4–6). In Bitter Springs cells, and in cells of other collections as old as 1800 million years, dark aggregates of matter appear that look like nuclei, or other intracellular structures typical of eukaryotic cells. The origin of these dark masses is controversial. When either eukaryotic or prokaryotic cells dry out, the cell contents shrink inside the cell membrane into a dark mass (Fig. 4–7). The dark structures of the Bitter Springs cells can be duplicated by drying modern prokaryotic cells; therefore their presence is not incontrovertible evidence that eukaryotes were present in Bitter Springs time. However, the large size of the cells is independent evidence of their eukaryotic nature.

Many black cherts containing microfossils occur

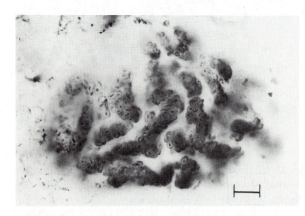

Figure 4-7   Coccoid cyanobacteria (*Eosynechococcus moorei*) showing pseudonuclei. From Early Proterozoic rocks 1900 million years old in the Belcher Islands, Canada. Scale bar 10 micrometers. (Courtesy of Hans Hofmann, The Paleontological Society, and *Journal of Paleontology*, v. 50, p. 1056.)

within and around layered, domed structures called stromatolites.

Stromatolites:

Finely laminated structures called stromatolites are built up of layer upon layer of mineral particles, commonly calcium carbonate, trapped or precipitated by bacteria or cyanobacteria (9). The thickness of the individual layers is usually a few millimeters, or fractions of a millimeter, but the layers can accumulate into structures many meters high (Fig. 4–8). In some stromatolites the layers are planar and build a tabular structure, but in most they are convex upward and build hemispherical, bulbous, or columnar forms. Modern freshwater stromatolites are formed of layers of calcium carbonate precipitated around the filaments of cyanobacteria. Modern marine stromatolites are communities of prokaryotic microorganisms whose surficial organic layers are stratified according to the light requirements of the organisms. Cyanobacteria that utilize light for photosynthesis and produce oxygen commonly form the upper layers; anaerobes that do not require light live in the basal part of the living layer. The sticky filaments of the bacteria trap a thin layer of calcium carbonate particles settling out of the water and, when covered with sediment, move through the accumulated layer to produce another sticky surface and trap the next layer. The layering may be a product of daily, monthly, or yearly cycles in the environment. Living stromatolites are uncommon but may be studied in western Australia at Shark Bay (Fig. 4–9), at Andros Island in the Bahamas, and in some lakes. When the microorganisms that form the stromatolite die and decay, little trace of their presence remains, except molds of the filaments. Stromatolites are identified on the basis of their growth form and microstructure. A single stromatolite may be formed by a community of several different bacteria, but the predominant builders are the cyanobacteria. Because most cyanobacteria are photosynthetic type 5 bacteria, the occurrence of stromatolites in ancient rocks suggests that photosynthesis was taking place at the time of their deposition and that oxygen was being produced.

Because many structures produced by microbes are not laminated, many paleontologists have expressed dissatisfaction with the term stromatolite. Robert Burne and Linda Moore (10) have suggested that the term microbiolite be used for all

Figure 4-8   Stromatolites from Manitounuk Island, Hudson Bay; the columns are about 2 cm across. (Courtesy of Hans Hofmann and Geological Survey of Canada.)

structures that formed from microbial community action in trapping, binding, and causing mineral precipitation, whether laminated or not.

The oldest stromatolites are found in the same beds with the oldest microfossils in the Warrawoona Group of western Australia. These are nodular, bulbous, and columnar forms much like those produced later by cyanobacteria (Fig. 4–10). A possible conclusion from this similarity is that oxygen-producing photosynthesizers were already present at this early date. However, some cyanobacteria can switch from photosystem II to photosystem I, which does not produce oxygen, when conditions change. Possibly these Archean stromatolite builders used photosystem I, and therefore the presence of stomatolites need not imply oxygen production. At least 12 other occurrences of stromatolites in Archean rocks are known. These fossils become increasingly abundant in Late Archean time.

During the Proterozoic Eon stromatolites reached their greatest abundance and diversity. During early Proterozoic time stromatolites in a host of different

Figure 4-9   Underwater photograph of living subtidal stromatolites about 40 cm high at Shark Bay, Western Australia. (Courtesy of A. E. Cockbain, Geological Survey of Western Australia, and Elsevier Science Publishers.)

growth forms constructed carbonate barriers at the margin of the North American continental platform. Soon after the beginning of Paleozoic time stromatolites became less abundant and were restricted to very shallow water, usually intertidal environments. The youngest extensive stromatolitic limestones in the North American succession are of Early Ordovician age. The stromatolites called *Cryptozoon* form the Fossil Gardens that are a tourist attraction in the Late Cambrian limestones of the Saratoga Springs area of New York State. The mid-Ordovician decline of stromatolites may have been caused by competition for space in shallow-water environments from more complex and faster growing organisms such as corals, sponges, stromatoporoids, and advanced algae, or by predation by the rapidly expanding mollusc and arthropod population.

Figure 4-10   Oldest known stromatolite from Warrawoona Group, North Pole, Australia. Scale is 10 cm long. (Courtesy of M. R. Walter, reprinted by permission from *Nature*, v. 284, p. 444, © 1980, Macmillan Magazines Ltd.)

Dispersed Organic Matter:

Pre-Paleozoic rocks contain carbon and carbon compounds dispersed in small quantities. Geologists speculate that these are fragments and degradational products of primitive organisms that remained trapped in the sediment after the organism itself was destroyed. Organic matter in older rocks is referred to as kerogen. It is defined as the organic matter in rocks that is insoluble in organic solvents. It has no definite chemical composition and exists in various states of metamorphism. In extensively metamorphosed rocks, organic matter is reduced to the mineral graphite, a form of elemental carbon. It is difficult to determine whether the graphite that is widely distributed in Archean rocks is the remnant of organisms or whether it was produced by metamorphic processes acting on carbonates. Biochemists can extract and analyse complex carbon compounds in small quantities and have discovered a wide range of compounds in pre-Paleozoic rocks that are at present produced only by organisms. These chemical fossils have been widely accepted as evidence of Archean life. Analyses of kerogen from the oldest rocks of the Isua Group have shown organic compounds, but these could have been produced either by primitive organisms or by inorganic processes acting in the primeval ocean as discussed earlier.

Stable Isotopes:

In order to distinguish compounds produced by life processes from those produced inorganically, biochemists have measured the proportions of the isotopes of carbon, nitrogen, sulfur, and oxygen that they contain. Nearly all the elements exist in various nuclear configurations called isotopes. Some isotopes are radioactive and can be used in dating rocks. The others are known as stable isotopes. Carbon has stable isotopes of atomic weights 12 and 13, which are symbolically represented as ^{12}C and ^{13}C. Oxygen has stable isotopes of atomic weights 16, 17, and 18; nitrogen, 14 and 15; and sulfur, 32, 33, 34, and 36. Although the isotopes of an element are identical chemically, they participate in reactions at slightly different rates owing to their different masses.

In chemical reactions that take place within organisms, the proportions between the isotopes in the organism and in the environment are changed, and the isotopes are said to be fractionated. A plot of the proportions of carbon isotopes of various organisms (Fig. 4–11) shows that organic carbons are enriched in ^{12}C over ^{13}C, compared to inorganic carbon of the atmosphere and oceans. However, the isotopic signatures overlap and a ratio is not specific to each organism.

The isotopic proportions of the kerogens in Archean rocks suggest, but do not prove, that they were produced by life. Kerogens 2000 to 3500 million years old have changes in ^{13}C values that range from −27 to −30 parts per thousand, in the range of many primitive organisms. Kerogens over 3500 million years old have values of about −16 parts per thousand, which may be interpreted as indicating that they are different and possibly inorganic in origin, although this value is within the range of the methanogens, cyanobacteria, and some plants.

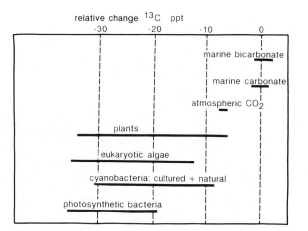

Figure 4-11 Carbon isotope composition of primitive organisms, the atmosphere, and the sea expressed as change of the carbon-13 isotope in parts per thousand (ppt) with respect to the carbon-12–carbon-13 ratio of a standard. Carbon-13 is depleted through fractionation by life processes.

OXYGEN LEVELS

Although oxygen is essential to cells that obtain energy by respiration, it is poisonous to anaerobes that are not equipped to use it. Eukaryotes have mechanisms for counteracting the deleterious effects of free oxygen but bacteria of types 1, 2, and 3 cannot live with it. In its presence no complex carbon compounds could have accumulated from inorganic processes before life began.

Although it could not have been generally available in the Early Archean atmosphere, oxygen may have been produced by photosynthesis or by the dissociation of water molecules by radiation from the sun. Some local, temporary concentrations, or oases, of oxygen could have existed where cyanobacteria formed stromatolites, but they would have been rapidly depleted in oxidizing organic matter and other compounds. Tracing the stages by which oxygen reached its present level of about 20 percent in the atmosphere has proved difficult, and geologists do not yet agree on what percentage was present at various times. Some evidence may be preserved in Archean and Proterozoic sedimentary rocks.

Evidence from Sedimentary Rocks:

In sediments that have been weathered from their parent rock and deposited under conditions where oxygen is present, the iron oxides are reddish

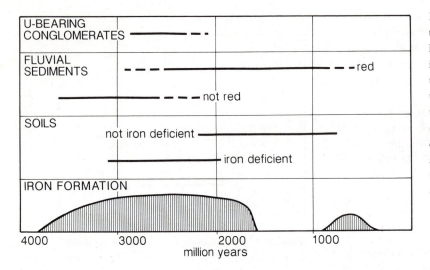

Figure 4-12 Sedimentary indicators of the oxidizing state of the atmosphere. The lines represent the range in time of the indicators; broken lines indicate uncertainty. Note that the indicators of oxidation (redbeds, iron-rich soils) overlap in time with those of a reducing atmosphere (uranium-bearing conglomerates, iron deficient soils) and a transition zone at about 2200 million years ago is suggested. (*Source:* Modified from J. W. Schopf and others, *Earth's Earliest Biosphere.*)

brown—they are in the oxidized, or ferric, state. Such sediments are called redbeds and they are common in Paleozoic and younger sedimentary successions. The oldest documented redbeds are between 2500 and 2000 million years old (Fig. 4–12). Red soils are also evidence that oxygen was available in the atmosphere when they were formed. The oldest red soils are about 2200 million years old. Banded iron formation (Fig 4.–13) is a sedimentary deposit that is confined to pre-Paleozoic rocks and is therefore believed to be a product of environmental conditions that disappeared at the end of Proterozoic time. Iron formation is composed of layers of iron oxides, such as hematite and magnetite, alternating with layers of iron silicates, iron-rich cherts, called jasper, and, locally, iron carbonate. The deposition of iron formation may have been controlled by periodic emissions from volcanic springs, by upwelling of deep iron-rich oceanic waters, or by blooming of microorganisms, but for all hypotheses ferric oxides in the deposits require that oxygen be available, at least locally, during deposition.

The oldest iron formation is found in the oldest rocks yet dated, the Isua Group of Greenland. Its presence suggests that this Early Archean atmosphere was not completely without oxygen. Deposits of this type reach their greatest abundance and widest distribution in rocks of Late Archean to Early Proterozoic age (3000 to 2000 million years old). Whether in their original state as silica-rich ores, or concentrated to nearly pure oxides by weathering, Early Proterozoic iron formations constitute nearly all the major iron ore deposits in the world.

Another sedimentary mineral deposit confined to the early part of Proterozoic time is uranium-bearing conglomerate. None is younger than 2200 million years. Commercially important deposits occur in the Elliot Lake mining area in Canada and in the Witwatersrand area of South Africa. The sedimentary particles of the mineral uraninite weather rapidly in the modern, oxygen-rich atmosphere, and their preservation through a cycle of erosion, transportation, and deposition in early Proterozoic time suggests an atmospheric oxygen level then of less than 1 percent of the present level. The restriction of these uranium ores to a particular time in the past suggests that deposition was terminated by a change in the worldwide environmental condition toward a more oxygen-rich atmosphere.

Eukaryotic Cells:

If eukaryotic cells could be easily recognized in the fossil record, their first occurrence would indicate when sufficient oxygen was available to support respiration. The oxygen level necessary to support respiration is about 1 percent of the present atmospheric level; this proportion is sometimes referred to as the Pasteur point. As already discussed, the identification of the first eukaryotes is a matter of disagreement, but William Schopf, one of the most experienced of pre-Paleozoic paleontologists, believes they can be recognized in rocks as old as 1500 million years.

These data suggest that progress toward the present oxygen-rich atmosphere was extremely slow. Some of the data seem contradictory. Oxygen was present only locally in the atmosphere and oceans

Figure 4-13 Contorted banded iron formation from Archean rocks, Beresford Lake, Manitoba. (Courtesy of Geological Survey of Canada.)

atmosphere prevents much of the ultraviolet radiation of the sun from reaching the earth's surface. This radiation is lethal to many microorganisms. Before oxygen was an important constituent of the atmosphere, no ozone shield would have formed, and the full strength of the lethal radiation must have forced microorganisms to live beneath the sediment surface, or in deep water where the water column would provide some protection. The relative insensitivity of cyanobacteria to ultraviolet radiation may explain their early success. The evidence that Early Archean life developed in shallow water, however, has led some geologists to suggest that the ozone shield was in place at this time and that oxygen was more abundant than previously thought (11). The increasing availability of oxygen to form the ozone shield would have progressively opened new environments in the oceans' surface waters to more sensitive microorganisms.

DEVELOPMENT OF PRE-PALEOZOIC LIFE

On the basis of the evidence that has been reviewed, a plausible sequence of events, a scenario, or model, can be formulated as to how life evolved in the three-quarters of earth history that are the Proterozoic and Archean eons.

1. In Early Archean time the oceans were habitable. They contained a mixture of inorganically produced carbon compounds, a great deal of carbon dioxide, and only local traces of oxygen produced by dissociation.
2. Before 3500 million years ago living cells originated in the oceans by the joining of carbon compounds into complex molecules that characterize all life-forms. Consideration of current hypotheses of the origin of life is out of place here, for it requires a background knowledge of microbiology and biochemistry. These simple cells were capable of building stromatolites. The primary production for this first community was inorganic, or abiotic, that is, cells were anaerobes that obtained energy by absorbing inorganically produced carbon compounds from the primeval ocean and processing them by fermentation.
3. The next step was the development of methanogenic bacteria (type 2). With them came the first photosynthetic bacteria that used

of Early Archean time. It was probably first produced in small amounts by photodissociation, but early photosynthesizers must have produced it in increasing quantities so that by Late Archean time when stromatolites and iron formation became widespread, oxygen oases were common. By mid-Proterozoic time when eukaryotes appeared, the oxygen level had reached about 1 percent of its present value. Some paleontologists have suggested that it did not reach its present value until the middle of the Paleozoic Era. From Hadean to Paleozoic time the atmosphere changed largely through the conversion of carbon dioxide to oxygen. The other major constituent, nitrogen, was derived either from early degassing or from oxidation of ammonia, and changed little in proportion after Hadean time.

Ozone Shield:

At present a layer of ozone (O_3 rather than the usual molecular form of oxygen, O_2) high in the

photosystem-I to form organic compounds (type 3). Some cells did not separate when they divided and formed long filaments.

4. By Late Archean time stromatolites built by anaerobic microorganisms were abundant enough to make significant oxygen oases and some types of cells (type 4 amphiaerobes) developed mediating mechanisms so that they could live in both oxygen-bearing and oxygen-deficient environments.

5. Some time in the Late Archean Era microorganisms (type 5) began using photosystem-II. Through photosynthesis, oxygen in quantity became available. The cyanobacteria, probably the first organisms to use this mechanism, left a record of their flourishing in the abundant Late Archean and Early Proterozoic stromatolites. Some oxygen produced was immediately combined with ferrous salts to form the extensive deposits of sedimentary iron oxides. Preston Cloud has suggested that the potentially lethal oxygen produced by early photosyntheizers was removed from the sea as it oxidized the abundant iron salts in the ocean, precipitating them as iron formation (12,13).

6. The major step in the history of pre-Paleozoic life was the appearance of eukaryotic cells about 1500 million years ago, after 2 billion years of prokaryotic evolution. The step was so large that it was probably made through several intermediate stages. Many microbiologists believe that the intracellular organelles that characterize eukaryotes, such as the nucleus, plastids, mitochondria, and other organelles, were once independant prokaryotic cells that entered a large prokaryotic host cell for their mutual benefit (14). The organisms that resulted not only could use the rapidly increasing oxygen supply by oxidizing carbohydrates but also could reproduce sexually.

All prokaryotes reproduce by simple division: the one copy of the genetic code, which forms a ring within the protoplasm, is divided into two identical daughter cells. Because the daugther cells will be identical to their parent unless there is some serendipitous accident (mutation) which changes the composition of the genetic material itself, the rate of evolution of prokaryotes is extremely slow. Some bacteria are capable of exchanging genetic material through a thin bridge between them in a type of sexual process. In eukaryotes the genetic material

is a sequence of genes arranged on paired chromosomes in the nucleus. In sexual reproduction the cells of both parents split into sex cells, or gametes, each of which takes one copy of the genetic material. The sex cells are dispersed and some eventually unite with sex cells from another individual of the same species to produce an individual that has some characteristics of each parent. By recombination of genes in sexual reproduction new variants of eukaryotic species are produced, some of which may be more advantageous to the survival of the species. Direct changes may also take place in the nature of the genes themselves through mutation. Innovations in the organism itself do not necessarily immediately follow mutation and recombination, as some of the new genes may be recessive, that is they are present in the chromosomes but not expressed in the body of the organism because their effect is overpowered by a dominant gene that determines the particular character or trait. The nature of dominant and recessive genes is discussed in elementary biology textbooks. The point to be made here is that sexually reproducing organisms can accumulate, in the form of recessive genes, a wealth of variation that only infrequently results in the character appearing in the body of the organism. However, this stored variability is available in the event of an environmental change that makes the common variant a less efficient competitor. The organism may have available a variant that can better cope with the changed circumstances and allow the species to persist in a modified form. Thus eukaryotes, in several ways, have potential for more rapid evolution than prokaryotes.

Despite this potential, the next step in the history of life—the appearance of many-celled organisms in which cells are specialized for various functions—did not take place for several hundred million years after the first appearance of the eukaryotes. The evolution of the first multicelled animals, called metazoans, took place at the close of the Proterozoic Eon.

REFERENCES

1. Palmer, A. R., 1985, The decade of North American Geology 1983 time scale: Geology, v. 11, p. 503–504.

2. Cairns-Smith, A. G., 1985, The first organisms: Scientific American, June 1985, p. 90–100.

3. Fox, G. E. et al., 1980, The phylogeny of prokaryotes: Science, v. 209, p. 457–463.

4. Woese, C. R., 1982, The primary lines of descent and uni-

versal ancestor. In Bendall, D.S. (ed.), Evolution from molecules to man: Cambridge, England, Cambridge Univ. Press, p. 209–234.

5. Schopf, J. W.(ed.), 1983, Earth's earliest biosphere: its origins and evolution: Princeton, N.J., Princeton Univ. Press, 343 p.

6. Aramik, S. M., Schopf, J. W., and Walter, M. R., 1983, Filamentous fossil bacteria 3.5×10^9 years old from the Archean of western Australia: Precambrian Research, v. 20, p. 357–374.

7. Schopf, J. W. and Packer, B. M., 1987, Early Archean (3.3 to 3.5 billion-year-old) microfossils from Warrawoona Group, Australia: Science, v. 237, p. 70–73.

8. Awramik, S. M., and Barghoorn, E. S., 1977, The Gunflint Microbiota: Precambrian Research, v. 5, p. 121–142.

9. Walter, M. R. (ed.), 1976, Stromatolites: New York, Elsevier, 790 p.

10. Burne, R. V., and Moore, L. S., 1987, Microbiolites: organosedimentary deposits of benthic microbial communities: Palaios, v. 2, p. 241–254.

11. Nagy, B. et al., (eds.), 1983, Developments and interactions of Precambrian atmosphere, lithosphere, and biosphere: Precambrian Research, v. 20(2–4), p. 103–585.

12. Cloud, P. E. 1973, Paleoecological significance of banded iron formation. In James, H. L., and Sims, P. K. (eds.) Precambrian iron formations of the world: Economic Geology, v. 68, p. 1135–1143.

13. Cloud, P. E., 1983, Banded iron formation: A gradualist's dilemma. In Trendall, A. F., and Morris, R. C. (eds.), Iron formation: facts and problems. Developments in Precambrian Geology, No. 6: New York, Amer. Elsevier, p 401–416.

14. Margulis, L., 1980, Symbiosis in cell evolution: San Francisco, Freeman, 419 p.

SUGGESTED READINGS

Hofmann, H. J., 1987, Precambrian biostratigraphy: Geoscience Canada, v. 14, p. 135–154.

Lipps, J. H. (ed.), 1987, Fossil procaryotes and protists: University of Tennessee, Studies in Geology 18, 303 p.

Margulis, L., 1982, Early life: Boston, Science Books International, 160 p.

Plumb, K. A. and James, H. L., 1986, Subdivisions of Precambrian time: recommendations and suggestions by the subcommission on Precambrian stratigraphy: Precambrian Research, v. 32, p. 65–92.

Schopf, J. W., l978, The evolution of the earliest cells: Scientific American, v. 239, no. 3, p. 110–134.

Schopf, T. J. M., 1980, Paleooceanography: Cambridge, Mass., Harvard Univ. Press, 341 p.

Chapter 5

Origin and
Diversification
of the Metazoa

METAZOANS AND PROTISTANS

Living organisms may be divided into five kingdoms (Fig. 5–1). The kingdom Monera consists of
prokaryotic organisms, such as bacteria, considered
in the last chapter. The eukaryotes that carry on
all life processes within a single cell are placed in
the kingdom Protista. Some of these single-celled
organisms live by photosynthesis, like plants, some
by digesting food particles, like animals, and some
by both. Organisms consisting of many cooperating
cells that live by photosynthesis constitute the Plantae. The primitive aquatic plants, generally known
as algae, are placed by some in a kingdom with the
Protista (the two together then constitute the kingdom Protoctista), by others in a kingdom of their
own, and by still others (as in this book) in the
plants. Fungi obtain food by absorbing organic matter through their cell walls, and the Animalia capture and digest organic matter produced by the
other kingdoms. Some biologists recognize the vi-
ruses as a separate kingdom and others split the
Monera into Archaebacteria (including the methanogens, sulfur-metabolizing and sulfate-reducing
bacteria) and Eubacteria (or true bacteria). This
chapter is concerned with the transition from sin-

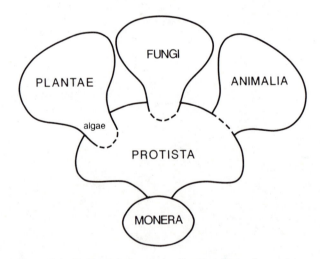

Figure 5-1   Relation between the five kingdoms of organisms
recognized in this book.

gle-celled eukaryotes (protistans) to animals composed of many cells that work together with specialized functions (metazoans).

Protistans are composed of single cells or colonies
of cells, but in either case, each cell is capable of all
of the life functions, such as ingestion, digestion,
respiration, reproduction, and excretion. In the

Animalia, cells are specialized for various functions and cooperate for the good of the whole organism. The Animalia are said to have reached the metazoan stage of organization. The specialized cells of metazoans are united into tissues, such as the exterior membrane or skin of an animal formed of flattened protective cells, or the digestive surface formed by the cells that line the gut. Tissues are united into organs, such as the heart, kidneys, and lungs which facilitate the circulation of fluids, the excretion of wastes, and the exchange of gases with the water or air. The simplest animals are those without backbones, known as invertebrates.

ANIMAL PHYLA IN THE FOSSIL RECORD

The invertebrates are an informal division of the Animalia that includes animals only slightly more complex than colonial protistans, and others, such as insects and sea urchins, with many intricately interacting organ systems. Most invertebrates, and all primitive ones, live in water and a vast majority live in the sea. The bodies of the simplest animals are small and composed of only a few cell layers. As a result, few of the cells are more than a millimeter or so from the water in which they live. The cells can obtain food and oxygen, and excrete wastes by diffusion through the cell walls that are directly in contact with the water, or through a layer of adjacent cells. As animals become larger and more active, with thicker bodies, interior cells become so separated from sources of food (the digestive tissue) and waste disposal (the exterior of the animal) that a system of plumbing is needed to circulate fluid bearing food and oxygen, and to remove waste. A pump, or heart, is needed to maintain flow in the pipes. Excretory organs, such as kidneys, are needed to dispose of the wastes in the fluid. A liver is needed to clean the circulating fluid, the blood. A whole range of organs was developed in the invertebrates as they grew in size, increased their activity, and invaded new environments.

The simplest invertebrates do not have organ systems, only sets of specialized cells united into tissues. (Some biologists would not consider the cell layers of the sponges to be tissues.) They consist of a cavity lined with digestive cells which is enclosed in a sheet of protective cells. When organ systems developed, they were located in cavities, called coe-

loms, in the body between the gut and the protective cells (Fig. 5–2). The primitive invertebrates, because they do not have coeloms, are said to be acoelomate; the rest are said to be coelomate.

The number of invertebrate phyla that are recognized varies from reference book to reference book. The variations depend largely upon the status given to a varied group of worm-like animals that have scanty fossil records. As this group is rarely encountered by paleontologists, it is given little attention in this book. Of the 50 or so invertebrate phyla recognized by some biologists, only eight have significant fossil records. The Protozoa are not invertebrates but are commonly grouped with invertebrates, because they are their immediate ancestors. The Chordates are by definition not invertebrates but those chordates that have close affinities with invertebrates are described in the following chapters with the invertebrates (Table 5–1).

Protozoa:

The protozoans are animal-like protistans, that is, they obtain food by capturing other organisms. Some

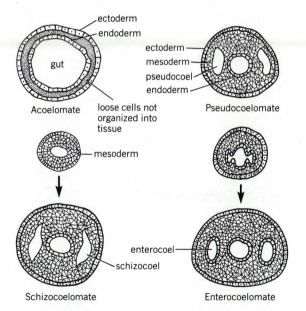

Figure 5-2 Diagrammatic representation of the development of the coelom in invertebrates. In acoelomates, there is no coelom in the unorganized cells between the endoderm and ectoderm. In pseudocoelomates, the pseudocoel is the space unfilled with mesoderm. The schizocoel develops by a split in the mesoderm. An enterocoel initially develops as pouches from the gut.

Table 5-1   PHYLA OF INVERTEBRATES AND PROTOCHORDATES WITH A MAJOR FOSSIL RECORD

	TISSUES AND ORGANS	TISSUE LAYERS	COELOM TYPE	METAMERISM	MOUTH
			Acoelomate–A		
		Endoderm–En	Coelomate–C	Oligomeric–O	
	Tissues–T	Ectoderm–Ec	Schizocoelomate–S	Metameric–M	Protostome–P
	Organs–O	Mesoderm–M	Enterocoelomate–E	Pseudometameric–P	Deuterostome–D
Protozoa	Neither	None	A	–	–
Porifera	?T	En, Ec	A	–	–
Cnidaria	T	En, Ec	A	–	–
Brachiopoda	T & O	En, Ec, M	C	O	P & D
Bryozoa	T & O	En, Ec, M	C	O	P & D
Mollusca	T & O	En, Ec, M	?	P	P
Annelida	T & O	En, Ec, M	S	M	P
Arthropoda	T & O	En, Ec, M	S	M	P
Echinodermata	T & O	En, Ec, M	E	O	D
Protochordata	T & O	En, Ec, M	E	O	D

of them secrete small shells of calcium carbonate or silica and have extensive fossil records. Since they are single-celled, they are neither metazoans nor invertebrates, but the more complex members reach organizational levels comparable to those of the Porifera.

Porifera:

The poriferans are the sponges. They probably developed from protozoans with whiplike extensions surrounded by a collar of fine hairlike processes. In the sponges, layers of these collar cells form digestive surfaces in internal cavities of the body through which water is impelled by the motion of the whips. Microorganisms in the circulating water are trapped by the collars and digested to produce food for the sponge.

Cnidaria:

This phylum comprises the corals, jellyfish, and their relatives. These animals are simple digestive sacks consisting of digestive tissue that lines the interior, protective tissue covering the outside of the animal, and a jellylike layer between the two. These animals capture small organisms in the water by means of tentacles equipped with stinging cells.

Brachiopoda:

The brachiopods are a small group of invertebrates enclosed in two shells. To feed they open their shells and filter micoorganisms and organic matter from the seawater with a comblike organ called a lophophore. They are more complex organisms than corals, and have organ systems for the circulation of fluids and the excretion of wastes. In brachiopods, and all more complex invertebrates, three layers of cells can be distinguished rather than the two of the lower invertebrates: the protective outer layers, called the ectoderm; the digestive surfaces, called the endoderm; and the cells between, which constitute the mesoderm.

Bryozoa:

Bryozoans are small colonial organisms that build a calcareous skeleton much like that of the corals. However, they have soft parts organized like those of the brachiopods. They have the same type of filtering apparatus for food gathering, the lophophore.

Mollusca:

The molluscs include the clams, squids, octopuses, and snails of the present day, their many fossil representatives, and several minor and extinct groups.

Many members of this phylum secrete a shell or pair of shells that enclose the soft parts. The molluscs have organs for respiration (gills), circulation (heart), digestion (digestive gland), etc., but do not have a well-developed body cavity or coelom.

Annelida:

The annelids are the segmented worms. In annelids the organ systems, appendages, and musculature are repeated in many almost identical segments along the axis of the body. As members of this phylum do not secrete skeletons, their fossil record is largely confined to their jaw parts, which are called scolecodonts, their burrows, the linings of the burrows, and their tracks in sediments.

Arthropoda:

This phylum includes not only the insects, spiders, and centipedes, but also the marine crabs, lobsters, and their minute floating and swimming relatives. The number of segments in the body is less than in the annelids, and different sets of segments are specialized for various functions, such as walking, swimming, or food handling. The segments have paired appendages that are modified to serve these functions. Arthropods secrete a rigid skeleton around the outside of the body (the exoskeleton) and therefore, to be efficient, these appendages must have flexible joints. The arthropods are among the most complex and successful of the invertebrate phyla.

Echinodermata:

Familiar members of this phylum of exclusively marine animals are the starfish and sea urchin. Other more obscure members of this group living today are the sea cucumber and sea lily (crinoid). Many members of the phylum secrete skeletons made up of plates of calcite fitted together, therefore they have a diverse and extensive fossil record. The echinoderms differ from other complex invertebrates in their fivefold symmetry and their intricate hydraulic water system that controls locomotion, food handling, and respiration.

Chordata:

By definition this phylum is not an invertebrate group because most of its members are character-
ized by a vertebral structure, a strengthening rod, or backbone. However, many of the primitive members, called protochordates, resemble invertebrates more than they do the advanced chordates called vertebrates. The modern representatives of the protochordates are marine organisms called sea squirts and acorn worms. The importance of the protochordates to the paleontologist lies in the information they give about the relations between the invertebrates and the vertebrates. The graptolites, a group of extinct, abundant fossils, probably were early protochordates.

All of these phyla, their living representatives as well as their fossil ones, are considered in greater detail in subsequent sections of this book.

RELATIONS BETWEEN THE PHYLA

One of the major unsolved problems of paleontology is the origin of the invertebrate phyla from advanced eukaryotic cells at the end of Proterozoic time. Two types of evidence provide information for the solution of this problem: the fossil record, and the comparative anatomy and embryology of living representatives. We will consider the latter first.

Development of the Coelom:

The relations between the invertebrate phyla are revealed by the origin of their coelomic cavities. The Cnidaria, Porifera, Protozoa, and Platyhelminthes (a group of flatworms that do not have a fossil record) are without a coelom and are said to be acoelomate. The molluscs have a poorly developed coelomic cavity and may be closely related to the acoelomates.

In coelomates the body cavity may form during the development of the embryo in various ways (Fig. 5–2). In the phyla grouped as the pseudocoelomates, it is the space between the gut (endoderm) and the external covering (ectoderm) that has not been filled with mesodermal cells during development. The phyla within this group (e.g., the rotifers and nematodes) are wormlike in form. Many are parasitic in habit, small in numbers, and most do not have a fossil record.

In embryos of echinoderms and chordates the coelom is formed when pouches from the gut extend into the mesodermal cells and are pinched off. These cavities are called enterocoels.

In the annelids and arthropods the coelom forms first as a split in the mesodermal tissue and is therefore called a schizocoel.

The way in which the coelom is formed (or not formed) divides the invertebrates into four groups: acoelomates, pseudocoelomates, enterocoelomates, and schizocoelomates (Table 5–1).

Metamerism:

Metamerism is a term zoologists prefer to the term *segmentation*. The phyla, such as the annelids and the arthropods, in which segmentation of the coelom and organs divides the whole body, are said to be metameric. In most of the other phyla the body cavity is divided into three parts from head to tail; these phyla are said to be oligomeric. In the Mollusca, the musculature of one of the primitive classes is repeated down the axis of the organism, but segmentation is not comparable to that of the annelids or arthropods. This condition is said to be pseudometameric. On the basis of segmentation the arthropods and annelids are separated from the rest of the phyla and the molluscs are distinct from both groups.

Location of the Embryonic Mouth:

In the early stages in the development of most invertebrates, the embryo is a hollow ball formed of a single layer of cells. One side of the ball pushes inward, is said to invaginate, to form a primitive digestive cavity and the cells that line the invaginated part become the digestive surface, or endoderm (Fig. 5–3). In the group of invertebrates that includes the molluscs, annelids, and arthropods, the entrance to this cavity becomes the mouth of the digestive system and the exit for the digestive tract (anus) develops by another infolding of the external cell layer that joins the opposite end of the digestive cavity. This group of phyla are called protostomes (Greek, *protos* = first, *stoma* = mouth). In the echinoderms and chordates, as the embryo develops beyond the invaginated ball stage, a new mouth forms opposite the aperture and the first aperture becomes the anus. These phyla constitute the deuterostomes (Greek, *deuteros* = second).

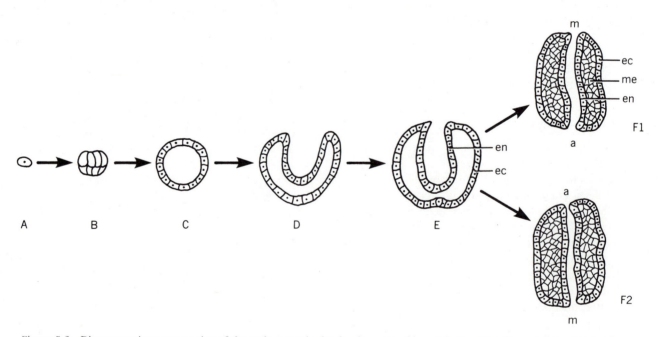

Figure 5-3   Diagrammatic representation of the early stages in the development of invertebrates. The diagrams are not to scale. A. Fertilized cell, B. Multiplication of cells by cleavage, C. Formation of a hollow ball of cells, D. Invagination of the hollow ball to form a digestive cavity, E. Differentiation of the lining of the cavity into endoderm (en) and the outer covering into ectoderm (ec), F. Filling of space between the endoderm and ectoderm with mesoderm. In protostomes (F1), the mouth (m) is at the site of the invagination; in deuterostomes (F2), the anus is in this position and a new mouth is formed.

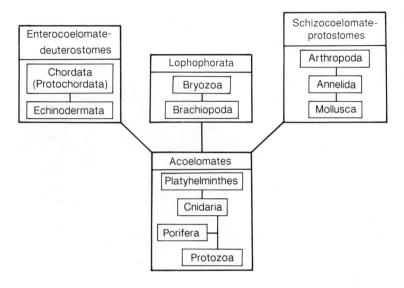

Figure 5-4 The major groups of invertebrate phyla. The lines joining the boxes suggest that the acoelomates are ancestral to the other groups.

Lophophorates:

The position of the embryonic mouth and the origin of the coelom separate the echinoderm–chordate group of phyla (deuterostomes, enterocoelomates) from the annelid–mollusc–arthropod group (protostomes, schizocoelomates). However, the brachiopods and the bryozoans do not separate neatly into these divisions. These phyla have characteristics in their embryonic development that suggest affinities with both the deuterostomes and the protostomes, and with both the schizocoelomates and the enterocoelomates. For this reason, when the relations between the phyla are expressed in a diagram (Fig. 5–4), these two phyla are commonly placed between the annelid–mollusc–arthropod group and the echinoderm–chordate group. Because both the bryozoans and the brachiopods possess a lophophore, the term lophophorates is used to refer to these phyla.

Embryology and Historical Relationships:

Why is the embryonic development of invertebrates so important to those who try to find relationships between phyla and the sequence in which they were derived from one another? The similarities between animals may not be obvious from a comparison of their adult forms, but the relations between them appear more clearly in their embryos. The principle that the embryos of animals resemble the embryos of other related animals much more than do adult forms was first recognized by the German biologist Karl Ernst von Baer (1792–1876). The pathways by which the fertilized cells of an invertebrate grow into an adult are relatively few. Some of the major differences in these pathways, such as the development of the coelom or the mouth, have been described above. The branchings of the developmental pathways are believed to preserve in the life history of the individual different adaptive choices made by ancestors at the time when the phyla were diversifying. For example, segmentation may have developed in primitive invertebrates to produce efficient burrowing movements, but many annelids that acquired segmentation early in their embryonic development are no longer burrowers. As the individual develops from the fertilized cell, the structures that are basic to the organization of its phylum, such as coelomic configuration, segmentation, and organ systems, are modified by the development of structures related to the specialized adaptations of that particular species. Therefore the structures that are essential to membership in a phylum are more clearly seen in embryos, and the relations between phyla are more clearly revealed by a comparison of embryos. We will return to this subject in Chapter 15 and discuss its relationship to the mechanisms of evolution.

By comparison of adult and embryonic anatomy of living invertebrates, the phyla can be placed in an organizational diagram (Fig. 5–4). The diagram represents the structural similarity between the invertebrates that are common in the fossil record and suggests how one might have been derived in time from another, but to produce a model of the course of differentiation of the phyla at the end of the Late Proterozoic Era, we must determine whether the conclusions from these comparative studies are supported by the fossil record.

PALEONTOLOGICAL RECORD OF THE EARLY METAZOA

The interval of earth history during which the diversification of the metazoans took place, at the end of the Proterozoic Eon, has been given various names, none of which has as yet received the approval of stratigraphic commissions or international acceptance. This interval lies between the end of the Late Proterozoic glaciation (approximately 680 million years ago) and the beginning of the Paleozoic Era, marked by the appearance of fossils of animals with hard skeletons. In this book, the term Ediacaran as defined by Preston Cloud and Martin Glaessner (1) is used, but the interval is considered to be part of the Late Proterozoic Era, not of the Paleozoic Era as they propose.

Tracks and Burrows:

The fossils of the earliest metazoans are not their shells nor the impressions of their bodies, but burrows that they have left in soft sediments. The oldest undisputed metazoan traces come from rocks of early Ediacaran age. Structures that have been described as metazoan traces have been recorded from older Late Proterozoic rocks (680 to 900 million years old) but the interpretation of these older structures is controversial. The oldest structures that may be metazoan burrows have been described by Erle Kauffman and James Steidtmann (2) from rocks as old as 2000 million years. A famous fossil of doubtful significance is *Brooksella canyonensis* (Fig. 5–5) from the middle Proterozoic beds of the Grand Canyon (1100 to 1300 million years old). It was originally described as a jellyfish impression and given the name of a jellyfish genus, but has now been interpreted as either the trace of a burrowing complex metazoan (3) or as a deformed sedimentary blister (4). Many paleontologists ascribe all pre–Ediacaran metazoan fossils to inorganic processes such as the escape of water or gas from consolidating sediments. The most convincing evidence for pre–Edicaran metazoan fossils are carbonaceous, transversely ridged tubes found below late Proterozoic tillites in China, but the affinity of these has also been questioned (5). At the beginning of the Ediacaran successions, indisputable traces of

Figure 5-5   A Middle Proterozoic structure of doubtful significance, *Brooksella canyonensis*, from the Grand Canyon. It has been interpreted as a jellyfish, a sedimentary blister, and a trace fossil. (Courtesy of Smithsonian Institution.)

metazoans that burrowed subhorizontally to shallow depths in the mud and sand appear in abundance. They are joined in late Ediacaran strata by the traces of animals that crawled along the surface, including traces that could only have been formed by such complex invertebrates as arthropods. Until the close of the Proterozoic, burrows were less than 2 cm deep but at that time deeper vertical burrows become common, possibly in response to the appearance of predators. The sequence from simple, shallow, horizontal burrows to horizontal, transversely annulated burrows, to horizontal burrows with upward prolongations, to vertical burrows, and finally to arthropod traces may be the key to biostratigraphic zonation of Ediacaran strata.

The Ediacaran Fauna:

A fauna of early metazoan body fossils (as opposed to trace fossils) is preserved in the Pound Quartzite in the Flinders Range of South Australia. The rock is a sandstone formed from sands on which the organisms were impressed as they lay on the seafloor. The soft bodies of the organisms themselves have decayed. The impressions have not preserved fine details of the animals' surfaces because the sand that has taken the impressions is relatively coarse (Fig. 5–6). In all, 28 genera of metazoans have been described from the Pound Quartzite.

Most of the organisms preserved are what are commonly known as jellyfish and more technically

Figure 5-6   Fossils of metazoans from the latest Proterozoic Ediacaran interval. A. *Spriggina floundersi* (×2), B. *Tribrachidium heraldicum* (×1.5), C. *Charniodiscus arboreus* (×0.25), D. *Dickinsonia costata* (×1), E. *Cyclomedusa radiata* (×0.7). (Courtesy of R. Jenkins, M. F. Glaessner and *Science*, v. 217, p. 783, © 1982 by the AAAS.)

as medusoids. They have been assigned to several classes of the Cnidaria, but some are difficult to place in any living group. In all, 12 genera of medusoids have been described so far. Some of the more common forms are illustrated in Figures 5–6 and 5–9. The second largest group of Ediacaran fossils are leaflike organisms as long as 1 m that were anchored to the seafloor (Fig. 5–6, 5–9). They are considered to be colonial cnidarians that were composed of many polyps housed in depressions on the leaflike structures. Modern relatives of these cnidarians are the sea pens and sea fans, which are referred to as the soft corals because they do not secrete a solid skeleton. Three genera have been

referred to the Annelida and two to the Arthropoda. *Tribrachidium* (Fig. 5–6) is a disklike impression with three symmetrical bent arms on its surface. Its resemblance to some primitive, pancakelike echinoderms of the early Paleozoic called edrioasteroids has suggested that it is an early representative of the phylum Echinodermata. However, the echinoderms characteristically have a fivefold symmetry; no other invertebrate has the threefold symmetry of *Tribrachidium*. A discoidal Ediacaran fossil with five radial grooves on the surface (*Arkarua*) has also been described as an ancestral edrioasteroid.

Since this remarkable fauna was found in the Flinders Range in the mid-1940s, similar fossils have

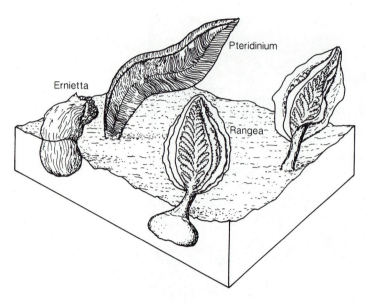

Figure 5-7 Reconstruction of primitive attached cnidarians (Petalonamae) from the late Proterozoic of Namibia, Africa. (*Source:* Based on reconstruction by Richard Jenkins.)

been found in beds of Ediacaran age in many places around the world. About 25 localities have been found and more are discovered every year. One of the richest faunas comes from the Nama Group of South Africa which overlies the Numees tillite, a deposit of the Late Proterozoic glaciation. This fauna was first described in the early years of this century. The tubular worm fossil, *Cloudina*, occurs at the base and in the middle of the group. In the middle part a diverse group of medusoids have been found, and many genera of leaflike colonial cnidarians including *Rangea* and *Petridinium* (Fig. 5–7). So widespread and varied are the attached cnidarians that H. D. Pflug (6) has suggested that they be recognized as a special order, the Petalonamae. Many genera of the Australian Ediacaran fauna occur in the Soviet Union where equivalent beds constitute the upper part of the Vendian System. Particularly rich finds have been made along the coast of the White Sea in arctic Russia and in the Ukraine. Other important Ediacaran faunas have been reported from England and China.

In North America the fauna appears in the Conception Group of eastern Newfoundland. It includes an enigmatic spindle-shaped organism (Fig. 5–8) whose surface is like the colonies of *Rangea*. The fossil impressions in Newfoundland are on beds below volcanic ash beds. Apparently the floating, fragile animals were killed and buried on the seafloor during eruptions of nearby volcanoes.

Figure 5-8 Spindle-shaped metazoan fossil from the latest Proterozoic Conception Group, eastern Newfoundland. (Courtesy of M. A. Anderson and S. B. Misra.)

Figure 5-9   Reconstruction of some components of the Ediacara fauna. The large leaflike organism in the foreground is *Charniodiscus*. Various medusoids are reconstructed in the water. A large *Dickinsonia* undulates through the water while another lies on the sediment surface near a *Tribrachidium*. (Courtesy of Mary Wade, Robert Allen, and the Queensland Museum, Brisbane.)

The soft-bodied fauna of medusoids, colonial cnidarians, annelids, arthropods, and enigmatic forms disappears before the end of the Late Proterozoic Era, and youngest Ediacaran rocks contain only trace fossils. In this fauna we glimpse a fleeting stage in the diversification of the metazoa, a record of animals that may have given rise to the more diverse and abundant invertebrates of the Cambrian and later periods, but which became extinct before the end of the era.

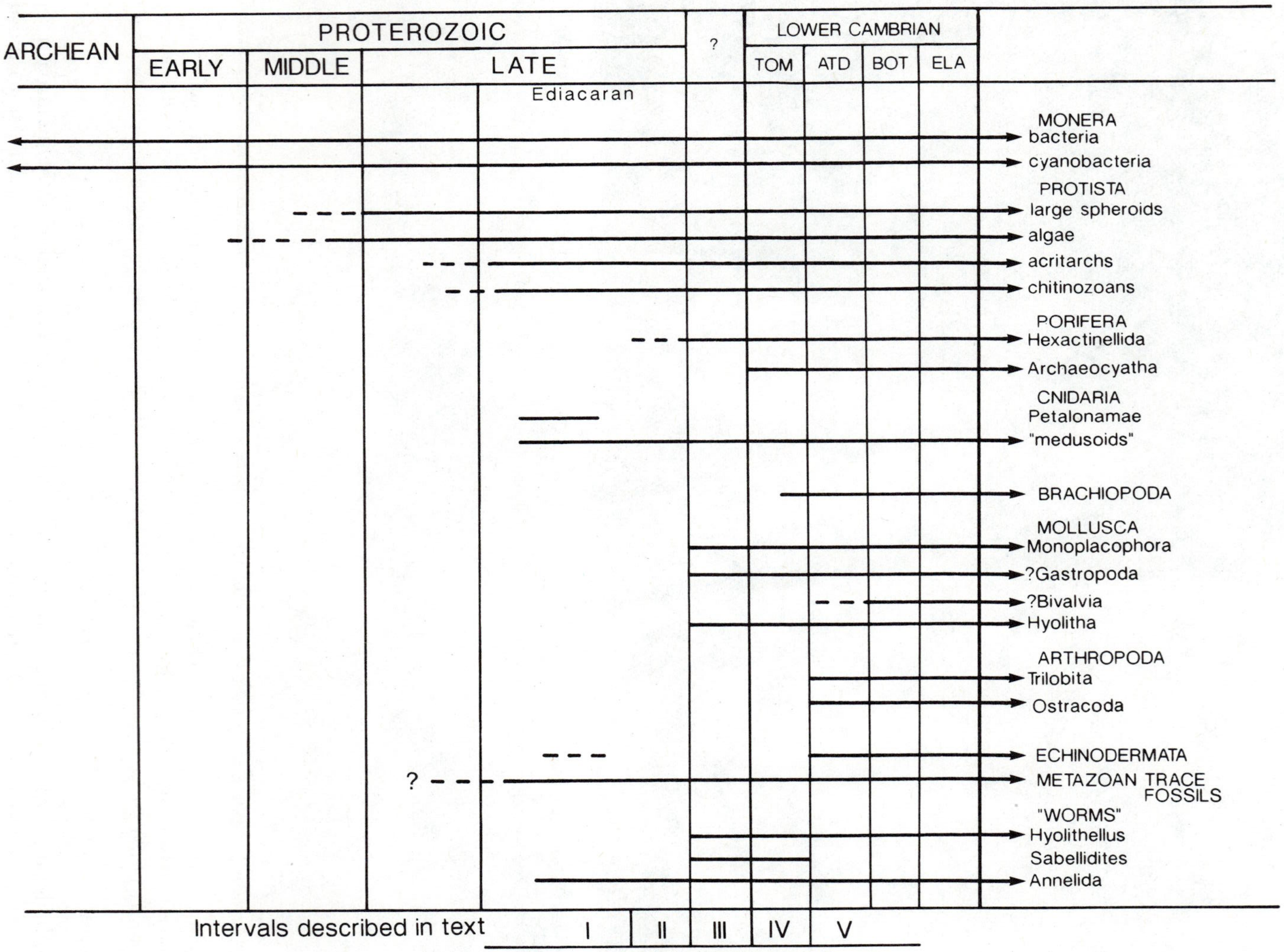

Figure 5-10 Ranges of major organisms across the Proterozoic–Cambrian boundary. The intervals described in the text are indicated at the bottom of the chart. The four stages of the Lower Cambrian recognized in the U.S.S.R. are abbreviated. (*Source:* Modified from Martin F. Brasier.)

The First Cambrian Fauna:

The base of the Cambrian System traditionally has been recognized by paleontologists at the appearance of fossils of animals that secreted a skeleton. In limestone successions of this age, the boundary was recognized by the appearance of an extinct group of calcareous sponges called archaeocyathids, which are described in the next chapter. The archaeocyathids were small, vase-shaped animals that grew in sufficient abundance to construct reef bodies in Early Cambrian rocks. In more shaly successions, the boundary was marked by the appearance of the chitinous trilobite arthropods, particularly those of the genus *Fallotaspis*. In recent years paleontologists have found increasingly older and older fossils of the skeletons of animals that were neither archaeocyathids nor trilobites in beds that underlie the first appearances of these two groups.

The newly discovered fossils have been found in the residues of limestones dissolved in acetic acid. They are small tubular and plate fossils composed of calcium phosphate. At this time in earth history the ocean appears to have been rich in phosphate for in many parts of the world at the base of the Cambrian System limestones are highly phosphatic. The phosphatic shells may have been originally secreted as phosphate, or they may have been calcium carbonate that was replaced in the phosphate-rich ocean.

The faunas near the Ediacaran–Cambrian boundary beds can be described in terms of five stratigraphic intervals that are coming to be recognized at many localities around the world (Fig. 5–10). The upper stages were first described in Siberia.

I. The lower part of the Ediacaran interval that follows the tillites of the Late Proterozoic Varangian glaciation is characterized by the soft-bodied Ediacaran fauna described above.

II. Late Ediacaran beds are barren of body fossils but contain the trails of metazoans, including trails that have been ascribed to arthropods, probably trilobites.

III. In Siberia the overlying beds constitute the Nemakit–Daldym Formation. They contain the first microscopic phosphatic fossils. *Anabarites*, *Protohertzina*, and *Hyolithellus* are typical examples (Fig. 5–11). The biological affinities of all these genera of tapering tubes are in doubt.

IV. The Tommotian Stage follows the Ne-

makit–Daldym zone in Siberia and contains a far more diverse fauna of phosphatic shells (Fig. 5–12). Included are sponges, brachiopods, gastropods, primitive molluscs, hyolithids (described later), and the plates and spines of organisms called tommotids. We know that these animals secreted a variety of hollow cones (*Lapworthella*) and plates around their bodies but we do not know to which phylum they belonged. The Tommotian fauna includes the calcareous archaeocyathid sponges. Recently the Tommotian has been shown to be of the same age as beds elsewhere that contain trilobites, and its value as a discrete stage has been put in doubt (7).

V. In Siberia the following stage, the Atdabanian, contains the first body fossils of trilobites and a host of the new invertebrate groups

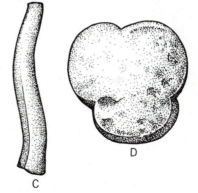

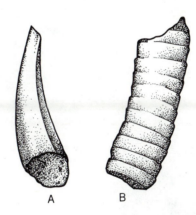

Figure 5-11   Phosphatic tubular fossils of the Cambrian–Ediacaran boundary beds. A. *Protohertzina*. B. *Hyolithellus*. C,D. *Anabarites*.

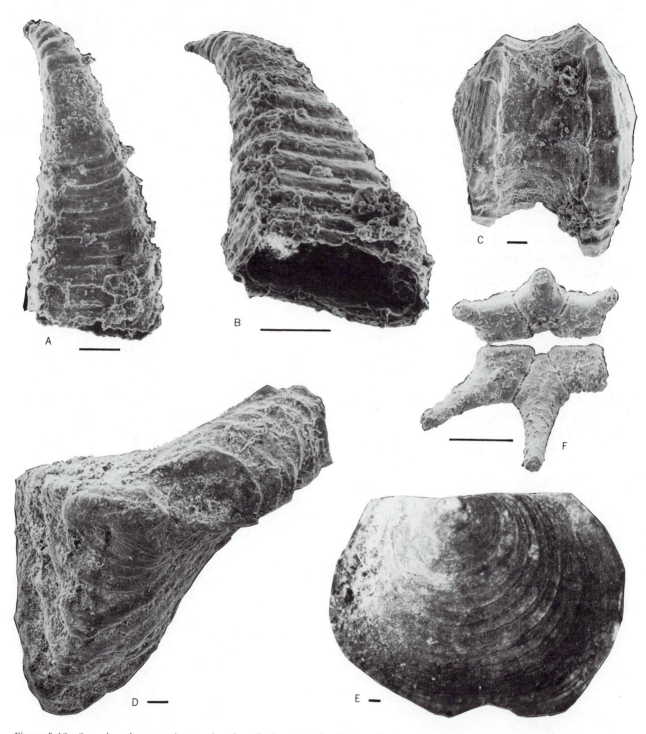

Figure 5-12 Scanning electron micrographs of small phosphatic fossils from the Lower Cambrian Tommotian stage. The scale bars are 0.1 mm long. A–D. Spines and scales of enigmatic organisms called tommotiids. A, B. *Lapworthella*, C. *Camenella*, D. *Sunnaginia*, E. *Paterina*, a brachiopod, F. *Chancelloria*, an organism of unknown affinities. (Courtesy of Godfrey Nowlan, Geological Survey of Canada, and *Canadian Journal of Earth Science* [A,B], M. F. Brasier, *Geology Today*, Sept. 1985, Blackwell Scientific Publication [C,E].)

that characterize the rest of the Lower Cambrian fauna.

Within this succession the position of the Ediacaran–Cambrian boundary has not been finally placed, but a commission of stratigraphers recently recommended that it be located between zones III and IV.

Hyolithids:

Small conical shells of uncertain affinities have been found in many early and middle Paleozoic shales and limestones in great abundance. The hyolithids are a group of bilaterally symmetrical, conical shells composed of calcium carbonate that appears at the base of the Cambrian System (Fig. 5–13). In cross section the cones are generally triangular, but may also be circular, oval, lenticular, or polygonal. The cones are open at their wide end, but this aperture could be closed by a trapdoor plate called an operculum. On the inside of the operculum are two or four paired scars where the muscles that opened and closed it were attached. The closed tip of the cone is commonly divided by a few partitions into small chambers but the rest of the cone, presumably occupied by the soft parts when living, is now empty. Rare specimens also show a pair of curved projections from the region of the aperture that point toward the apex of the shell (Fig. 5–13). The shells are from about 1 to 150 mm in length. The typical genus is *Hyolithes*.

The hyolithids thrived in the Cambrian and declined steadily through the Paleozoic, until they became extinct in the Permian. They are usually considered to be a class of the Mollusca but their soft parts and way of life cannot be reconstructed with confidence. Some of them may have been light enough to float near the surface, like the pteropods, which are modern aberrant molluscs with thin calcareous shells. Others could have lived with the point of the shell stuck in the seafloor sediment. Others may have swum or propelled themselves on the seafloor. No agreement has been reached on the function of the thin arms. They may have been used to balance the organism, to pull it around on the seafloor, or to support a fleshy swimming membrane that has not been preserved.

The hyolithelminthes, small shells found in Tommotian rocks, have been compared to hyolithids because they have a conical form and an operculum. However, the shell is entirely without internal

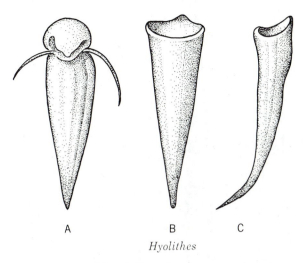

Hyolithes

Figure 5-13 *Hyolithes.* A. Exceptionally preserved specimen showing the operculum and projections or arms. B, C. Two views of a normally preserved specimen of the shell. About three times natural size.

partitions, is composed of phosphate, and is circular in cross section. In addition, the muscle scars on the underside of the circular operculum are in five pairs. The affinities of *Hyolithellus* (Fig. 5–11) and several related genera are unknown but they have been placed customarily in the Annelida, or some other worm phylum.

Cricoconarids:

A second group are known as cricoconarids (Fig. 5–14) and, like the hyolithids, are generally considered to belong to an extinct class of the Mollusca. The slender cones of the members of this group are composed of calcium carbonate, and are round in cross section, commonly marked by rhythmically spaced transverse rings, rarely by fine longitudinal grooves. Transverse partitions may form a series of chambers across the tip of the cone. No operculum closing the aperture of these shells has been found.

The cricoconarids are rare in Ordovician rocks but increase in abundance through the Silurian and into the Devonian. They became extinct near the close of Devonian time. The genus *Tentaculites* is common in Silurian and Devonian rocks, and locally their tiny cones cover bedding planes in the Late Silurian limestones of New York State. *Styliolina* is a fragile cone with fine transverse grooves that is found in Late Silurian and Devonian shales.

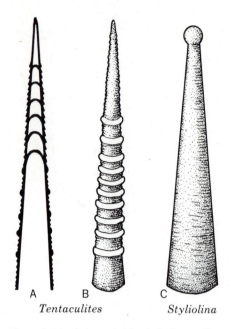

A   B   C
Tentaculites   *Styliolina*

Figure 5-14   Cricoconarids. A, B. *Tentaculites*, exterior and cross section to show the internal partitions, C. *Styliolina* showing the initial bulbous part of the cone that is rarely preserved. The shells are about 5 mm long.

Some cricoconarids have been found to be valuable in determining the division and correlation of Devonian successions.

Cricoconarids probably lived floating near the surface of the ocean. This conclusion is based on (1) the delicacy of their shells, which appear to be unsuitable for the rough and tumble of life on the sea floor; (2) the presence of chambers that were presumably gas-filled and buoyant; (3) their widespread distribution geographically; and (4) their occurrence in abundance in beds containing few or no fossils of bottom-living organisms, which suggests that they drifted down after death into an inhospitable bottom environment.

A MODEL OF METAZOAN ORIGIN AND DIVERSIFICATION

Diversification:

By the middle of Early Cambrian time most of the major groups of invertebrates that will be described in detail in the next chapters had appeared in the sea. The sponges are represented in the fossil record by spicules, assigned to several different or-

ders, as well as by archaeocyathids; the cnidarians, both as medusoids and attached soft corals, had been abundant in Ediacaran seas; primitive shells assigned to the gastropods and monoplacophorans represent the beginnings of the molluscan classes; both inarticulate and articulate brachiopods had appeared; annelids were represented by a variety of tracks and burrows as well as such Ediacaran fossils as *Dickinsonia*. The trilobites were represented by body fossils, but trace fossils suggesting their presence are known from older Ediacaran strata, and primitive echinoderms herald the coming of the vast gardens of crinoids and cystoids that carpeted Ordovician and later Paleozoic seas.

The rapid appearance of most of the invertebrate phyla and their major classes and orders in Early Cambrian time is a striking feature of the paleontological record. The diversification of life has been documented by John Sepkoski and his co-workers (8). Figure 5–15 shows the number of families of marine invertebrates that have been identified as belonging to the Ediacaran–Cambrian interval. Similar diagrams can be constructed for orders or genera. All show an interval of about 80 million years, during which the variety of life increased exponentially. The exponential nature of the increase is best shown if the scale of the number of fossil groups is logarithmic (Fig. 5–16), so that the curve plots as a straight line. Sepkoski has pointed out that such a pattern would result from repeated branching from a stem. If each group originated by the simple splitting into two of an ancestral group, then the rate of increase in the number of groups is exponential, two groups give rise to four, four to eight, eight to sixteen, and so on.

Obviously this rate of increase in diversity could not continue indefinitely. In Ordovician time the diversity curve levels off and, with minor perturbations, continues at this level until near the end of the Paleozoic. An equilibrium was reached between the appearance of novel forms and the extinction of old ones.

Although the rocks of the Ediacaran interval preserve a record of early metazoans, the record does not reveal how or when the simpler phyla gave rise to more complex phyla. Reconstructions of this most important period for the evolution of life between 680 and 570 million years ago must be based largely on comparative anatomy and embryology of living invertebrates, but must be compatible with the fossil evidence. The widely accepted views of R. B. Clark (9) are used as a basis for the following discussion.

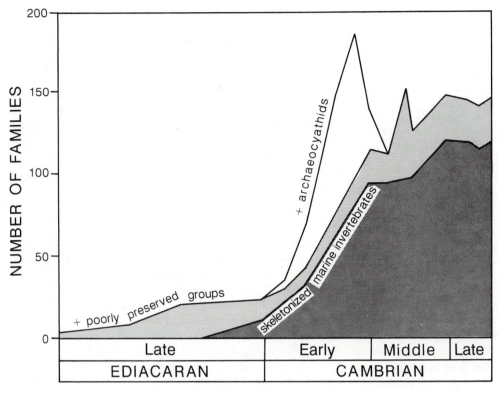

Figure 5-15 Exponential expansion in skeletonized metazoans at the beginning of Cambrian time. The archaeocyathids are separated from the rest of the metazoans as they appear to have undergone a separate expansion and decline. Note that the increase in the number of families levels off after Early Cambrian time. (Courtesy of J. J. Sepkoski and The Paleontological Society, *Paleobiology*, v. 5, p. 229.)

Acoelomates:

The first metazoan was probably a wormlike organism resembling the planula larva of modern cnidarians (Fig. 5–17). Zoologists believe that it evolved from a colonial form of the most advanced of the protozoans, the flagellates. It was covered with cilia, by which it swam or worked its way through the sediment. It consisted of a gut lined with endoderm and an outer ectodermal layer of cells. It had no intermediate layer of cells (mesoderm) or coelom. The original metazoan gave rise to the Cnidaria, in which an anus does not develop in the gut and a simple digestive sack makes up most of the body of the adult polyp or medusa. The abundance of both mobile and fixed cnidarians in the Ediacara fauna supports the conclusion that cnidarians were among the most primitive metazoans. The crawling and swimming way of life of the original planula is maintained in the modern free-living flatworms which belong to the phylum Platyhelminthes. The only organ system of these simple worms is the reproductive system. Platyhelminthes rely on diffusion to transport nutrients from the many-branched gut to interior cells. The free-living flatworms are believed to be similar to the ancestors of the higher invertebrates, but unfortunately the phylum has no fossil record to confirm their presence in Ediacaran seas. Many members of the phylum Platyhelminthes have become parasitic organisms.

The Porifera probably arose directly from colonial, collar-celled protozoans. In many ways their organization is only slightly more advanced than that of a colonial protozoan.

Coelom Development:

In the simplest coelomates (pseudocoelomates), the nematodes, rotifers, acanthocephala, and others, the coelom acts as a hydrostatic skeleton. The pressure

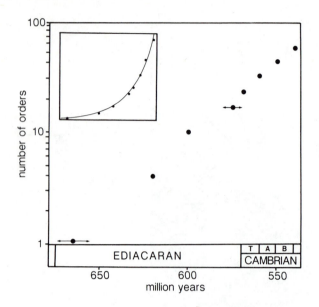

Figure 5-16   Increase in the number of orders at the end of Proterozoic and beginning of Paleozoic time on a semilogarithmic plot to show its exponential nature. Arrows extending from points indicate range of the uncertainty of age assignment. T, A, and B are abbreviations for Lower Cambrian Stages. (Courtesy of J. J. Sepkoski and The Paleontological Society., *Paleobiology*, v. 4, p. 228.)

of the fluid inside the coelom is slightly higher than in the surrounding water or air. It gives the body rigidity so that when it is distorted by muscles it will resume its shape. The muscles of all animals are capable of contraction only, and in pseudocoelomates are reextended after contraction by internal pressure in the animal. The sealed cavity of the coelom allowed wormlike invertebrates to increase their efficiency of movement in burrowing by making it possible for the animal to return to its original shape after flexing, like a bent elongate balloon.

Metamerism:

Although a single coelom allows the whole worm to change its proportions and resume its shape, the changes are of limited value in burrowing. An efficient burrower, such as the earthworm, an annelid, has longitudinal muscles whose contraction thickens the body, and encircling muscles whose contraction thins it. The worm progresses by making certain zones, in which the body is thickened, pass backward, thereby pressing against the wall of the burrow and forcing the animal forward (Fig. 5–18). Such a mechanism requires that different

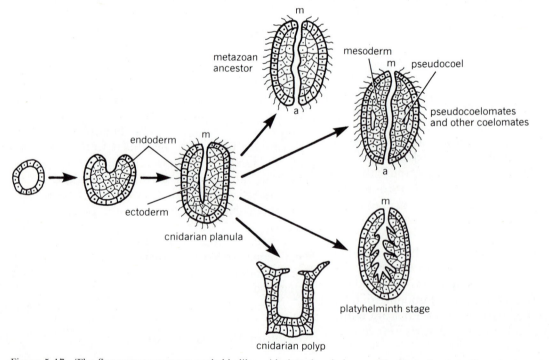

Figure 5-17   The first metazoans were probably like cnidarian planula larvae. The larva develops from the invaginated ball stage into a coral polyp. Flat worms (Platyhelminthes), which have a blind gut, are at the cnidarian planula stage. The development of mesoderm and cavities in it leads to coelomate invertebrates. m = mouth, a = anus.

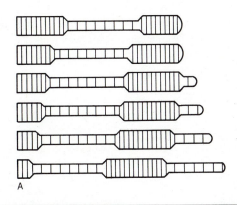

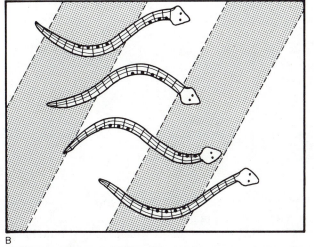

A

B

Figure 5-18 Locomotion and metamerism in metazoans.
A. The earthworm progresses by passing sections of the body
that are thickened by the successive contraction of muscles
backward along the body. B. Swimming invertebrates pass
lateral waves that press back on the water down the body by
contraction and relaxation of muscle blocks on either side of
the body. Contracting muscle blocks are marked by a dot.
(*Source:* Modified from M. Wells, *Lower Animals.*)

zones in the body respond to muscular contractions
and relaxations in an independent, yet coordinated,
way. The body must be divided into many small
cavities along its length, and each division must
have its own muscle system. The major segmented
phyla are believed to have arisen as adaptations to
more efficient locomotion on, and within, the seafloor sediments. The segmented muscle systems
could then be adapted to a swimming motion, in
which lateral waves pass down the length of the
body by the alternate contraction and relaxation of
lateral muscle blocks (Fig. 5–18). That metamerism
was an early adaptation of metazoans is shown by
the early appearance of annelids, such as *Dickin-*

sonia; arthropods, such as *Parvancorina*; and their
tracks in Ediacaran sediments.

The oligomerous phyla, which include the lophophorates and the deuterostomes, did not develop full metamerism. The relations between these
phyla are unclear, and their coelomic cavities may
have formed in response to different needs. The
lophophorates were probably first represented by
worm-like forms that resembled living members of
the phylum Phoronida (Fig. 5–19). These lophophore-bearing worms secrete tubes more than a
meter long in seafloor sediments. Paleontologists
have tentatively identified some enigmatic tube fossils of the earliest Paleozoic as phoronids. By the
beginning of Paleozoic time, animals like phoronids
may have given rise to the Brachiopoda, but the
second major group of the lophophorates, the Bryozoa, are not clearly represented in the fossil record
until the Ordovician Period.

Deuterostomes:

R. B. Clark believes that the echinoderms developed from tube-dwelling worms, and the chordates
from swimming ancestors. In the chordates the
muscles for locomotion pull against a rigid rod along
the back, which in advanced members of the phylum becomes a segmented, bony vertebral column.
The segmentation of the musculature was devel-

Phoronis

Figure 5-19 The phoronid *Phoronis*.

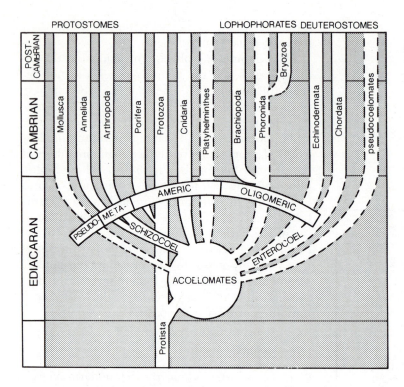

Figure 5-20   Model of diversification of the invertebrate phyla at the beginning of the Paleozoic Era. pseudo = pseudomeric, meta = metameric.

oped to facilitate the passage of lateral waves down the length of the body, producing the swimming motion seen in fish or snakes (Fig. 5–18). The deuterostomes were probably later developments in the diversification of the metazoans since chordates do not appear until the middle of the Cambrian Period, and echinoderms (unless *Tribrachidium* is considered to be an echinoderm) not until the beginning of the Cambrian Period.

Figure 5–20 is a summary of these relationships. Much remains to be learned about this period of life history when the basic architecture of the animal phyla was determined, but much progress has been made in the last 20 years through the combined efforts of paleontologists and zoologists.

CAUSES OF METAZOAN DIVERSIFICATION

Dramatic events, such as the diversification of the invertebrate phyla at the beginning of the Paleozoic Era illustrated in Figures 5–10, 5–16, and 5–20, demand an explanation. The 500 million year interval between the evolution of the eukaryotic cell, the previous major evolutionary step, and the first metazoan fossils suggests that the microorganism community had acquired great stability, and re-

quired some internal or external stimulation to proceed to another major innovation. The problem can be stated in another way—what changes could have caused the unicellular eukaryote community that had dominated the oceans for 1 billion years to have given rise to all the major groups of complex animals within about 100 million years?

Environmental Factors:

M. J. S. Rudwick (10) suggested that the diversification was related to the ending of the Varangian glaciation in latest Proterozoic time. The Varangian glacial deposits are widespread in latitude, suggesting a particularly cold climate during this time. The improvement of the climate accompanying the waning of the glaciers about 700 million years ago is postulated to have expanded the environments in the sea that were hospitable to life, thereby triggering the expansion of metazoan diversity.

Several hypotheses postulate an increase in availability of shallow-water marine environments favorable to animals that lived on or in marine sediments at the beginning of the Paleozoic. Michael LaBarbera (11) suggested that the extensive continental shelves that now underlie shallow seas came into existence at the beginning of Paleozoic time and that the increased area and diversity of shallow-water environments on such shelves was the trigger for metazoan diversification. M. D. Brasier (12) showed that the diversification accompanied the rise of sea level that caused the ocean to extend over the continental platforms in shallow seaways in which many new habitats were opened. James Valentine and E. W. Moores (13) related the diversification to the availability of new habitats as the continents, which had been united into a supercontinent in late Proterozoic time, broke into smaller masses. This resulted in a great increase in coastline lengths and a rapid increase in the availability of habitats for shallow-water organisms.

Most hypotheses of chemical change have suggested that an increase in oxygen in the seas and atmosphere controlled the diversification. Preston Cloud (14) has estimated that in Ediacaran time, attainment of an oxygen content of 8 percent of present atmospheric level stimulated rapid diversification. Bruce Runnegar (15) suggests that the first metazoans appeared when oxygen level was still only 1 percent of present atmospheric level, and that it had only risen to 6 to 10 percent by the time the Ediacaran fauna appeared. These estimates are based on the respiration requirements of such large (up to 1 meter) and flat animals as *Dickinsonia*, which are thought to have been adapted to obtaining oxygen entirely by diffusion.

Biological Factors:

Many of the invertebrates that originated in the Ediacaran interval were suspension feeders. This means that they extracted microorganisms from seawater by passing the water through various forms of filtering devices. Their sudden diversification suggests that the availability of floating and swimming microorganisms increased rapidly at this time, and that prior to this time the microorganisms lived on the surface of, or within, sediments. With the exception of the enigmatic acritarchs (see Chapter 10), microorganisms are not generally preserved in rocks of this age, making this hypothesis difficult to test.

Once the process of diversification started, additional environments would be created by the organisms themselves, because most organisms depend on other organisms for food or homes. Although this concept of rapidly expanding, biologically controlled habitats helps explain the snowballing effect of diversification, it does not explain why it started.

Secretion of Skeletons:

The secretion of skeletons of phosphate or hard organic compounds, such as proteins, appears to have accompanied or closely followed the diversification of invertebrates into phyla. External skeletons are usually a defense from predators, and many hypotheses formulated to explain their acquisition in Cambrian time have speculated that predators first became abundant at this time. However, hard parts also serve as supports to bring organisms above the smothering effects of bottom sediments, as solid surfaces for the attachment of muscles, as guides for feeding currents, and for many other purposes. Calcium and phosphate ions are both essential to the cells of metazoans: the first is involved in the control of muscle contraction, and the second in energy transfer within cells. Skeletal parts may at first have been reservoirs of these valuable materials for the animal. The skeletal parts of most early Paleozoic animals were composed of calcium phosphate or hard organic compounds, although some animals (such as archaeocyathids) secreted calcium carbonate as early as earliest Cambrian time. The animals that are heavy users of calcium carbonate did not appear in abundance until mid-Ordovician time. At present calcium carbonate is not secreted by animals living in zones of the ocean where oxygen content is lower than 1 milliliter per liter. This would be the maximum concentration of oxygen in an ocean in equilibrium with an atmosphere containing 16 percent of the present atmospheric level of oxygen. This suggests that oxygen level controls on skeleton secretion may have been important in the early Paleozoic also. Calcium carbonate skeletons probably could not have been formed until oxygen content reached the 10 percent level in Cambrian time. The approach of oxygen levels to those of the present in Ordovician time may have allowed the proliferation of such carbonate-secreting animals as the corals, sponges, brachiopods, bryozoans, molluscs, and echinoderms. No single factor is likely to provide a complete explanation for the proliferation of skeletons from Early Cambrian to Ordovician time.

Compared to the 3.5 billion years of life history, the diversification of the metazoa and the acquisition of skeletons by the invertebrates appear to be relatively rapid events. Before the details of the Ediacara and lowest Cambrian faunas were known, these events were characterized as explosive evolution. However, we now know that they took place over a period of several hundred million years and, against any scale other than a cosmic one, were extremely slow.

REFERENCES

1. Cloud, P. E., and Glaessner, M. F., 1982, The Ediacaran Period and System: the metazoa inherit the earth: Science, v. 217, p. 783–792.

2. Kauffman, E. and Steidtmann, J., 1981, Are these the oldest metazoan trace fossils?: Journal of Paleontology, v. 55, p. 923–947.

3. Glaessner, M. F., 1983, The emergence of the metazoa in the early history of life. Precambrian Research, v. 20, p. 427–441.

4. Cloud, P. E., 1973, Pseudofossils: a plea for caution. Geology, v. 1, p. 123–127.

5. Sun, W-G. Wang, G-X., and Zhou, B-H., 1986, Microscopic wormlike body fossils from the Upper Precambrian (900–700Ma) Huainan district, Anhui, China, and their stratigraphic and evolutionary significance: Precambrian Research, v. 31, p. 377–403, and following discussion.

6. Pflug, H. D., 1972, The Phanerozoic–Cryptozoic boundary and the origin of the metazoa: 24th International Geological Congress, Montreal, Section 1, p. 58–67.

7. Moczydlowska, M. and Vidal, G., 1988, How old is Tommotian?: Geology, v. 16, p. 166–168.

8. Sepkoski, J. J., 1978, A kinetic model of Phanerozoic taxonomic diversity. I. Analysis of marine orders: Paleobiology, v. 4, p. 223–251.

9. Clark, R. B., 1979, Radiation of the metazoa. in House, M. R., (ed.) Origin of the Major Invertebrate Groups: Systematics Association Special Paper 12, p. 55–102.

10. Rudwick, M. J. S., 1964, The Infracambrian glaciation and the origin of the Cambrian fauna. in Nairn, A. E. M. (ed.), Problems in Paleoclimatology, New York, Wiley Interscience p. 150–155.

11. LaBarbera, M., 1978, Precambrian geological history and the origin of the metazoa: Nature, v. 273, p. 22–25.

12. Brasier, M. D., 1982, Sea level changes and the late Precambrian–Early Cambrian evolutionary explosion: Precambrian Research, v. 17, p. 105–123.

13. Valentine, J. W. and Moores, E. M., 1972, Global tectonics and the fossil record: Journal of Geology, v. 80, p. 167–184.

14. Cloud, P. E., 1976, The beginnings of biospheric evolution and their biochemical consequences: Paleobiology, v. 3, p. 351–382.

15. Runnegar, B., 1982, The Cambrian explosion: animals or fossils?: Journal of the Geological Society of Australia, v. 29, p. 395–411.

SUGGESTED READINGS

Brasier, M. D., 1985, Evolutionary and geological events across the Precambrian–Cambrian boundary: Geology Today, Sept./Oct., p. 141–146.

Conway Morris, S., 1985, The Ediacaran biota and early metazoan evolution: Geological Magazine, v. 122, p. 77–81.

Conway Morris, S., 1987, The search for the Precambrian–Cambrian boundary: American Scientist, v. 75, p. 157–167.

Glaessner, M. F., 1984, The Dawn of animal life: a biohistorical study: London/New York, Cambridge Univ. Press, 244 p.

Seilacher, A., 1984, Late Precambrian and Early Cambrian Metazoa: preservational or real extinctions. in Holland, H. D. and Trendall, A. F. (eds.) Patterns of change in earth evolution: Berlin/New York, Springer-Verlag, p. 159–168.

Chapter 6

The Invertebrates Dominate the Seas:

ACOELOMATES AND LOPHOPHORATES

By the end of Early Cambrian time all the major groups of invertebrates were represented in the marine fauna, and the exponential increase in diversity that characterized the early Paleozoic was well under way. Many of these groups had acquired carbonate or phosphate skeletons. They had radiated into the major ecological roles, such as suspension feeder, deposit feeder, scavenger, and predator. In this and the following chapter each of the invertebrate groups that are important to the paleontologist is described, and its history in the Paleozoic summarized. A large number of these groups were decimated by events that ended the era, and Mesozoic and Cenozoic marine faunas are much different from those of the Paleozoic. For this reason the story of this second wave of invertebrates is reserved for Chapter 10, after the consideration of the plants and vertebrates of the Paleozoic Era.

Protistans and animals secrete calcium carbonate skeletons of two different minerals: calcite and aragonite. These minerals have the same chemical composition but the calcium, carbon, and oxygen atoms are arranged in different ways so that the minerals have different physical properties. How organisms control the nature of the carbonate min-

eral deposited is not clear, and many deposit both minerals at the same time in different locations. The pearly inner layer of a clam shell is made of aragonite, the chalky outer part of calcite. Aragonite is less stable and more soluble than calcite under conditions at the earth's surface. Fossils (or parts of fossils) that were made of aragonite tend to dissolve or be replaced by calcite when the shell is buried. Only under extraordinary conditions has aragonite resisted replacement or dissolution in fossils of Paleozoic age.

PROTOZOA

Protozoan Biology:

The protozoans are members of the single-celled kingdom, the Protista, that obtain energy by ingesting other cells. Like all protistans, they are capable of all life functions within a single cell, but the independant cells may be united into large colonies. The phylum contains many groups of great diversity and astronomical numbers of individuals that are important in the modern environment, many as parasites on people and animals. However, such protozoans as the flagellates, ciliates, and sporozoans are of little importance to the paleontologist as they do not secrete preservable hard parts, and have little or no fossil record. The two groups that are well represented in the record, the Foraminiferida and the Radiolaria, belong to the subphylum Sarcodina, which is characterized by extensions from the body called pseudopods. The radiolarians have stiff pseudopods extending radially from the body and secrete a delicate skeleton of opaline silica (Fig. 10–7). Because their skeletal material is relatively soluble, they have a poor Paleozoic record, but are important for dating deep sea deposits in Cenozoic sediments and contribute their fragile tests to make extensive deposits of siliceous ooze in equatorial belts of the modern Pacific Ocean. They are further described in Chapter 10.

Biology of Foraminiferida:

These are the most abundant and diverse group of unicellular organisms that secrete preservable hard parts. In modern seas they are abundant, although rarely conspicuous, floating in the water or living on and in the bottom sediment at all water depths. Nearly all foraminiferans are marine. Distinctive features of the forams, as they are commonly called, are the skeletons called tests they secrete or construct from particles, and the nature of the thread-like pseudopods. Unlike those of other protozoans, the pseudopods of the foraminiferans form a net around the cell and are full of granules which stream up one side of the threads and down the other (Fig. 6–1). Food particles, such as algal and bacterial cells, are trapped in the net, transported towards the center, broken up, and digested.

The foraminiferan test may be a single chamber or a series of chambers that are added as the cell grows. The test has an aperture where most of the pseudopods are concentrated but also may be perforated by pores (the *foramina* that give the group its name). The function of these pores is problematic. In some recent foraminiferans, the pores are covered with a membrane. A considerable part of the cell is outside the test, and in certain planktonic species the pseudopod net is ten times the diameter of the test. When the cell has grown too large for the test a new chamber is formed. The pseudopods first extend far out from the aperture, and their ends form a protective cyst consisting of a thin membrane with debris cemented to it. The protoplasm of the central cell then bulges out of the

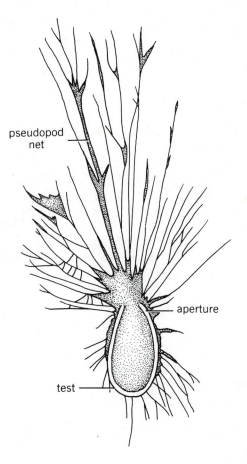

Figure 6-1   The simple living foraminiferan *Allogromia*, showing the relation between the soft tissue and the test. The cell is about 0.5 mm in width.

aperture within the protective cyst to the shape of the next chamber. This bulge is bounded by another membrane. Small crystals of calcium carbonate appear on its surface and eventually fuse to form a new chamber wall. The series of chambers formed by the repetition of this process may be linear, biserial, coiled planispirally, coiled in a trochoid spiral, or in a variety of complex forms (Fig. 6–2).

The reproductive cycle is important to the paleontologist because it results in dimorphism in the tests of many species. Unfortunately, the life cycle of only about 20 of the approximately 4000 living species is known. Although these cycles differ in detail, they all involve alternation of sexual and asexual reproductive phases (Fig. 6–3). In the asexual phase an individual completely divides into many smaller foraminiferans that leave the old test and start independent lives. This process may be repeated several times without the intervention of the sexual phase, whose onset may be triggered by environmental conditions such as increase in temperature. Then an individual cell breaks down completely into gametes that may be released freely into the water or may be exchanged when two individuals come together. The union of the gametes produces an individual that grows to adult size and then reproduces asexually. The initial chamber of the sexually produced foram is usually smaller than that of the asexually produced one although the test of the sexually produced foram may be larger. The dimorphic forms are referred to as microspheric (small initial chamber) and megalospheric (large initial chamber) tests. Because several asexual generations may intervene between sexual ones,

megalospheric tests usually far outnumber microspheric ones.

The division of the order Foraminiferida into suborders is based on the nature of the wall of the test. In the most primitive suborder, the Allogromiina, the test is an uncalcified sack formed of proteins. The members of the second suborder, the Textulariina, form a test of particles of sediment, usually silt, held together by an organic or mineral cement. Such tests are said to be agglutinated. They are granular in appearance, and may incorporate such particles as mica flakes, sponge spicules, and carbonate grains. Forams of the suborder Fusulinina have walls composed of microgranular calcite. The wall of the Miliolina is calcite that has a matte surface like unglazed porcelain and, in adult tests,

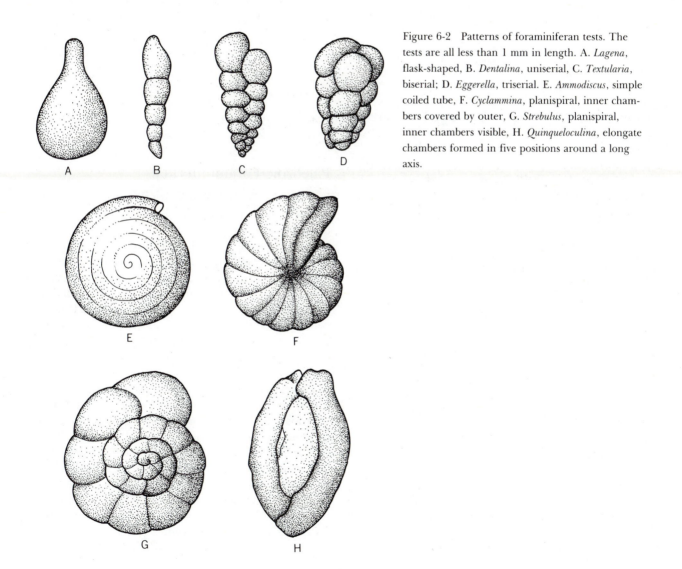

Figure 6-2   Patterns of foraminiferan tests. The tests are all less than 1 mm in length. A. *Lagena*, flask-shaped, B. *Dentalina*, uniserial, C. *Textularia*, biserial; D. *Eggerella*, triserial. E. *Ammodiscus*, simple coiled tube, F. *Cyclammina*, planispiral, inner chambers covered by outer, G. *Strebulus*, planispiral, inner chambers visible, H. *Quinqueloculina*, elongate chambers formed in five positions around a long axis.

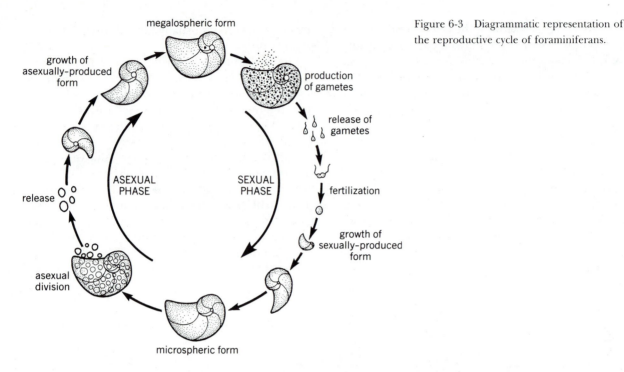

megalospheric form

growth of asexually-produced form

production of gametes

release of gametes

ASEXUAL PHASE

SEXUAL PHASE

fertilization

release

growth of sexually-produced form

asexual division

microspheric form

Figure 6-3 Diagrammatic representation of the reproductive cycle of foraminiferans.

is imperforate. The Lagenina have walls that are a single layer of radially arranged calcite crystals. The largest suborder, the Rotaliina, have walls that appear to be glassy (hyaline) and are perforated by many pores. Nearly all foram tests are composed of calcite, but a few are aragonite and one minor group secretes silica.

Kingdom—Protista
  Phylum—Sarcomastigophora
    Subphylum—Sarcodina
      Class—Granuloreticulosa
        Order—Foraminiferida
          Suborder—Allogromiina
          Suborder—Textulariina
          Suborder—Fusulinina
          Suborder—Miliolina
          Suborder—Lagenina
          Suborder—Globigerinina
          Suborder—Rotaliina
              + 5 minor
              suborders

The Paleozoic history of the foraminiferans is dominated by the agglutinated Textulariina and the calcareous Fusulinina. The suborders Miliolina, Lagenina, and Rotaliina, which dominate modern

seas, become abundant in Mesozoic times. Their expansion is described in chapter 10.

Paleozoic Foraminiferida:

Because the Foraminiferida are the protozoans with the best fossil record, nineteenth century paleontologists expected that their fossils would be the first to appear in the paleontological record. Some pre-Paleozoic organic structures were then identified as foraminiferans, not because they resembled living ones, but because foraminiferans were expected to be the oldest fossils. Since that time many Proterozoic unicellular fossils have been discovered but none has been a foraminiferan.

The oldest foraminiferans are straight or coiled agglutinated tubes associated with archaeocyathids in Lower Cambrian rocks of the southwestern states. Members of the order without agglutinated or calcareous tests may have existed earlier, but their organic capsules have not yet been discovered. Most of the approximately 15 Ordovician genera that have been described consist of a single subspherical, or tubular, chamber, or of a few connected chambers, and have been discovered by examination of insoluble residues of limestones dissolved in acid. By Silurian time more complex genera of irregularly, planispirally, and trochoidally coiled single

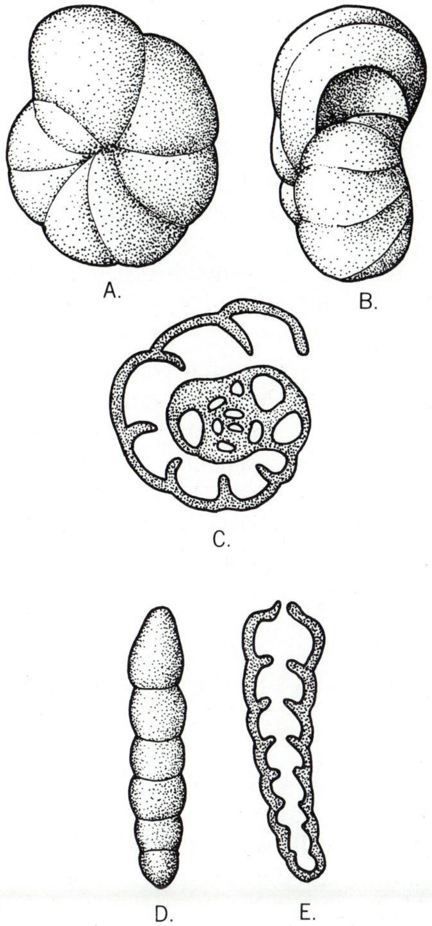

6–4A, B, C), became so abundant that they were important contributors to limestones in the upper Mississippi valley. These granular limestones are extensively quarried in Indiana and Illinois for building stone. The stone is suitable for carvings and facing of buildings because it is uniform in texture and available in large blocks from very thick beds. Late in the Early Carboniferous the discoidal endothyrids gave rise to spindle-shaped foraminiferans that dominated the rest of the Paleozoic Era, the fusulines. These became so abundant that many late Paleozoic limestones are composed almost entirely of their remains (Fig. 6–5).

Fusulines are externally similar and must be thin sectioned along two planes at right angles to reveal their internal structure. Externally the tests are shaped like footballs or grains of wheat, but are usually less than 5 mm long. However, some members of this group grew to be giants up to 10 cm

Figure 6-4   Mid-Paleozoic calcareous foraminiferans. A, B, C. *Endothyra* (0.4 mm), side and apertural views and cross section. D, E. *Nodosinella* (2 mm long), side view and cross section.

chambers had appeared (such as *Ammodiscus*, Fig. 6–2E) and the number of species had tripled. By Carboniferous time a great variety of coiled, uniserial, biserial, and triserial tests were being secreted by the Textulariina (Fig. 6–2C) and this variety of form persists today.

The oldest calcareous foraminiferans have been found in Ordovician rocks, but calcareous tests were not abundant until the Devonian Period. Many of the early members of the suborder Fusulinina, which includes the primitive calcareous forams, secreted a linear series of chambers (Fig. 6–4D, E). By Devonian time multiple-chambered tests coiled in a single plane were becoming prominent. By Early Carboniferous time (Mississippian) these minute calcareous tests, typified by the genus *Endothyra* (Fig.

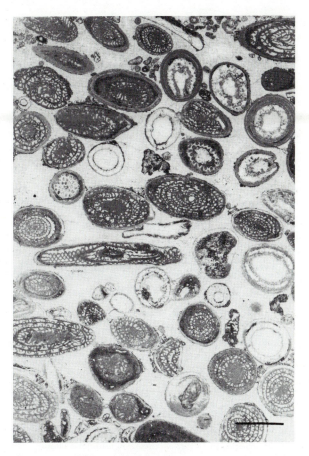

Figure 6-5   Thin section of fusuline limestone from Permian rocks of Texas. The scale bar is 2 mm. (Courtesy of C. W. Stearn.)

long. The group evolved rapidly and spread widely and is therefore useful for correlating late Paleozoic limestones around the world.

The test consists of chambers arranged in a planispiral coil, but, unlike the endothyrids in which the axis of coiling is the smallest dimension, in the fusulines the axis is the largest dimension. The chambers are bounded above and below by a spiral wall that is wrapped around the axis of coiling and they are separated laterally by septa, which are partitions that run from pole to pole (Fig. 6–6). The conti-

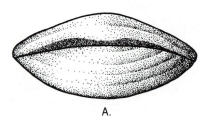

A.

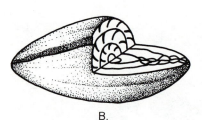

B.

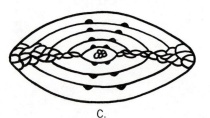

C.

D.

Figure 6-6 Late Carboniferous fusulinids. A, B. *Triticites*, view of the aperture and a specimen cut away to show the internal structure, C. *Profusulinella*, longitudinal section showing the initial chamber, spiral wall, and folded septa, D. *Profusulinella*, a saggital section across the coiling axis showing the spiral wall and septa.

nuity of the spiral wall is best shown by sections through the initial chamber at right angles to the axis of coiling. Sections parallel to the axis are also parallel to the septa but repeatedly cut the spiral wall as it winds around the axis. Evolutionary trends in the fusulines include the folding of the septa, the filling of the test with secondary deposits, the resorption of spiral passages (tunnels) through the septa, changes in the nature and thickness of the chamber walls, and changes in the shape of the test and the height of the chambers.

Although they were extraordinarily abundant in Permian seas and reached their greatest size in this period, fusulines became extinct at the close of the Permian, a time of crisis for many invertebrate groups.

Fusulines lived in shallow, warm seas as part of the bottom fauna. They appear to have been confined to tropical waters.

SPONGES: THE PHYLUM PORIFERA

Biology of Living Sponges:

Sponges have no organs and the body is composed of only two layers of cells separated by a jellylike substance containing amoeboid cells. These mobile cells are capable of many functions including digestion, transfer of nutrients and waste products, reproduction, and secretion of the skeleton (Fig. 6–7).

Nearly all sponges are marine but a few genera live in fresh water, and encrust rocks and vegetation in lakes. A typical sponge is shaped like a vase, jar, cylinder, or chimney attached to the seafloor, but many form irregular incrustations on rocks and other organisms. The living tissue of a typical sponge surrounds a central cavity called the spongocoel. Both tissues and skeleton are highly porous and penetrated by numerous canals. The sponge draws water from outside the body cylinder through fine pores on its surface into internal cavities where the fine organic particles, mostly bacteria, are trapped and digested. The filtered water follows canals through the cylinder wall into the spongocoel and flows out through its orifice which is called the osculum (Fig. 6–7C). In encrusting sponges, the small pores through which water is admitted (incurrent pores) and the larger pores through which it is released (excurrent pores) share the upper surface.

The filtering and feeding processes take place in

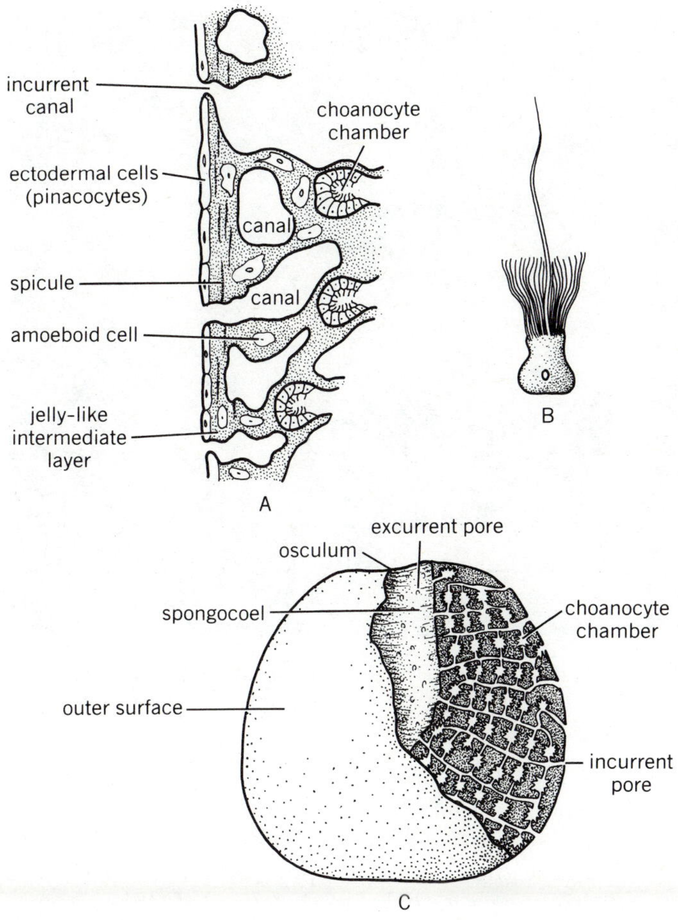

incurrent canal

choanocyte chamber

ectodermal cells (pinacocytes)

canal

spicule

canal

amoeboid cell

jelly-like intermediate layer

A

B

Figure 6-7 Anatomy of living sponges. A. Diagrammatic cross section of the outer surface showing the cell layers and canals. The cells are a few micrometers across, B. An isolated choanocyte showing its flagellum and collar of fine cilia, C. Small typical demosponge cut away to show the canal systems and choanocyte chambers. These are much enlarged for clarity.

excurrent pore

osculum

spongocoel

choanocyte chamber

outer surface

incurrent pore

C

internal chambers connected to the incurrent and excurrent canal systems. These minute chambers are lined with collar cells. Each collar cell has a single flagellum (a whiplike extension) pointing toward the exit from the cavity. The flagellum is surrounded at its base by a ring of very fine, hairlike tentacles which form the collar. The motion of the flagellum generates a current which draws water and its suspended microorganisms through the collar, where the bacteria are trapped and passed into the interior of the cell for digestion. The collar cells pass particles to the amoeboid cells for further digestion and distribution of the nutrients throughout the body. Large thick-walled sponges have hundreds of thousands of chambers lined with collar cells and connected by canal systems of great complexity. The flow of water through the sponge is maintained by the action of all the flagella. Some of the food requirements of sponges are also met by absorption of dissolved organic matter in seawater.

Sponges secrete skeletons of (1) organic fibers,

(2) individual elements of mineral matter called spicules, and (3) layers and plates of calcium carbonate. The organic fibers are a type of collagen. Collagen is an important fibrous protein of both invertebrates and vertebrates and forms the connective tissue outside the cells. In mammals the thick inner skin is largely composed of this polymer. Sponges strengthened only by organic fibers are relatively soft and flexible and some are sold for cleaning. Paleontologists are more interested in sponge skeletons formed of mineral matter because these are more likely to be preserved. Each spicule of the sponge skeleton is secreted by a single cell, or a group of cells, in the jellylike intermediate layer. Most such spicules are formed of hydrous silica (Fig. 6–8), but in one class of sponges they are calcite. The simplest spicules are needlelike and lie unconnected to one another in the soft tissue, then fall apart on the death of the sponge. More complex spicules have several radiating shafts or axes (Fig. 6–9). Many are shaped like tuning forks; others have highly irregular forms. Complex spi-

Figure 6-8   Scanning electron micrographs of the spicules of two living demosponges. The scale bars are 10 micrometers. A. *Agelas*. B. *Leucetta*. (Courtesy of C. W. Stearn.)

cules may interlock with adjacent ones to produce a rigid framework that may be preserved in the fossil record. In place of or in addition to a spicular skeleton, some living and many fossil sponges secrete (or secreted) a solid, continuous skeleton of either calcite or aragonite laid down by ectodermal cells at the base of the animal.

Phylum—Porifera
    Subphylum—Symplasma
        Class—Hexactinellida
    Subphylum—Cellularia
        Class—Calcarea
            Order—Pharetronida
            Order—Heteractinida
                + others
        Class—Stromatoporoidea*
        Class—Archaeocyathida*
        Class—Chaetetida*
        Class—Demospongia
            Order—Lithistida
            Order—Sclerospongea
            Order—Sphinctozoa
            Order—Keratosa
                + others
*These fossil groups may belong in the Demospongia or Calcarea

The Hexactinellida are placed in a subphyllum separate from the rest of the Porifera because the boundaries between their cells are poorly defined, unlike those of other sponges. Members of this group secrete delicate skeletons of siliceous spicules of six rays which may interlock to produce a three-dimenional grid. They are commonly known as glass sponges and are now largely confined to deep-water environments.

The Demosponges secrete skeletons of siliceous spicules and include about 95 percent of all sponges. Those demosponges in which the spicules have complex processes that interlock to form a semicontinuous skeleton are known as lithistid sponges and have the most extensive fossil record. Some demosponges have only organic fibers as skeletons (the Keratosa). The sclerosponges secrete a basal calcareous skeleton of aragonite or calcite. Modern sclerosponges are rare but are closely related to the fossil chaetetids and stromatoporoids.

The Calcarea secrete calcite spicules that are basically three-rayed but are usually complex in shape.

The Stromatoporoidea are an extinct group of Paleozoic reef-building sponges that secreted a continuous basal skeleton of calcium carbonate.

The archaeocyathids are another extinct group that are now placed in this phylum.

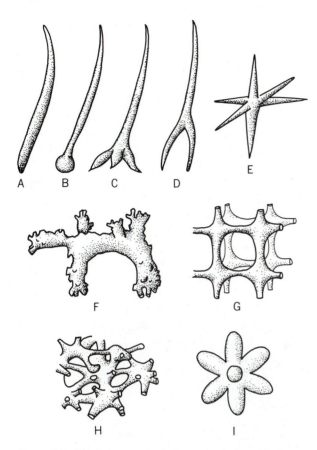

colonies. In the *Treatise on Invertebrate Paleontology* published in 1972, about 250 genera of archaeocyathids are listed; probably several tens of additional genera have been described since then.

The living tissue of the animal is believed to have occupied the space between the cones, leaving the central cavity empty like the spongocoel of a modern sponge. The porosity of the whole skeleton suggests that the archaeocyathid lived like a sponge, filtering seawater that passed through the body from outer to inner cone. However, no soft parts have been found with these fossil skeletons to confirm that this was their way of life. Because the shape, radial partitions, tabulae, and lack of spicules of archaeocyathids also suggest a relationship with corals, paleontologists have commonly placed them between the phyla Porifera and Cnidaria in a phylum of their own. The recent discovery of many living sponges that secrete a solid, non-spicular, calcareous skeleton like that of the archaeocyathids has removed most of the objections to uniting them with the Porifera (1).

In Early Cambrian time the archaeocyathids joined forces with an enigmatic, calcium carbonate secreting organism called *Renalcis* (Fig. 8–2), which was probably an alga, to build mounds tens of me-

Figure 6-9 Typical sponge spicules and spicule networks. A, B. Simple needlelike spicules called monaxons, C. Four-rayed spicule called a tetraxon, D. Tuning fork spicule, E. Six-rayed spicule typical of hexactinellids, F. Isolated irregular spicule of a lithistid, G. Interlocking spicule network of a lithistid, H. Spicule network of a hexactinellid, I. Heteractine spicule.

Archaeocyathida:

Archaeocyathids formed a skeleton composed of two cones, one inside the other, separated by radial partitions or septa (Fig. 6–10). The space between the cones was usually divided by horizontal plates called tabulae. The cones, partitions, and tabulae were perforated by pores, tubules, or slits of various shapes and sizes. The animal grew up from the seafloor like a vase from its point, which was embedded in the sediment. Many secreted skeletal tissue consisting of overlapping curved plates called cysts outside the cone at the base to stabilize themselves or to cement themselves to other skeletons. Most of the conical animals were solitary, but some stayed together after budding from a parent to produce

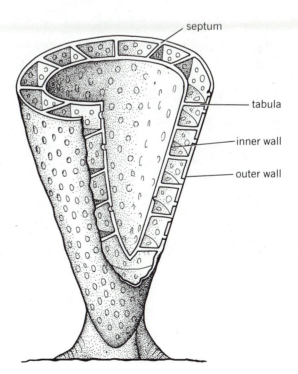

Figure 6-10 Diagrammatic reconstruction of a simple archaeocyathid.

ters high (Fig. 6–11). These are the first reefs built by metazoans. Some of the great carbonate masses found in Proterozoic rocks can also be described as reefs, but they were formed by cyanobacteria, not metazoans. Archaeocyathids are found in all continents but are particularly abundant and widespread in Siberia, southern Australia, and the Appalachian and Rocky Mountain regions of North America.

The archaeocyathids were the first metazoans to use calcium carbonate extensively to secrete skeletons. Their time of greatest success lasted for 30 million years. With the exception of a few genera that survived into Middle Cambrian, and one that may have survived into Late Cambrian time, the group became extinct at the end of the Early Cambrian Epoch (Fig. 6–12). They make excellent in-

dex fossils because they are restricted to a narrow time slot and can be easily recognized by the geologist in the field.

Hexactinellida

These sponges have six-rayed spicules which, in more advanced forms, fuse to make a three-dimensional rectilinear network (Fig. 6–13) like the steel framework of a modern office building. Isolated spicules of these sponges occur in Lower Cambrian rocks, but the oldest complete fossils are of Middle Cambrian age. These were shaped like a vase formed of a single sheet of spicules and were attached to the seafloor by a long tuft of straight spicules. By Middle Ordovician time they had a

Figure 6-11   Small archaeocyathid reef in Lower Cambrian rocks of southern Labrador, photographed from above. Note that the unbedded limestone of the reef in the center is more resistant that the surrounding bedded rocks and as the overlying sedimentary beds have been removed by erosion, the original relief of the reef has been revealed. (Courtesy of N. P. James and *Sedimentology*, v. 25, p. 7.)

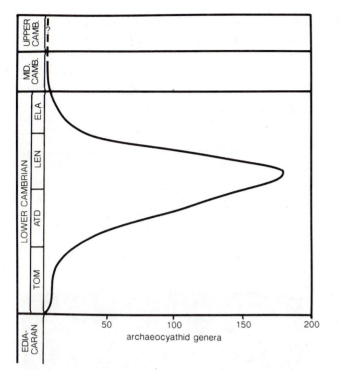

Figure 6-12   Generic diversity of the archaeocyathids. The names of the stages of the Lower Cambrian are abbreviated. (*Source:* Modified from M. F. Glaessner, *The Dawn of Animal Life.*)

double layer of spicules. The class became abundant and diverse in Middle Devonian time. Some beds in the Devonian sandstones of central New York State are famous for the many well-preserved hexactinellid sponge fossils they contain (Fig. 6–14). The class was again abundant in Jurassic and Cretaceous time. The sponges appear to have lived then in shallow water, but today their survivors have retreated into parts of the ocean deeper than 100 meters.

Demospongia:

Most complete demosponge fossils belong to the order Lithistida (Fig. 6–15) and have complex knobby siliceous spicules fused to one another. Scattered demosponge spicules are also preserved. The oldest demosponges occur in Middle Cambrian rocks. In Ordovician time the lithistids increased greatly in numbers and diversity and became prominent members of shallow-water marine communities. Important families in this period and the following Silurian Period are the astylospongiids and hindiids (Fig. 6–15A). Like the hexactinellids, the demosponges were not as widespread in the late Paleozoic

and early Mesozoic as they were in early Paleozoic and late Mesozoic time. Mesozoic demosponges are more abundant in Europe than in North America. Although the Cenozoic record of this group is not good, they thrive today in many environments.

Calcarea:

The Calcarea comprise those sponges that secrete calcium carbonate spicules and, like the demosponges, they include one group, the pharetronids, that built rigid skeletons and are therefore much better represented in the fossil record than the others. The first pharetronids, from Lower Permian rocks, are part of a reef fauna, and succeeding members of the group continued to prefer the reef environment. In Triassic reef limestones of the Cassian Formation of northern Italy, the calcareous sponges are so faithfully preserved that the spherulitic microstructure and aragonite mineralogy of their skeletons is still intact. Pharetronids thrived in Jurassic and Cretaceous reefs of central Europe where they are represented by such genera as *Stellispongia* and *Peronidella* (Fig. 6–16). Although only a few genera of the Pharetronida persist to the present, other

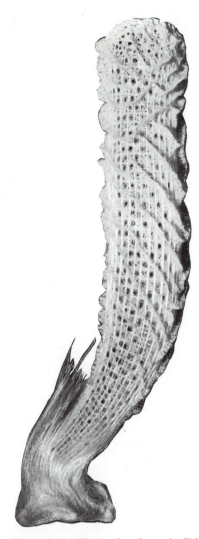

Figure 6-13   The modern hexactinellid sponge *Euplectella*, called Venus' flower basket. The sponge is attached by the fine, hairlike spicules at the base. The specimen is about 35 cm long. (Courtesy of H. Reiswig and C. W. Stearn.)

confined to Paleozoic rocks. The best-known North American genus is *Astraeospongium*, a saucer-shaped sponge with star-shaped spicules common in Middle Silurian rocks of the eastern states (Fig. 6–16D).

The second group, the Sphinctozoa (sometimes placed in the Calcarea but here considered demosponges), secreted a calcite skeleton that was not composed of spicules but consisted of a line of chambers arranged vertically and connected axially by a canal (Fig. 6–16A). The chamber walls were pierced by many pores through which water reached the axial canal and flowed out the osculum at the top. The sphinctozoans first appear in Cambrian rocks and persist as a single genus, *Vaceletia*, to the present. They were important contributors to the

Figure 6-14   The fossil hexactinellid sponge *Hydnoceras* from Upper Devonian rocks of New York. The specimen is about 20 cm long. (Courtesy of C. W. Stearn and Redpath Museum, McGill University.)

orders of the Calcarea are more abundant.

Two other groups of sponges that secreted calcareous skeletons but are excluded from the Calcarea by some paleontologists are of importance geologically. The Heteractinida secreted spicules that typically have six rays in one plane and two shorter rays at right angles to these (Fig. 6–9I). They are

building of reefs in Permian time in the Guadalupe Mountains of West Texas, and in Triassic time in the Alps.

Stromatoporoids:

The stromatoporoids are an extinct group of Paleozoic animals whose position in the invertebrate phyla has been controversial for over a century. Most paleontolgists now place them in the Porifera, but the older view that they are fossil hydrozoans (Cnidaria) is still held by some.

The stromatoporoids built skeletons of calcium carbonate that are hemispherical, tabular, laminar, columnar, bulbous, or dendroid (Fig. 6–17). Their fossils are now composed of calcite but the inferior preservation of many stromatoporoid fossils compared to that of calcite-secreting corals and brachiopods entombed with them, suggests that most stromatoporoids secreted aragonite or some combination of the two carbonate minerals.

The skeleton is composed of five main structural elements:

1. Dissepiments—thin, overlapping, upwardly convex plates.
2. Pillars—rods, round or irregular in cross section, perpendicular to the growth surface.
3. Laminae—laterally persistent plates, usually perforated, formed at more or less regular intervals parallel to the growth surface.

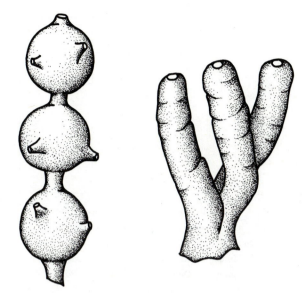

A.   *Girtycoelia*   B.   *Peronidella*

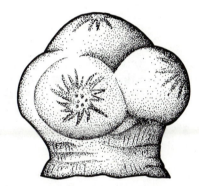

C.   *Stellispongia*

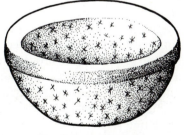

D.   *Astraeospongium*

Figure 6-16   Sponges with calcareous skeletons. A. The Permian sphinctozoan *Girtycoelia* (Class Demospongia). B. *Peronidella* (Class Calcarea, Order Pharetronida) (Triassic–Cretaceous). C. *Stellispongia* (Class Calcarea, Order Pharetronida) (Permian–Jurassic). D. *Astraeospongium* (Order Heteractinida) (Silurian).

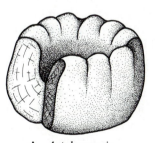

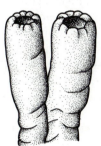

A.   *Astylospongia*

B.   *Cylindromphyma*

Figure 6-15   Lithistid sponges. A. *Astylospongia*, cut away to show the radial and concentric texture reflecting the canal trends (Silurian). B. *Cylindromphyma* (Jurassic).

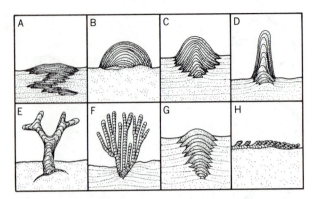

Figure 6-17 Growth forms of reef-building organisms and their relationship to sedimentation rates. A. Laminar or tabular, slow sedimentation, B. Hemispherical, no sedimentation, C. Hemispherical, slow sedimentation, D. Columnar or digitate, E. Dendroid, branching, F. Phacelloid, G. Bulbous, fast sedimentation, H. Creeping, no sedimentation.

4. Coenosteles—vertical elements, perpendicular to the growth surface, forming in cross section a maze or labyrinthine network.
5. Coenostroms—short, thick elements parallel to the growth surface connecting coenosteles (see Fig. 6–18D).

Study of the internal structure of these fossils is essential for their identification. The skeleton is cut along planes tangential and perpendicular to the growth surface, and slabs are ground thin enough to transmit light for microscopic examination.

Astrorhizae are small radiating sets of grooves on the upper surface that are incorporated as tubes when the skeleton grows upward around them (Fig. 6–19). Usually the centers of these canal systems keep the same location from one growth stage to another, forming tiers of superposed canals through the whole skeleton. The radial astrorhizal canals are parallel to successive growth surfaces, but the centers of the superposed sets may be connected by vertical canals as the skeleton is extended upward. The surface of many stromatoporoids rises into regularly spaced small elevations.

The living tissue of stromatoporoids must have been confined to the surface and the upper part of the skeleton (Fig. 6–20). The astrorhizal systems are nearly identical to the canals that collect the outgoing current in modern encrusting sponges. Unlike most modern sponges, the stromatoporoids of the Paleozoic are not composed of spicules and no trace of spicules has been found in their skeletons. Several modern demosponges secrete a basal skeleton that so closely resembles the skeleton of stromatoporoids that some zoologists believe they should be recognized as living stromatoporoids, and the stromatoporoids as fossil demosponges.

The Order Labechiida includes stromatoporoids whose horizontal structural elements are largely dissepiments. In the typical genus *Labechia* (Fig. 6–18A), these are combined with pillars. The Labechiida are the first stromatoporoids (Middle Ordovician) and dominate early faunas. They persist as minor members of later communities and are among the last of the Paleozoic stromatoporoids to die out at the end of Devonian time.

The labechiids gave rise to the Clathrodictyonida in Middle Ordovician time. This order comprises forms in which persistent, simple laminae are separated by pillars confined to the space between two laminae (Fig. 6–18B). In the Actinostromatida, pillars are the dominant structural element, and laminae are formed by rodlike processes extending horizontally from them at intervals (Fig. 6–18C). In the Stromatoporellida the axes of the laminae are occupied by a zone of small voids or cells. The order Stromatoporida includes stromatoporoids composed of coensteles and coenostroms. The genera of this order have thick structural elements that are minutely porous or cellular in microstructure (Fig. 6–18D).

The labechiid-clathrodictyonid faunas of the Ordovician are followed in Silurian time by more diverse assemblages of clathrodictyonids, actinostromatids, and stromatoporids. The greatest diversity of the stromatoporoids was reached in Middle Devonian time when they built great reefs in the tropical zones that are now Europe, the Soviet Union, the Middle East, North Africa, China, Australia and western North America. They continued to flourish into early Late Devonian time but suffered a rapid decline in the middle of the epoch. A few genera survived to make a comeback near the boundary between the Devonian and Carboniferous periods before the whole class disappeared for the rest of the Paleozoic Era.

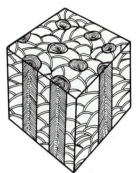

A. *Labechia*

B. *Clathrodictyon*

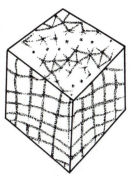

C. *Actinostroma*

Figure 6-18   Stromatoporoids and a chaetetid. These are continuous structures but
are illustrated as if the skeleton was cut in
a cube to show the internal structure in
three dimensions. The blocks are about 2
mm square. A. *Labechia* (Ordovician–
Devonian). B. *Clathrodictyon*, (Ordovician–
Devonian). C. *Actinostroma* (Silurian–
Devonian). D. *Stromatopora* (Silurian–
Devonian). E. *Stromatoporella* (Devonian).
F. *Chaetetes* (Devonian–Permian).

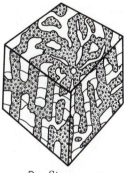

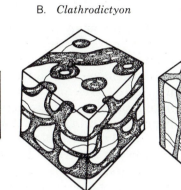

D. *Stromapora*

E. *Stromatoporella*

F. *Chaetetes*

Figure 6-19   Astrorhizal canal systems on the growth surfaces
of a stromatoporoid from the Devonian of Michigan. One
layer has been broken away to show the layer beneath. The
canals are about 0.5 mm in diameter. (Courtesy of C. W.
Stearn and Redpath Museum, McGill University.)

In Triassic rocks a group of reef organisms appears that in structure is nearly identical to Paleozoic stromatoporoids but its members are distinctly
fibrous in microstructure. This group of Mesozoic
"stromatoporoids", also called sphaeractinoids or
neostromatoporoids, continued in reef rocks into
Cretaceous time before becoming extinct. These
Mesozoic fossils contain remnants of spicules in the
calcareous skeletal elements showing without a doubt
that they were sponges, probably demosponges. If
these fossils represent descendants of Paleozoic
stromatoporoids, why is the lineage not represented in the Late Paleozoic fossil record? Bruno
Mistiaen (2) has suggested that stromatoporoids lived
during this period of 100 million years as soft-bodied organisms, and in Triassic time again began to
secrete preservable skeletons. Another possibility is
that the Mesozoic stromatoporoids evolved from a
Paleozoic group of encrusting demosponges and
evolved structures similar to the Paleozoic stromatoporoids as adaptations to the same environments.

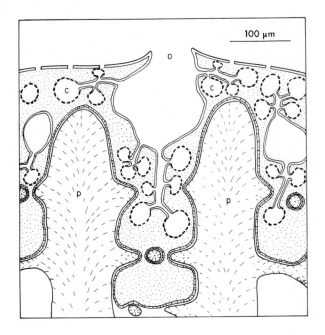

Figure 6-20   Reconstruction of the soft tissue of the stromatoporoid *Actinostroma*. The soft tissue occupies only the top surface of the skeleton which, in the reconstruction, is dominated by two pillars (p). The choanocyte chambers (c) are diagrammatically represented by cavities (delineated by thick, dashed lines) and are connected through the astrorhizal canals to an osculum or excurrent pore (o) at the surface. (Courtesy of C. W. Stearn and *Lethaia*, v. 8, p. 89.)

Chaetetida:

The skeletons of this group of extinct sponges are composed of many polygonal tubes less than 1 mm across and divided by remotely spaced tabulae (Fig. 6–18F). For many years these fossils were considered to be corals but recent discoveries of the remains of spicules in the tube walls, of astrorhizal canal systems on their surfaces, and of living sponges that secrete skeletons identical to fossil chaetetids have resulted in the recognition of their poriferan nature. The living genus *Acanthochaetetes* (Fig. 6–19) has all the skeletal features of fossil chaetetids and the soft tissue of a sponge.

Chaetetids occur sparingly in rocks of early Paleozoic age but become abundant in some Upper Carboniferous limestones, particularly those of the central and southwestern states in which they form small reefs. They are also found in many Mesozoic reefs usually as a minor element of the fauna.

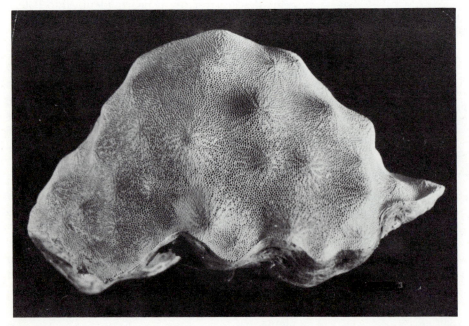

Figure 6-21   The skeleton of the living chaetetid *Acanthochaetetes* from the western Pacific Ocean. Note the astrorhizal canal systems on the surface. The specimen is 8 cm across. (Courtesy of Kei Mori and Tohoku University Science Reports, v. 46, p. 11.)

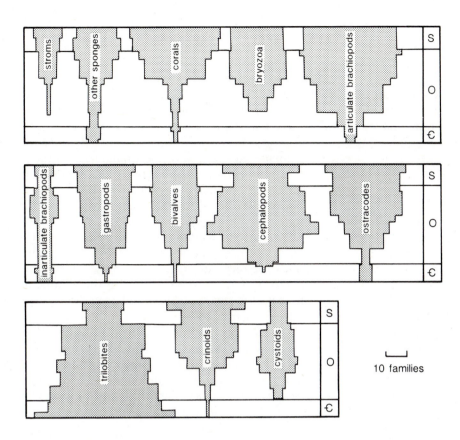

Figure 6-22 Rapid increase in diversity of animals secreting calcareous skeletons at the beginning of Ordovician time measured by the number of families. Stromatoporoids are abbreviated as stroms. (*Source:* Modified from J. J. Sepkoski and P. M. Sheehan, *Biotic Interactions in Recent and Fossil Benthic Communities*, p. 677, 678.)

10 families

RISE OF THE MID-PALEOZOIC REEF COMMUNITY

Middle Ordovician Changes:

During Middle Ordovician time the number of marine organisms that secreted carbonate skeletons increased greatly. Invertebrate diversity tripled during the period as the animal communities that were to dominate the rest of the Paleozoic were established. The trilobites that had dominated communities of the Cambrian Period did not disappear but declined steadily in importance. John Sepkoski and Peter Sheehan (3) showed that trilobites moved from shallow inshore environments to the deeper water of the continental slope. Inshore environments became the home of animals that secreted heavy calcareous shells and obtained food by filtering seawater. The groups of brachiopods that formed carbonate shells increased greatly in abundance. The cephalopods and gastropods increased in number and diversity at the beginning of the

Ordovician Period and bivalves rose to prominence, particularly in estuarine environments, near the end of the period. In addition, reef-building organisms such as stromatoporoids and corals became dominant faunal elements in shallow-water communities. Sepkoski (4) groups all these skeleton-forming invertebrates as the Paleozoic fauna and notes that their increase in the Ordovician Period was exponential, just as was the increase of the Cambrian fauna from late Proterozoic into Cambrian time (Fig. 6–22). At the end of the Early Ordovician Epoch the seas withdrew from the continental platform of North America, resulting in a prominent gap in the stratigraphic record. The seas returning at the beginning of the Middle Ordovician Epoch brought with them faunas enriched in animals that secreted thick calcareous skeletons. The dominance of animals secreting calcite over those secreting phosphatic or chitinous shells was then complete, and their expansion resulted in a significant change in the nature of shallow-water limestones. Younger limestones are composed almost entirely of the shells and skeletons of these animals. A few older limestones have similar fossil fragment textures, but most are composed of fine crystalline grains that appear to have been precipitated from seawater by biochemical processes associated with algae.

The returning seas of Middle Ordovician time introduced animals that would for the next three periods produce reefs in shallow continental seas—the stromatoporoids and two extinct groups of corals, the tabulates and rugosans.

What is a Reef?:

A reef is a mound on the seafloor built under the influence of organisms. Some geologists would add modifying phrases requiring that the mound be wave resistant or contain a certain proportion of organisms, but a broader definition of reefs is adopted here. Geologists also use the terms bioherm and buildup.

The term reef was first used for any obstruction to shipping, that is, for a rock near the surface of the sea. Geologically it has come to mean an organically built elevation on the seafloor ranging in size from patch reefs a few meters across (Fig. 6–23), to fringing reefs along shorelines (Fig. 6–24), to great walls of limestone hundreds of kilometers long, such as the outer edge of the Great Barrier Reef of Australia. They are locations of intense calcium carbonate production—carbonate factories of the sea. Modern reef builders, principally corals and algae, grow best near sea level in the zone agitated by waves where food, oxygen, and light are easily available. Builders of fossil reefs also appear to have preferred shallow water and therefore were subjected, like their modern counterparts, to the destructive action of the surf in storms. Because the carbonate skeletons of the reef builders are broken up by wave action and the action of bioeroders, the reef acts as a source of carbonate particles ranging in size from blocks several meters across to fine mud.

The steep-sided edifices of modern reefs are supported by a framework of intergrown coral and algal skeletons. Although these rigid constituents may form only 10 to 20 percent of the volume of the reef, they, like the pillars and beams of a building, are essential to the stability of the reef. Much of the space between their framework is filled with broken skeletons and shells of animals living on the reef. These loose materials may range from fine sand to massive boulders dislodged by hurricanes. They are bound into the reef framework by invertebrates and algae that encrust and coat grains and by chemical and biochemical processes that precipitate calcium carbonate from seawater cementing the whole into solid rock. Geologists call reef car-

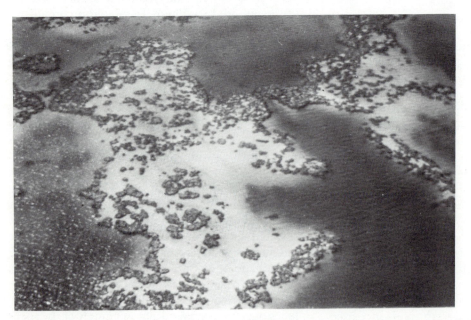

Figure 6-23 Aerial view of patch reefs growing in shallow water in the North Lagoon, Bermuda. Notice the light-colored sand that mantles the shallow sea floor around the reefs. (Courtesy of C. W. Stearn.)

Figure 6-24   Aerial view of the Bellairs fringing reef, Barbados. (Courtesy of C. W. Stearn.)

bonates dominated by framebuilders, framestones; those dominated by the sediment binders are bindstones.

Filling of the space between the framebuilders by reef debris and binding and cementing are rarely complete enough to eliminate cavities produced by the irregular growth. As a result, limestones of ancient reefs are porous and act as traps for oil and gas. In North America the Silurian reefs of the Great Lakes region, the Devonian reefs of western Canada, the Permian reefs of West Texas, and the Cretaceous reefs of Texas and Mexico are examples of such reservoirs. An understanding of the ecological requirements of the organisms that formed these bodies is essential to their discovery and exploitation.

Mid-Paleozoic Reefs:

After the rapid decline of the archaeocyathids at the close of the Early Cambrian Epoch, no invertebrates had the ability to construct reefs, and through the rest of the Cambrian Period only stromatolites built mounds on the seafloor. In Early Ordovician time sponges and algae built small reefs in combination with an encrusting organism of doubtful affinity called *Pulchrilamina* (Fig. 6–25), which may have been the first stromatoporoid.

The first reefs to be built by the highly successful combination of stromatoporoid sponges and corals occur in limestones of early Middle Ordovician age in the northeastern states. In building these mounds, these reef builders had help from the demosponges and bryozoans. Throughout the Ordovician Period reefs were small, at most a few meters across (5). By Silurian time the reef-building community was capable of forming barrier reefs hundreds of kilometers long in the Great Lakes area and the Arctic Islands. When the continents are restored to their mid-Paleozoic positions with the aid of paleomagnetism studies, these ancient reefs are seen to have been distributed within, or near, the tropics, just as are modern reefs, and the occurrence of reef builders at latitudes now high in the Arctic is understandable. The Silurian equator ran from the present position of California to that of Hudson Bay. Silurian reefs are also widespread in England and Scandinavia.

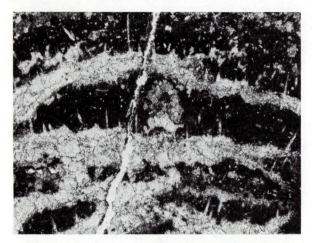

Figure 6-25   The Early Ordovician reef-building organism *Pulchrilamina* in thin section. The upper lamina is about .5 mm thick. This fossil may be the oldest stromatoporoid. (Courtesy of D. F. Toomey, M. Nitecki, and The Paleontological Society, *Journal of Paleontology*, v. 41, p. 984.)

The greatest success of the mid-Paleozoic reef community came between late Early and early Late Devonian time. In this interval patch reefs and barrier reef systems were built in western and arctic North America, Australia, central Europe, the Middle East, Afghanistan, and the central Soviet Union. This great reef-building episode lasted 100 million years, but ended abruptly in the middle of the Late Devonian Epoch. In latest Devonian time only remnants of the stromatoporoid–coral fauna continued to form small reefs until the end of the period. Although catastrophic and extraterrestrial mechanisms have been suggested to explain this episode of relatively rapid extinction (see Chapter 16), a cooling of the sea, caused by radical change in ocean currents as continents changed their positions, seems more likely.

Shapes of Reef Builders:

Many organisms that are attached to the seafloor secrete a calcareous skeleton at the base of the soft tissue and live on top of it. Such organisms include some calcareous sponges and algae, bryozoans, colonial corals, and stromatoporoids. Upward growth is necessary to keep the organism above the fouling effects of accumulating sediment. Although these organisms are dissimilar in their soft tissues, they grew to similar shapes. The basic growth form is hemispherical or tabular (Fig. 6–17). If it is nec-

essary for the organism to direct much of its growth energy to keeping up with accumulating sediment, it may grow to a columnar or bulbous form that appears to penetrate through the sediment (Fig. 6–17G). Columnar fossils tens of centimeters high may never have had more than a few centimeters of relief during life. Branching or dendroid forms raise the feeding surface high above the bottom.

The rise of the mid-Paleozoic reef community coincided with the diversification of organisms that are typically colonial or clonal, such as corals and bryozoans. Clonal organisms can grow rapidly to large size by asexual reproduction, or by the budding of one sexually identical individual from another, and thus compete efficiently for space on the ocean floor. Although most are capable of sexual reproduction, most rely almost completely on asexual reproduction to produce long-lived, large colonies and to propagate new colonies.

CORALS

The Biology of Cnidaria:

The Cnidaria include the corals, jellyfish, and hydrozoans, but the last two groups are of little interest to paleontologists. The phylum Coelenterata included the Cnidaria and the Ctenophora when they were placed in the same phylum, but the two groups are now recognized by most biologists as separate phyla, and the term Coelenterata has fallen into disuse. The Cnidaria are primarily digestive sacks composed of two cellular layers separated by a jellylike layer in which is embedded a primitive nerve network. The inner layer is the digestive surface, or endoderm. When food is in the sack, the endoderm cells engulf and absorb nutrients, sending them to the rest of the body by diffusion. The cells of the outer layer, or ectoderm, enclose the body, secrete a carbonate skeleton (if there is one), effect the muscular contraction of the body when it is disturbed, and include the stinging cells, or nematocysts. These stinging cells scattered in the ectoderm contain a poison sack and a minute, coiled, dart-tipped appendage that can be rapidly ejected to poison potential food or predators. The nematocysts of corals and jellyfish can inflict painful rashes on swimmers who touch them and, when the stinging effect is massed, as in such animals as the Portugese man-of-war, nematocysts are capable of inflicting life-threatening wounds. The nematocysts

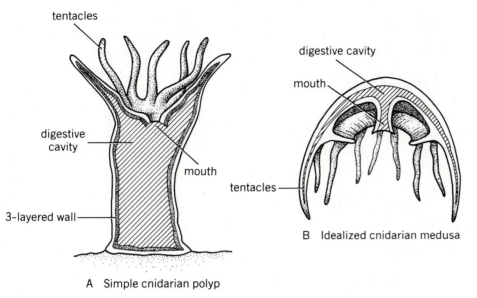

tentacles

digestive
cavity

mouth

3-layered wall

A Simple cnidarian polyp

digestive cavity

mouth

tentacles

B Idealized cnidarian medusa

Figure 6-26 Two forms of the cnidarian individual.

are concentrated on tentacles that surround the mouth, which is the only opening to the digestive cavity (Fig. 6–26). The tentacles immobilize small organisms through the action of the nematocysts and bring them to the mouth. After the enzymes secreted by the endoderm have broken down the prey and ingestible particles have been taken up, waste material is ejected from the mouth.

The digestive sack may be attached to the bottom at its closed base, or it may float in the water with its mouth downward (Fig. 6–26). The attached form is called a polyp and the floating form, a medusa. In some cnidarians medusa and polyp alternate from generation to generation; in others, one or the other phase may be more important.

The major divisions of the Cnidaria are the classes Hydrozoa, Scyphozoa, and Anthozoa.

Hydrozoans are the simplest of the Cnidaria. The digestive cavity is a simple cylinder, and massive skeletons are the exception rather than the rule. In many hydrozoans a polyp generation alternates with a medusa generation. In colonial hydrozoans specialized polyps form periodically and release small medusae. After dispersing, the medusae release sex cells into the water, and the union of these cells produces a swimming larva. When the larva attaches itself to a hard surface, it grows into a polyp. Most hydrozoan polyps live in small treelike colonies with delicate, horny, organic skeletons. However, some secrete calcareous skeletons and one genus, *Millepora*, the fire coral, is an important

contributor to the framework of modern reefs. The hydrozoa are not well represented in the fossil record but some of the Ediacaran leaflike fossils may belong in this class, and several groups of enigmatic Permian and Triassic corallike calcareous fossils have been assigned to it.

The class Scyphozoa includes cnidarians in which the medusa generation is the most conspicuous, and in many representatives, the only generation. These large jellyfish were only rarely preserved when their insubstantial bodies settled on the ocean floor, leaving impressions on the soft sediment. Some of the jellyfish of the Ediacara fauna may have belonged to the Scyphozoa.

The Anthozoa are the corals. They have no medusa stage. Sex cells are released from the polyp and unite in the water to form a new polyp. Radial divisions of the digestive cavity, called mesenteries, increase the area of the digestive surface (Fig. 6–27). Most of the anthozoans secrete an aragonite skeleton, but anemones are composed entirely of soft tissue. The simplest form of coral skeleton is a cup that encloses the bottom and sides of the digestive sack. As the polyp grows, the walls extend upward forming a tube that is called a corallite. If budding takes place, a colony composed of many tubelike corallites each containing a polyp is formed. As the corallite grows, the polyp lifts its base and secretes a plate beneath it for support. If the base is lifted as a whole, a flat plate called a tabula (plural = tabulae) is formed across the tube (Fig. 6–27).

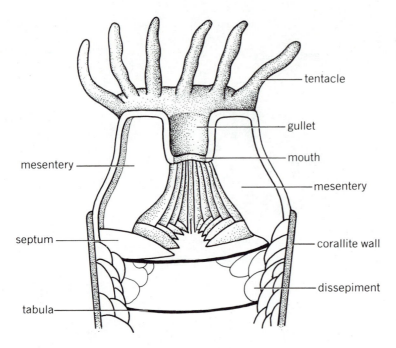

Figure 6-27   Diagram of an anthozoan polyp cut across the axis to show the relation between the soft tissue and the skeleton.

If the base is lifted in sections, one edge at a time, each part is supported by a small convex plate that section off a small part of the corallite periphery. These overlapping cysts are called dissepiments.

The base of the polyp is folded upward radially between the mesenteries. These folds secrete radial plates called septa (singular = septum) in the base of the cup. The major elements of the anthozoan skeleton are septa, dissepiments, tabulae and the corallite wall that encloses the polyp.

```
Phylum Cnidaria
    Class—Hydrozoa
    Class—Scyphozoa
    Class—Anthozoa
        Subclass—Octocorallia
        Subclass—Zoantharia
            Order—Tabulata
            Order—Rugosa
            Order—Scleractinia
```

The Anthozoa are divided into the hard corals (subclass Zoantharia) and soft corals (subclass Octocorallia). Only a few of the latter secrete a continuous skeleton; most have isolated calcareous spicules embedded in the soft tissue. The octocorals have a poor fossil record but are well represented in modern seas by the sea fans and sea pens, the precious coral of the Mediterranean (*Corallium rub-*

rum), and the organ-pipe and blue corals of Pacific reefs (*Tubipora musicum, Heliopora coerulea*). The Zoantharia have left an excellent fossil record. The subclass is divided into three orders: Tabulata, Rugosa, and Scleractinia. The first two were confined to the Paleozoic Era and are considered in this chapter. The Scleractinia, the corals of modern seas, are considered in Chapter 10.

Tabulata:

The tabulates are a group of extinct, exclusively colonial corals composed of tubes in which tabulae are usually prominent and septa usually poorly developed. Although their skeletons resemble to those of calcareous sponges, the recent discovery of polyps preserved in the corallites of a few specimens confirms their affinity to the Anthozoa (6).

Several small fossils composed of clusters of thin, simple tubes have been described from Cambrian rocks as possible tabulates, but all are of doubtful, affinity. The oldest fossils generally accepted as tabulates occur in rocks of early Middle Ordovician age and belong to the genus *Lichenaria* (Fig. 6–28A). They consist of closely packed polygonal tubes about 1 mm in diameter, without septa, and with rare tabulae and pores connecting the tubes. The tubes unite in a colony of hemispherical or tabular form tens or hundreds of centimeters across. Representatives of this ancestral genus *Lichenaria* be-

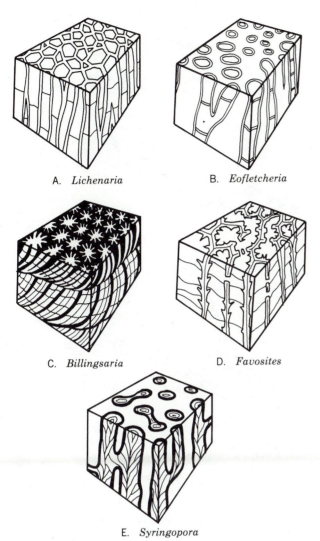

A. *Lichenaria*   B. *Eofletcheria*

C. *Billingsaria*   D. *Favosites*

E. *Syringopora*

Figure 6-28   Block diagrams illustrating the internal structure of tabulate corals. A. *Lichenaria* (Ordovician), B. *Eofletcheria* (Ordovician), C. *Billingsaria* (Ordovician), D. *Favosites* (Late Ordovician–Devonian), E. *Syringopora* (Ordovician–Permian).

came rapidly abundant in the Middle Ordovician, and gave rise to other pioneering genera such as *Eofletcheria* and *Billingsaria* (Fig. 6–28B,C) that were ancestral to several new suborders of the Tabulata.

The largest of these suborders is the Favositida. The typical genus, *Favosites* (Fig. 6–28D), derives its name from the Latin (*favus*) for honeycomb. Its skeleton looks much like a fossil honeycomb because the corallites are closely packed, hexagonal, and about the same size as honeycomb cells. The corallites of the favositids are connected by open-

ings called mural pores. In *Favosites* these are in the walls of the corallites, but in its predecessor, *Paleofavosites*, they are in the corners of the corallites. The septa are six or twelve in number and are not radial plates, as in most corals, but spines projecting from the walls. In the few specimens in which tentacles are preserved (6), there are twelve of them. Favositids are the most common corals in mid-Paleozoic reefs and grow to diameters of several meters.

Eofletcheria, which is composed of simple cylindrical, isolated tubes, is believed by Colin Scrutton (7) to be ancestral to the suborders Auloporida and Halysitida. The auloporids comprise tabulates composed of tubes that are round in cross section and united into a colony by horizontal tubes. Septa are absent and tabulae have the form of funnels (Fig. 6–28E). *Syringopora*, the most common genus, is abundant in Silurian and Devonian reefs, and continued to flourish into the late Paleozoic. In the Halysitida, corallite tubes are joined along their edges to two or more neighbors in order to produce colonies that are chainlike in cross section (Fig. 6–29A). In the common Silurian tabulate, *Halysites*,

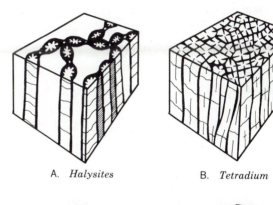

A. *Halysites*   B. *Tetradium*

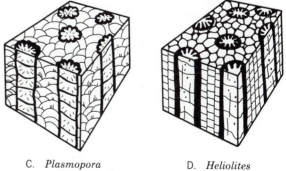

C. *Plasmopora*   D. *Heliolites*

Figure 6-29   Internal structure of tabulate corals. A. *Halysites* (Silurian). B. *Tetradium* (Ordovician). C. *Plasmopora* (Ordovician–Silurian). D. *Heliolites* (Ordovician–Devonian).

the large corallites are separated by thinner, more finely tabulated tubes that may have housed a smaller polyp specialized for some unknown function. The halysitids can be traced back to Middle Ordovician ancestors and became extinct at the close of the Silurian Period.

The Sarcinulida and Heliolitida comprise tabulates in which the corallite tubes are separated by skeletal tissue of various types. In the former this tissue is an outgrowth of the tabulae and septa, and forms horizontal plates connecting the corallite tubes. In the latter the skeleton is composed of a mass of cysts enclosing the tubular corallites (*Plasmopora*), or of finely tabulated polygonal tubes (*Heliolites*, Fig. 6–29C,D).

A group of puzzling, coral-like organisms typified by the genus *Tetradium* (Fig. 6–29B) joined the mid-Ordovician tabulate expansion. These fossils are sets of tabulated tubes that are square in cross section and divided by four septa growing from the middle of the tube walls. As the tubes grow in width, the septa meet in the center dividing each tube into four new tubes. Septa form on the walls of these tubes, and as the process is repeated the size of the whole mass increases. Although the tubes appear to be analogous to corallites, the method of fourfold division is unique to the tetradiids. They may be allied to the chaetetid sponges previously discussed rather than to the tabulate corals.

The tabulate corals are abundant in limestones deposited in shallow water environments in the Ordovician, Silurian, and Devonian Periods but declined in the late Paleozoic and became extinct at the end of that era.

Rugosa:

While the affinity of some tabulates is suspect, the rugose corals were without doubt true corals. Like modern corals, they have well-developed septa, but the septa are not arranged in sets of six. Instead they have a bilateral symmetry and the septa are arranged in sets of four. Because the septa are so arranged, this group is also referred to as the tetracorals. The other name of the order, the Rugosa, is derived from the coarse ridges, called rugae, that form on the outer wall of corallites owing to changes in the rate of growth (Fig. 6–30). Rugose corals were both solitary and colonial animals. The solitary polyp grew in diameter as it built its cuplike skeleton upward, forming a cone embedded in the sediment or cemented to a hard object. If the rate

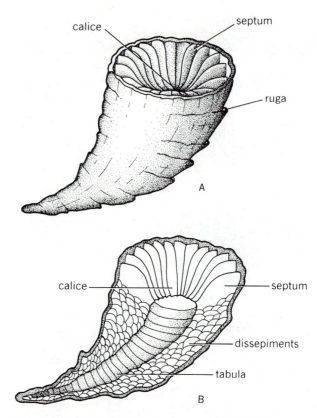

Figure 6-30   Typical solitary rugose coral. A. Side view showing calice, septa, and growth ridges or rugae, B. Cutaway view showing the relationship of septa, tabulae, and dissepiments.

of expansion of the polyp was relatively slow, the cone may resemble a cylinder. Most solitary rugose corals did not form symmetrical cones, but were curved slightly so that they resembled the horn of a cow; hence their common name, horn corals (Fig. 6–30).

Many rugose corals were colonial. They formed colonies in many of the patterns adopted by the tabulates. In some, the corallites are pressed against one another and are polygonal (Fig. 6–31C,D,E); in others they are separate like subparallel pipes (Fig. 6–31A,B). The term corallum is used to refer both to the hornlike skeleton secreted by a single polyp and to the mass of corallites secreted by colonial forms.

When the rugosan polyp was alive, a depression at the top of the corallum, called the calice, enclosed its base. The edges of the calice are lined by septa, and its bottom is divided radially by them. The center of the calice may be flat and formed by the

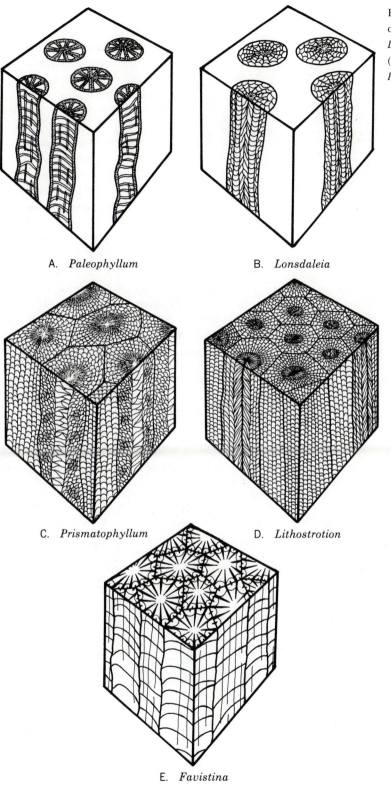

Figure 6-31   Block diagrams of colonial rugose corals. A. *Paleophyllum* (Ordovician–Silurian), B. *Lonsdaleia* (Carboniferous), C. *Prismatophyllum* (Devonian), D. *Lithostrotion* (Carboniferous), E. *Favistina* (Ordovician).

A. *Paleophyllum*

B. *Lonsdaleia*

C. *Prismatophyllum*

D. *Lithostrotion*

E. *Favistina*

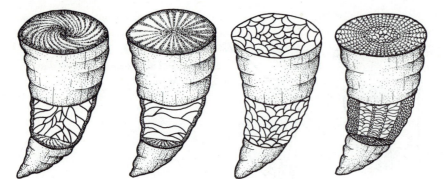

A. *Dinophyllum*   B. *Streptelasma*   C. *Cystiphyllum*   D. *Aulacophyllum*

Figure 6-32   Solitary rugose corals cut to show the relationship of the skeletal elements in longitudinal and cross sections. A. *Dinophyllum* (Silurian) showing the septal swirl forming a columella, B. *Streptelasma* (Ordovician) showing tabulae crossing the whole width of the corallum, C. *Cystiphyllum* (Ordovician–Devonian) showing the whole corallum filled with dissepiments, D. *Aulacophyllum* (Devonian) showing peripheral dissepiments, central tabulae, and a complex colummellar structure.

last secreted tabula (Fig. 6–30B), or it may be elevated into a knob. These central elevations are the ends of axial structures that pass through the whole length of the corallum. Some of these are made of a solid rod of calcite called a columella (Fig. 6–31D); others are formed by the swirling union of the inner edges of the septa (Fig. 6–32A). Some late Paleozoic colonial rugose corals have complex axial structures consisting of radial plates and cysts (Fig. 6–32D). Commonly the edges of the calice are floored by dissepiments and the central part by tabulae (Fig. 6–30B). The arrangement of these horizontal structures results from the polyp lifting itself in small steps around its periphery and in larger steps at its axis. The proportion between the two structures varies widely. In some rugose corals of the order Cystiphyllida, the corallum is completely filled with dissepiments (Fig. 6–32C), while in others dissepiments are absent and tabulae extend from wall to wall. The floor of the calice may be marked by one or three radial depressions called fossulae (singular fossula). These mark places where the septa are introduced as explained later. Details of the calice are preserved only in fossils collected from shales and shaly limestones. To identify and study rugose corals enclosed in limestones, paleontologists must cut the corallum transversely and longitudinally to reveal its internal structure.

Septa are best studied in transverse sections. In most rugose corals short, second-order septa alternate with major septa, and in some a third order of even shorter septa may appear between those of the second order. By making successive, or serial, sections across the corallum from its tip to the calice, the paleontologist can determine how the septal

pattern originated because through its life the coral retains the record of its early development (Fig. 6–33). Early in life six septa were secreted as follows: the cardinal, and on the opposite side, the counter septa; on either side of the cardinal, two alar septa; and on either side of the counter, two counterlateral septa (Fig. 6–33C). As the polyp grew in diameter, additional major septa were added in four sectors—on either side of the cardinal septum and on the counter side of the alar septa (Fig. 6–33D–I). With each major septum a shorter second-order septum is introduced beside it. The insertion of additional septa continues until the adult compliment is reached. The insertion of septa is the result of the need to increase the area of the digestive surface

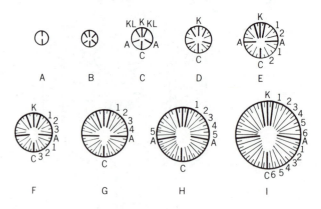

Figure 6-33   Diagrammatic serial sections of a solitary rugose coral to show the order of insertion of the septa. Septa are inserted on either side of the cardinal septum and between the alar and counterlateral septa. A = alar septum. KL = counterlateral septum. K = counter septum. C = cardinal septum. Major septa are numbered 1 to 6.

of the polyp as it grew, by the successive introduction of additional mesenteries each of which, we assume, was separated by a septum as in modern corals. The cardinal septum is commonly shortened in adult stages, and the prominent fossula that develops in the cardinal position is shaped partly by this shortening and partly by the shortness of the new major septa introduced on either side of it. In some calices fossulae also developed where insertion took place near the alar septa. The plane of bilateral symmetry passes through the cardinal and counter septa.

In most colonial rugose corals the corallites are separated by a discrete wall, and the septa of each corallite are independant. Many colonial genera have a thin, wrinkled layer of skeletal material (holotheca) that envelopes the base of the corallum, isolates the initial polyps from sediment, and protects the abandoned parts of the corallites from parasitic boring organisms.

The classification of the rugose corals is based on the microstructure of the fibrous elements that form the septa and on the arrangement of the septa, tabulae, and dissepiments. The suborders and families cannot be briefly characterized, and are therefore not described in this summary.

The oldest rugose corals occur at the base of the Middle Ordovician Series in North America in beds that contain the record of the first diversification of the mid-Paleozoic reef community. *Lambeophyllum*, the first genus, has a small conical corallum with a deep calice that extends to the tip of the cone, leaving no room for tabulae or dissepiments. Horn corals rapidly became abundant, and *Streptelasma*, which has a thick outer wall and septa twisted into an axial vortex, and its relatives are common fossils in Middle and Upper Ordovician shaly limestones. Colonial genera, such as *Paleophyllum* and *Favistina* (Fig. 31), also appear in Middle Ordovician rocks. In Silurian time the diversity continued to increase and both solitary and colonial forms are common in the reef facies. The cystiphyllids, in which septa are inconspicuous and dissepiments plentiful, are particularly abundant (Fig. 6–32C). Following a setback, possibly caused by the regression of the seas from the continents in latest Silurian and earliest Devonian time, the rugose corals reached their greatest abundance during the reef-building period of late Early Devonian to Late Devonian time. Particularly characteristic of the Devonian Period were the colonial genera *Hexagonaria* and *Philipsastrea* and their relatives. Carboniferous rugose corals developed complex axial and confluent septal structures. Typical genera of this period were *Lonsdaleia* and *Lithostrotion* (Fig. 6–31B,D). The Rugosa became less abundant in Permian time and finally became extinct in the last stages of the period as the Paleozoic Era ended.

BRYOZOA

The bryozoans were among the pioneer reef builders during the rise of the mid-Paleozoic reef community. Their hemispherical, encrusting, and branching colonies grew in such abundance that they were capable of forming mounds in Ordovician seas, either alone or with the flourishing corals and stromatoporoids. Alone they did not form such large structures as the corals, but they were important contributors to most Paleozoic shallow-water communities. Their skeletons share a common architecture with the corals but the individuals of the colonies were smaller and lack the radial septa characteristic of coral skeletons.

Biology of Bryozoa:

The animals of this phylum are inconspicuous but common and are widely distributed in modern marine environments. They live in habitats under rocks and overhangs and in caves in modern tropical reefs. However, they also live in polar and in deep waters, and one group is common in fresh water. Because many of the marine species secreted a calcite skeleton, the phylum is well represented in rocks of Ordovician to recent ages (Fig. 6–34). The phylum Bryozoa has been given several other names; the most common are Polyzoa and Ectoprocta.

All but one genus of the phylum are colonial. The forms of bryozoan colonies are similar to those of other colonial animals such as the corals. Encrusting (Fig. 6–35), hemispherical, dendroid, sticklike, and vinelike colonies are all common bryozoan forms. Erect sheet-like colonies attached to the seafloor by one corner or edge are more common among the bryozoans than among the corals, where such colonies are formed only by the soft corals (Octocorallia).

The individuals are small, rarely more than 1 mm across, and are housed in tubes or boxes called zooecia (singular = zooecium). With the exception of the first, which grows from a larva, each individual in a colony is budded from another and is

Figure 6-34   Bedding plane of Middle Ordovician limestone with many sticklike trepostome bryozoans and many netlike fenetrate bryozoans. The prominent trilobite cephalon is of *Flexicalymene* and measures about 2 cm across. (Courtesy of C. W. Stearn and Redpath Museum, McGill University.)

undergoes a complete change of form. The larva breaks down into what appears to be an unorganized mass of cells and from this mass the first individual of the new colony grows.

The tentacles of the Bryozoa are covered with fine cilia that sweep microorganisms towards the mouth at their base. They are not armed with stinging cells like the tentacles of corals. In addition to food gathering, the circle of tentacles is responsible for respiration. This structure, which is common to the phyla Bryozoa and Brachiopoda, is called a lophophore.

Bryozoans have no excretory organs to rid the body of the nitrogenous waste products of metabolism. Periodically individuals regenerate themselves. The body disintegrates into apparently unorganized cells, and a bud emerges from them to grow into a new individual. This regenerative process leaves behind a mass of brown organic matter called a brown body. Biologists have suggested that this extraordinary rejuvenation may be caused by a need to deposit accumulated waste.

When disturbed, the tentacles are rapidly pulled into a sheath and, in forms that secrete a skeleton, into the protection of the skeleton. This withdrawal

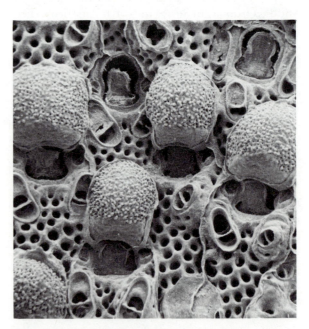

Figure 6-35   Surface of the skeleton of the living cheilostome bryozoan *Hippopetraliella* from Ghana. Several of the zooecia have brood pouches. The zooecia are about 0.5 mm wide. (Courtesy of Patricia Cook, P. J. Chimonides, and the British Museum (Natural History).)

assumed to be genetically identical to the others. All individuals are clones and the animal is said to be clonal. The only permanent internal organ is a U-shaped digestive tract (Figs. 6–36,6–37). The mouth is surrounded by a circle of 8 to 35 tentacles and the anus is placed outside, but close to the tentacle circle. The gut is suspended in the coelomic cavity in which, at certain seasons, both male and female reproductive cells are formed in the same individual. They are released when mature through a break in the wall of the coelom. After the cells are fertilized, the embryos are retained in brood pouches formed on the outside of specially modified zooecia (Fig. 6–35). When the larva is released from the pouch, it has a relatively short-lived free-swimming stage before it attaches to a substrate and

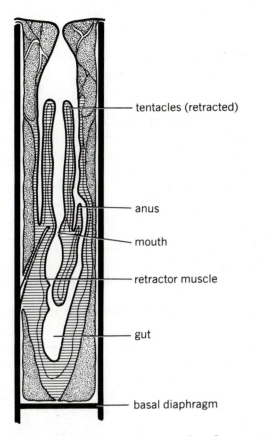

Figure 6-36   Diagrammatic cross section of a trepostome bryozoan retracted into its zooecium, based on the anatomy of living bryozoans. (*Source:* Modified from *Treatise on Invertebrate Paleontology*, pt. G (revised).)

is performed by a retractor muscle (Figs. 6–36, 6–37) that stretches the full length of the individual. The same muscle cannot push the tentacles out when danger has passed, as muscles can only retract. The extrusion mechanism depends on increasing the hydrostatic pressure inside the coelom thus forcing the tentacles out of their sheath into the water (Fig. 6–37). The pressure may be increased by muscles pulling inward on a flexible outer wall of the housing or by a sack within the coelom inflating with water, which displaces the tentacles from their sheath (see Chapter 10).

Some zooecia may be modified by a brood pouch. Others may grow into elongate whips and still others look like birds' heads with small, snapping beaks. These serve to sweep away sediment that might cover the feeding surface and to repel larvae of other marine organisms that might try to colonize the surface.

Modern classifications recognize three classes in the phylum Bryozoa. The Phylactolaemata include the freshwater bryozoans which do not secerete a mineral skeleton and (with the exception of some dubious reproductive bodies) are unknown as fossils. The Stenolaemata are a largely Paleozoic group in which the zooids secrete a tubular housing. One order persists to the present day, and these representatives serve as models for the reconstruction of the soft parts of the rest of this class. The Gymnolaemata comprise most modern bryozoans. Their

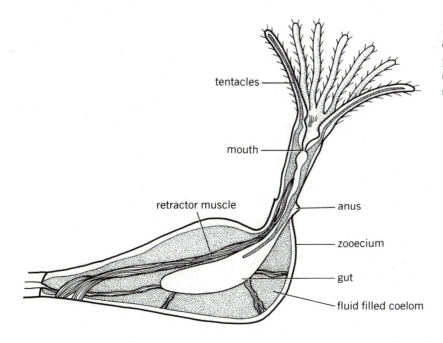

Figure 6-37   Diagrammatic cross section of a living bryozoan individual in feeding position with lophophore extended. (*Source:* Modified from Treatise on Invertebrate Paleontology, pt. G (revised).)

skeleton is composed of boxlike zooecia that are united into a sheetlike colony. The seven orders that are grouped into the last two classes are discussed more fully in the following section and in Chapter 10.

Phylum Bryozoa
    Class—Phylactolaemata
    Class—Stenolaemata
        Order—Tubuliporata
        Order—Trepostomata
        Order—Cystoporata
        Order—Fenestrata
        Order—Cryptostomata
    Class—Gymnolaemata
        Order—Cheilostomata
    Order—Ctenostomata

Paleozoic Bryozoa:

Of the seven orders now recognized, six originated in the Ordovician and the seventh, which comprises most of the modern forms, did not appear until Late Jurassic time.

The members of the Order Tubuliporata, commonly called cyclostomes, have colonies consisting of a set of simple cylindrical tubes with open, round orifices (Fig. 6–38). The tubes may be undivided or crossed by plates that are like tabulae, but are called diaphragms in bryozoans. The cyclostomes are bit players in the Paleozoic drama, but when most of the Paleozoic bryozoans died out in Permian time, they briefly became stars and were abundant and diverse in the early part of the Mesozoic Era. They were eclipsed by the rise of modern bry-

ozoans (order Cheilostomata) in Cretaceous time but still play minor roles in modern seas.

The most prominent bryozoan members of mid-Paleozoic shallow water communities had a massive calcareous skeleton and grew in thick branching, hemispherical, and encrusting colonies. They are placed in the order Trepostomata. Zooecia are long, cylindrical tubes crossed by transverse diaphragms. The diaphragms may be incomplete (Fig. 6–39B) or replaced in part by curved plates called cystiphragms, which look like dissepiments of a coral (Fig. 6–39A). Between the larger zooecia (autozooecia), there may be smaller, polygonal, closely spaced tubes called mesozooecia (Fig. 6–39A) that presumably housed individuals of modified, but unknown, form. Rodlike acanthostyles located in the zooecial walls may have been bases for still another type of individual (Fig. 6–39C). The structure of the colony in many trepostomes changes near its surface. In this mature zone walls thicken, diaphragms are more closely spaced, and such features as acanthostyles and mesozooecia appear between the autozooecia (Fig. 6–40A,B). The surface of trepostomes may be marked by raised areas, as in *Hallopora* (Fig. 6–40A), or by smooth areas of darker appearance where mesozooecia are concentrated. For identification, the internal structure of trepostomes must be studied with the aid of thin sections.

The members of the order Cystoporata, also known as fistuliporids, secrete heavy colonies in a variety of forms. In this group the tubular zooecia are isolated from each other by cystose tissue, a structure similar to that of the heliolitid tabulate corals, but on a smaller scale (Fig. 6–39D). *Constellaria*, a distinctive Ordovician representative of

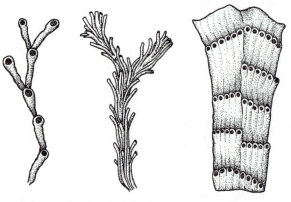

A. *Stomatopora*   B. *Entalophora*   C. *Idmonea*

Figure 6-38 Byrozoans of the order Tubuliporata. Zooecia are about 0.1 mm. A. *Stomatopora* (Ordovician–recent). B. *Entalophora* (Jurassic–recent). C. *Idmonea* (Jurassic).

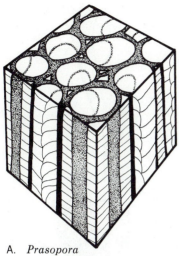

A. *Prasopora*

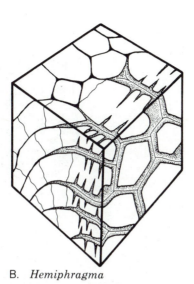

B. *Hemiphragma*

Figure 6-39 Block diagrams to show the internal structure of three hemispherical trepostome and a cystoporate bryozoans (zooecia about 0.2 mm). A. *Prasopora* (Ordovician). B. *Hemiphragma* (Ordovician). C. *Stigmatella* (Ordovician). D. *Fistulipora* (Silurian–Permian).

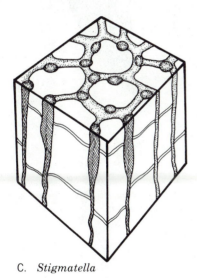

C. *Stigmatella*

D. *Fistulipora*

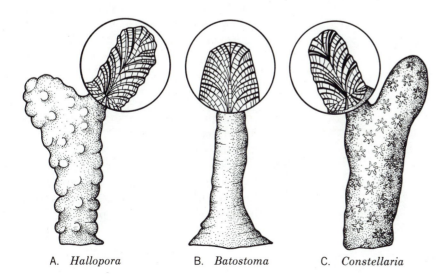

A. *Hallopora*   B. *Batostoma*   C. *Constellaria*

Figure 6-40 Sticklike trepostome and cystoporate bryozoans with a longitudinal section of a branch enlarged to show the internal structure. A. *Hallopora* (Ordovician), B. *Batostoma* (Ordovician). C. *Constellaria* (Ordovician).

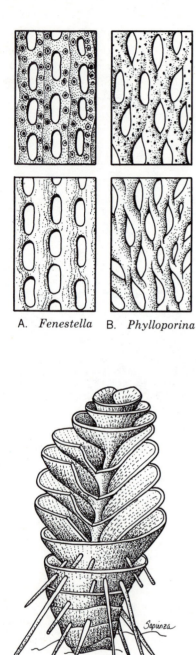

A. *Fenestella*   B. *Phylloporina*

C. *Archimedes*

Figure 6-41 Bryozoans of the Order Fenestrata. A. Front and back views of a frond of *Fenestella* (Ordovician–Permian), B. Front and back views of a frond of *Phylloporina* (Ordovician). In A and B the branches are about 1 mm across. C. A complete specimen of *Archimedes* (Early Carboniferous) showing the spiraling frond cut away in the center to reveal the screwlike axis. The axis is about 5 mm in diameter.

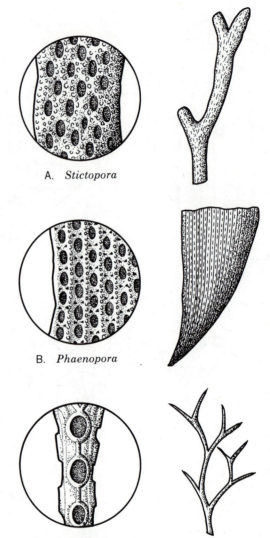

A. *Stictopora*

B. *Phaenopora*

C. *Arthrostylus*

Figure 6-42 Bryozoans of the Order Cryptostomata. On the right, complete specimens about natural size; on the left, the surface enlarged to show the zooecia. A. *Stictopora* (Ordovician), B. *Phaenopora* (Ordovician–Devonian), C. *Arthrostylus* (Ordovician).

this group, has prominent star-shaped elevations covering its surface (Fig. 6–40C).

Many bryozoans secreted short, boxlike zooecia. Such short zooecia were united into sheetlike colonies that would encrust a surface or stand vertically in the water like a wall. Other colonies were made in the form of sticks, trees, or planks with the individuals opening on one or both sides. All these forms evolved in order to place the animals in a

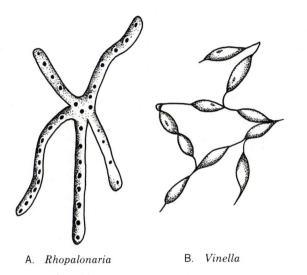

A. *Rhopalonaria*   B. *Vinella*

Figure 6-43 Bryozoans of the Order Ctenostomata. Width of impressions is about 0.1 mm. A. *Rhopalonaria* (Ordovician–Permian). B. *Vinella* (Ordovician–Cretaceous).

favorable position for filtering out microorganisms drifting in the sea currents.

Many erect colonies were sheets perforated by holes through which nourishing currents flowed. They belong to the Order Fenestrata. Their form is illustrated by the genus *Fenestella* (Fig. 6–41). On the downcurrent surface of these lacy fronds the minute apertures of the boxlike zooecia can be seen but the upcurrent surface is without zooecia. One of the most complex of the fenetrates is *Archimedes*, a fossil characteristic of Lower Carboniferous rocks in the United States. In this genus the perforated sheet is wound spirally about a solid calcareous axis that stood, supported by spines, in the seafloor (Fig. 6–41C). The part of *Archimedes* that is usually preserved is the solid axis, which looks much like a petrified screw. The fenestrate bryozoans reached their greatest abundance in Devonian and Carboniferous time when they were important rock-building organisms covering bedding planes with their fragile fronds. They became extinct at the end of the Paleozoic Era.

Erect planklike or straplike colonies typified by genera such as *Phaenopora* and *Stictopora* (Order Cryptostomata) have boxlike zooecia opening on both sides. (Fig. 6–42). The individual zooecia are arranged in rows and the walls between them are thickened near their apertures, which are narrowed by these deposits. Sticklike bryozoans like *Arthrostylus* have apertures opening on all sides.

One group of bryozoans that ranges from Ordovician to recent is known from the fossil record, not from the skeletons they secrete, but by excavations they make in shells. *Vinella* and *Rhopalonaria* (Fig. 6–42) are Paleozoic genera typical of this group (Order Ctenostomata). Although long ranging, the group was never a prominent one.

BRACHIOPODS: SEASHELLS OF THE PALEOZOIC ERA

Biology of Brachiopods:

The brachiopods are exclusively marine invertebrates that enclose their bodies within two shells and obtain food by means of a lophophore, in the form of a coiled and folded ciliated ribbon. The shells are called valves and are positioned at right angles to the plane of bilateral symmetry of the animal, that is, in a position above and below the body. In typical brachiopods each valve is bilaterally symmetrical but one of the valves is considerably larger than the other (Fig. 6–44). Through an opening in the larger valve a muscular stalk called the pedicle protrudes and attaches the animal to the seafloor, or to another organism. This larger valve is called the pedicle valve. The smaller valve contains the supports for the lophophore, which was originally thought of as a sort of arm (Latin = *brachium*) and gave its name to the phylum (literally, arm-foot). This valve is therefore called the brachial valve. The exterior of the valves is marked by concentric growth lines which show that the valves started to grow near the pedicle opening where the valves are pointed, or beaked. Brachiopods may live in various positions, but for descriptive purposes the area of the pedicle opening is considered to be posterior and is placed at the top in illustrations, and the edge along which the valves open is considered to be anterior.

The space within the two valves is divided into a mantle cavity and a body cavity (Fig. 6–45) by the body wall. The valves are covered on the inside by a fleshy layer called the mantle which is continuous with the body wall. In the lining of the mantle cavity, the mantle is doubled over and the body cavity extends in branching canals between the double layers. These canals may leave impressions on the inside surfaces of the valves. Respiration takes place in the mantle aided by these canals, and in the

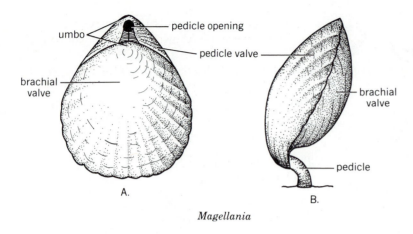

Figure 6-44   The living brachiopod *Magellania*. A. Dorsal view of the shell. B. Side view showing the pedicle and brachial valve and the position of the pedicle.

Magellania

lophophore. The mantle also secretes the mineral matter of the two valves which is deposited at the anterior and lateral edges extending the valves, and to a lesser extent along the inner surface, thickening the valves. Variations in the deposition at the growing shell margins form the concentric growth lines.

The lophophore is a ribbon of tissue attached through the body wall to the back of the brachial valve that may be supported by a calcareous band. The ribbon is commonly folded, looped, or wound in a spiral so that a great length is enclosed in a modest space. The lophophore is equipped with two lines of closely spaced, staggered filaments like the teeth of a comb, and these are covered with fine, hairlike cilia. When the valves are open to feed, the movements of the cilia normally bring a current of seawater bearing microorganisms in from two sides of the valve and force it out the front of the valves. The lophophore and filaments are so arranged that the water must pass through the comb as it flows through the mantle cavity. Organisms trapped in the cilia are passed to a groove along

the lophophore axis to the mouth situated in the body wall where the strands of the lophophore ribbon converge.

The brachiopods are divided into three classes. In the largest class, the Articulata, the valves are composed of calcium carbonate and interlock in teeth and sockets along the hinge, the pedicle is functional, the gut is blind (without an anal opening), and the coelom is an enterocoel. In the Class Lingulata, the shell is calcium phosphate, the gut is U-shaped opening into the lophophore cavity, and the coelom is a schizocoel. In the Class Inarticulata the shell is carbonate, the teeth, sockets and pedicle are poorly developed, and the gut opens toward the beak region.

Waste products from the blind gut of articulate brachiopods are regurgitated as small pellets and expelled from the mantle cavity by the rapid closing or snapping of the valves. In addition to the gut, the body cavity contains one or two digestive glands, gonads for reproduction, a contractile sack or heart that circulates fluids, and two excretionary pores for ridding the body of nitrogenous wastes and, at

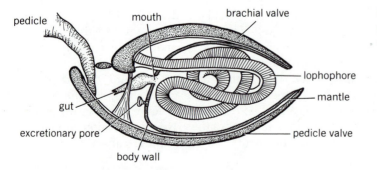

Figure 6-45   Cross section of a brachiopod showing the relationship of the internal organs to the two valves.

appropriate times, for releasing male or female reproductive cells from the gonads.

Nerves run from the edge of the mantle to a central ganglion around the mouth and from there to the muscles controlling the valves. However, brachiopods have no specific sense organs. The individual cells at the mantle edge are apparently capable of sensing a variety of subtle changes in the environment that cause the animal to close the valves rapidly in a defensive posture.

The opening and closing of the valves are controlled by two pairs of muscles in articulate brachiopods. The pair of muscles that stretch between the valves and whose contraction closes them, are called adductor muscles. The valves are opened by a set of muscles called diductors (Fig. 6–46). These pass from the floor of the pedicle valve to an knob on the brachial valve that extends behind the hinge-line. Contraction of these diductor muscles pulls down on this knob, called the cardinal process (Latin, *cardo* = a hinge), and tilts the brachial valve up opening the shell so that the brachiopod can feed. The attachment surface of the muscles inside the valve produces two heart-shaped scars which are preserved in fossils. The musculature of lingulate brachiopods is more complex and allows lateral, as well as vertical, movement of the valves.

Cambrian Brachiopods:

The brachiopods must have evolved rapidly from one or more ancestral lophophorates at the end of Ediacaran time, but if the first brachiopod had ancestors in the Ediacaran faunas, they are not like the typical brachiopod. The first lophophore-bearing animals may have been some of the tube-dwellers of Proterozoic and earliest Cambrian time that have been assigned to the phoronid worms, but the transition from these to the brachiopods of

Early Cambrian time is not preserved in the fossil record. The extent of the differences between the brachiopod classes in embryology, shell composition, and organ systems suggests that their common ancestry in some ancestral lophophorate is remote and the phylum is polyphyletic. In Lower Cambrian rocks all orders of lingulate brachiopods appear, the oldest in the Tommotian Stage. The oldest articulates appear in the overlying *Fallotaspis* zone.

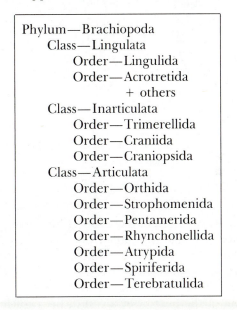

```
Phylum—Brachiopoda
    Class—Lingulata
            Order—Lingulida
            Order—Acrotretida
                    + others
    Class—Inarticulata
            Order—Trimerellida
            Order—Craniida
            Order—Craniopsida
    Class—Articulata
            Order—Orthida
            Order—Strophomenida
            Order—Pentamerida
            Order—Rhynchonellida
            Order—Atrypida
            Order—Spiriferida
            Order—Terebratulida
```

Of the four orders of the class Lingulata only two have extensive fossil records; the other two are confined to Cambrian rocks and are relatively unimportant. The Lingulida secrete elongate, biconvex, phosphatic shells shaped like a tongue (Latin, *lingula* = tongue) (Fig. 6–47). The pedicle emerges between the two valves rather than through an opening in one of them. The other important order, the Acrotretida, secrete subcircular, cone-shaped shells and the pedicle emerges from a hole in the

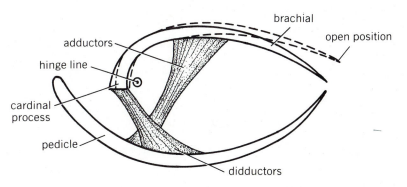

Figure 6-46   Cross section of a brachiopod showing the muscles for opening and closing the valves.

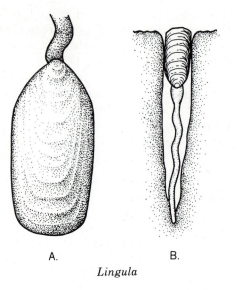

A.   B.

Lingula

Figure 6-47 *Lingula*, the shell is about 3 cm long. A. Dorsal view. B. The brachiopod at the top of its burrow attached to the bottom by a long pedicle.

pedicle valve. Both of these orders have living representatives that differ little from their early Paleozoic ancestors and are in this sense "living fossils."

Lingula live at present in vertical burrows mostly in the tidal zone. Each individual is attached to the bottom of the burrow by a contractile pedicle several times the length of the valves (Fig. 6–47). This burrowing lifestyle is rare in brachiopods but *Lingula* adopted it in Ordovician time and the adaptation has been so successful that the genus has

persisted essentially unchanged in this ecologic niche for nearly 500 million years.

Brachiopods of the order Orthida appear in Early Cambrian rocks and are ancestral to all other orders of the Articulata. They are biconvex shells, that is, each valve is convex, and the hinge is long and straight (Fig. 6–48). The opening for the pedicle in the pedicle valve and the notch that may develop in the brachial valve opposite it are open in the orthids, indicating that the pedicle was functional and attached to the seafloor. The only skeletal supports for the lophophore are small, rodlike extensions from the hinge of the brachial valve. Many shells have coarsely ribbed, or folded, shell surfaces. The orthids reached their greatest abundance in the Ordovician Period, but continued in abundance through the Paleozoic until they became extinct in the Permian crisis that affected many groups of this phylum. The orthids were the only articulates in Early Cambrian time but they were joined in Middle Cambrian time by a second order of the articulates, the pentamerids.

Ordovician Brachiopod Explosion:

Early in the Ordovician Period the calcareous brachiopods increased greatly in abundance and diversity. Their shells are the most common fossils found in mid-Paleozoic rocks. On modern beaches the clams and snails are the common sea shells, but collectors on Paleozoic beaches (if shell collectors had lived then!) would have found brachiopods.

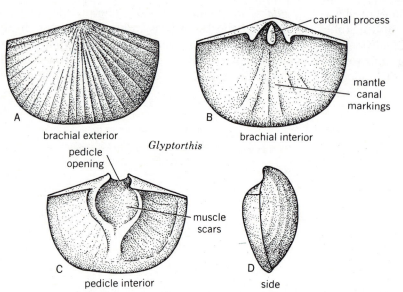

brachial exterior   brachial interior

Glyptorthis

pedicle opening

muscle scars

pedicle interior   side

— cardinal process

mantle canal markings

Figure 6-48 The orthid brachiopod *Glyptorthis* (Ordovician). A. Exterior of pedicle valve. B. Interior of brachial valve. C. Interior of pedicle valve. D. Side view.

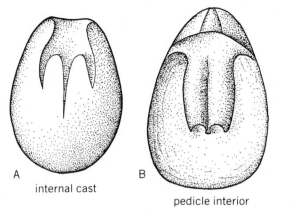

Trimerella

Figure 6-49   The inarticulate brachiopod *Trimerella* (Silurian).
A. Preserved in the usual manner as an internal mold with
the vaults under the muscle platforms filled with sediment,
B. Reconstruction of the pedicle valve from the mold,
showing the muscle platforms. The shell is about 6 cm long.

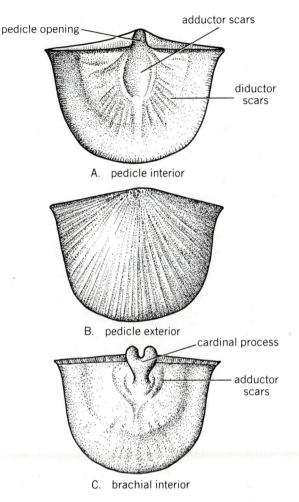

Figure 6-50   The strophomenid brachiopod *Strophodonta*
(Devonian), about 3 cm wide. A. Pedicle valve interior,
B. Pedicle valve exterior, C. Brachial valve interior.

Brachiopods of the inarticulate order Trimerellida grew to a relatively large size and secreted heavy shells, apparently of aragonite, the more soluble form of calcium carbonate. Paleontologists deduce that the shells were aragonite because they are usually dissolved away and the brachiopods are known largely from molds, casts, and replacements. The efficiency of the muscles in this and other brachiopod groups was increased by shortening the distance over which they were stretched. The trimerellids evolved platforms supported by calcareous plates above the inner surfaces of both valves (Fig. 6–49). This structure was also developed in different forms in a group of orthids and, more notably in the pentamerids described later. A group of small inarticulates comprising the order Craniida lived with the pedicle valve cemented to another shell or to rocks, and they are commonly found plastered onto the valves of other brachiopods.

By Early Ordovician time the lingulate and inarticulate brachiopods were completely dominated by the rapid diversification and increase in numbers of the articulates. The orthids thrived and gave rise to one of the most successful of the Paleozoic brachiopod orders, the Strophomenida. Strophomenids have thin shells with long straight hinge lines, resembling a knight's shield in shape (Fig. 6–50). Usually one valve is flat and the other is slightly convex, but in some genera one of the valves is concave and the space between the valves is reduced

to a cavity a few millimeters high (Fig. 6–51). In the strophomenids the pedicle opening is usually restricted, and in many it is completely covered by calcareous plates. When the opening was covered, the pedicle was obviously nonfunctional and these brachiopods must have lay unattached on the sediment surface, or were partly buried in sediment. In several Silurian and Devonian strophomenids the pair of teeth at the hingeline was replaced by a set of small teeth along the whole length of the hinge (Fig. 6–50).

Typical brachiopods of the Order Pentamerida are large, biconvex shells with short hinge lines and prominent beaks (Fig. 6–52). In order to reduce the length of muscles in their inflated shells the pentamerids developed a spoon-shaped platform

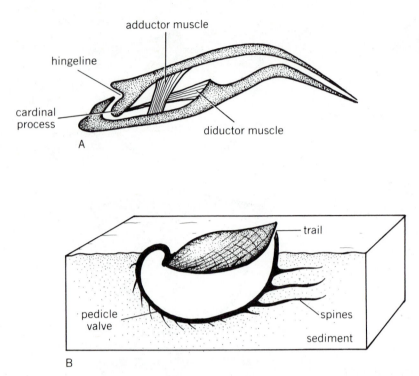

Figure 6-51 Cross sections of stropho-menid brachiopods. A. *Rafinesquina* (Ordovician) showing the connection of muscles to the cardinal process and the thinness of the space between the valves, B. A productid brachiopod embedded in sediment showing the function of the trail.

in the pedicle valve. The platform was supported below and attached to the floor of the valve by a vertical plate. Some pentamerids had a similar structure in the brachial valve so that the muscle bases at both ends were elevated above the floors of the valves. Large pentamerids were particularly abundant in Silurian time when they inhabited reef environments in concentrations resembling those of modern oyster banks.

Brachiopods of the Order Rhynchonellida evolved from the pentamerids in Ordovician time. They have short hinge lines, prominently beaked posteriors (Greek, *rhynchos* = snout), and, typically, strongly folded biconvex shells that interlock at the front in a tongue-and-groove structure (Fig. 6–53). This interlock was developed to a lesser extent in

many other brachiopod orders. Although the structure probably strengthened the shell so that it could not be twisted open by a predator, it also separated the incoming lateral currents from the outgoing central current when the animal was feeding. The lophophore was supported at its base by small hook-like extensions from the hinge area of the brachial valve. Members of this order were abundant in mid-Paleozoic time and survived both Late Devonian and terminal Permian crises in brachiopod history. They persist in modern seas.

New Silurian and Devonian Orders:

Late in Ordovician time the rhynchonellids gave rise to the Order Atrypida in which the lophophore

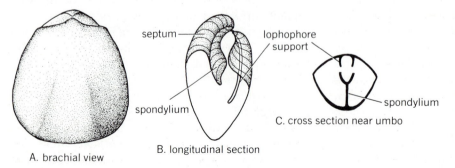

A. brachial view

B. longitudinal section

C. cross section near umbo

Figure 6-52 The pentamerid bra-chiopod *Pentamerus* (Silurian), about 5 cm long. A. Dorsal view, B. Cross section through the pedicle opening showing the muscle platforms, C. Cross section perpendicular to that in B, showing the spoon-shaped spondylium in the pedicle valve.

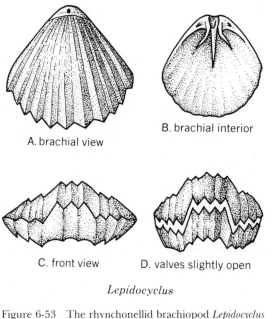

A. brachial view

B. brachial interior

C. front view

D. valves slightly open

Lepidocyclus

Figure 6-53 The rhynchonellid brachiopod *Lepidocyclus* (Ordovician). A. Dorsal view, B. Interior of brachial valve, C. Front view with valves closed showing the tongue-and-groove interlock, D. The same with the valves open showing the division of the opening into three.

was coiled in the form of two spirals shaped like antique beehives (Fig. 6–54). The axis of the two spirals was directed upward into the higly convex brachial valve. The feeding current was apparently drawn into the axis of the spiral from the edges of the valve and filtered as it passed through the many coils of the lophophore. Because the lophophore was reinforced by a ribbon of calcite, these elegant structures are usually preserved. Some paleontologists classify all brachiopods with spiral lophophore supports in the order Spiriferida; others consider the Atrypida to be a separate order.

In the Spiriferida the spirals are directed laterally rather than upward. As a result many of the shells, but not all, are wide and have a long hinge line. Typical spiriferids have shells that are coarsely ribbed or folded and interlock at the front in a tongue-and-groove (Fig. 6–54). However, some have short hinge lines, spirals with few coils, and smooth shells that in many ways resemble those of the terebratulids described later. Although spiriferids range from Middle Ordovician to Jurassic in age, they were particularly abundant in the Devonian Period.

The oldest brachiopods of the Order Terebratulida appear in Lower Devonian rocks. They are the last major group of brachiopods to appear. They

have smooth, elongate shells with short hinge lines and a functional pedicle (Fig. 6–54E,F). The lophophore was supported by a calcareous ribbon in the form of a simply or complexly folded loop. The terebratulids diversified rapidly in Early Devonian time but did not became major constituents of Paleozoic brachiopod faunas. They survived the Permian crisis and became the most abundant brachiopods in Mesozoic and Cenozoic seas.

Late Paleozoic Brachiopods:

Late Devonian was a difficult time for brachiopods. The Atrypida and Pentamerida became extinct and

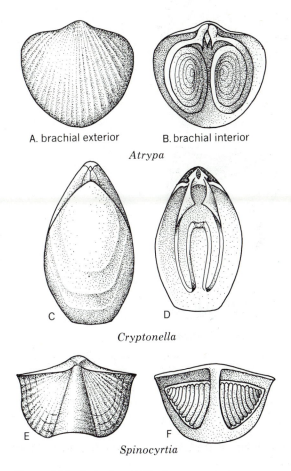

A. brachial exterior

B. brachial interior

Atrypa

C

D

Cryptonella

E

F

Spinocyrtia

Figure 6-54 A, B. *Atrypa* (Silurian–Devonian), an atrypid, pedicle valve and brachial valve interior, showing the spiral lophophore support, C, D. *Cryptonella* (Devonian–Permian), a terebratulid, dorsal view and brachial interior, showing the looped lophophore support, E, F. *Spinocrytia* (Devonian), a spiriferid, dorsal view and brachial interior, showing the spiral lophophore support.

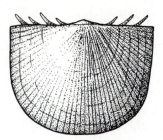

A.   *Chonostrophia*

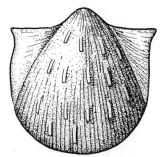

B.   *Cancrinella*

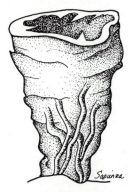

C.   *Prorichtofenia*

Figure 6-55   Late Paleozoic strophomenid brachiopods. A. *Chonostrophia* (Silurian–Devonian), B. *Cancrinella* (Carboniferous–Permian) showing spine bases, C. *Prorichtofenia* (Permian) showing the coral-like form.

major groups of the Strophomenida and Orthida died out.

With the setbacks suffered by these groups, the strophomenids that survived came to dominate Carboniferous and Permian brachiopod faunas. In Devonian time several strophomenids had developed spines on the outside of the valves (a group called the Productida) or along the hinge (a group called the Chonetida) and these spiny brachiopods

diversified rapidly during the Carboniferous (Figs. 6–51B,6–55). As in many other strophomenids, the pedicle was nonfunctional, the pedicle opening was closed, and the shell lay on the sediment surface or was partly buried by sediment. The spines are thought to have kept the shell from sinking into muddy sediment on which these brachiopods lived. Most productids have a flat valve and a highly convex one. The front margins of both valves are extended forward to form a narrow slit (Fig. 6–51B). This structure, called a trail, allowed the animal to be largely covered with sediment and still let in a feeding current when the valves were slightly opened. In some productids the pedicle valve was cemented to hard objects. This valve then grew much larger than the brachial valve into the form of a horn anchored at its tip. The brachial valve became a subcircular cap on the open end of the horn (Fig. 6–56C). These aberrant brachiopods of the family Richtofeniidae resemble corals and lived like corals in reef habitats.

At the end of Paleozoic and the beginning of Mesozoic time the brachiopods suffered a major setback from which the phylum has not fully recovered. The following did not survive the end of the Paleozoic Era: (1) major groups of the strophomenids, such as the productids and chonetids; (2) all of the orthids; (3) all of the spiriferids except a single superfamily which did not survive the Early Jurassic. The extinctions left only the terebratulids, rhynchonellids, and a small obscure group of deep-sea brachiopods of the superfamily Thecideacea which are thought by some to be survivors of the strophomenids, but differ from Paleozoic strophomenids in the microstructure of the shell.

REFERENCES

1. Debrenne, F., and Vacelet, J., 1984, Archaeocyatha: is the sponge model consistent with their structural organization?: Paleontographica Americana, v. 54, p. 358–369.

2. Mistiaen, B., 1984, Disparition des stromatopores ou survie du groupe: hypothèse et discussion: Bulletin Société Géologique de France. sér. 7, v. 24(6), p. 1245–1250.

3. Sepkoski, J. J., and Sheehan, P. M., 1983, Diversification, faunal change and community replacement during Ordovician radiations. In Tevesz, M. J. S., and McCall, P. L., (eds.), Biotic Interactions in Recent and Fossil Benthic Communities: New York, Plenum, p. 673–717.

4. Sepkoski, J. J., 1981, A factor analytical description of the Phanerozoic marine fossil record: Paleobiology, v. 7, p. 36–53.

5. Webby, B. D., 1984, Ordovician reefs and climate: a review. in Bruton, D. L. (ed.) Aspects of the Ordovician System: Oslo University, Contributions to Palaeontology, v. 295, p. 89–100.

6. Copper, P., 1985, Fossilized polyps in 430-Myr-old *Favosites* corals: Nature, v. 316(6024), p. 142–144.

7. Scrutton, C. T., 1984, Origin and early evolution of Tabulate Corals: Palaeontographica Americana, v. 54, p. 110–118.

SUGGESTED READINGS

Boardman, R. S., and et al., 1983, Bryozoa. Part G (revised), v. 1. In Robison, R. A. (ed.), Treatise on invertebrate paleontology: University of Kansas and Geological Society of America. 625 p.

Brasier, M. D., 1980, Microfossils: London, Allen & Unwin, 193 p.

Culver, S. J., 1987, Foraminifera. In Lipps, J. H. (ed.), Fossil Prokaryotes and Protists: University of Tennessee Studies in Geology, No. 18, p. 169–212.

Dutro, J. T., and Boardman, T. W. (eds.), 1981, Lophophorates: University of Tennessee Studies in Geology, No. 5, 253 pp.

Fagerstrom, J. A., 1987, The evolution of reef communities: New York, Wiley, 600 p.

Haynes, J. R., 1981, Foraminifera: New York, Wiley, 433 pp.

Hill, D., 1972, Archaeocyatha, Part E (revised), in Teichert, C. (ed.), Treatise on Invertebrate Paleontology. Boulder, Colorado and Lawrence, Kansas, Geological Society of America and University of Kansas, 158 p.

Oliver, W. H., et al. (eds.), 1984, Recent advances in the paleobiology and geology of Cnidaria: Palaeontographica Americana v. 54, 557 pp.

Rudwick, M. J. S., 1970, Living and Fossil Brachiopods: London, Hutchison Univ. Press, 119 pp.

Stearn, C. W., 1975, The stromatoporoid animal: Lethaia, v. 8, p. 89–100.

Chapter 7

The Invertebrates Dominate the Seas II:

SCHIZOCOELOMATES AND ENTEROCOELOMATES

BIOLOGY OF LIVING MOLLUSCS

In modern seas, molluscs are the most conspicuous shell-bearing animals and are outnumbered only by the forams. Many molluscs secrete calcareous shells suitable for preservation, but many do not. The fossil record of the phylum extends back to the beginning of the Cambrian Period. The molluscan architecture of the soft body enclosed in one or two hard shells proved to be adaptive not only to marine, but also to freshwater and terrestrial environments.

Molluscs are unsegmented (or pseudosegmented), coelomate metazoans. The repetition of gills and/or muscle scars in the Monoplacophora and Polyplacophora, two minor groups, suggests that primitive molluscs shared a segmented ancestor with annelids and arthropods, but opinion is divided on the significance of these features. In molluscs, a muscular extension of the body, commonly referred to as a foot, was modified for burrowing, creeping, and grasping food. Molluscs have well-developed digestive, circulatory, reproductive, nervous, and sensory systems. In all groups except the bivalves, scaphopods, and some cephalopods, the food is torn into pieces for digestion by a unique rasplike ribbon behind the mouth called a radula.

The phylum is usually divided into eight classes, five of which are of minor importance to the paleontologist, although all but one have a fossil record. The minor classes are: (1) the Polyplacophora (chitons, most of which cling to rocks in the surf zone and secrete spicules and a covering of eight plates); (2) the Monoplacophora (single saucer-shaped shell, possibly segmented); (3) the Rostroconchia (bivalved shells joined permanently across the top); (4) the Scaphopoda (tooth or tusk shells), and (5) the Aplacophora (without shells). Small conical fossils such as *Hyolithes* and *Tentaculites* may represent a ninth class (Chapter 5), but some paleontologists regard them as a separate phylum. Only the three large classes are described separately in the following discussion.

Phylum—Mollusca
    Class—Aplacophora
    Class—Polyplacophora
    Class—Rostroconchia
    Class—Scaphopoda
    Class—Bivalvia
    Class—Monoplacophora
    Class—Gastropoda
    Class—Cephalopoda

The first molluscs probably evolved from flatworm (phylum Platyhelminthes) ancestors in Ediacaran time. Three of the six classes of molluscs with readily preservable shells appear in earliest Cambrian (Tommotian) beds, but all their shells are minute. Runnegar (1) believes that the Ediacaran ancestors of molluscs were a millimeter or less across and of very simple structure. The first shelled molluscs may have been simple conical monoplacophorans (Fig. 7−15), and these gave rise rapidly at the beginning of Cambrian time to the Rostroconchia and Gastropoda, and slightly later to the Bivalvia. High conical monoplacophorans with septa across the apex probably evolved into the cephalopods in Late Cambrian time.

Class Bivalvia:

This class includes the clams, mussels, scallops, and oysters. As the name implies, the Bivalvia have two valves, but they are not the only invertebrates with two valves. Other scientific names that have been applied to them include Lamellibranchia and Pelecypoda. The two shells, or valves, are mirror images of each other laterally placed on either side of the body. Like brachiopod shells, they have a bluntly pointed apex called the beak. This is the position of the larval shell where growth begins. Subsequent shell growth usually produces growth lines parallel to the margin of the larval shell (Fig. 7−1A). A single valve is not bilaterally symmetrical like that of a brachiopod, but the area of maximum curvature of the shell near the beak, called the umbo, is usually inclined forward. Just below the beak the valves usually interlock where the teeth in one valve fit into sockets in the opposite one (Fig. 7−1C). The teeth are at the hinge line along which the valves pivot to open the lower edge. The valves are drawn together by adductor muscles that leave scars on each valve (Fig. 7−1C). Oysters and scallops have only one muscle; most bivalves have two. When these muscles relax, the valves spring apart by the action of organic ligaments. These ligaments are located either above the hinge line, in which case the ligament is under tension when the valves close, or in a triangular cup inside the hinge line, in which case it is under compression.

The valves are secreted by the mantle that lines them. The mantle is usually attached to the shell at a scar near the shell's edge called the pallial line. The pallial line usually connects the two muscle scars (Fig. 7−1C). At the back the mantle lobes may

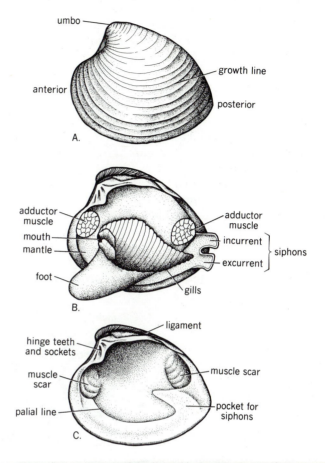

Figure 7-1 Anatomy of the living bivalve *Mercenaria*, about 10 cm long. A. Exterior of the left valve. B. Left valve and mantle removed to show the internal organs. C. Interior of the right valve.

be joined to produce two tubes, or siphons, one of which brings water into the shells and the other expels it. These siphons may be short, extending only to the edge of the valves, or they may be several times the length of the valves in species that burrow deep into the sediment. Small siphons can be retracted into a pocket when the valves close but larger ones cannot and the valves must gape at the back when closed to accomodate them. The existence of a pocket for retraction of the siphons is indicated by a bend in the pallial line (Fig. 7−1C).

Most bivalves are filter feeders and the water delivered by the siphon brings microorganisms for food. Between the viscera and the mantle lobes hang the gills, which filter water for food and extract oxygen for respiration. Small food particles collected on the gills are moved forward by cilia to the mouth which lies just below the front adductor muscle (Fig. 7−1B). Waste is extruded into the outgoing water stream. Bivalves do not have a head.

The heart surrounds the gut and pumps a fluid into open cavities in the body which act as a hydrostatic skeleton. The lower part of the body is a wedge-shaped foot that contains one of these large cavities. Most bivalves live partly or wholly within sediment. When they move forward, the valves are opened, and the foot is thrust forward into the sediment and expanded by fluid which is pumped into its cavity. When the foot is anchored, the muscles in it can then drag the valves toward it. The foot–anchor is then deflated by withdrawal of fluid, and thrust forward again. Progress is slow. Not all bivalves push through sediment; some stay in one place, some are cemented to hard objects, some attach to objects by threads, some bore into rock or wood, others swim through the water. The basic pattern of soft and hard parts previously described was much modified as bivalves adapted to these varied ways of life.

Class Gastropoda:

Gastropods are snails and slugs. They are presently the most diverse and abundant of the mollusc classes and successfully occupy marine, terrestrial, and freshwater environments in large numbers. The vast majority secrete a single shell, usually coiled in a conispiral, like a conical corkscrew. The shell is basically a cone translated down the axis of coiling as it turns. If the young, pointed part of the shell is oriented upward (Fig. 7–2C), in most gastropods the aperture is to the right of the axis of coiling. These shells are called dextral. Those few in which the opening is to the left are sinistral.

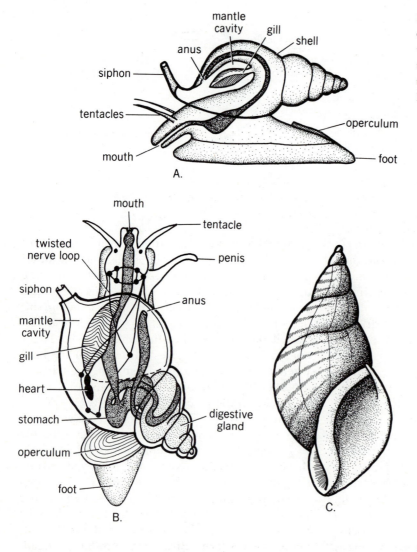

Figure 7-2   Gastropod anatomy. A. Side view of a typical gastropod in cross section. B. Top view of a mesogastropod with the shell made transparent. C. Dextral gastropod as conventionally oriented in illustrations (*Calicantharus*, Pleistocene).

Snails have distinct heads commonly equipped with eyes and feelers (Fig. 7–2A). They creep on a muscular foot. The body and foot can be retracted into the shell of most snails and many secrete a plate, or operculum, that closes the opening. During embryological development most snails undergo torsion, that is, the back of the embryo twists 180 degrees so that it comes to lie above the head. Both the end of the digestive tract and the mantle cavity containing the gills are twisted into this position. In this manoeuver, most snails lose one of the two gills and one of the two kidneys, and the nerve loop is twisted into a figure eight. In one subclass of gastropods, detorsion brings the mantle cavity and the anus to the back again but with only one gill and kidney. Although several theories have been proposed to explain the advantages of torsion to gastropods, no one theory is widely accepted. The gastropods that left the sea to live on land dispense with the remaining gill and use the mantle cavity as a lung by multiplying the number of blood vessels on its surface to absorb oxygen. Those that live in fresh water must rise to the surface periodically to trap air in order to supply the respiratory needs of their lung.

Many gastropods feed by scraping organic films and algae off the bottom. Others are deposit feeders and filter feeders. Some are parasitic and many are predatory. Some members of five different families drill neat holes in the shells of bivalves and other gastropods to feed on the soft tissue inside.

Gastropod classification is based largely on features of the soft parts, and paleontologists are forced to infer the nature of the anatomical features that are critical for classification from a study of their influence on the shell. Three subclasses are generally recognized: (1) the Prosobranchia, which are largely marine gastropods with viscera twisted by torsion; (2) the Opisthobranchia, in which detorsion has taken place and the viscera are no longer twisted; and (3) the Pulmonata, which are snails in which the mantle cavity is modified into a lung and the gill is lost. Opisthobranch gastropods with shells sufficiently thick for preservation are rare in the fossil record. An order of this subclass, the Pteropoda, reduced the shell to a thin semitransparent envelope, modified the foot into a pair of beating "wings", and joined the plankton floating near the surface of the sea. Their insubstantial shells accumulate on modern oceans floors locally to form pteropod oozes.

Class Cephalopoda:

The approximately 400 living species of cephalopods are a small fraction of the thousands that lived in the past and are known to us through their fossils. All living, and we suspect all fossil, cephalopods are, or were, marine. Bivalves and gastropods are basically adapted to living in and crawling on top of the sediment respectively, but cephalopods are adapted to swimming. The living squid, octopuses, cuttlefish, and their fossil representatives, the belemnites, are members of the subclass Coleoidea. *Nautilus* is the last living genus of the subclass Nautiloidea which has many fossil members. A vast number of fossils are included in the extinct subclass Ammonoidea. In each of these the head is equipped with tentacles for grasping food and bringing it to the mouth. The cephalopods are predators and scavengers and the largest are capable of defending themselves against sperm whales. Below the tentacles the mantle margin is shaped into a tube through which the water in the mantle cavity can be jetted by rapid muscular contraction (Fig. 7–3B). In reaction to this jet the cephalopod moves in the direction opposite to that in which the jet is directed. In *Nautilus* the backward motion is relatively slow but in the streamlined squid the repeated jetting of water from the mantle may be powerful enough to lift the body from the sea and meters into the air. The squid (Fig. 7–4) and octopus have largely dispensed with skeletons but *Nautilus* secretes an external shell made buoyant by the gas that fills its chambers (Fig. 7–3A,B). Although the shell is coiled, it differs fundamentally from gastropod shells in its division by partitions, or septa, with a central canal. The septa mark successive back walls of the body chamber secreted as the animal enlarges its shell, abandoning and sealing off the early sections. The animal removes liquid from these chambers and substitutes gas through a thread of soft tissue containing veins and arteries that passes in a tube through each septum and chamber to the initial one.

PALEOZOIC BIVALVES

John Pojeta and Bruce Runnegar (2) have presented evidence that *Fordilla* and *Pojetaia* from Early Cambrian rocks are the first bivalves (Fig. 7–5). They believe they are descended from the primitive

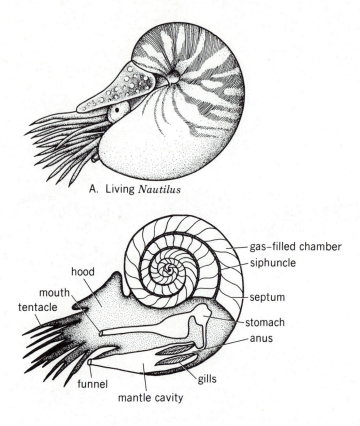

A. Living *Nautilus*

hood
mouth
tentacle
funnel
mantle cavity
gas-filled chamber
siphuncle
septum
stomach
anus
gills

Figure 7-3   *Nautilus* (about 1.2 cm in diameter). A. Side view of exterior of the animal in living position. B. Cross section to show the relationship of the internal organs to the shell.

molluscan class Rostroconchia. The rostroconchs have two valves that gape at the front and back and are joined rigidly along the top by calcareous layers, so that they do not open and close as bivalves do. They lived from Cambrian to Permian time and in certain places were sufficiently abundant to form shell beds. The transition from rostroconchs to bivalves required the replacement of the rigid calcareous bridge between the valves by a flexible protein ligament and development of shell closing muscles. The diversification of bivalves, like that of several other groups secreting heavy calcareous skeletons, did not occur immediately after their ap-

pearance; in fact, no Middle or Upper Cambrian bivalves have yet been found. However, by Middle Ordovician time, all major subclasses of the bivalves had appeared.

Bivalves and other molluscs are not as well preserved in most Paleozoic sediments as are brachiopods. Usually bivalve fossils are molds of the interiors or exteriors of single valves. Possibly their shells contained a higher proportion of the less stable carbonate, aragonite, than those of later bivalves which are better preserved. When bivalves die and the adductor muscles relax, the ligaments spring the valves open and they are usually preserved sep-

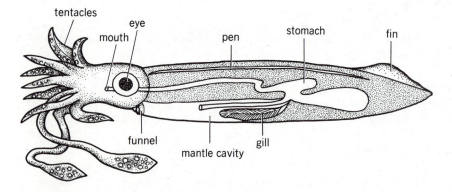

tentacles
mouth
eye
pen
stomach
fin
funnel
mantle cavity
gill

Figure 7-4   The living squid *Loligo* cut away to show some of the internal organs.

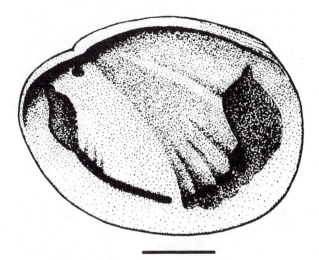

Figure 7-5   Reconstruction of the interior of the right valve of one of the first bivalves, *Fordilla*, from the Lower Cambrian rocks of New York. The two adductor muscle scale are more darkly shaded than the rest of the shell. The scar bar is 1 mm. (Courtesy of J. Pojeta and *Science*, v. 180, p. 867, © 1973, by the AAAS.)

arately whereas brachiopods, which are not spring loaded, are most often preserved with the valves in the closed position. Bivalves did not compete directly with brachiopods but many lived in nearshore, shallow environments and were capable of

tolerating much suspended sediment (3). As a result they are commonly found in siltstones and fine sandstones deposited in estuarine environments. Through the Paleozoic they expanded into offshore environments and developed forms that could burrow deeper into soft sediment.

Classification of bivalves by zoologists is based on features that are not preserved in fossils, such as gill structure. Paleontologists use a classification that combines features of fossil shell, such as teeth and sockets at the hinge line, ligaments, shell microstructure, and shape. The close relation between bivalve shape and habitat in modern bivalves allows paleontologists to deduce habitats of fossil species (Fig. 7–6).

Many early Paleozoic bivalves moved about within, or on top of, soft sediment in the sea and fed by filtering microorganisms from the water or eating dead organic matter on the seafloor. Some stayed buried in one place; others attached themselves to hard substrates with horny threads, called a byssus, secreted by a gland in the foot. Because shells of early Paleozoic bivalves show no evidence that they had long siphons, paleontologists deduce that they were unable to burrow deep in sediments. Efficient burrowers are characterized not only by extensive siphons, but also by streamlined shells. Later Paleozoic rocks contain fossils of the forerunners of

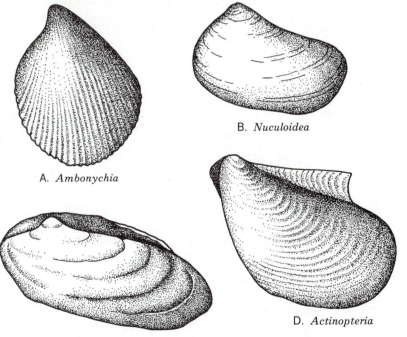

B. *Nuculoidea*

A. *Ambonychia*

C. *Archanodon*

D. *Actinopteria*

Figure 7-6   Paleozoic bivalves. Approximately natural size. A. *Ambonychia* (Ordovician). B. *Nuculoidea* (Devonian). C. *Archanodon*, nonmarine (Carboniferous). D. *Actinopteria* (Devonian).

the bivalves that swam and cemented themselves to hard substrates. These ecological roles became important in Mesozoic time. Devonian reefal limestones are characterized by large, thick shells of the bivalve *Megalodon*, which was the ancestor of the coral-like rudist bivalves of the Mesozoic described in Chapter 10.

In Middle Devonian time bivalves were able to invade freshwater environments, and their shells appear in post-Devonian coal-bearing successions. Modern freshwater clams disperse their young by attaching the larvae to the gills and fins of fish for a short time. Although there is no evidence, it is presumed that the evolution of this curious piggyback adaptation allowed some marine Paleozoic bivalves to dispatch their young upstream against river currents to lakes on the gills of fish, such as rhipidistians, which, like modern salmon, may have migrated between marine and freshwater environments.

FIRST CEPHALOPOD SUCCESS

Living *Nautilus*:

For fossil groups that do not have living representatives, paleontologists may have great difficulty relating shell and skeleton, which are preserved, to soft tissue, which is not. In the absence of knowledge of the living animal, the relationship of the hard parts to the behavior of the animal remains speculative. Fortunately, remnants of both of the major subdivisions of the cephalopods are alive today but the shell-bearing branch has been reduced to a single genus. Because the pearly nautilus species of the western tropical Pacific are the sole living

representatives of a host of fossil cephalopods with external shells, their shells and behavior have been extensively studied, particularly within the last decade (4,5,6).

Most species of *Nautilus* secrete a conical shell that is closely coiled in one plane so that succeeding whorls embrace the previous whorl and only the final one can be seen (Fig. 7–3A). When the shell is cut across the axis of coiling, or X-rayed (Fig. 7–3B,7–7), the septa that divided the coiled cone into gas-filled chambers are revealed. The body occupies the last, or living, chamber and, as it grows, it moves forward periodically in the shell, lets water into the space behind the body, secretes a septum on its back surface, pumps the water out, and separates itself from the abandoned chambers. The addition of chambers stops when the shell has reached mature size in five to ten years. The chamber formed last is filled with water and the recently formed chambers are partly filled, but most of the chambers are filled with a gaseous mixture rich in nitrogen that diffuses into them from the animal. *Nautilus* slowly replaces the fluid in the last-formed chambers and introduces the gas by means of the siphuncle, a string of tissue in a calcareous sheath that extends back from the body through perforations in each septum. The sheath of the siphuncle is formed of two parts: the edges of the septa around the perforations turned backward as necks, and cylinders of poorly calcified tissue called connecting rings that join the free end of the septal neck to the front of the next septum (Fig. 7–7A). *Nautilus* is able to pump fluid out of the chambers using an osmotic mechanism through the permeable connecting rings, and it controls its buoyancy as the body weight grows by the addition of new buoyant chambers and the gradual emptying of the newly

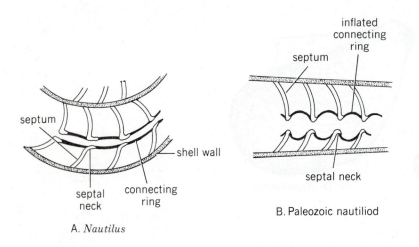

Figure 7-7   Siphuncle structure of modern *Nautilus* and Paleozoic straight nautiloids.

A. *Nautilus*

B. Paleozoic nautiliod

formed chambers. As a result it maintains a slight negative buoyancy.

The shell of *Nautilus* is formed entirely of aragonite. When the outer porcellaneous layer is artifically polished off, the beauty of the mother-of-pearl interior layers is revealed. Application of studies of these shells to the interpretation of the habitats of fossil cephalopods is discussed further in Chapter 13.

Early Paleozoic Cephalopods:

The essential features of the shell of *Nautilus*, the chambered cone and siphuncle of septal necks and connecting rings, are also present in its Paleozoic ancestors. In most, but not all, of these early cephalopods the shell is not coiled but straight or slightly curved. Some of these straight cones were 9 meters long—the largest shells ever secreted by invertebrates. Their internal structure can be understood in terms of the problems of stabilizing such shells.

In coiled cephalopods the gas-filled chambers that exert a buoyant force upward are above the body, which exerts a gravitational force downward, and the animal is suspended in the water in the position shown in Figure 7–3A. The center of gravity is directly below the center of buoyancy, a situation in which the shell is stable and the body chamber is more or less horizontal. A long straight cone can only be propelled parallel to the seafloor through the water if the cone is horizontal. However, the buoyancy of the chambers of a simple straight cone will bring the shell into a vertical position and the heavy body face down (Fig. 7–8A,B). To bring the center of buoyancy and the center of gravity together so that the shell will be horizontal, the center of gravity must be moved back by weighting the apical end, and/or the center of buoyancy must be moved forward by bringing the gas-filled chambers above the heavy body (Fig. 7–8C).

Phylum—Mollusca
  Class—Cephalopoda
    Subclass—Nautiloidea
    Subclass—Coleoidea
    Subclass—Ammonoidea
      Order—Anarcestida
        Suborder—Clymeniina
        Suborder—Bactritina
      Order—Goniatitida
          + others (Ch. 10)

Paleozoic cephalopods deposited calcium carbonate around the margins of the chambers in the apical parts of their shells. How the deposition took place is not clear. It has been proposed that the deposit was made from a mantle lining left behind when the body abandoned the chamber, but such a lining could not have been connected to the siphuncle and it has no counterpart in modern *Nautilus*. Possibly the carbonate was deposited on organic membranes supplied by diffusion through the siphuncle when the chambers were full of water. Paleozoic straight cephalopods also formed heavy deposits of six main types in their siphuncles. The most common were annular masses, kidney-shaped in cross section, located on the septal necks (Fig. 7–8C,D,E). In the actinoceratoids these are associated with inflated connecting rings and grew to fill the space of the siphuncle completely, leaving only a few canals. Such siphuncles look like a long row of toy tires (Fig. 7–9). They filled most of the apical part of the shells and are usually the only parts of the shell preserved because they are more massive than the walls or septa. In the endoceratoids, the siphuncle was weighted with a series of nested cones whose apexes were pointed toward the apex of the shell and pierced by the siphuncle (Fig. 7–10A). The siphuncles of other straight cephalopods were weighted with radial plates extending from the periphery to just short of the axis. The effect of the weighted siphuncles was to counterbalance the weight of the body and move the center of gravity toward the back of the shell. They are thickest near the apex where their effect as a counterbalance would be greatest. As the body grew heavier and the shell longer, new deposits were secreted near the living chamber and the old ones were thickened near the apex (Fig. 7–8C). In many straight nautiloids the heavy siphuncle was not central but near the lower margin to resist the tendency of the shell to roll.

A group called the ascocerids added a few large chambers above the body chamber rather than behind it. Their adult shells were short, small, and the apertures of some are partially closed over the body chamber leaving holes for the funnel, tentacles, and eyes (Figs. 7–10B, 7–11). Many early Paleozoic cephalopods balanced their shells, as does *Nautilus*, by coiling the buoyant chambers above the body.

Slightly curved cephalopods (Fig. 7–11) must have had difficulty swimming and are assumed to have floated near the seafloor or crawled on it in a more or less vertical position.

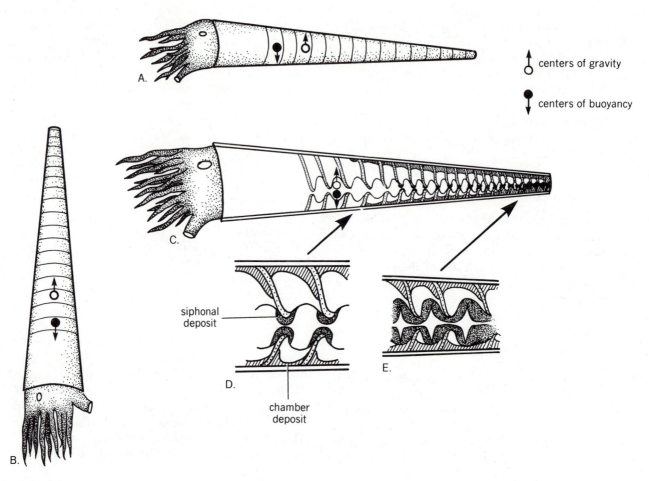

centers of gravity

centers of buoyancy

siphonal deposit

chamber deposit

Figure 7-8   Buoyancy control in Paleozoic straight nautiloids. A, B. In a simple unweighted cone the center of buoyancy is at about .75 of the length, the center of gravity is near the front, and the position in the water that results is shown in B. C. Siphuncular and chamber deposits bring the centers of gravity and buoyancy together and the shell floats horizontally. D. Detail of deposits near the living chamber. E. Detail of deposits near the apex showing the development of a heavy siphuncle and the position of the siphuncle near the ventral side of the cone.

Straight cephalopods differ little in exterior form although a few have distinctive ridges and nodes. They may be cut along the siphuncle to reveal the interior structure for identification.

From Paleozoic time to the present, one group of cephalopods replaces another in four periods of rapid expansion, each followed by slow decline or catastrophic disaster. They evolved from monoplacophoran ancestors in Late Cambrian time. Early Ordovician time was the period of greatest expansion for the Nautiloidea. By mid-Ordovician time ten orders were represented in the fossil record and their long, straight cones were abundant fossils in limestones and shales. From that time, the number of orders decreases to nine in mid-Silurian, to six in mid-Devonian, and to four in Early Carboniferous time. By mid-Paleozoic time the endoceratoids

and actinoceratoids were extinct. The orthoceratoids reached the Triassic but by middle Mesozoic time only the ancestors of the modern *Nautilus* remained of the early Paleozoic cephalopod fauna.

In Silurian time the nautiloids gave rise to a transitional group, the Bactritina, which in turn began the late Paleozoic expansion phase of the cephalopods, in which the ammonoids were the principal players. The bactritines were small, straight cephalopods, commonly with a bulbous initial chamber and small marginal siphuncle on the lower edge of the cone.

In most animals the lower side is technically called the ventral and the top side is referred to as the dorsal (from Latin words for belly and back). In a coiled shell the lower and upper sides of the coiling tube change places along its length and these terms

Figure 7-9 *Armenoceras* (Ordovician), a straight nautiloid with a heavy siphuncle (light-colored). The specimen has been cut longitudinally and polished so that the siphuncle is seen in section. The specimen is 16 cm long.

are confusing. The outside edge of a coiled cephalopod shell, which is in the ventral position at the living chamber, is called the venter; the inside edge, the dorsum.

Late Paleozoic Cephalopods:

The second major cephalopod expansion began in Devonian time from bactritine ancestors and involved the subclass Ammonoidea. The following features are typical of this group: (1) shells that are coiled planispirally; (2) cylindrical, unweighted, small marginal siphuncles located on the venter; (3) septa folded radially; (4) small, spherical, embryonic shell in the center of the coil.

Of these features, the folded septa are most distinctive of ammonoid cephalopods. In most early Paleozoic cephalopods, the septa are parts of spherical shells and concave toward the aperture. The line along which the septum meets the shell wall is

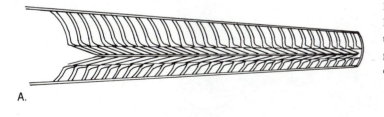

A.

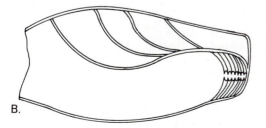

B.

Figure 7-10 Balancing straight nautiloid shells. A. Diagrammatic cross section of *Endoceras*, showing the invaginating cones in the siphuncle. B. Diagrammatic cross section of an ascocerid showing the chambers formed above the living chamber.

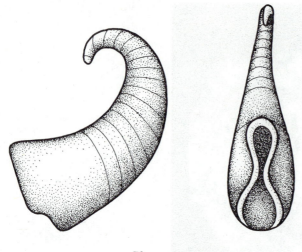

Phragmoceras

Figure 7-11 *Phragmoceras* (Silurian), side and front views.
Note the curved shell and the restricted aperture.

suture line will reflect the complexity of that fold-ing. The lines are commonly illustrated as if re-moved from the curved surface of the shell and flattened (Fig. 7–13). The folding of the septa in-creased their strength without increasing their weight, so that they would not break from pressure changes when the animal changed swimming depth (see also Chapter 13).

Ammonoids of the Devonian, Carboniferous, and Permian belong largely to the order Goniatitida. The suture lines of this group are composed of simple open or sharp curves indicative of simple folding of the septa. The goniatites are excellent index fossils for late Paleozoic strata for they, like most of the ammonoids, evolved rapidly and were distributed widely. Not only were they capable of swimming to extend their range, but after death their shells must have remained buoyant and were distributed by currents.

called the septal suture. It is invisible on the exterior surface of shells in which the wall is preserved, but in fossils the wall is often partially dissolved leaving the edges of the septa exposed where they meet the wall (Fig. 7–12). If the septa are simple concave plates, the suture will be straight or slightly curved, but if the septa are fluted at their margins, the

PALEOZOIC GASTROPODS

The first gastropods appear in the Tommotian Stage at the base of the Cambrian System and belong to the genus *Aldanella* (Fig. 7–14). These trochoid shells are accompanied by simple, low, conical shells like

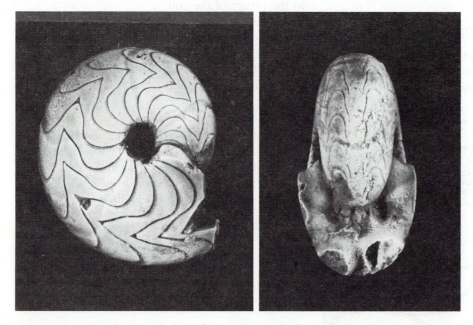

Figure 7-12 Side and apertural views of the goniatite *Bolandiceras* (Carboniferous). The specimen is 2.6 cm across. Coiled cephalopods are conventionally illustrated with the aperture at the top, a completely unnatural position. In this book the aperture is placed at the bottom to approximate a living position. (Courtesy of N. J. Riley and British Geological Survey.)

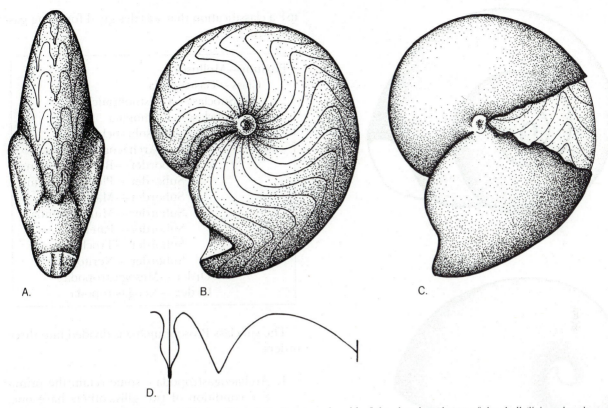

Figure 7-13   The goniatite *Imitoceras* (Carboniferous). A, B. Internal mold of the chambered part of the shell (living chamber missing) from the front and side showing the folded face of the first septum. C. Exterior of the shell broken to show the septal sutures. D. Suture line as commonly illustrated.

those of limpets that cling to rocky, wave-washed shores. These conical shells are members of the Monoplacophora, the group from which the gastropods evolved (Fig. 7–15). The elevation of the cone to accomodate an enlarging visceral mass is believed to have led to the coiling of the shell in a plane (planispiral) so that it was more compact and less likely to be knocked off the rocks by waves. Such planispirally coiled monoplacophorans were probably ancestors of both gastropods and cephalopods (Fig. 7–16). Cephalopods continued to use the planispiral coil extensively but most of the gastropods adopted the asymmetrical trochoid spiral with the shell slung over the shoulder.

The first gastropods were small and most did not reach a diameter of over a centimeter until the Ordovician Period when they first became abundant. The diversity of gastropods did not wax and wane in cycles like that of the cephalopods, but all orders increased or maintained their diversity through Devonian, Permian, and Cretaceous crises that decimated other invertebrates.

The common types of shell form evolved early in gastropod history: high spired trochoid forms of many whorls, low squat trochoid forms of few whorls, and planispirally coiled forms (Fig. 7–18). Analysis of gastropod shell form by computer is discussed in Chapter 13. One type of coiling evolved in which the spire was depressed below the last whorl instead of being elevated above it as in most gastropods. Such shells have been called both hyperstrophic and ultradextral (*Maclurites, Lecanospira*, Fig. 7–17E,F).

The orders of gastropods are recognized on the basis of their different gill structures and by the modifications of the mantle to supply them with water—features that are not preserved in fossils. However, clues to the plumbing of the mantle cavity are preserved in the margins of the aperture. In advanced gastropods the mantle folds into tubes to control the currents, like the siphons of bivalves. The front siphon brings water to the gill, or gills, above the head, and a rear one empties it behind the animal. Usually the margin of the shell is folded

A. *Aldanella*

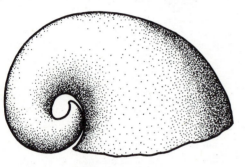

B. *Pelagiella*

Figure 7-14   Internal molds of the first gastropods (Cambrian). A. *Aldanella* (0.5 mm). B. *Pelagiella*, possibly a monoplacophoran (1 mm).

use a classification that was designed for living gastropods.

Phylum—Mollusca
    Class—Gastropoda
        Subclass—Opisthobranchia
        Subclass—Pulmonata
        Subclass—Prosobranchia
            Order—Archaeogastropoda
                Suborder—Bellerophontina
                Suborder—Pleurotomariina
                Suborder—Macluritina
                Suborder—Murchisoniina
                Suborder—Patellina
                Suborder—Trochiina
                Suborder—Neritopsina
            Order—Mesogastropoda
            Order—Neogastropoda

The subclass Prosobranchia is divided into three orders:

1. Archaeogastropoda—some retain the primitive condition of two gills; others have one. As they are without siphons, the water for the aeration of the gills enters the mantle cavity beneath the edges of the shell.
2. Mesogastropoda—have an improved gill structure and one of the gills is always lost during the torsion process. Advanced members have siphons.
3. Neogastropoda—have a still different gill structure, a single gill, and all have siphons. Many are carnivorous. As this group did not appear until the middle of the Cretaceous Period, their history is described in Chapter 10.

The archaeogastropods are the snails typical of

over the siphon to support it, or the siphon occupies a notch in the margin that leaves behind interruptions in the growth lines on the shell (Fig. 7–17A,D). The influence of siphons and water flow on the form of the aperture allows the paleontologist to

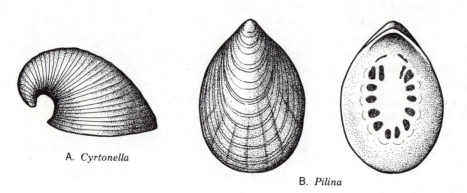

A. *Cyrtonella*

B. *Pilina*

Figure 7-15   Monoplacophorans. A. *Cyrtonella* (Silurian–Devonian) (25 mm), an advanced curved shell. B, C. Exterior and interior of *Pilina* (Ordovician–Silurian) (35 mm), showing the paired muscle scars.

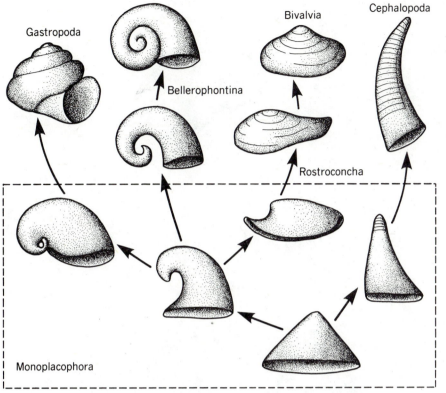

Figure 7-16 Origin of the major molluscan groups from various types of monoplacophorans. (Modified from B. Runnegar, *University of Tennessee, Geology Studies* 13.)

the Paleozoic Era but their descendants are still common in modern oceans. They reached their maximum diversity in the Ordovician Period. Robert Linsley (7) noted that most of them have radial apertures (Fig. 7–18). A radial aperture is one whose plane passes through the axis of coiling of the shell and as a result the shell cannot be carried parallel to the rock, or sediment, on which the snail is crawling. Snails with this type of aperture cannot clamp their shell by means of the muscular foot onto the seafloor when they are attacked but must rely on an operculum plate carried on the back of the foot to close the aperture. To clamp to a surface for protection, the gastropod must have an aperture that is tangential to the circumference of the last whorl.

The bellerophontine archaeogastropods have planispiral shells that increase rapidly in diameter during coiling (Fig. 7–17H,I). In this they resemble the shells of advanced monoplacophorans and some paleontologists believe they should be transferred from the gastropods to this class. Supporting their assignment to the monoplacophorans are paired muscle scars on the inside of the shell that imply that the animal had not undergone torsion; supporting the assignment to the gastropods are notches at the aperture margins of some that imply a circulation of water over a gill above the head which is found only in gastropods. Linsley suggests that the group is not homogeneous and includes both classes of molluscs. The pleurotomarines have low-spired shells with radial apertures and a deep slit in the outer edge of the aperture to channel the outflow of water from the mantle cavity (Fig. 7–17D). The group is common in shallow-water deposits of early Paleozoic age but modern representatives are found only in deep water. The macluritines are large, hyperstrophic shells with depressed spires and flat bottoms that were abundant in Middle and Late Ordovician seas. They are believed to have been sedentary filter feeders (Fig. 7–17E). Many of the high spired forms, like the common Ordovician genus *Hormotoma*, are included in the suborder Murchisoniina. Another group, of which the genus *Platyceras* is representative lived in a parasitic or symbiotic relationship with stalked echinoderms sitting near their anal openings and feeding from their waste (Fig. 7–17G).

Mesogastropods appear in Ordovician time but the order did not begin to diversify rapidly until the end of the Paleozoic and the beginning of the

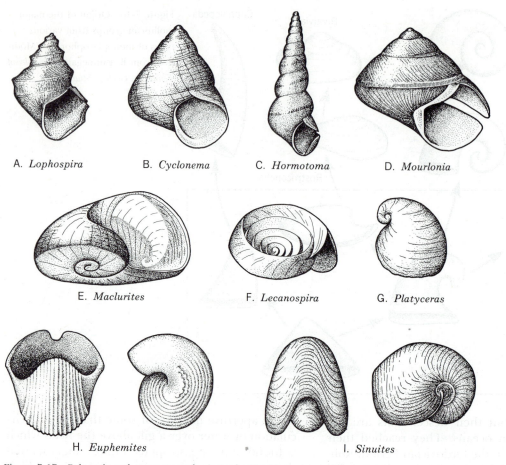

Figure 7-17　Paleozoic archaeogastropods. A. *Lophospira* (Ordovician–Silurian) (Pleurotomariina). B. *Cyclonema* (Ordovician–Devonian) (Trochina). C. *Hormotoma* (Ordovician) (Murchisoniina). D. *Mourlonia* (Ordovician–Permian) (Pleurotomariina). E. *Maclurites*, a hyperstrophic shell (Ordovician) (Macluritina). F. *Lecanspira* (Ordovician) (Macluritina). G. *Platyceras* (Silurian–Permian) (Trochina). H. *Euphemites* (late Paleozoic) (Bellerophontina). I. *Sinuites* (Ordovician) (Bellerophontina).

Mesozoic. Typical mesogastropods are illustrated in Figure 7–19.

The terrestrial and fresh water gastropods of the order Pulmonata appear first in rocks of Carboniferous age. Thereafter they steadily increased in abundance and today are found in many terrestrial environments. In Pleistocene successions of the Great Plains these gastropods have been useful in dating and correlating beds.

ARTHROPODA: BIOLOGY

Arthropods far outnumber the members of other invertebrate phyla combined and are the most successful animal group living today, whether success is measured in numbers of individuals or species, diversity of form, or territory occupied. The term Arthropoda refers to the jointed legs of this large and complex group.

Common Characteristics:

All arthropods are covered with a tough outer sheathing that is called an exoskeleton or carapace. Chitin forms the exoskeleton of most terrestrial arthropods. It is a complex polysaccharide protein which has a structure like that of cellulose and forms long chains of great strength and stability. In marine arthropods it is extensively reinforced with calcium carbonate and/or phosphate. The exoskeleton is divided into segments and the segments are grouped into regions that define units of the body, such as the head, abdomen, thorax, tail, etc. At the sutures between the segments the exoskeleton may be flexible, allowing movement, or the boundaries

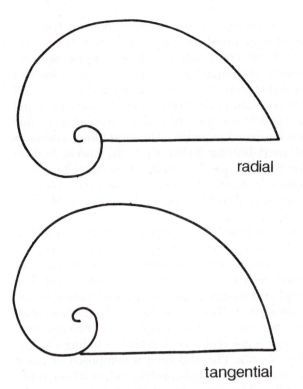

Figure 7-18   Radial and tangential apertures in gastropods. (Courtesy of R. L. Linsley, and The Paleontological Society, *Paleobiology*, v. 3, p. 196.)

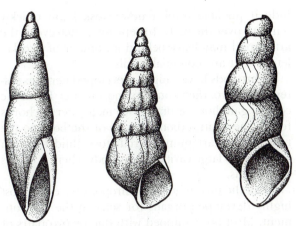

A. *Subulites*   B. *Stephanozyga*   C. *Loxonema*

Figure 7-19   Mesogastropods from Paleozoic Rocks. A. *Subulites* (Ordovician–Devonian). B. *Stephanozyga* (late Paleozoic). C. *Loxonema* (Ordovician–Carboniferous).

between several segments may be fused, making sections of the body rigid.

Typically each segment has a pair of appendages composed of up to nine rigid components separated by joints where the exoskeleton is flexible. The movement of the appendages is controlled by muscles attached to the inside of the exoskeleton. Generally the appendages are specialized for various functions, such as feeling, catching food, bringing food to the mouth, swimming, walking, etc., and the appendages in any one region of the body are similar (Fig. 7–20). Most appendages are simple linear sets of components but in some marine arthropods each consists of two branches of which one is featherlike and acts as a gill.

In marine arthropods respiration may be by means of such appendages, or by gills on the underside of the body. Arthropods that live in the air respire through pores in the exoskeleton that lead into branching tubes within the body.

The exoskeleton gives the arthropod a protective covering and the rigidity required for efficient working of muscles, but it also has disadvantages. The animal cannot grow within the armor plating

and, in order to grow, all arthropods periodically shed their exoskeletons and secrete larger ones— a process called molting. Before molting the inside layers of the chitinous exoskeleton are dissolved by body fluids so that it separates from the body within. The body takes on fluid so that it swells and cannot fit within the old shell, splitting it apart along special weak lines called sutures or along flexible joints. The soft animal then crawls out of the old exoskeleton which it discards or eats. At this stage in the life cycle it is vulnerable to predators and must

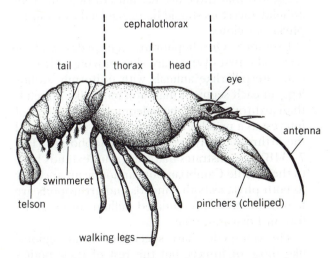

Figure 7-20   Modern crayfish, a typical aquatic arthropod, showing the division of the animal into segments and regions and the specialization of the 21 pairs of appendages.

hide if possible until a new exoskeleton quickly hardens over the body. In repeating such cycles the arthropod may leave behind a number of exoskeletons that are potential fossils.

Arthropods have a highly developed nervous system to control their complex movements. From the central ganglion, or brain, two major nerve chords follow the ventral (bottom) side of the body to the rear. The heart pumps circulatory fluid through arteries into large cavities in the body where it bathes the organs.

Of all the invertebrates, arthropods have the most highly developed organs for sensing the environment. Most are equipped with one or two pairs of antennae or feelers and one, or several, sets of eyes. The body is usually covered with minute hairs, or setae, by which it can sense movement in the medium around it.

Classification:

Segmentation and organ systems indicate that the Arthropoda evolved from annelidlike animals, but the diversity of the arthropods suggests that they are not all descended from a single annelid ancestor. If all the arthropods were descended from a single annelid that evolved across the boundary between annelidness to arthropodity, then the phylum Arthropoda is monophyletic. If two or more annelids gave rise to different groups of arthropods, then the phylum is polyphyletic. Many zoologists believe the latter, and S. M. Manton (6) has suggested that three distinct lines of descent from annelid ancestors should be recognized as separate phyla as follows:

Uniramia: Onychophorans, centipedes, and insects. The first of these are slow moving, soft-bodied, caterpillarlike animals with uniformly stubby legs on each segment that look more like annelids than arthropods. For the paleontologist they are of interest as a possible link between the annelids and the arthropods. *Aysheaia* (an onychophoran) (Fig. 7–34B) demonstrates that the group extends back to the Middle Cambrian but cannot be an ancestor to both phyla as both annelids and arthropods are known to have differentiated millions of years before in Ediacaran time.

The centipedes have specialized head regions, like those of insects, but the rest of their bodies consist of almost identical segments with similar appendages like the annelids. They have a poor fossil record.

The insects apparently evolved from the centipedes in mid-Paleozoic time. Their bodies are clearly differentiated into head, thorax, and abdomen. Although they are now the most abundant of arthropods, the insects have left a sparse fossil record. Some of the best preserved insects are found trapped in the gum of tropical trees that in time has hardened to amber. The most famous deposits of insect-bearing amber are in the Eocene beds of the Baltic region, but similar occurences have been found in North America in Cretaceous rocks of California, Alaska, and Canada, and in the Miocene rocks of Mexico.

Chelicerata: This group includes the horseshoe crabs, eurypterids, scorpions, and spiders (Fig. 7–21). They do not have antennae and the first appendages are pincher claws for grasping food. The head and thorax form a single unit called the cephalothorax. The horseshoe crabs (Xiphosura), common on most American east coast beaches, have a large semicircular cephalothorax hiding the legs (Fig. 7–22). The abdomen is composed of fused segments bearing gills, and the tail, or telson, is like a spike. The ancestors of the Xiphosura can be traced back to Silurian time.

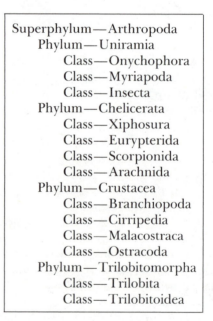

Superphylum—Arthropoda
  Phylum—Uniramia
    Class—Onychophora
    Class—Myriapoda
    Class—Insecta
  Phylum—Chelicerata
    Class—Xiphosura
    Class—Eurypterida
    Class—Scorpionida
    Class—Arachnida
  Phylum—Crustacea
    Class—Branchiopoda
    Class—Cirripedia
    Class—Malacostraca
    Class—Ostracoda
  Phylum—Trilobitomorpha
    Class—Trilobita
    Class—Trilobitoidea

The eurypterids are an extinct group of large aquatic, Paleozoic arthropods that are described more fully later. The scorpions are adapted to a terrestrial existence and probably evolved from eurypterid stock in early Paleozoic time.

The oldest known spiders (Arachnida) come from

Figure 7-21   Wolf spider preserved in amber (Eocene), Baltic region (length about 5 mm) (Courtesy of Redpath Museum).

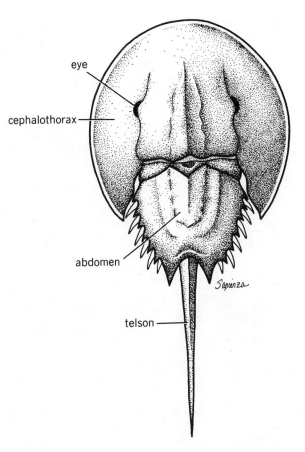

Figure 7-22   *Limulus*, the horseshoe crab (recent). Total length about 50 cm.

Devonian rocks. Their fossil record, like that of the insects, is poor and the details of their evolution, therefore, obscure.

Crustacea: Paleontologically significant groups include the branchiopods, ostracodes, barnacles, crabs, and lobsters. The trilobites are considered to be crustaceans by some paleontologists; others regard them as a phylum comparable to the Crustacea. Crustaceans are largely marine animals but some can live in fresh water and a few on land. Most have two pairs of antennae and appendages with two branches. Ostracodes and trilobites, which have excellent fossil records, are discussed later.

The branchiopods are small, shrimplike crustaceans that secrete two shells, or a single shell, enclosing most of the body. That they have flourished in great numbers in freshwater ponds and lakes since late Paleozoic time is shown by deposits in which their shells cover the bedding planes, but, as their shells are lightly calcified, they do not have an extensive fossil record.

The barnacles live attached to rocks in the intertidal zone encased in many calcareous plates which retain moisture when the tide is out and open to allow feeding when it is in. Plates of barnacles are uncommon fossils but have been found in rocks as old as mid-Paleozoic age.

Although crabs and lobsters (Fig. 7–20) have well-calcified carapaces, they have not left an extensive

fossil record. The elliptical body of a crab that is visible from above is the fused head and thorax region. Crabs carry the abdomen tucked under this carapace. They, and the lobsters, have developed large pinching appendages for attacking food and for defense. The oldest crabs are found in Jurassic rocks.

TRILOBITES

Of the fossil arthropods, the trilobites have been given the most attention because they are abundant in Paleozoic rocks, useful for correlation, and attractive in appearance (Fig. 7–23). Complete specimens of these fossils are rare for, like all arthropods, they were composed of rigid segments of the exoskeleton joined by flexible organic connections that decayed on the death of the animal. Although complete exoskeletons are uncommon, the separated parts of the carapace are common fossils in

Figure 7-23   The Early Cambrian redlichiid trilobite *Olenellus* 5 cm long. (Courtesy of V. J. Okulitch.)

Cambrian, Ordovician, and Silurian rocks and are proof of the abundance of these animals in early Paleozoic seas. The carapaces of trilobites were highly calcified, in fact, as preserved, they are entirely calcite but are assumed to have contained an organic component, probably chitinous, during life. Except in extraordinary conditions, the only part of the trilobite that is preserved is the exoskeleton that covered the back or dorsal side. The ventral side and the appendages were not calcified and are known only from a few specimens in which this soft tissue was preserved as a thin film or was replaced by pyrite.

The Dorsal Carapace:

In both the head and tail regions several segments fused together to make a single unit. The head is called the cephalon and the tail the pygidium (Fig. 7–24). The thorax between the cephalon and pygidium is composed of segments that articulate with each other. In addition to this threefold division

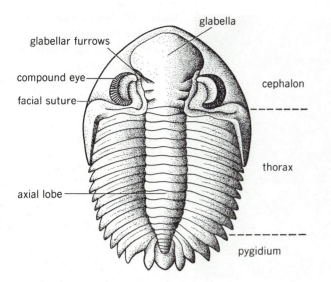

Figure 7-24   Major features of the dorsal carapace of a trilobite illustrated by *Greenops* (Devonian).

into cephalon, thorax, and pygidium, the trilobite body is divided longitudinally into three lobes. The name trilobite was suggested by this longitudinal division.

The axial section of the cephalon is the glabella (Fig. 7–24). It is usually divided transversely by up to four furrows of which the one at the back is usually complete and the others, incomplete. These furrows are relicts of the sutures between the original segments that were fused to form the cephalon. Exceptionally preserved specimens show that the glabella overlay the stomach and that the mouth was beneath its back end (Fig. 7–25).

The lateral lobes of the cephalon are divided in most trilobites into two parts by a groove called the facial suture. When the carapace of the trilobite was periodically molted, the cephalon split along the facial suture and the animal crawled out. Along this

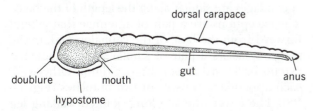

Figure 7-25   Diagrammatic reconstruction of a cross section from front to back of a trilobite to show the position of the gut, mouth, and hypostome.

suture is a prominence that carried the compound eye. The location of the eye on the outer side of the suture apparently insured that this delicate organ separated from the old carapace easily during molting and that the trilobite was deprived of sight for a minimum of time during this dangerous period. The suture does not cross the top of the cephalon in all trilobites; in some it follows the front margin and in others it is absent. Cephalons left behind when the trilobite molted are missing the sections outside the suture because these sections are separated from the rest of the cephalon in the molt.

Trilobites were the first organisms, as far as we know, to have eyes. In many Cambrian trilobites the eyes are not well preserved and their nature is obscure. They had a suture below the ocular surface which easily separated from otherwise well-preserved specimens. In those Cambrian trilobites that have preserved eyes and in all younger trilobites, the eyes are compound, that is, composed of many individual lenses, like those of living insects (Fig. 7–26). The lenses may be adjacent to one another (in which case the eyes are said to be holochroal) or separated by opaque exoskeleton (schizochroal). In both forms, the lenses are composed of crystalline calcite with the c-axis of the crystal oriented perpendicular to the surface. Whether the images produced by schizochroal and holochroal eyes were different, is not known. Several trilobites, such as *Cryptolithus* (Fig. 7–30D), do not have eyes, presumably because they lived below the photic zone on the ocean floor and had no need of sight. In these trilobites the facial suture follows the front edge of the cephalon.

In most trilobites the thorax consists of about ten segments which moved on articulating surfaces along their forward and backward edges giving the animal flexibility. This flexibility was so great that the animal was capable of bringing the pygidium underneath the body so that it was pressed against the underside of the cephalon (Fig. 7–27). In this enrolled position all the underside and the appendages were safely protected by the tough dorsal carapace and safe from all but the largest predators. The thoracic segments are similar in most trilobites but one or more sets may be elongated backwards at their tips. The lateral segments of the thoracic carapace covered most of the walking and swimming legs on the ventral surface.

In many trilobites, the pygidium is about the same size as the cephalon so that it can shield the soft

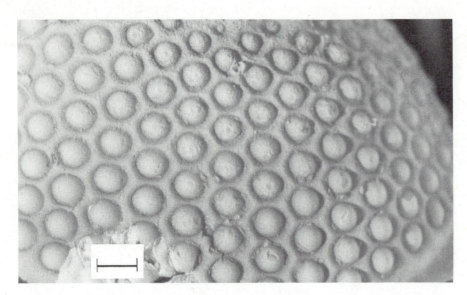

Figure 7-26   Schizochroal eye of the trilobite *Phacops* (Devonian), Silica Shale, Ohio. The scale bar is 1 mm (C. W. Stearn.).

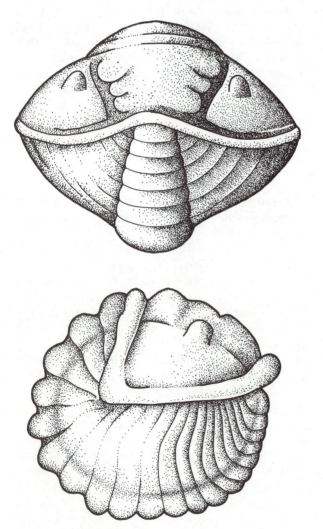

Figure 7-27   Front and side views of the enrolled trilobite *Flexicalymene* (Ordovician).

parts effectively. Other trilobites have smaller pygidia. The segments that are fused to form them can be distinguished in most pygidia by furrows. However, in some genera the surface of the pygidium is smooth.

Ventral Surface:

In the pygidium and cephalon, and to a lesser extent in each thoracic segment, the dorsal carapace is folded over onto the ventral surface to form a ledge called the doublure (Fig. 7–25). Hinged to the inner edge of the doublure at the front of the cephalon is a shield-shaped plate called the hypostome that partly covers the area beneath the glabella where the stomach was located. Apart from the hypostome and doublure, the ventral surface must have been composed of tough, soft tissue.

Appendages have been observed in about 20 species of trilobites. The appendages of most arthropods are specialized for different functions in different parts of the body, but in the trilobites most appendages are similar along the length of the body. The exceptions are a pair of antennae that extend forward from the base of the cephalon and, in the genus *Olenoides* a pair of similar appendages that extend backwards from the pygidium. Otherwise, each appendage consists of two branches (Fig. 7–28). The lower one is a long, jointed walking leg and the upper a featherlike gill that was probably also used as a paddle to help the trilobite swim. The appendages beneath the pygidium were much smaller than those of the thorax.

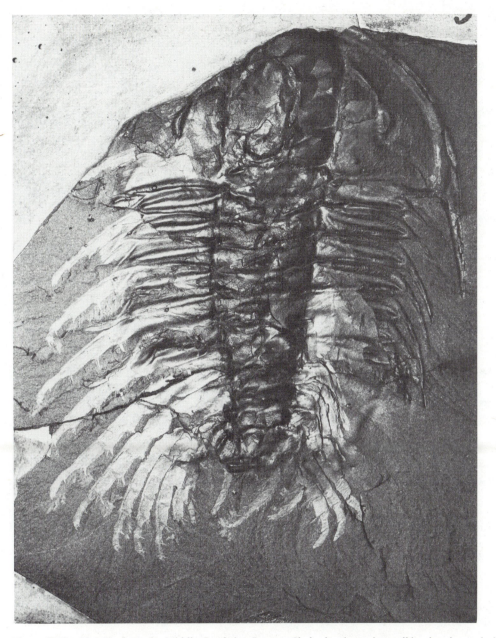

Figure 7-28 *Olenoides* from the Middle Cambrian Burgess Shale showing spiny walking legs beneath the dorsal carapace (7 cm). (Courtesy of H. B. Whittington.)

Ontogeny:

The growth stages of trilobites are best known from specimens that have been replaced by silica or phosphate and can be dissolved out of limestone with acid. From these minute fossils paleontologists can reconstruct the developmental stages of several species. The initial stage is a disklike, usually spiny carapace which has a segmented axial lobe. The eyes and the suture are located at the front margin (Fig. 7–29). As the trilobite grows the facial suture moves back over the top of the carapace and the eyes move to the top surface with it . Another suture forms across the posterior of this growth stage. As successive molts take place, segments of the thorax are formed at this suture. The segments take form in the posterior part of the larval carapace, move forward, and eventually, when separated from this posterior section by a suture, become free. Each stage in the addition of segments is separated by a

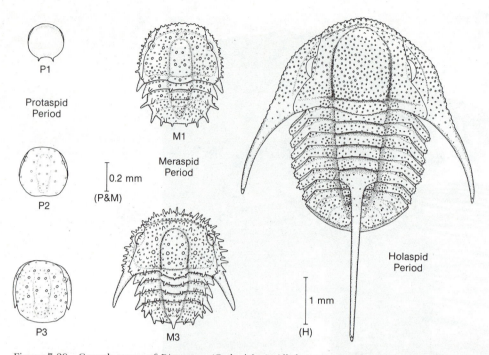

Figure 7-29   Growth stages of *Dimeropyge* (Ordovician). All three protaspid growth stages (P1-P3), two of the seven meraspid stages (M1, M3), and one of an indeterminate number of holaspid stages are shown. The protaspid period consists of larvae whose head is still joined to the tail. The meraspid period encompasses those growth stages where the head and the tail are separated but not all of the thoracic segments have been released from the tail portion. The holaspid period has all of the thoracic segments released as separate units. Most of the growth of this trilobite took place during the holaspid period, although it has already molted its shell ten times before reaching this stage. The smaller scale (P-M) is for the protaspid and meraspid stages: the larger (H), for the holaspid stage. (Courtesy of Brian Chatterton.)

molt of the whole carapace. After the adult complement of thoracic segments has been added, the basic shape of the trilobite does not change but it may continue to grow in size as successive molts take place. The pygidium is what is left of the posterior part of the carapace after all the thoracic segments have separated from it.

Superphylum—Arthropoda
Phylum—Trilobitomorpha
Class—Trilobita
Order—Agnostida
Order—Redlichiida
Order—Corynexochida
Order—Ptychopariida
Order—Proetida
Order—Phacopida
Order—Lichiida
Order—Odontopleurida

Classification:

The trilobites can be divided into eight groups which have been recognized as either orders or classes. Representatives of these groups are illustrated in Figure 7–30.

1. The agnostids are small Cambrian and Ordovician trilobites with subequal cephala and pygidia, no eyes (except in one family), and only two or three thoracic segments.
2. The redlichiids are large Early and Middle Cambrian trilobites having semicircular cephalons with spines at the posterior corners, and small pygidia. They are the oldest group of the trilobites and have primitive features such as many segments and a marginal facial suture.
3. The corynexochids are smaller Cambrian trilobites with boxlike glabellas, seven or eight

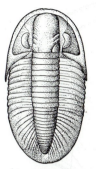

A. *Phillipsia*

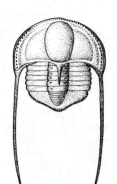

B. *Cryptolithus*

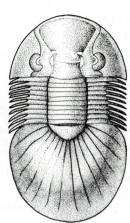

C. *Scutellum*

D. *Arctinurus*

E. *Homagnostus*

F. *Odontopleura*

G. *Olenellus*

H. *Ceraurus*

I. *Eoharpes*

Figure 7-30   A. *Phillipsia*, proetid, one of the last trilobite genera (Permian) (25 mm). B. *Cryptolithus*, blind ptychopariid (Ordovician) (30 mm). C. *Scutellum*, ptychopariid (Silurian) (11 cm). D. *Arctinurus*, lichid (Silurian) (14 cm). E. *Homagnostus*, agnostid (Upper Cambrian) (7 mm). F. *Odontopleura*, odontopleurid (Silurian) (40 mm). G. *Olenellus*, redlichiid (Early Cambrian) (10 cm). H. *Ceraurus*, phacopid (Ordovician) (40 mm). I. *Eoharpes*, ptychopariid (Ordovician) (25 mm).

thoracic segments, and large pygidia. *Olenoides* is a typical genus.

4. The ptychopariids comprise the largest group of trilobites and are difficult to characterize in a few words but typically have facial sutures and pygidia of moderate size. They are Early Cambrian to Devonian in age and include such genera as *Eoharpes*, *Cryptolithus*, *Triarthrus* and *Scutellum*.

5. The proetids lived from Ordovician to Permian periods and are the last group of the trilobites to die out. *Phillipsia*, a Late Carboniferous trilobite, is typical.

6. The phacopids are an Early Ordovician to Late

Devonian group that includes some of the most common middle Paleozoic trilobites such as *Greenops Flexicalymene* (Fig. 7–27), and *Ceraurus*.

7. The lichids are large mid-Paleozoic trilobites with flat cephalons and pygidia. *Arctinurus* is typical of this small group.

8. The odontopleurids are Early Ordovician to Late Devonian trilobites characterized by their spiny carapaces, such as that of *Odontopleura*.

Geological History:

Trilobites first appear in basal Cambrian rocks above the Tommotian Stage. In Lower Cambrian rocks the dominant group is the large primitive Redlichiida, typified by the genus *Olenellus*. They are joined in the Early Cambrian by the agnostids, corynexochids, and ptychopariids. Species of these trilobites are used to subdivide Cambrian rocks into zones and to correlate them around the world. All but some agnostids and ptychoparids died out at the end of the Cambrian Period and were replaced by mid-Paleozoic groups such as the proetids, phacopids, lichids, and odontopleurids. The trilobites reached their greatest diversity and abundance in Ordovician time and declined through the Silurian and Devonian periods. They suffered a second major setback in Middle and Late Devonian time which only the proetids survived. Trilobites are relatively rare in Carboniferous and Permian sedimentary rocks, and the last expired as the Paleozoic Era drew to a close.

OSTRACODA

The ostracodes are a stratigraphically useful but inconspicuous group in the fossil record. Numerically their fossils far outnumber those of other arthropod groups. Most of their shells are less than 1 mm across and therefore easily escape the notice of the casual collector. Ostracodes are arthropods that enclose their bodies in two valves held on either side of the body and hinged along the back of the animal. This bivalved carapace is composed of chitin reinforced by calcite fibers at right angles to the surface. Inside the valves the body is not obviously divided into head, thorax, and abdomen (Fig. 7-31). It is equipped with a variety of jointed appendages for sensory, feeding, swimming, crawling, and cleaning functions. When the valves are slightly open the appendages can be extended into the water or bottom sediment to carry out their functions. A median muscle that extends from valve to valve opens and closes them, and other muscles attach the body and appendages to the inner valve walls. The attachment points of these muscles may be visible on the inner and outer surfaces of the valves. Each valve may also be marked by a spot or area that transmits light to the ostracode eye beneath. Like all arthropods, each ostracode individual forms many exoskeletons (as many as ten) throughout its life by molting.

The surface of the valves may be smooth, or it may be sculptured with nodes, knobs, depressions, flanges, reticulate patterns, pores, and so on (Fig.

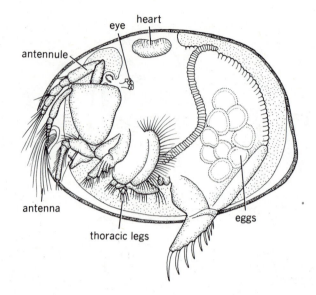

Figure 7-31   The living ostracode *Cypridina* with one valve removed to show the body and appendages. About 3 mm in width.

7–32). In some the hinge line is a straight, smooth junction along which the two valves are joined by an elastic membrane. In others it is equipped with teeth, sockets, flanges, and troughs along which the valves interlock. In those groups that have female and male individuals, the valves may show sexual dimorphism. The female valves resemble those of the males but may have a pouch at the front, a widened or subdivided frill along the free margin, or they may be enlarged at the back to provide for the incubation or protection of the young.

Ostracodes are confined to aquatic habitats although a few species are capable of living in wet vegetation. They live in marine water up to 2,800 meters in depth and are abundant in all freshwater habitats. The eggs of freshwater species may survive dessication for prolonged periods. Some species survive in water of intermediate salinity of estuaries. Most of the freshwater and estuarine forms have smooth or only slightly ornamented valves. Ostracodes swim and crawl in all aquatic environments but are most common on and within the bottom sediment in shallow water. The simple, smooth, lightly calcified valves of the first ostracodes appear in Early Cambrian sediments. In the middle Paleozoic the smooth leperditiids grew to large sizes up to 30 mm across. Their abundance in Paleozoic carbonate rocks has been interpreted as an indicator of nearshore, lagoonal–estuarine

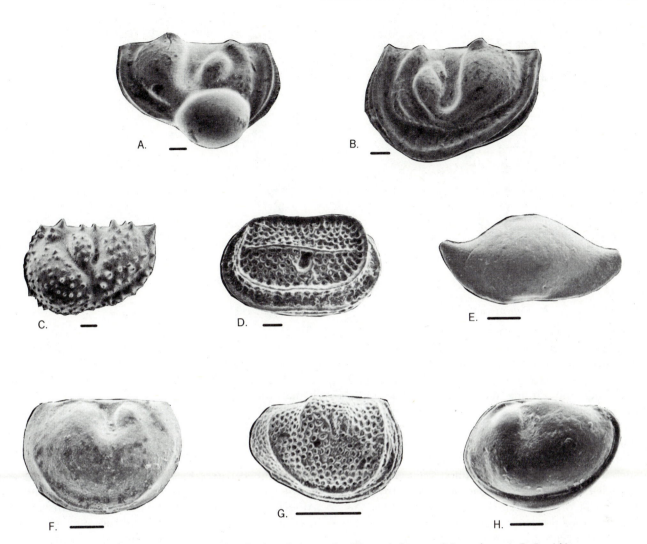

Figure 7-32   Scanning electron micrographs of Paleozoic ostracodes. The scale bars are 0.2 mm long. A, B. *Beyrichia (Mitrobeyrichia)* (Silurian), female with brood pouch and male. C. *Beyrichia (Beyrichia)* (Devonian), female with brood pouch. D. *Amphizona* (Devonian), left valve. E. *Bairdia* (Devonian), left valve. F. *Milleratia* (Ordovician), right valve. G. *Ulrichia* (Devonian), right valve. H. *Sansabella* (Carboniferous), left valve. (Courtesy of Murray Copeland and the Geological Survey of Canada.)

conditions. Freshwater ostracodes first appeared in Carboniferous rocks of the coal swamps. Three of the five orders of ostracodes did not survive the end of the Paleozoic, but Mesozoic and Cenozoic faunas are abundant and diverse.

Because the ostracodes are small, they may be identified in the fragments of cuttings a few millimeters in diameter brought to the surface in the drilling of wells for oil, gas, or water. Because many are rapidly evolving and the ranges of species are short, they make excellent fossils for the dating of Phanerozoic sediments.

EURYPTERIDS AND SCORPIONS

The eurypterids are a group of large Paleozoic arthropods whose chitnous carapaces are found in marine, brackish, and freshwater deposits. The head region, or prosoma, is generally rectangular in shape and the appendages attached to its ventral surface extend beyond its margin (Fig. 7–33). Six appendages were attached to the head. In *Pterygotus* the first is equipped with pinchers for defense and food-handling. The four appendages behind these are walking and balancing legs which are usually cov-

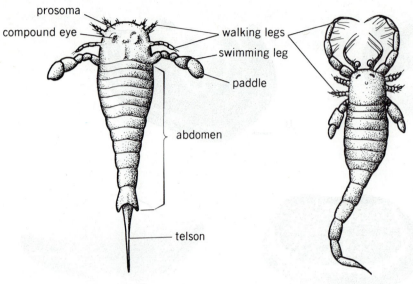

prosoma

compound eye

walking legs

swimming leg

paddle

abdomen

telson

A. *Eurypterus*

B. *Mixopterus*

Figure 7-33   Eurypterids. A. *Eurypterus* (Silurian) (20 cm long). B. Reconstruction of *Mixopterus* (Early Devonian) (90 cm long).

ered with spiny processes. The last appendage is large in most eurypterids and ends in a broad paddle apparently used in swimming or digging. The head region was equipped with large compound eyes and other simple light receptors. The abdomen behind the head was divided into twelve segments of which the first seven were usually wider than the last five. At the end of the abdomen the final segment, or telson, was in many eurypterids, such as *Eurypterus*, a long, pointed spike. In some, such as *Pterygotus*, it was a paddlelike structure like the telson of a lobster. The gills opened on the underside of the front segments of the abdomen. The eurypterids are believed to have walked along the bottom and swam for short distances. Most are between 10 and 20 cm long but, from fragments of its carapace, a specimen of *Pterygotus* has been reconstructed as 250 cm long. Tracks from Carboniferous rocks of Pennsylvania, which are believed to have been made by a eurypterid, suggest an even larger animal. Eurypterids were the largest arthropods ever known.

The ancestors of the eurypterids are found among Cambrian arthropods but the oldest fossils recognized as members of the group occur in Lower Ordovician rocks. The order became extinct in Permian time.

Eurypterids are uncommon fossils but scorpions are rarer still. The scorpions have large pinchers on the first appendage and a clearly defined tail that ends in a stinger. At present all scorpions are terrestrial in habit but they are clearly descended from aquatic ancestors which resembled eurypter-

ids. The first scorpion, *Paleophonus*, of Silurian age, was originally described as an airbreather but later investigators have suggested it was aquatic and that fully terrestrial scorpions did not appear until Devonian time.

THE BURGESS SHALE

Above the Burgess Pass near the town of Field in the southeast corner of British Columbia, the American paleontologist Charles Walcott discovered a thin lens in Middle Cambrian beds of the Stephen Formation that contained fossils of soft-bodied organisms. Between 1910 and 1917 he directed the excavation of a quarry in this lens and from the material collected he described the extraordinary fauna and flora. In the last 20 years investigation of the fossils from this locality has been continued by Harry Whittington and Simon Conway Morris (10,11,12) and colleagues at Cambridge University. Other occurrences of similar deposits have been found at different stratigraphic levels in the mountains around Field by Desmond Collins and his colleagues at the Royal Ontario Museum.

The fossils are preserved as films of aluminosilicates in the dark shale (Fig. 7–34). Most are not pressed on to a single bedding plane but cross the fine laminations in the shale. The organisms lived near an abrupt transition from muddy sediments, in which they were entombed, to a thick limestone to the east. This abrupt transition has been inter-

Figure 7-34 Fossils from the Burgess Shale fauna. A. *Marrella*, an arthropod. The dark stain at the back is caused by decay products leaking out of the body into the surrounding sediment (16 mm wide). B. *Aysheaia*, an onychophoran, (40 mm long). C. *Canadia*, a polychaete annelid (35 mm long). D. *Burgessochaeta*, a polychaete annelid (28 mm long). (Courtesy of H. B. Whittington [A, B] and S. Conway Morris [C, D].)

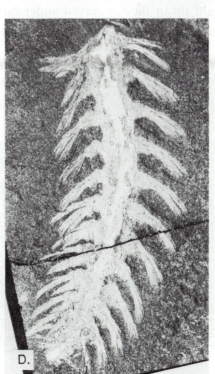

preted as a submarine escarpment down which a sudden turbid cascade of mud flowed to cover the Burgess Shale organisms (Fig. 7–35). An alternative interpretation of the environment casts doubt on the prominence of the escarpment during deposition. The conditions under which preservation took place are still unclear. The lack of damage to the organisms by bacteria or scavengers suggests that once they were covered with mud they were protected from decay and predation by lack of oxygen in the sediment.

At present 119 genera and 140 species have been identified from this locality and most of these have not been collected elsewhere. The major groups represented are: trilobitoid arthropods (25 percent), sponges (18 percent), trilobites (14 percent), priapulid and polychaete annelids (14 percent), and lophophorates, largely brachiopods (6 percent). The remaining groups which make up less than 5 percent each include chordates and hemichordates, coelenterates, molluscs, echinoderms, and a large group (16 percent) of organisms that are difficult to place in any phylum. Nearly all these animals lived on the bottom or burrowed into the sediment. A few swimming animals were caught up in the catastrophic flow and preserved with the bottom fauna. On the evidence of their gut contents and the nature of their appendages, a large number of the arthropods can be shown to have been predatory.

The Burgess deposit invites questions about the nature of the fossil record. Why, in all the rest of the paleontological record, have not comparable deposits been found? Why have no medusoids been found in this soft-bodied fauna even though they are abundant in the Ediacaran fauna and at present? Is the proportion of soft-bodied to skeletonized animals in the fauna (5 to 1) representative of Cambrian faunas as a whole? Is the proportion applicable to invertebrate faunas from Paleozoic to recent time, or was the Cambrian Period unique as a time when only a small proportion of marine animals were skeletonized? The low diversity of Cambrian skeletonized faunas suggests that this may be true. Whether or not the proportion is widely applicable, the Burgess Shale fauna reminds paleontologists that past seas were populated by many animals that may never appear as fossils because they are without skeletons and can only be preserved under exceptional conditions.

Figure 7-35　Reconstruction of the Burgess Shale environment and fauna at the National Museum of Natural History, Washington, D.C. (Courtesy of Smithsonian Institution, photo 84-4711.)

BIOLOGY OF ECHINODERMATA

The echinoderms are among the most numerous, complex, and diverse of living invertebrates and have an extensive fossil record stretching back to the beginning of the Cambrian Period and perhaps into Ediacaran time. Most of the members of the phylum secrete calcareous skeletons composed of plates. Echinoderms are exclusively marine animals and most cannot tolerate salinities less than that of normal sea water. Typical echinoderms live on the seafloor or within the sediment and either prey on other invertebrates, eat organic detritus and algae, or filter microorganisms from the water. Three features characterize all living members of this phylum: (1) adult five-fold symmetry that is superposed on primitive or embryonic bilateral symmetry, (2) a skeleton composed of individual calcite plates honeycombed with organic matter, (3) a unique hydrostatic system called a water vascular system.

```
Phylum—Echinodermata
    Subphylum—Blastozoa
            Class—Blastoidea
                    Cystoidea
            Class–Diploporita
            Class—Rhombifera
    Subphylum—Crinozoa
            Class—Crinoidea
    Subphylum—Asterozoa
            Class—Asteroidea
            Class—Ophiuroidea
    Subphylum—Echinozoa
            Class—Edrioasteroidea
            Class—Holothuroidea
            Class—Echinoidea
```

The phylum is divided into four subphyla and as many as 21 classes in some classifications. Many of these classes are established for a small number of rare genera that are unlikely to be encountered outside a museum. Only five classes include common fossils: most of the Cystoidea, Blastoidea, and Crinoidea are, or were, attached to the seafloor by a stalk; the Echinoidea (sea urchins) move freely within or on top of the seafloor sediments; and the Edrioasteroidea are an extinct group of pancakelike attached echinoderms. Three other classes have many representatives in modern seas but few fossils. The Asteroidea (sea stars or starfish), Ophiuroidea (brittle stars), and the Holothuroidea (sea cucumbers) have poor fossil records because their small plates are isolated in soft tissue and usually fall apart when the animal dies. The abundance of both these classes in modern seas and of their scattered plates in sediments suggests that they were more abundant in the past than indicated by their described fossil representatives.

The free-swimming larvae of echinoderms resemble those of cephalochordates, urochordates, and hemichordates and are much different from those of annelids, arthropods, and molluscs. The echinoderms and these chordate and hemichordate relatives are enterocoelous deuterostomes but unlike the chordates, the echinoderms do not have a part of the body differentiated as a head.

The plates of echinoderms are distinctive. They are each secreted as a single crystal of calcite containing up to 15 percent magnesium carbonate. Since they are single crystals, the plates can readily be identified as of echinoderm origin in thin sections of sedimentary rocks by means of the petrographic microscope, even when the form of the animal that secreted them is completely broken up by waves and currents. The plates are secreted within the mesoderm and are penetrated throughout by a network of soft tissue (Fig. 7–36). This allows them to grow, to be resorbed, or to be repaired after they have been damaged. The echinoderms therefore have many of the advantages of an internal skeleton but, unlike that of the chordates, it is located at the periphery of the body. In classes with a good fossil record the plates form a rigid box around the animal.

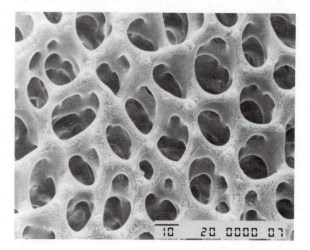

Figure 7-36 Scanning electron micrograph of a plate of the recent echinoid *Strongylocentrotus*. The bar scale at the base is 10 micrometers.

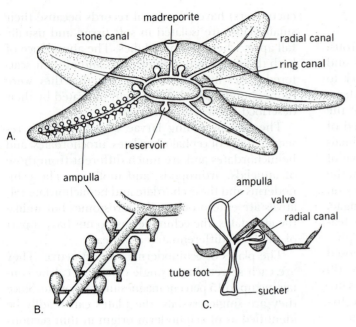

madreporite
stone canal
radial canal
ring canal
A.
reservoir
ampulla
ampulla
valve
radial canal
tube foot
sucker
B.
C.

Figure 7-37 Water vascular system of an asteroid. A. General view. B. Detail of a radial canal, ampullae, and tube feet. C. Cross section of an arm showing the relation between the radial canal, ampulla, and tube foot.

The water vascular system consists of a series of pipes, reservoirs, pressure chambers, and valves used by echinoderms for locomotion, food gathering, respiration, sensing, and tube building. In many of the echinoderms the system is open to sea water through a porous plate called the madreporite (Fig. 7–37). Its fine openings prevent unwanted particles, or organisms, from entering the system. The stone canal leads from the madreporite to a ring canal that encircles the mouth. On this canal are sacklike bodies that may act as reservoirs to supply fluid rapidly to the system in case damage leads to large leaks. From the ring canal five radial canals extend outward into feeding arms or just beneath the body surface. These canals repeatedly and alternately give off short lateral canals that in many echinoderms lead to small muscular bulbs (ampullae) and to muscular, closed tubes (tube feet) (Fig. 7–37B). In some echinoderms the radial canals and ampullae are within the skeleton; in others the canal is outside and the ampullae inside. Crinoids do not have ampullae. A tube foot is extended when its ampulla squeezes water into it by contracting and a valve closes isolating it from the radial canal. Tube foot motion is controlled by muscles along its sides. The end of each tube foot is flattened into a sucker which, with the help of a secretion and contraction of the central part to produce suction, attaches itself to objects. Contraction of the tube foot muscles can then pull the object toward the animal, or the an-

imal toward the object. The tube feet act with extraordinary coordination to perform their various functions. The sectors of the echinoderms underlain by the radial canals and along which the tube feet emerge, are called ambulacral areas and the areas between them are interambulacral areas.

Echinoderms have a poorly developed circulation system consisting of rings of canals around the mouth and anus and radial canals, but nutrients and waste products are also transported through the large open coelom by currents driven by cilia. Exchange of gases takes place through the tube feet and extensions of the coelom, either through small pores in the skeleton or as branched gills outside the body. The echinoderms have a complex nerve system to coordinate the intricate movement of the tube feet. They are covered with sensory cells that are sensitive to touch, radiation, and the chemical conditions of their environment. On the tips of the arms of sea stars and the upper surfaces and tube feet of echinoids are concentrations of light sensitive cells that act as simple eyes.

In each class of the phylum these organ systems are expressed differently. Unlike the sea stars and the echinoids, the crinoids live with their mouth upward. The gut makes a U-turn and both the anus and the mouth are located on the upper surface (Fig. 7–39). The body is a plated calyx (Greek = cup), from which arise featherlike branching arms in multiples of five. The arms are equipped with

Figure 7-38   Living crinoids. A. Isocrinoid at approximately 500 m depth, Strait of Florida. Note the horizontal position of the axis of the arms facing into the current and the cirri on the stem grasping the rocky sea floor. (Courtesy of A. Conrad Neumann.) B. Comatulid crinoid (*Dichrometra*) at about 2 m depth, Palau Islands, with arms forming an arcuate filtration fan across the current flowing into the photograph. Cirri at the base grasp the top of a coral. (Courtesy of David Meyer.)

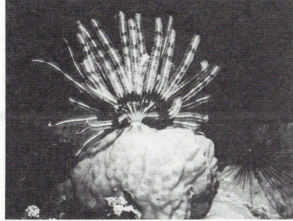

rows of tube feet specialized to collect microorganisms from the water and transfer them to a food groove that lies at their bases. Cilia in the grooves pass these particles along to the mouth which is at the convergence of the grooves. Modern crinoids that live in deep water are attached to the bottom or other hard objects by a stem and holdfast system composed of buttonlike plates (Fig. 7–38A). The feather stars that live in shallow water are crinoids that have abandoned their stems and can float freely in the sea but, by means of the cirri at their base, they ordinarily rest grasping a ledge or another organism (Fig. 7–38B). In asteroids, ophiuroids, echinoids, and holothurians the mouth is either at the front or on the bottom where the ambulacral areas converge, and the anus is on the back or at the top.

The echinoids give their name to the phylum. The Greeks used the same word, *echinos*, for both the hedgehog and the sea urchin for both are covered with spines. The spines are attached to the plates by a ball-and-socket that allows them to be moved by a circlet of surrounding muscles (Figs. 10–26, 10–30). The ambulacral areas occupy five sectors of the skeleton, or test, and the plates in these sectors are pierced by rows of pores for the tube feet. Because echinoids are rare fossils in Paleozoic rocks, and become abundant only in Mesozoic time, their paleontological record is considered in Chapter 10.

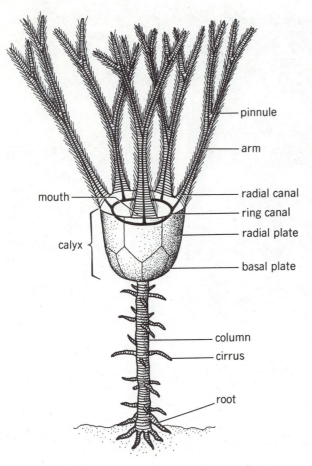

Figure 7-39   Diagram of some features of the anatomy of a stalked crinoid.

ATTACHED ECHINODERMS

The common Paleozoic echinoderms lived attached to the bottom. Of the four important classes only the crinoids survive to the present; the cystoids, blastoids, and edrioasteroids were extinct before the end of the Paleozoic Era. The carpoids, considered by some to be a fifth subphylum, may or may not be echinoderms. They are considered in the next section.

Crinoids:

Modern crinoids can be divided into two groups on the basis of their habitats: deep-living stalked forms, called isocrinoids, and free-living shallow-water forms, called comatulids. Both types belong to the subclass Articulata which includes all post-Triassic crinoids. Three other subclasses (with the exception of one Triassic genus) are confined to Paleozoic

rocks and nearly all the crinoids in them were attached but, unlike modern isocrinoids, they grew in vast undersea gardens in shallow water.

The calyx of crinoids consists of a cup of regularly arranged plates and a top, or tegmen, which may be flexible and leathery or solidly roofed with small plates. Both mouth and anus may open directly on the top surface or the mouth may be roofed over by the small plates of the tegmen. The plates of the cup of all crinoids, except those of the oldest (Middle Cambrian) genus, are arranged in regular cycles most of which consist of five plates (Fig. 7–40). The simplest cups are composed of two cycles of five plates and the arms are attached to the upper cycle, the radial plates (Fig. 7–40B). The lower cycle of five plates, called basal plates, alternates with the radials. In some cups a third cycle of plates, called infrabasals, underlies the basal cycle (Fig. 7–41C). This regular pattern of plates is modified in three ways. (1) Plates above the radials that would normally be part of the arms (brachial plates) were incorporated in the cup, apparently to increase its size (Fig. 7–40A,C). In some cups the pattern of these brachial plates indicates that parts of the arms that have branched once or twice have been taken into the cup. The spaces between the brachial plates are filled by interbrachial plates. (2) The basal and infrabasal cycles may contain plates that are fused so that only three plates occupy the cycle, or in rare cups, only one at the top of the column. (3) Additional plates may be introduced into the radial or basal cycles in the sector that on the upper surface contains the anus (Fig. 7–40C).

The arms of the crinoid contain extensions of the coelom, the radial water canals, nerve system, and circulatory system, as well as the food grooves. In feeding, the arms are held at right angles to prevailing currents, that is, with the axis of the circlet of arms in a horizontal position to form a filtration net (Fig. 7–38A). Minute animals in the water are trapped in mucus on the tube feet that border the food grooves, are pushed into the groove and carried by ciliary action down to the base of the arm. From there the groove may follow the surface of the tegmen to the central mouth, or may go beneath the tegmen if the mouth is covered by tegmen plates. The arms may branch many times and each branch may be bordered by a line of fine pinnules that give it a featherlike appearance. As the food grooves extend on to all branches and pinnules, the length of the food gathering surface can be enormous. The soft tissue of the arms is supported by calcareous plates in a single or double series held together

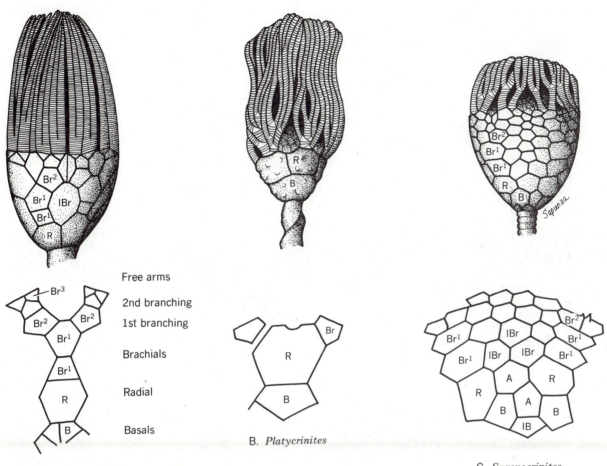

Free arms

2nd branching

1st branching

Brachials

Radial

Basals

A. *Eucalyptocrinites*

B. *Platycrinites*

C. *Sagenocrinites*

Figure 7-40 Paleozoic crinoids. The arrangement of plates in the calyx is shown diagrammatically below each drawing. Br = brachial, IBr = interbrachial, R = radial, B = basal, A = anal. A. *Eucalyptocrinites* (Silurian), three orders of brachials are incorporated in the calyx. B. *Platycrinites* (Carboniferous), the arms are free above the radial plates. C. *Sagenocrinites* (Carboniferous), several brachials and many interbrachials and two anal plates are incorporated in the calyx.

by ligament and muscles. The arms can be moved to take advantage of currents or rolled up during the day when many modern crinoids hide in crevices. Since the cylindrical arm plates fall apart when the muscles and ligaments decay, the whole filtration apparatus is rarely preserved but the individual plates are abundant fossils.

The column is composed of a single line of buttonlike plates (columnals) which may be several meters, but is usually a few centimeters in length (Fig. 7–39). The string is held together by ligament fibers that run through small holes in the columnals. Radial ridges on the surfaces of columnals interlock like a stack of poker chips. A muscular cord runs down an opening in the center of the plates to a holdfast or anchor system at the base. The hole in

the plate may be circular or star-shaped and in one Early Devonian genus it is double. Most crinoid columns appear to have been capable of flexing so that the calyx and arms could be advantageously positioned. Some column plates form the bases of circlets of short appendages called cirri that grasped adjacent organisms for support.

Crinoid History:

Complete crinoids are uncommon fossils in early Paleozoic rocks. Only under exceptional circumstances, such as in catastrophic burial, did calices and arms remain together after death. Where Carboniferous crinoid "gardens" were overwhelmed by sediment, as they were at Crawfordsville, Indiana,

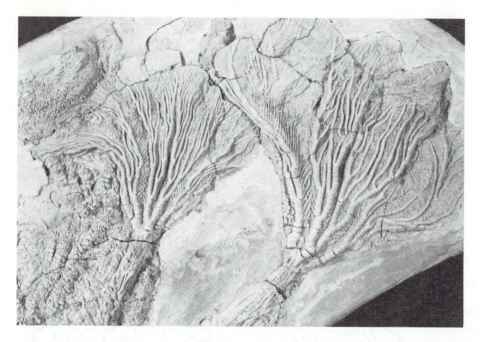

Figure 7-41   Two specimens of *Pentacrinites* from Jurassic limestone, England, showing the fine preservation of the arms and their pinnules. The slab is 16 cm across. (Courtesy of Redpath Museum.)

and LeGrand, Iowa, the bedding planes have been quarried and many specimens have been shipped to museums around the world (Fig. 7–41).

The oldest known crinoid comes from the Burgess Shale and has an irregular arrangement of calyx plates. The next oldest specimens come from

Early Ordovician rocks and the main expansion of the class coincides with that of many other organisms with carbonate skeletons, in Middle Ordovician time. By early in Middle Ordovician time grainstone and packstone limestones were dominated by the plates of crinoids and other attached

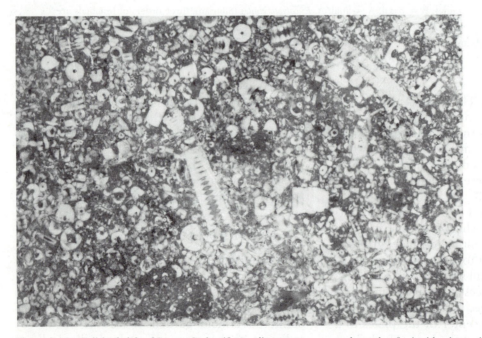

Figure 7-42   Polished slab of Lower Carboniferous limestone composed mostly of crinoid columnals. About 16 cm in width.

echinoderms. In mid-Paleozoic time the camerate crinoids, which had rigid calices and lived in turbulent waters close to reefs, were abundant. The acme of crinoid diversity and abundance was in Early Carboniferous time when they thrived in the shallow epicontinental sea that covered central North America and contributed their plates in astronomical numbers to the accumulating limestones (Fig. 7–42). At the close of Permian time the three Paleozoic subclasses became extinct, and the modern subclass, the Articulata, replaced them in Triassic time.

Cystoids:

Like crinoids, most cystoids are attached to the bottom but some have no columns and probably moved about on the seafloor. Unlike those of crinoids, the plates of the cystoid body are pierced by pores that are believed to have had respiratory functions but may have been related to locomotion or food gathering also. Rarely the body is symmetrical and formed of regular cycles of plates but, more often than not, the plates are irregular in size and shape and the body is without obvious symmetry (Fig. 7–43). Some cystoids are flattened and the plates are differently arranged on the bottom and top sides. These are believed to have lived lying on their sides on the seafloor.

The pore systems are of two types and this difference is used to distinguish two classes (Fig. 7–44). In one (Diploporita) the pores are scattered over the surface of the plates usually in pairs. In the other (Rhombifera) they occur in pore-rhombs that cross the boundaries between the plates. Pore-rhombs appear to have been infolds of the plates into the coelom or outfolds of the coelom into the seawater but their internal structure is poorly preserved. In some genera the pores are covered partly or completely by a thin calcareous plate.

The mouth is at the center of the upper surface and food grooves radiate from it over the plates. The grooves are usually bordered by, and extend on to, biserial unbranched appendages called brachioles (Fig. 7–45). The number of brachioles ranges from two, in *Pleurocystites*, to over a hundred (Fig. 7–43). The anus on the upper or side surface is surrounded by a pyramid of plates.

Cystoids first appear in Lower Ordovician rocks and by Middle Ordovician time had reached their maximum diversity. Their numbers dwindled through Silurian and Devonian time and by the end of the Devonian Period they were extinct.

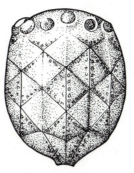

A. *Caryocrinites*

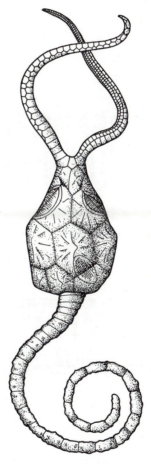

B. *Pleurocystites*

Figure 7-43 Cystoids. A. *Caryocrinites* (Silurian), calyx only showing the rows of pores at the margins of the regularly arranged plates, B. *Pleurocystites* (Ordovician), a complete specimen showing the upper surface, 2 arms, and 3 pore-rhombs. Both specimens about 3 cm long.

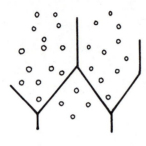  A. Simple pores

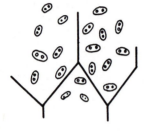 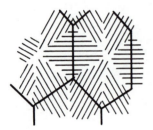 B. Diplopores

 C. Pore–rhombs (internal hydrospires)

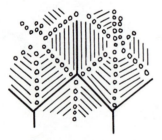 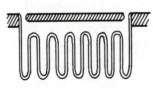 D. Covered hydrospires

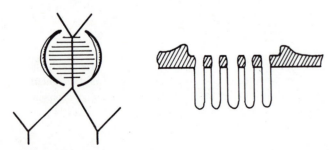 E. Pectinirhomb

Figure 7-44   Pore systems of cystoids. A. Simple pores. B. Double pores or diplopores. C. Pore-rhombs. D. Covered pore-rhombs. E. Pectinirhombs.

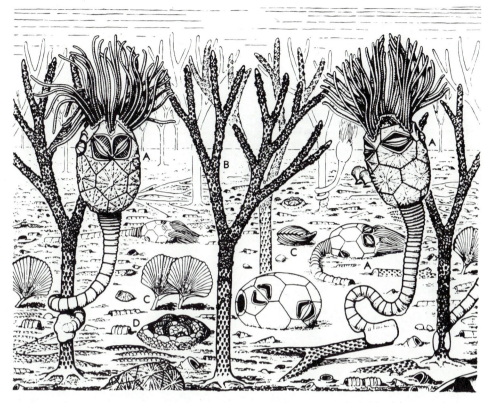

Figure 7-45   Reconstruction of the seafloor in Indiana during late Ordovician time, showing the cystoid *Lepidocystis* (A) attached to sticklike trepostome bryozoans (B). Also shown are brachiopods (C) and, in the foreground, an edrioasteroid (D). (*Source:* From R. V. Kesling and L. W. Mintz, used by permission of the University of Michigan Museum of Paleontology.)

Blastoids:

The blastoids are a late Paleozoic group of attached echinoderms descended from the cystoids. They are characterized by a respiration system that opens on either side of the ambulacral areas. Like many attached echinoderms they consist of a column, a body or cup, and small armlets called brachioles. Their structure is conveniently described by using *Pentremites*, the most common representative, as an example.

The calyx of this genus has clearly defined ambulacral areas with food grooves and pores (Fig. 7–46). The cup outside the ambulacral areas is formed of thirteen plates arranged in three cycles: three basal plates around the column, five radial plates notched to accommodate the ambulacral system, and five deltoid plates. The mouth is centrally located on the top surface and is surrounded by five large apertures. These apertures, called spiracles (Fig. 7–46B), lead down into the interior of the calyx where each opens into a pair of pleated organs called hydrospires. These organs are believed to

have been respiratory in function and to have been derived from the pore-rhomb systems of the cystoids. Because one of the spiracles, which is larger than the rest, incorporates the anus, we deduce that they all must have acted as exits for water circulating through the hydrospires. Water entered the hydrospires through many small pores that line either side of the ambulacral areas (Fig. 7–46C). The axes of these areas contain a food groove that leads to the mouth. It is covered, like those of many echinoderms, with small hinged plates and it was bordered in life by two lines of unbranched biserial brachioles. The brachioles themselves are rarely preserved but the sockets for their bases are evident at the margins of the ambulacral areas. In other blastoids, such as *Codaster* (Fig. 7–47A), the hydrospire folds are exposed as slits at the margins of the ambulacral areas and resemble the pore-rhombs of cystoids.

The first blastoids appear in Middle Ordovician rocks. They increased in number until Carboniferous time and reached their greatest abundance in the Lower Carboniferous limestones of the central United States. They are rare in Upper Carbonifer-

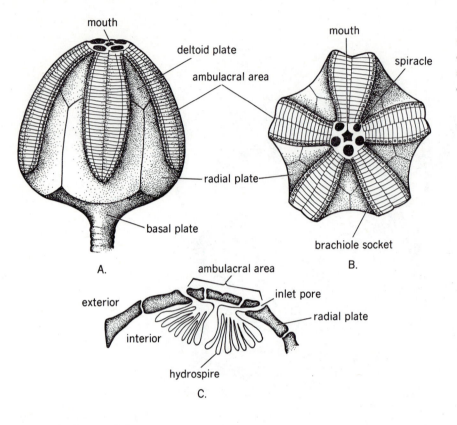

mouth
deltoid plate
ambulacral area
radial plate
basal plate
A.

mouth
spiracle
brachiole socket
B.

ambulacral area
exterior
inlet pore
radial plate
interior
hydrospire
C.

Figure 7-46   The blastoid *Pentremites* (Carboniferous). A. Side view of calyx. B. Top view of calyx. C. Cross section of the ambulacral area. The calyx is about 1 cm wide.

ous rocks but make a strong comeback in Permian time (Fig. 7–47B) before their extinction at the end of the Paleozoic Era.

Edrioasteroids:

These are a group of extinct Paleozoic echinoderms that fixed themselves to hard objects, such as rocks or shells, by their whole undersurfaces. Most are disks less than 1 cm across. The top surface, which is the only one usually observed, shows five ambulacral areas composed of regular lines of plates covering food grooves (Fig. 7–48). The areas between the ambulacra are covered with an irregular mosaic of plates. The ambulacral areas may be sinuously curved like the arms of a brittle star or they may be straight like those of a sea star, but, in either case, they are immobile. The anus is located between the ambulacral areas and is covered by a pyramid of plates, and the mouth is at the junction of the ambulacra beneath covering plates. Another opening on the surface is thought to be the entrance to a water vascular system. The first edrioasteroid is of Early Cambrian age but the group is only common in Middle and Upper Ordovician rocks where their circular tests commonly encrust brachi-

opod shells. The Ediacaran fossils, *Tribrachidium* and *Arkarua* may be edrioasteroid ancestors.

Other Echinoderm Groups:

Many other subclasses, classes, and orders of echinoderms have been established to receive a small number of genera with features that are transitional between the groups described above. Recently the tendency of echinoderm paleontologists has been to give these taxa higher positions in the classification hierarchy so that some primitive echinoderm classes now contain only a few species. Such groups as the Eocrinoidea, Paracrinoidea, Helicoplacoidea, Edrioblastoidea, Ophiocystoidea, Cyclocystoidea, Lepidocystoidea, and Somasteroidea are unlikely to be encountered outside a museum.

CARPOIDS, HOMALOZOANS, OR CALCICHORDATES

One group of fossils that have long been considered to be primitive echinoderms deserve further consideration, not because they are common fossils, but because they have many structures that have been

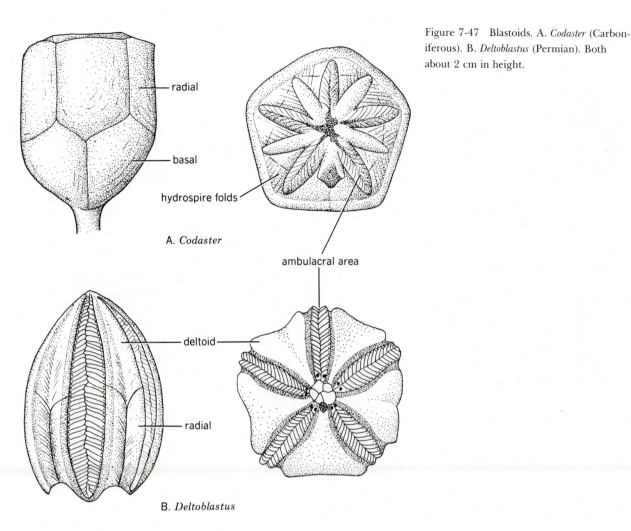

Figure 7-47   Blastoids. A. *Codaster* (Carboniferous). B. *Deltoblastus* (Permian). Both about 2 cm in height.

A. *Codaster*

B. *Deltoblastus*

interpreted as being comparable to those of primitive chordates, and some paleontologists claim that they are the link between the invertebrate phyla and the chordates. Others disagree, not only on the position of the group as chordate ancestors, but also on its name and on which end of the animals is the front. Until about 20 years ago these fossils were considered to be a division of the Cystoidea, the order Carpoidea. However, they differ from cystoids in their complete lack of symmetry and in the absence of pore systems penetrating their plates. For those who do not accept the chordate connection these fossils constitute the subphylum Homalozoa of the Echinodermata; for those who do, they constitute the subphlyum Calcichordata of the Chordata.

Certain anatomical features are diagnostic of primitive chordates, such as the Cephalochordata and Urochordata. These animals have a strength-

ening rod along the back called a notochord which in more advanced chordates becomes surrounded by bony elements called vertebrae. The principal nerve chord runs along the back side of this notochord. Water is passed into the mouth and passes out through a series of gill slits in which blood is aerated and, in some primitive chordates, food is filtered from the water. The skeletal material of chordates is calcium phosphate. The degree of similarity of the carpoids to chordates can be assessed in terms of notochord, dorsal nerve chords, gill slits, and phosphatic skeleton.

Ceratocystis is a carpoid that has been interpreted as showing many of the chordate features (Fig. 7–49A,B). It consists of a plated body flattened to lie on the seafloor, and a long stem, or arm, with complex plate structure. The outline of the body is highly irregular and it extends away from the stem in three lobes. The plate arrangements on top and bottom

Figure 7-48 The edrioasteroid *Agelacrinites* (Devonian). Note the pyramid of plates covering the anus between the two curving posterior ambulacral areas. About 15 mm in diameter. (Courtesy of the Smithsonian Institution.)

are different and both lack symmetry. On the top surface are a row of seven elliptical openings. According to R. P. S. Jefferies these are gill slits used by the animal in filter feeding (13,14). According to Georges Ubaghs (15) they are pores leading into a respiratory system similar to that of many cystoids. On the underside between two of the lobes is an aperture surrounded by small plates which Jefferies identifies as the mouth, and Ubaghs identifies as the anus. Opposite this opening, near the base of the stem, is a bilobed opening that Jefferies says is the combination of anus and genetic pore, and Ubaghs says is the mouth. What is called here the stem is called the aulacophore by Ubaghs and the tail by Jefferies. Ubaghs thinks it was held up in the water in front of the animal and brought microorganisms to the mouth by means of a food groove on its surface. Jefferies reconstructs the stem as a tail that was largely embedded in soft sediment and was used to pull the body backward along the sediment surface. In other carpoids, such as *Enopleura* (Fig. 7–49C), the stem does appear to have elements suitable for anchoring it in soft sediment. Jefferies believes that the canals in the structure he calls the tail can be interpreted as passageways for the notochord and the dorsal nerve chord. Ubaghs attributes these passageways to the water vascular or nerve system of an echinoderm.

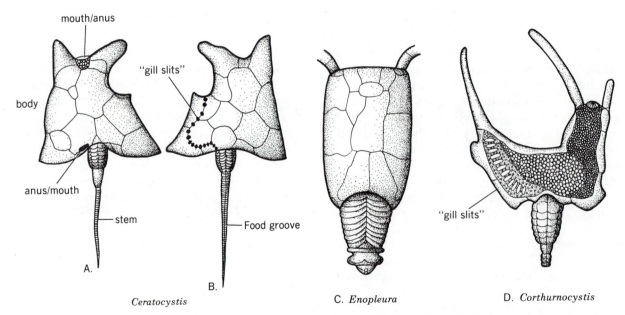

Figure 7-49 Ordovician carpoids. A, B. Top and bottom surfaces of *Ceratocystis*. C. *Enopleura*. D. *Corthurnocystis*.

Cothurnocystis is even more bizarre than *Ceratocystis* (Fig. 7–49D). Its body is framed by heavy, marginal plates within which was a flexible covering studded with small, round plates. Sixteen elliptial openings on the top surface may represent gill slits. The mouth (or anus) is in a similar position to that of *Ceratocystis*, between two extensions backwards (or forwards) of the body.

Paleontologists agree that at some remote time echinoderms and chordates arose from a common ancestor but disagree on whether the carpoids are in the direct line of descent. Not only is the embryological development of echinoderms and chordates similar, but the microstructure of their skeletons is comparable. Both echinoderm plates and bones are enclosed in soft mesodermal tissue that permeates the mineral matter completely and can modify the skeleton after it has been laid down. However, the skeleton of echinoderms is calcium carbonate and that of chordates, calcium phosphate. Jefferies believes that the descendants of the carpoids (calcichordates) at a certain stage abandoned a carbonate skeleton and later secreted a phosphate one.

The problem of the evolutionary position of these complex fossils remains but may be clarified as new species are found.

BIOLOGY OF HEMICHORDATA

The phylum Chordata comprises animals with a backbone or a longitudinal strengthening rod called a notochord. In the middle ground between the chordates and the invertebrates lie several groups of animals that have been either placed in the Chordata, placed collectively in the single phylum Protochordata, or distinguished as several different phyla. They are all enterocoelous deuterostomes linked by similarities in embryonic development, but very different in adult form. Three groups are usually distinguished: the Cephalochordata, the Urochordata, and the Hemichordata. The first two, either in embryonic or adult stages, possess a notochord and are therefore closely related to the chordates. The hemichordates do not have a notochord and may be considered to be a phylum of the invertebrates. However, the presence of a dorsal nerve chord and gill slits in some members of this group is evidence of their close relationship to the other two groups of primitive chordates.

Two classes of living animals that are dissimilar in adult form are united in what is here considered to be the subphylum Hemichordata. The Enteropneusta, commonly called acorn worms, live within burrows in shallow-water marine sediment and have no fossil record. Of more interest to the paleontologist are the Pterobranchia, comprising three living genera of encrusting colonial organisms that play a minor role in modern oceans but had abundant relatives called graptolites in early Paleozoic seas. The oldest rhabdopleurid is *Rhabdotubus* from Middle Cambrian beds of Sweden.

Phylum—Chordata
   Subphylum—Cephalochordata
   Subphylum—Urochordata
   Subphylum—Hemichordata
      Class—Enteropneusta
      Class—Pterobranchia
      Class—Graptolithina
         Order—Dendroidea
         Order—Graptoloidea

Rhabdopleura is a typical living pterobranch. It is a small encrusting marine organism that spreads threadlike tubes on shells and other hard objects (Fig. 7–50). The colony consists of creeping tubes that branch over the surface and erect tubes that house the individuals. The individual animals are attached to the base of their tubes by a stalk that can contract to bring them into the safety of the tube. They are about one-half millimeter across and equipped with a lophophore composed of two hollow ciliated arms. A disklike pre-oral lobe, which overhangs the mouth, secretes the tubes and is used to creep up the tube. The alimentary canal leads from the mouth to the stomach and through the gut in a U-shaped bend to the anus below the lophophore (Fig. 7–50C). Circulatory and nerve systems are present but rudimentary. The sexes are separate.

The tubes are composed of hard protein, but not of chitin. The creeping tube is secreted as a series of half-rings that join along two zigzag sutures. Inside the tube is a dark rod, or stolon, which branches into the base of each erect tube where it is attached to the contractile stalk. The rings of protein that form the erect tubes are complete, not half-rings like those of the creeping tubes.

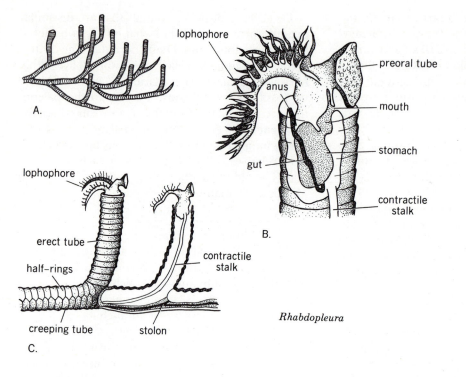

Figure 7-50 The living hemichordate *Rhabdopleura*. A. Fine tubes of the whole organism encrusting a hard surface. B. Detail of creeping tube, half rings, and stolons within the tubes. C. Individual at the top of its tube. The tubes are about 0.2 mm across.

GRAPTOLITES

The graptolites are considered by most paleontologists to be an extinct class of the phylum Hemichordata because they secreted tubes of protein which, in primitive representatives, are formed of half-rings and contain a system of branching stolons like those of *Rhabdopleura*. They were colonial marine organisms whose fossils are abundant in Silurian and Ordovician sedimentary rocks, particularly in dark shales. The colonies were linear arrangements of short tubes that were connected by a common canal through which, presumably, the individual animals shared food and coordinated their movements. They were called graptolites because they look like the markings of a lead (graphite) pencil on rocks (*graphos* (Greek) = writing). In graptolites, but not in *Rhabdopleura*, the inner layer formed of rings and half-rings is overlain by a laminar protein layer called the cortex which may have been secreted by a soft tissue that completely covered the outside of the skeleton, or may have been secreted by the animal on the inside of the tube (Fig. 7–51A). The exact composition of the skeleton is unknown but it was neither chitin nor collagen. During preservation both layers are usually extensively carbonized so that much of the detail of their structure is lost. They are also compressed

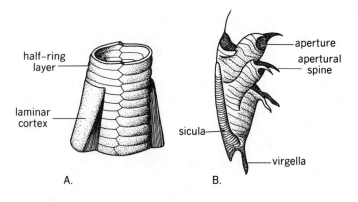

Figure 7-51 A. Two layers of the graptolite skeleton. B. Intial part of the graptolite *Saetograptus*.

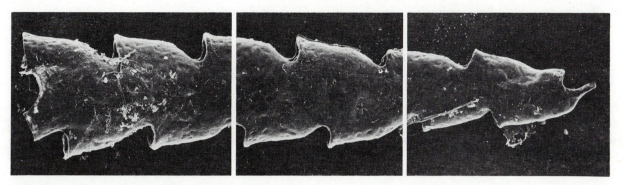

Figure 7-52   Three-dimensional diplograptid graptolite dissolved from a concretion, Silurian, Cornwallis Island. Length 5 mm. (Courtesy of Alfred Lenz, University of Western Ontario.)

onto the bedding planes of shales. Where it can be released from the shale, the carbonized skeleton can be bleached with potassium chlorate and, when it becomes translucent, details of the structure can be studied. Studies of the growth and wall structure have been made also on specimens preserved in three dimensions in siliceous and carbonate concretions from which they may be released by acid-leaching (Fig. 7−52).

Dendroidea:

The class Graptolithina is divided into two main orders and a number of minor ones. The order

Dendroidea comprises bushlike graptolites with many branches bearing tiny openings for housing the individuals. The colony was either attached to the ocean floor or hung from floating seaweed by the apex from which the branches diverge. The openings on the branches are of two sizes, that is, the individuals making up the colony were dimorphic (Fig. 7−53B). Although the soft tissue structures of graptolites that have been preserved are all of dubious identity, the individuals are assumed to have been similar to those of *Rhabdopleura* (Fig. 7−50), but the nature of the dimorphism is unknown. The small openings may have housed male individuals and the larger ones female, or the

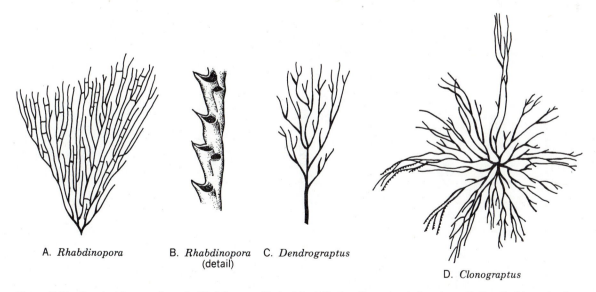

A. *Rhabdinopora*   B. *Rhabdinopora* (detail)   C. *Dendrograptus*   D. *Clonograptus*

Figure 7-53   Dendroid graptolites. A. *Rhabdinopora* (Ordovician−Carboniferous), whole colony. B. Detail of branch of *Rhabdinopora* showing the dimorphism. C. *Dendrograptus* (Ordovician). D. *Clonograptus* (Ordovician). All except Figure B natural size.

small ones may have been individuals specialized for cleaning.

A common dendroid genus is *Dendrograptus* (Fig. 7–53C). Its treelike colonies with branches that are free from their bases are abundant in Late Cambrian and Early Ordovician black shales. Some dendroids lived on into Carboniferous time after the most advanced graptolites had become extinct. The advanced dendroids, called anisograptids, have partially lost the dimorphism of the more primitive members of the order and branch in regular patterns (Fig. 7–53A,D). In *Rhabdinopora* (formerly called *Dictyonema*) the branches are joined together at intervals into a net by connecting cortex tissue. Through reduction in the number of branches, conversion to regular branching, elimination of dimorphism and stolonal systems, and transition to a free-floating or swimming lifestyle, the dendroid graptolites gave rise in Early Ordovician time to the second major order, the Graptoloidea. Some paleontologists suggest that the anisograptids are better placed in this more advanced order (16).

Graptoloidea:

Early graptoloids had 16, or in some more, separate branches (Fig. 7–54A) formed by four orders of branching (2^4) from the initial conical chamber which is referred to as the sicula (Fig. 7–51B). The classification of these graptolites is based on the order in which individuals are branched from this sicula and on whether or not it has a downward-pointing spine called a virgella (Fig. 7–51B). Within Early Ordovician time genera with eight, four, and 2 branches appeared (Fig. 7–54). In some of these forms the branches hung down from the sicula, in others they were on the same level, and in still others they grew above it. By Middle Ordovician time graptolites in which the apertures of the individuals opened on either side of a central thread (Fig. 7–54F,H) had evolved. Because a typical member of this group is *Diplograptus*, this Middle and Late Ordovician fauna is referred to as the diplograptid fauna. In the Silurian Period the dominant graptoloids had a single line of apertures on one side

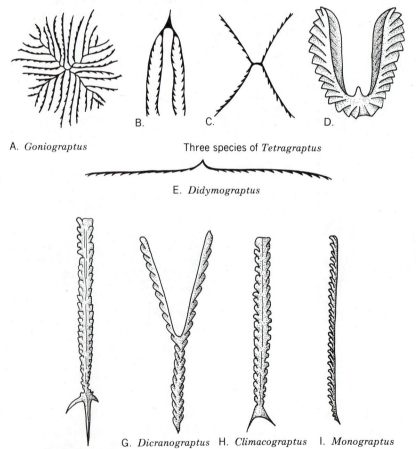

A. *Goniograptus*

B. C. D.

Three species of *Tetragraptus*

E. *Didymograptus*

F. *Orthograptus*

G. *Dicranograptus* H. *Climacograptus* I. *Monograptus*

Figure 7-54 Graptoloid graptolites. A. *Goniograptus*. B, C, D. Three species of *Tetragraptus*. E. *Didymograptus*. F. *Orthograptus*. G. *Dicranograptus*. H. *Climacograptus*. I. *Monograptus*. All Ordovician except *Monograptus*, which is Silurian.

of the colony. Many of these are assigned to the genus *Monograptus*. Their apertures were extended into tubes opening downward, upward, or sideways, and were bordered by flanges and spines. *Monograptus*, the last of the graptoloids, became extinct near the end of Early Devonian time for reasons that are not apparent. No other group became extinct at this time. Possibly predation from newly evolving fish caused extinction of the graptolites that were already in decline.

Although paleontologists generally agree that graptoloid graptolites lived in the open ocean, they disagree on whether they floated passively, buoyed up by gas trapped in their soft tissue, or kept themselves up and migrated vertically by the concerted swimming action of their lophophores. Nancy Kirk (17,18) believes that migration of the graptolites into more turbulent water near the surface accompanied the transition from multibranched forms to those with few branches.

The graptolites evolved rapidly and spread widely, features that have made them useful for dating and correlating sedimentary rocks. The oldest graptolites come from Middle Cambrian rocks but the group was not abundant until the beginning of the Ordovician Period. Dendroids characterize earliest Ordovician faunas; anisograptids, later Early Ordovician faunas; diplograptids, Middle and Late Ordovician faunas; and monograptids, Silurian and Early Devonian faunas. From Middle Ordovician time onward, the trends in evolution changed from reduction of the number of branches to modification of the apertures of the individuals. Many genera now recognized are polyphyletic and represent stages of evolution. The tracing of phylogenies on the basis of early stages in colony development is now in progress (17,18). The rocks of Ordovician, Silurian, and Early Devonian ages that contain graptolites have been divided in Europe into 53 zones, each characterized by an assemblage of species. Unfortunately, most of these fine divisions of the rock sequence can only be used for correlating black shales where graptolites are abundant. Limestones and sandstones rarely contain these fossils.

CONODONTS

Conodonts have been called "fascinating little whatzits" (19). This unspecific name was suggested by uncertainties about the biological relationships of these beautiful, minute, toothlike fossils that are common in Paleozoic rocks (Figs. 7–55, 7–56). Many specialists have placed them in no living phylum but believe them to be representatives of an extinct phylum, the Conodonta.

Conodonts are microfossils up to 8 mm, but usually less than 1 mm, long composed of a single cusp, linear sets of small conical teeth attached at their bases, or plates and platforms that are sculptured with ridges and knobs. They are composed of calcium phosphate with carbonate, sodium, and water present in the crystalline structure in small amounts. The composition of conodonts allows paleontologists to release them from enclosing carbonate rocks by dissolving the rock in weak acids, such as acetic, in which the phosphate is insoluble. The acid residue is further concentrated by placing it in a liquid of density greater than 2.8 in which the heavy conodonts (specific gravity 3.1) sink to the bottom and lighter insoluble minerals float to the top.

Conodonts are light amber, black, gray, or clear. Their color is an indicator of the temperature that the fossil was subjected to after it was buried in sediment. Such heating is usually caused by deep burial by sediments that accumulate in a basin of deposition but could be caused by nearness to an igneous intrusion. Experiments in which conodonts are heated indicate that golden ones darken progressively as the temperature is raised above the 50 to 80°C. range. They change through brown to black at 300°C., and through gray and white to clear between 300 and 600°C. The color change is not reversed on cooling. The color alteration index of conodonts, therefore, gives paleontologists a method of determining the thermal history of the rock in which they are found and, since this is related to the rock's ability to release and preserve petroleum, of estimating the likelihood of finding oil in the rocks. Because the rate at which the temperature of the earth increases with depth is known, the color alteration index can also be used to determine the depth to which sediments containing conodonts have been buried even if they have been subsequently raised to the surface. Similar methods of assessing burial depth employ the color of graptolites and spores.

The elements (as the individual fossils are called) have been divided into three basic forms but a great variety exists within each group (Fig. 7–57). Coniform elements are curved, fanglike forms resembling a single tooth. Ramiform elements have a large cusp from which processes bearing rows of smaller teeth extend. Pectiniform elements are sculptured

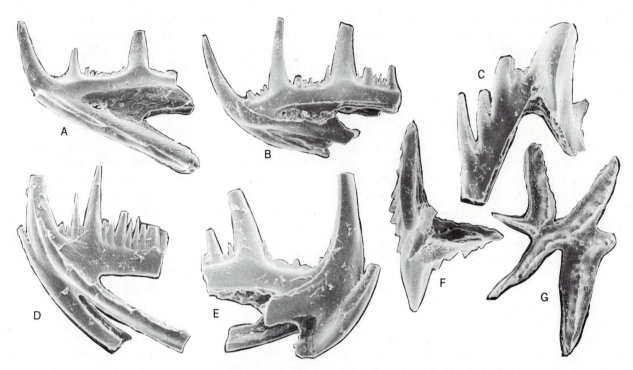

Figure 7-55   Scanning electron micrographs of the seven elements of *Amorphognathus ordovicicus*. A. ×200. B. ×300. C. ×180. D. ×300. E. ×210. F. ×180. G. ×110. (Courtesy of Godfrey Nowlan and the Geological Survey of Canada.)

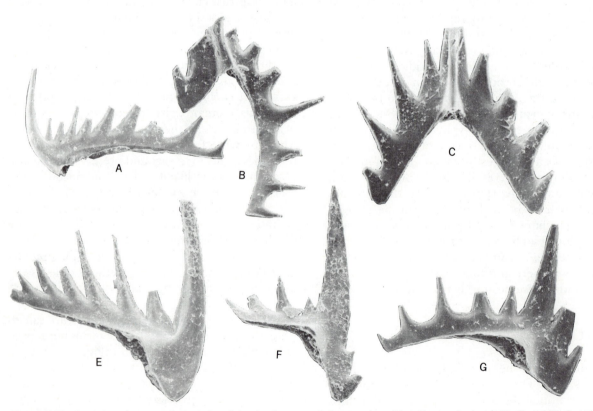

Figure 7-56   Scanning electron micrographs of the six elements of the conodont *Plectodina tenuis*. A. ×180. B. ×200. C. ×300. E. ×210. F. ×170. G. ×180. There is no d element. (Courtesy of Godfrey Nowlan and the Geological Survey of Canada.)

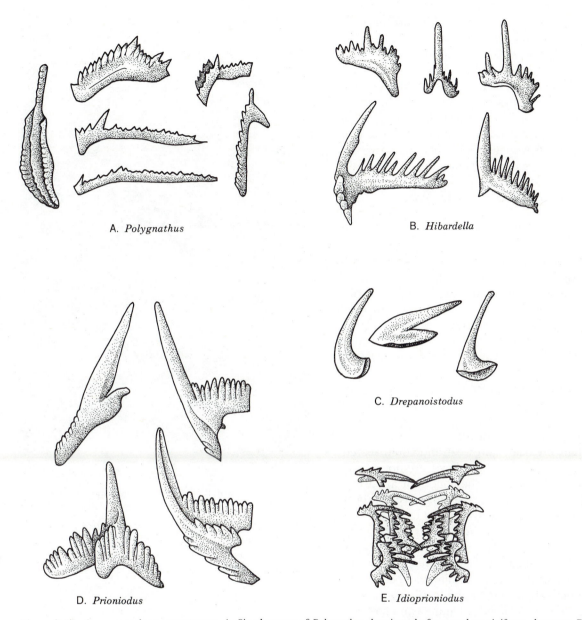

Figure 7-57   Some conodont apparatuses. A. Six elements of *Polygnathus* showing platform and pectiniform elements. B. Five element apparatus of *Hibardella*, most are ramiform elements. C. Three element apparatus of *Drepanoistodus*. D. Four element apparatus of *Prioniodus*, composed of ramiform elements. E. Reconstruction of the elements in the apparatus of *Idioprioniodus*.

platforms and denticulate bars. Rarely, conodont elements are found regularly arranged in assemblages or fused clusters, usually on bedding surfaces of shales. These assemblages show that the conodont animal was equipped with several different kinds of elements. The basic number of different types of elements is six and assemblages that contain this number of elements are said to be seximembrate. Five different types of assemblages have been distinguished: (1) unimembrate with coni-

form elements only; (2) multimembrate, including only coniform elements; (3) ramiform and pectiniform elements only; (4) pectiniform and coniform elements only; (5) several types of ramiform elements.

Although about 500 conodont assemblages have been described, most conodonts are known only from acid residues in which the elements are isolated. Each element was given a specific name when it was first described and each was assumed to belong to a different animal. This older taxonomy is revised as the different elements are found in association with one another in assemblages which must have belonged to a single animal. So much is now known about the recurrent associations of elements that assemblages are reconstructed from isolated elements in acid residues. Conodont paleontologists face a problem in taxonomy similar to that of paleobotanists who describe roots, stems, leaves, and flowers with scientific names but find eventually that they belong to a single organism which has been given four different specific names.

Biological Affinity:

Although conodonts look like teeth, and most paleontologists have assumed that they were used for capturing and breaking up food, their internal structure and lack of wear do not strongly support this interpretation. The internal structure of conodont elements shows that they have grown by addition of layers to the outside surface. In order to grow, the conodont elements must have been covered, at least periodically when a new layer was laid down, by soft tissue that secreted the phosphate. If conodont elements acted as teeth, they must have been withdrawn into a sheath lined with cells that could repair breakage and wear and could secrete another layer at times when they were not in use.

In what organisms did the conodont apparatus function? Although many assemblages have been found, none, except the specimens considered in the following discussion, show traces of the organism that bore them. The conodont animal must have been soft-bodied and disintegrated rapidly after death. One specimen, at first claimed to be the conodont animal, appears to be the fossil of an animal that swallowed the conondont animal and had conodonts in its gut. Conodont assemblages that are clearly in place in the preserved body of an animal come from Early Carboniferous shales in Scotland (Fig. 7–58) (20,21,22). The conodont elements are arranged behind a bilobed head of a wormlike organism about 40 mm long, compressed on the bedding plane. The body appears to have had a long fin supported by rays. A separate ray-supported fin may have been present along the back of the animal. The body of the fossil also shows faint traces of transversely segmented muscles. In Lower Silurian rocks of Wisconsin, elements of the genus *Panderodus* have been found at the head of a wormlike body that shows few details of its structure.

Figure 7-58   A conodont animal, *Clydagnathus?*, from the Carboniferous of Scotland. The specimen is about 4 cm long. The head (H) is bilobed and contains a group of conodont elements. The oblique lines on the body toward the tail (T) suggest segmented muscles. (Courtesy of E. N. K. Clarkson and *lethaia*, v. 16, p. 4.)

The Scottish fossils of the conodont animal most closely resemble two groups of living animals: the chaetognaths and the primitive chordates. The phylum Chaetognatha comprises a few genera of organisms called arrowworms that live in the open ocean, locally in large numbers. At the front of the head a series of chitinous grasping spines and teeth are used for capturing food. These can be covered by a fleshy hood or sheath. The body is equipped with lateral fins for propulsion. The primitive chordates (Cephalochordates), typified by *Brachiostoma* (also called amphioxus) are small, elongate animals that embed themselves in sand or swim short distances. Like the conodont animal and the chaetognaths, they have longitudinal fins along a wormlike body. Transverse segmentation of muscles is evident along the sides of the body. No feature of these primitive chordates, however, corresponds to the conodont elements themselves. Conodont phosphate does more closely resemble bone secreted by chordates than the chitin secreted by chaetognaths. However, it has been suggested that some simple cone-shaped elements classified as protoconodonts are chaetognath spines. The paleontologists studying the Scottish specimens believe that they represent a primitive chordate. Others have suggested an affinity with aplacophoran molluscs should be considered. Despite the discovery of fossils of the conodont animal, controversy about their affinity continues.

Stratigraphic Distribution and History:

Much attention has been paid to conodonts because they are so useful for correlation of Paleozoic strata. Some conodont species have relatively short stratigraphic ranges. Conodont elements are found in many different kinds of marine sedimentary rocks and were widely distributed geographically. Ordovician and Silurian limestone successions have been divided into about 40 successive conodont zones, and specialists are able to place samples containing these fossils in the time scale with a precision of a few million years. For example, the 21 million years of Early Devonian time represented in the rocks of western North America can be divided into nine different conodont zones.

Although conodonts are widely distributed and the fossils are valuable for dating rocks, they are not entirely independent of sedimentary facies. Conodont elements are distributed in many types of sedimentary rocks but some are restricted to certain lithologies. Certain genera appear to have been restricted to certain depth zones but not all of the factors that affected the distribution of conodonts have yet been assessed.

Conodonts appear in rocks near the Ediacaran–Cambrian boundary as simple cone-shaped elements, but these are rare. They diversified rapidly early in the Ordovician Period and again near its close. With the exception of a brief recovery in the Middle Silurian, they declined in diversity through the rest of the Paleozoic. Conodonts survived the Late Permian crisis that decimated invertebrate ranks but became extinct at the boundary between the Triassic and Jurassic periods.

PALEOZOIC INVERTEBRATE HISTORY

In examining the history of individual classes, orders, and phyla of invertebrates in the Paleozoic, an appreciation of the "forest" may be lost in looking at the "trees". In this section some of the major trends and events of Paleozoic invertebrate life are reviewed. Quantitative conclusions about such events have been based on changes with time in the numbers of species, genera, families, etc. Estimates of changes in the number of individuals, or of the biomass, in time are more difficult to determine and so far have been purely subjective.

Diversification:

The rate of diversification of life in Ediacaran and Cambrian time is exponential as documented in Chapter 6. By early in the Cambrian Period all the major invertebrate phyla and most of the major classes were represented in the fossil record. The production of new taxa slowed by Ordovician time and thereafter the rise of one group of invertebrates appears to match the fall of another. For example, the decline of the trilobites in the Ordovician coincides with the rise of the corals, brachiopods, and stalked echinoderms. Several groups that secreted heavy calcareous skeletons, such as the stalked echinoderms, bivalves, and articulate brachiopods, have representatives in Lower and Middle Cambrian rocks, are rare or absent in Upper Cambrian rocks, and do not become abundant until the middle of the Ordovician Period. Only the trilobites, inarticulate brachiopods, and archaeocyathids are abundant in Cambrian rocks. The delay in the spread of carbonate-secreting invertebrates to Ordovician time may have been related to the attainment of a critical level of oxygen in the atmosphere (see Chapter 6) or to the spread of vast epicontinental seas across the continents in the Ordovician, offering greater opportunity for the diversification of shallow-water reef builders and dwellers. Some idea of the rise and fall of invertebrates in the Paleozoic can be obtained from Figure 7–59, although the data plotted are somewhat out-of-date. John Sepkoski (23) has divided the Paleozoic faunas into Cambrian, Paleozoic, and modern components (Fig. 7–60). The first two faunas show a rapid rise to maximum diversity and a slow decline. For the Cambrian fauna this decline was completed in the late Paleozoic, for the Paleozoic

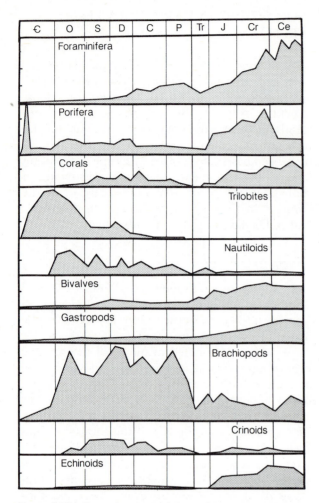

Figure 7-59   General pattern of rise and fall of invertebrate groups during the Phanerozoic. Abbreviations for the geological periods are across the top. The major divisions on the left each represent 50 genera.

fauna it continues to the present, and for the modern fauna it has not yet started.

Limitations of Diversity Curves:

The significance of plots of diversity, such as Figures 5–15, 7–59, and 7–60, is difficult to assess. The number of species, genera, orders, etc. of a fossil group that have been described from rocks of a particular age is not solely dependent on the number of such taxa that lived at that time. The number of recorded fossils of a particular age is also dependent on the area of exposure of those beds, which is an important control on the opportunity for a collection to be made from them. Other factors that may influence the number of families,

genera, or species counted include the volume of sediments entering the geological record during each period and the number of paleontologists working on fossils of this age (Fig. 7–61). The shape of the diversity graph also depends on the time of publication of major monographs relative to the time of compilation. For example, the publication of a large monograph on Permian brachiopods containing descriptions of a hundred new species and the establishment of new families and genera would have significant effect on the height of the diversity curve at the Permian Period and might not be balanced by studies in similar detail from other periods for many years. Graphs showing diversity changes for the total invertebrate fauna (for example, Fig. 7–59), are based on the assumption that the taxonomic groupings being summed (such as families) are comparable in diversity. The assumption is that the total diversity of invertebrates can be obtained bt treating one order of brachiopods as if it were equivalent to one order of sponges. This assumption is difficult to justify because the orders are subjective concepts of different paleontologists using anatomies of very different organisms to arrive at convenient divisions. In addition, James Valentine (24) has pointed out that for such graphs to be valid for all life, the proportion of organisms with preservable soft parts must be constant in time. No data is available to assess the possible variations in this proportion.

Estimates of the diversity of marine organisms from the beginning of the Paleozoic to the present (Fig. 7–60) show a general increase in early Paleozoic time, stability in mid to late Paleozoic time, decreased diversity in the Permian and Triassic periods, and a rapid rise in diversity through late Mesozoic to the Recent. Despite the resemblance between diversity plots such as Figure 7–60 and the biasing factors (Fig. 7–61), Phillip Signor (25) concludes that the pattern of early Paleozoic and late Mesozoic increases, and late Paleozoic and early Mesozoic decreases, is real.

Crises:

The development of the Paleozoic biota was not without interruptions; at least three periods of major crisis have been distinguished: end of Ordovician, within Late Devonian, and at the end of Permian time. Just before the end of the Ordovician about 22 percent of the families of invertebrates became extinct. The groups that were particularly

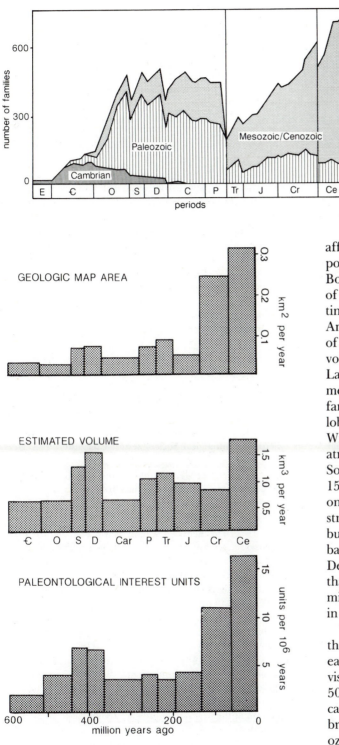

Figure 7-60   Variations in the diversity of the three major Phanerozoic faunas distinguished by Sepkoski. (Courtesy of J. J. Sepkoski, and The Paleontological Society, *Paleobiology*, v. 10, p. 249.)

Figure 7-61   Biasing factors in measurement of Phanerozoic diversity. The paleontological interest units are derived from the number of paleontologists working on the fauna of a given interval. (Courtesy of Phillip Signor and the Geological Society of America, *Geology*, v. 10, p. 626.)

affected were the trilobites, graptolites, cephalopods, articulate brachiopods, and crinoids. Arthur Boucot (26) has stated that 60 percent of the species of Late Ordovician brachiopod genera became extinct. Many of these were species endemic to North America and they were replaced at the beginning of Silurian time by European migrants. The Devonian extinctions took place near the middle of Late Devonian time between the Frasnian and Famennian stages. About 30 percent of invertebrate families became extinct and brachiopods, corals, trilobites, and ammonoids were seriously affected. Within the brachiopods the orthids, pentamerids, atrypids, and strophomenids became extinct. James Sorauf and Allen Pedder have calculated that of 159 species of shallow water Frasnian rugose corals, only ten survived into the Famennian stage. The stromatoporoids, which were the principal reef builders of the mid-Paleozoic, suffered a major setback and formed only small patch reefs in latest Devonian time. George McGhee (17) has estimated that 70 percent of the species of larger (that is, not microscopic) invertebrates disappeared at this time in the Applachian area.

The Late Permian life crisis is so conspicuous in the paleontological record that it was recognized as early as the early 19th century as the basis for division of the Paleozoic from the Mesozoic eras. About 50 percent of the animal families in the world became extinct at this time. Major divisions of the brachiopods, crinoids, echinoids, corals, and bryozoans and all the trilobites, blastoids, fusulines, and goniatites died out at this time. The new marine fauna of the Mesozoic was dominated by the expansion of the gastropods, bivalves, and new types of ammonoids, new forms of echinoids and corals, and new calcareous forams. This new fauna is discussed further in Chapter 10. The causes of these periods of extinction will be considered in Chapter 16 after the accounts of life in the Mesozoic and Cenozoic eras.

REFERENCES

1. Runnegar, B., 1985. Origin and early history of Mollusks. In Bottjer, D. J., Hickman, C. S, and Ward, P. D. (eds.) Mollusks: University of Tennessee Studies in Geology, v. 13, p. 17–32.

2. Pojeta, J., and Runnegar, B., 1974, *Fordilla troyense* and the early history of the pelecypod molluscs: American Scientist v. 62, p. 706–711.

3. Gould, S. J., and Calloway, C. B., 1980, Clams and brachiopods—ships that pass in the night: Paleobiology, v. 6, p. 383–396.

4. Saunders, W. B., and Landman, N. H., 1987, *Nautilus*, biology and paleobiology of a living fossil: New York, Plenum, 632 p.

5. Ward, P., 1987, The natural history of *Nautilus*: London, Allen & Unwin, 192 p.

6. Westermann, G. E. G., 1982, The connecting rings of *Nautilus* and Mesozoic ammonites: implications for ammonite bathymetry: Lethaia, v. 15, p. 373–384.

7. Linsley, R. M., 1977, Some "laws" of gastropod shell form: Paleobiology, v. 3, p. 196–206.

8. Linsley, R. M., 1978, Shell form and evolution of gastropods: American Scientist, v. 66, p. 432–441.

9. Manton, S. M., 1977, The arthropoda: habits, functional morphology, and evolution: London, Oxford Univ. Press, 527 p.

10. Conway Morris, S. and Whittington, H. B., 1985, Fossils of the Burgess Shale: Geological Survey of Canada Miscellaneous Report 43, 31 pp.

11. Conway Morris, S. and Whittington, H. B., 1979, The animals of the Burgess Shale: Scientific American, v. 241, July, p. 127–133.

12. Whittington, H. B., 1985, The Burgess shale: New Haven, Conn. Yale Univ. Press, 151 p.

13. Jefferies, R. P. S., 1975, Fossil evidence concerning the origin of the chordates: Zoological Society of London, Symposium 36, p. 253–318.

14. Jefferies, R. P. S., 1986, The ancestry of vertebrates: London/New York, Cambridge Univ. Press, 376 p.

15. Ubaghs, G., 1968, Stylophora and Homostelea. Treatise on Invertebrate Paleontology: Geological Society of America and University of Kansas Press, part S2, p. 495–581.

16. Fortey, R. A. and Cooper, R. A., 1986, A phylogenetic classification of graptoloids: Palaeontology, v. 29, p. 631–654.

17. Kirk, N. H., 1980, Controlling factors in the evolution of graptolites: Geological Magazine, v. 117(3), p. 277–284.

18. Bates, D. E. B., and Kirk, N. H., 1985, Graptolites: a fossil case history of evolution from sessile animals to automobile superindividuals: Proceedings Royal Society of London, v. B228, p. 207–224.

19. Sweet, W. C. 1985, Conodonts—those fascinating little whatzits: Journal of Paleontology, v. 59, p. 485–494.

20. Briggs, D. E. G., Clarkson, E. N. K., and Aldridge, R. J., 1983, The conodont animal: Lethaia, v. 16, p. 1–14.

21. Aldridge, R. J., Briggs, D. E. G., Clarkson, E. N. R., and Smith, M. P., 1986, The affinities of conodonts, new evidence from the Carboniferous of Scotland: Lethaia, v. 19, p. 279–292.

22. Smith, M. P., et al., 1987, A conodont animal from the Lower Silurian of Wisconsin, U. S. A., and the apparatus architecture of panderodontid conodonts. in Aldridge, R. J. (ed.), Paleobiology of conodonts: Chichester, Ellis Horwood, p. 91–104.

23. Sepkoski, J. J., 1981, A factor analytical description of the Phanerozoic marine fossil record: Paleobiology, v. 7, p. 36–53.

24. Valentine, J. W., 1970, How many marine invertebrate fossil species, a new approximation: Journal of Paleontology, v. 44, p. 410–415.

25. Signor, P. W., 1982, Species richness in Phanerozoic: compensating for sample bias: Geology, v. 10, p. 625–628.

26. Silver, L. T., and Schultz, P. H. (eds.), 1982, Geological implications of large asteroid and comets on the earth: Geological Society of America Special Paper 190, 528 p.

SUGGESTED READINGS

Aldridge, R. J., (ed.), 1987, Palaeobiology of conodonts. British Micropalaeontological Society series: Chichester, Ellis Horwood, 189 p.

Austin, R. L. (ed.), 1987, Conodonts: investigative techniques and applications: Chichester, Ellis Howard, 422 p.

Bell, B. M., 1976, A study of North American Edrioasteroidea: New York State Museum Memoir 21, p. 1–447.

Bottjer, D. J., Hickman, C. S., and Ward, P. D. (eds.), 1985, Mollusks: notes for a short course: University of Tennessee Studies in Geology 13, 306 p.

Broadhead, T. W., and Waters, J. A. (eds.), 1980, Echinoderms: notes for a short course: University of Tennessee Studies in Geology 3, 235 p.

Bulman, D. M. B., 1970, Graptolithina, Treatise on Invertebrate Paleontology, part V, revised: Boulder/Lawrence Geological Society of America and Univ. of Kansas Press. 163 p.

Clark, D. L., et al., 1981, Conodonts. Treatise on invertebrate paleontology, Part W, Supplement 2: Boulder/Lawrence Geological Society of America and Univ. Kansas Press, 202 p.

Jollie, M., 1982, What are the 'Calcichordata'? and the larger question of the origin of chordates: Zoological Journal of the Linnean Society, v. 75, p. 167–188.

Pojeta, J., 1978, The origin and taxonomic diversification of the pelecypods: Philosophical Transactions of the Royal Society of London. v. 284B, p. 225–246.

Stanley, S. M., 1970, Relation of shell form to life habits of the Bivalvia (Mollusca): Geological Society of America Memoir 125, p. 1–496.

Chapter 8

Land Plants and Their Ancestors

ALGAE

The photosynthetic eukaryotes can be divided into two groups: those that live on land and require water-gathering and circulating systems, and those that live in water and damp soil and need neither. The water supply system comprises conduits made of vascular tissue and the plants that have it are vascular plants. The Bryophyta, mosses and liverworts, occupy a middle ground between vascular plants and algae; they live on land but only grow actively in damp places where they do not need a vascular system. The largest group of nonvascular plants are the algae.

Soft Algae:

Proterozoic algal cells were only preserved under exceptional circumstances in shales or when impregnated with silica in cherts. In general, algae that did not secrete a covering or support of silica or calcium carbonate did not leave a fossil record. Possibly the seas since Proterozoic time have had a flora of unicellular and more complex soft green and brown algae comparable in diversity and abundance to that of present seas, but impressions of these seaweeds have been left as fossils only in those rare deposits where soft tissue is preserved.

In the Mesozoic Era several groups of unicellular algae became planktonic, abundant, and diverse in the ocean. The dinoflagellates secreted cellulose tests; the coccoliths, calcium carbonate; and the diatoms, silica. The history of these groups is briefly reviewed in Chapter 10.

Kingdom—Plantae
  Subkingdom—Algae
    Division—Chlorophyta (green)
    Division—Chrysophyta (diatoms and others)
    Division—Phaeophyta (brown, soft)
    Division—Rhodophyta (red)
    Division—Pyrrhophyta (dinoflagellates)
  Subkingdom—Bryophyta
  Subkingdom—Tracheophyta

Red Algae (Rhodophyta):

The red algae have been important limestone makers since the early Paleozoic. In northern seas they thickly encrust rocks and shells with calcium carbonate, and in tropic seas they bind fragments of corals and shells together in reefs. Some form bush-like plants and contribute their jointed stems to the accumulating calcareous sediment when they die.

The carbonate secreted by most algae is on the outside of their tissues but the red algae secrete calcite around the margins of their cells, producing a three-dimensional network structure (Figs. 8–1, 8–2A). The most important superfamily, the Corallinaceae, form the ocean-facing algal ridges in Pacific reefs that take the brunt of the surf. The near-shore zones of many Caribbean reefs are completely covered with these coralline algae. Internally, the skeletal tissue is composed of minute cells, usually in two sizes—an internal coarse layer and an ex-

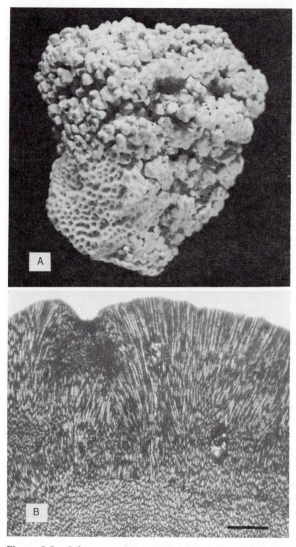

Figure 8-2   Calcareous algae. A. The crustose coralline alga *Lithophyllum* growing in a knobby mass on the scelractinian coral *Agaricia* on a recent fringing reef, Barbados. The specimen is about 9 cm in width. B. *Solenopora* (Ordovician), a fossil commonly assigned to the red algae illustrated in thin section showing the fine tubular structure. The scale bar is 1 mm. (C. W. Stearn.)

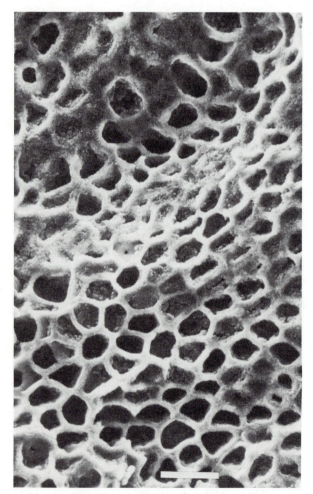

Figure 8-1   Scanning electron micrograph showing the cellular structure of an encrusting coralline red alga from a recent reef in Barbados. The scale bar is 20 micrometers. (C. W. Stearn.)

ternal fine layer containing cavities to house the reproductive spores. The coralline algae first appear in Jurassic rocks, but some late Paleozoic genera have been tentatively assigned to the group.

The Solenoporaceae (Ordovician to Miocene) are a superfamily, probably ancestral to the coralline algae. Their fossils consist of fibrous rows of cells coarser than those of the true corallines but not differentiated into the two layers (Fig. 8–2B). The

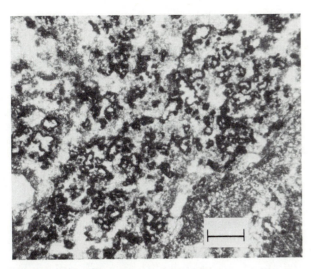

Figure 8-3 *Renalcis*, a thin section of the opaque-walled, irregular chambers of this enigmatic reef-building fossil. Devonian of Alberta. Scale bar is 1 mm. (C. W. Stearn.)

group reaches its greatest abudance in Jurassic reef limestones, but the small nodular fossils are also common in Paleozoic reef limestones.

Two widespread reef-forming organisms of the early Paleozoic may belong in the red algae but paleontologists are uncertain of their classification and some place them in the cyanobacteria. The genera *Renalcis* and *Epiphyton* are aggregates and branching series of rounded to crescentic chambers defined by walls of fine crystalline carbonate that is opaque in thin section (Fig. 8–3). The clumps seem to grow downward from the top of cavities. The two genera are abundant in reefs from Early Cambrian, when they consorted with the archaeocyathids, to Late Devonian time, when they filled spaces between stromatoporoids and corals. In the shallow Carboniferous seas of the central and southwestern United States, leaflike red algae grew abundantly as vertical sheets and formed small banks in the sea. These phylloid algal banks are important petroleum reservoirs in the Four Corners region of Utah and Colorado.

Green Algae (Chlorophyta):

Although most of the marine and freshwater species of this group do not secrete hard tissue, some of them are, and have been, important contributors to carbonate sediments. The codiacean green algae include the genus *Halimeda* (mid-Jurassic to recent),

whose heavily calcified plates secreted in branching series (Fig. 8–4) fall apart on death and supply a large part of the carbonate sands around modern reefs. Banks up to 19 meters thick, formed by *Halimeda*, occupy a large part of the platform behind the northern Great Barrier reef near Australia. The plates are 97 percent aragonite. Other codiaceans, such as the living genus *Penicillus*, secrete fine aragonite needles between their filaments and make major contributions to the deposition of lime muds when the soft tissue decays. In the Florida reef area *Penicillus* contributes 25 grams per square meter per year to the accumulating lime muds. Many lime mudstones of the past were probably formed of such fine particles precipitated by algae whose soft parts have completely decayed.

Ischadites (Fig. 8–5A) and *Receptaculites* (Fig. 8–5B) are representatives of an enigmatic group of fossils that are now generally assigned to the green algae. They are pillow or flask-shaped fossils up to tens of centimeters across composed of dumbell-shaped calcareous elements with rhombohedral ends arranged in a spiral pattern. *Receptaculites* is a common fossil in Late Ordovician limestones in western

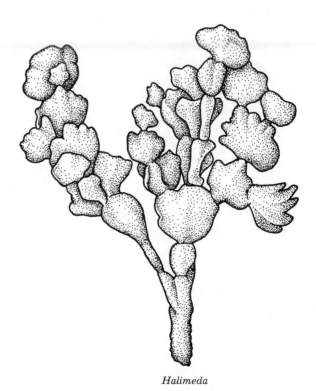

Halimeda

Figure 8-4 The green alga *Halimeda* (Cenozoic) about natural size.

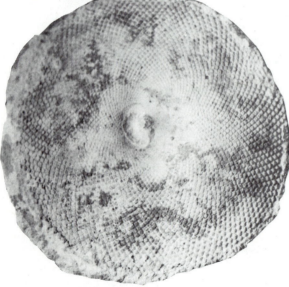

Figure 8-5   Receptaculitids. These sacklike fossils are commonly assigned to the algae. A. *Ischadites* (Silurian). Two specimens about 3 cm across. (Courtesy of Seigfried Reitschel and Forschungsinstitut Senckenberg [*Senckenbergiana lethaea*, v. 50, pl. 3.]) B. *Receptaculites* (Ordovician), the sunflower fossil, about 30 cm in diameter. (Courtesy of R. J. Elias, University of Manitoba.)

North America where it is known as the sunflower fossil because the pattern of plates resembles that of the seeds in a sunflower. The genus ranges from Middle Ordovician to Late Permian but most specimens come from pre-Middle Devonian rocks.

In the organization of their life cycle, their photosynthetic pigments, and biochemistry, some of the green algae appear to be the most likely ancestors of the land plants whose fossils first appear in Silurian rocks. Before this time the land surface was barren of vegetation and the processes of weathering must have been different from those of the present day, for some of the processes are controlled by soil acids derived from decaying plants. Possibly some rocks were covered with lichens, symbioses of fungus and algae, that are still widespread. The fossil record of the lichens is insignificant but a questionable fossil from the Proterozoic of South Africa has been attributed to this group.

BASIC FEATURES OF VASCULAR PLANTS

Because none of the cells of algae is far from water, the plant does not need organs to obtain moisture or to exchange metabolic gases. To live on land, plants had to evolve a system of gathering water and nutrients from the soil and pipe-like cells in the stem to conduct these nutrients to all parts of the body. How this vascular tissue evolved from the tissue of green algae is not clear because we have not yet discovered complete fossils that illustrate intermediate steps by which adaptation to land life was made. Scraps of cyclindrical cells that might have served to conduct water and nutrients have been recovered when sedimentary rocks as old as Early Silurian in age have been dissolved in hydrofluoric acid.

Marine plants withdrawn from water soon dry and die. The first land plants must have had the waxy covering for retaining moisture that characterizes modern land plants. However, the surface, while resisting drying, had to allow the passage of metabolic gases, oxygen, and carbon dioxide and regulate the transpiration of water vapor. For these purposes it was pierced by pores called stomates which are capable of opening and shutting in response to changes in environmental conditions. Algae are supported by water but the first vascular plants also required rigid cells that could support the weight of the plant above the ground.

Alternation of Generations:

All vascular plants have similar life cycles but there are many variations. In the cycle's simplest form as illustrated by ferns, the two phases or generations are separate plants. The sexually produced generation has two sets of chromosomes (diploid condition) and is called the sporophyte. The asexually produced generation has one set (haploid) and is called the gametophyte. The large plant (the fern) produces a single type of haploid spore by meiosis in dotlike cases on the leaves. The spores are dispersed and grow on damp ground into minute plants (gametophytes) that release male (sperm) and female (egg) cells from their lower surfaces. When the female and male elements combine along the base of the gametophyte, the fertilized cells produce a new fern plant (sporophyte). This alternation of generations is similar to that of the foraminiferans—one generation is produced asexually by spores, and the next sexually by fusion of sex cells. In more advanced vascular plants two kinds of spores are released, one that forms the female sex cells, and the other, the male cells. In the most advanced vascular plants, the flowering plants and gymnosperms, the transition from asexually produced spores to sperm cells that are contained in the pollen takes place within the pollen-bearing organs (anthers) of the flower and the transition from spores to eggs or ovules takes place in the ovary of the flower. In a sense the gametophytes have become part of the sporophyte.

The bryophytes (mosses, liverworts, and hornworts) are land plants without stiffened vascular tissue for conducting water and nutrients from the soil. They have a small fossil record that begins in the Devonian System. In bryophytes the plant that produces male and female cells (the gametophyte) is the conspicuous one in the life cycle; in vascular plants this phase, if separate, is minute and the large plant represents the spore-bearing phase (the sporophyte).

The first well-preserved land plant, *Cooksonia*, from Middle Silurian rocks records the success of the adaptation of plants to land.

EARLIEST VASCULAR PLANTS

Paleobotanists are not agreed on the time of appearance of the first vascular plants because they do not agree on the significance of fossil trilete spores. These are spores with three scars on their surfaces. Some have regarded these spores as diagnostic of vascular plants and, since they have been reported from rocks as old as Proterozoic in age, have placed the origin of land plants in that eon. However, others have pointed out that some algae and bryophytes produce spores that are much like trilete spores of higher plants and that the possibility of contamination of the Proterozoic samples by modern spores has not been eliminated. Trilete spores, probably of land plants, are known from the Late Ordovician of Libya, but scraps of vascular tissue are not found in rocks older than those of Early Silurian age. The great diversification of trilete spores in Middle Silurian rocks may reflect the rise then of vascular plants.

Cooksonia and other primitive plants are placed in the division Rhyniophyta. The name is derived from the town of Rhynie, Scotland, where a fossil deposit has provided much of our knowledge of the simplest land plants. Here, in Lower Devonian beds, the detailed structures of the earliest members of several plant groups were preserved impregnated with silica, forming chert. The rhyniophytes were short-stemmed plants branching upward from a horizontally creeping root system. The stems were smooth, without leaves, and carried spore cases at their tips (Fig. 8–6). *Psilophyton*, one of the first Devonian land plants to be described, was more complex than the rhyniophytes. Small branches emerged from the main stem and the spore cases grew on the ends of these side branches (Fig. 8–6). In the Zosterophyllophyta the spore cases are arranged along the main stem and many genera have spiny surfaces. Plants of this last division were the ancestors of the major coal-forming plants of the late Paleozoic, the Lycophyta.

Sub-Kingdom—Tracheophyta (vascular plants)
   Division—Rhyniophyta
   Division—Zosterophyllophyta
   Division—Trimerophytophyta
   Division—Lycophyta
   Division—Sphenophyta
   Division—Pteridophyta (ferns)
   Division—Progymnospermophyta
   Division—Pteridospermophyta (seed ferns)
   Division—Cycadophyta (cycads)
   Division—Cycadeoidophyta
   Division—Gingkophyta
   Division—Coniferophyta
     (conifers)
   Division—Anthophyta (flowering
     plants)(angiosperms)

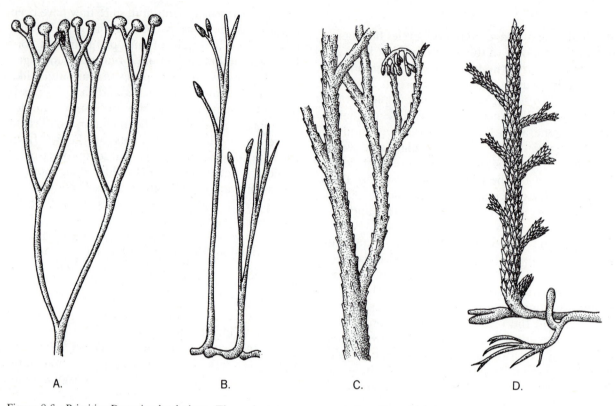

A.   B.   C.   D.

Figure 8-6   Primitive Devonian land plants. These plants were commonly 5 to 20 cm high but *Psilophyton* may have grown as high as 60 cm. A. *Cooksonia*. B. *Rhynia*. C. *Psilophyton*. D. *Asteroxylon*.

LATE PALEOZOIC FLORA

Lycophyta:

In the extensive coal swamp forests of the Carboniferous Period, trees of this division were the largest and most abundant. Coal was formed from partly decayed vegetation slowly transformed by the pressure of overlying sediment into the black rock that is one of our major sources of energy. Most coal today is mined from rocks of Carboniferous and Cretaceous ages and is burned in thermal power stations to make electricity. The details of its plant structure have been obliterated in the decay of plants before burial, during the preservational process, and by compression as the original organic mush was pressed into coal, but where the rock has been infiltrated by calcium carbonate to produce a concretion called a coal ball, the fine structure of the lycophyte trees that formed much of it is well preserved (Fig. 8–7).

The lycophyte trees of the coal swamps grew to heights of 38 meters and to diameters of 2 meters. Their spreading root systems, known by the generic name *Stigmaria*, extended 15 meters from the trunk. Unlike modern trees in drier climates whose roots reach deep to tap the water table, lycophytes had roots that spread out horizontally because the trees grew in swamps.

The trunk of the common lycophytes of the coal swamps had diamond-shaped scars on the surface in a pattern like reptilian scales and for that reason are called *Lepidodendron* (scale tree) (Figs. 8–8, 8–10) and *Lepidophloios* (scale bark). The scars, which are confined to the upper part of the trunk, mark places where leaves have fallen off. The leaves were unlike those of modern deciduous hardwood trees in that they were attached directly to the main trunk and branches. Most of the leaves of lycophytes were short and bladelike but some of *Lepidodendron* were 1 meter long. The tree branched a few times near the top to form a crown, and the spores were carried in cones that hung among the crown branches. In the genus *Sigillaria* the leaf scars appear to be in vertical lines (Figs. 8–8, 8–9, 8–14). Actually, the arrangement of leaf scars in both *Lepidodendron* and *Sigillaria* is helicoidal, that is, like a spring coiling up the trunk. Not all lycophytes were high trees;

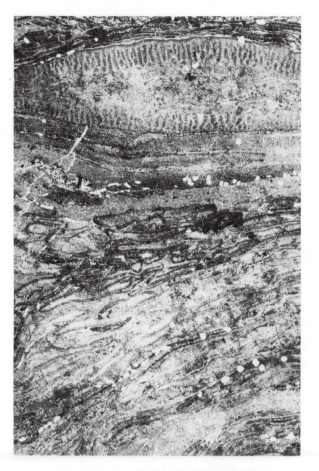

Figure 8-7   Cellular structure of the plant matter making up coal, revealed in a coal ball where the plants have been empregnated with calcium carbonate. The lenslike structure at the top is the stem of a leaf of the seed-fern *Medullosa* and is about 4.5 mm thick. (Courtesy of Redpath Museum, C. W. Stearn.)

some were bushes growing in environments away from the coal swamps.

Lycophytes arose from the Zosterophyllophyta in the Early Devonian and some fossils from the Rhynie chert are believed to be the forerunners of these large trees. They persist to the present day as five genera of clubmosses of which the ground pine, *Lycopodium*, a creeping plant of northern forests, is the most familiar. As the Paleozoic Era closed the lycophytes declined, perhaps owing to the cooler and drier climate of that time and after the Paleozoic were insignificant members of the world flora.

Sphenophyta:

Growing with the towering lycophytes of the coal forests were 20 meter trees that looked like giant relatives of the common horsetail or scouring rush that grows now in marshy places. The horsetail (*Equisetum*) is a living fossil in the sense that it is the only surviving genus of a group of plants that were common in late Paleozoic and early Mesozoic landscapes. These plants that belonged to the division Sphenophyta are characterized by stems regularly interrupted by nodes or joints from which other stems or small leaves, like those of the lycophytes grew. The stem between the nodes was vertically ridged (Fig. 8–11).

The most common genus of coal forest sphenophytes is *Calamites* (Figs. 8–10, 8–11, 8–14). Many of the fossils of this tree are fillings of the central hollow in the stem. The circulets of leaves that grew at the joints of *Calamites* stems have been given the generic names *Annularia* or *Asterophyllites* (Fig. 8–12). Different parts of fossil plants are commonly separated before preservation and when found are given different names until they can be associated with a species or genus already named. Intermediate forms between sphenophytes and primitive plants of the Early Devonian have been found in Middle Devonian rocks.

Ferns (Pteridophytes):

The leaves of ferns bear spores, and the two generations are separate plants. Most ferns are small plants but ferns today in the tropics grow to treelike proportions and grew in Carboniferous coal swamps to similar sizes. They have been common plants since the Late Carboniferous but are confined to damp environments because the sperm require a film of water to reach the eggs at the base of the gametophyte. Not all the fernlike leaves preserved in sediments associated with coals belonged to true ferns—several different groups had fernlike leaves in the Carboniferous swamps.

Progymnosperms:

This group is believed to be ancestral to the conifers because the wood of the stems is much like that of conifers, but the leaves are not like the needles of conifers. The wood of the Middle Devonian tree *Archaeopteris* (Fig. 8–13A) was first classified as a conifer and the leaves as a fern before it was realized that they belonged to the same plant that shared these features. In Middle Devonian sandstones of New York, stumps of the progymnosperm *Aneurophyton* that indicate a tree 12 m high have been excavated at the Gilboa reservoir (Fig. 8–13B).

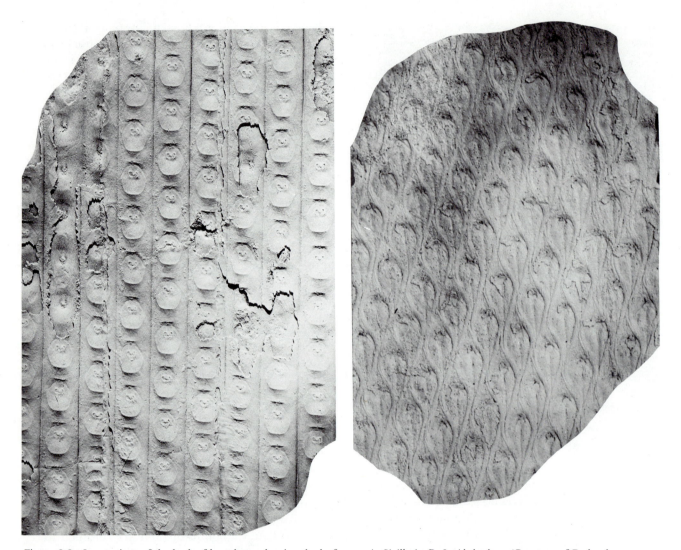

Figure 8-8   Impressions of the bark of lycophytes showing the leaf scars. A. *Sigillaria*. B. *Lepidodendron*. (Courtesy of Redpath Museum.)

Seed Ferns (Pteridospermophytes) and the Origin of Seeds:

All the plants so far discussed distributed small spores that grew into a minute separate plant in which both types of sex cells were formed. Spores that have been separated by acid solution of sediments show that many plants were dispersing two kinds of spores by later Devonian time, large megaspores that gave rise to female gametophytes and small microspores that gave rise to male gametophytes. Seeds evolved when the formation of the eggs and sperm no longer took place outside the tissue of the parental sporophyte. Spores have a single set of chromosomes (haploid) and are formed by meiosis in the sporophyte. In seed plants (the gymno-

sperms and angiosperm) the microspores develop into haploid pollen grains and the megaspores develop into female sex cells in ovules while still attached to the plants (the sporophyte). In most, both types of sexual organs are attached to the same sporophyte (for example, the male and female cones of conifers), in others the sporophytes themselves may be male or female. When the pollen reaches the ovule and the sperm in the pollen fertilizes the female gametophyte, it develops into a seed. The seed in its capsule then may enter a resting stage lasting many months before it gives rise to an new plant. The oldest fossil seeds come from the Upper Devonian of West Virginia and Belgium.

The seed ferns were fernlike in foliage but seeds hung from the tips of their branches. They are

Figure 8-9   Reconstruction of the lycophyte *Sigillaria*.

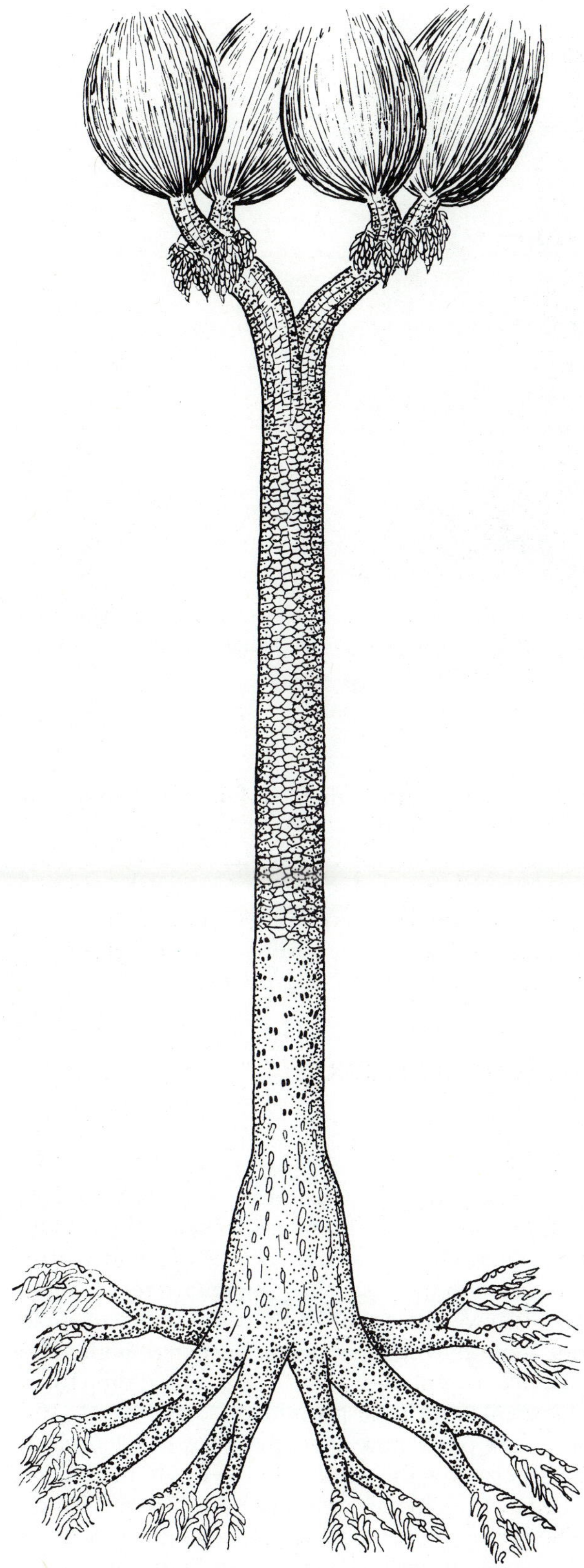

Sigillaria

Figure 8-10   Reconstruction of the coal swamps of the Carboniferous Period. Tall branched *Lepidodendron* trees dominate the scene. *Calamites*, with whorls of leaves and many fernlike plants, are prominent in the swamp. (*Source:* Mural by G. A. Reid, courtesy of the Royal Ontario Museum.)

difficult to distinguish from ferns on the basis of foliage alone. They were important contributors to the undergrowth of the coal swamps and were the dominant plants of floodplains and uplands. They survived into mid-Mesozoic time.

Conifers and Cordaites:

Conifers are the familiar cone-bearing evergreens of northern forests, the pines, spruces, and hemlocks. Their leaves are typically needlelike or scalelike. The plants have small cones that release pollen to the wind and larger cones in which the ovules are fertilized by that pollen and grow into seeds that are released when the cone opens. The reproductive cycle is a slow one in some species, taking up to two years to complete. The first conifers appeared in Late Carboniferous time and the group became abundant and diverse in late Paleozoic and early Mesozoic time (Fig. 8–14). The most familiar conifer fossils (*Araucarioxylon*) are the silicified logs of the Petrified Forest in Arizona (Fig. 8–15). These Triassic trees were apparently rafted to the site of their burial, for no traces of leaves or cones have been found in the sediments that enclose them. By mid-Jurassic time all the major families of conifers

that are found in the modern fauna had appeared in the fossil record.

The Cordaites are Carboniferous and Permian cone-bearing trees descended from early gymnosperms. The cones were carried on small branches among the leaves which were up to 1 meter long and shaped like straps (Fig. 8–14).

MESOZOIC FLORA

Cycads:

The cycads are squat palmlike trees that grow in Central America and the southern hemisphere. Only ten genera are living but their leaves and trunks are common in Mesozoic sedimentary rocks. Some cycads have short trunks shaped like a pineapple but some have trunks up to 10 meters high. The palmlike compound leaves are attached directly to the trunk and leave prominent scars (Fig. 8–16). The male plants have cones that release pollen which is dispersed by the wind or by beetles to the female plants whose ovules are borne on different cones. The cycads are descendants of seed ferns of the late Paleozoic but did not give rise to higher plants.

Figure 8-12   *Asterophyllites*, the leaves of an arthrophyte such as *Calamites*. Carboniferous. The specimen is about 15 cm long. (C. W. Stearn, Redpath Museum.)

Figure 8-11   Jointed stem of *Calamites*. This is a sediment filling the hollow interior of the stem. (Courtesy of the Field Museum of National History (Geo5373), Chicago.)

Cycadeoids:

The cycadeoids are an extinct group with squat trunks covered with leaf scars similar to those of the cycads. These silicified trunks are common in Cretaceous sediments of the Black Hills of South Dakota. The cones are embedded in the trunks and contain both the male and female organs like the flowers of more advanced plants. Some reconstruc-tions show the cycadeoids covered with flowers but there is little evidence that they flowered and some evidence that the cones were closed at the top and the plant practiced self-fertilization. The cyca-deoids became extinct at the close of the Cretaceous without giving rise to higher plants.

Flowering Plants (Anthophyta or Angiosperms):

In the seed plants considered so far, the egg is fertilized directly by sperm derived from the pollen and is not encapsulated in an ovary of a flower. Such seed plants are called gymnosperms because the seed is naked (Greek—*gymna* = naked). The

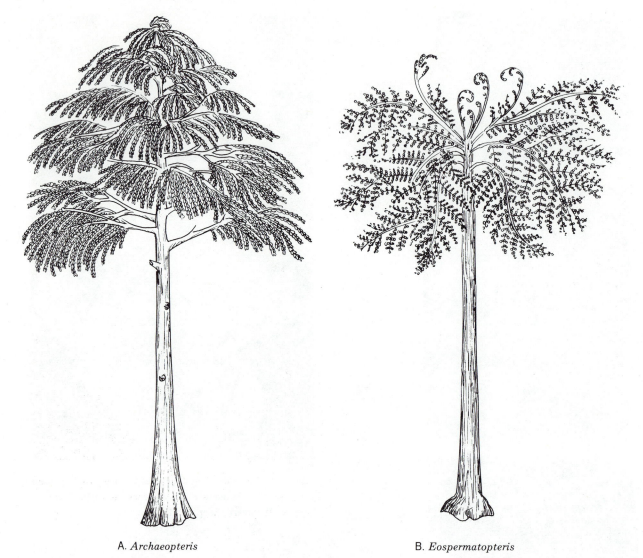

A. *Archaeopteris*   B. *Eospermatopteris*

Figure 8-13 Reconstructions of the Late Devonian progymnosperms. A. *Archaeopteris*. B. *Eospermatopteris* and *Aneurophyton*. The trunk of this plant was given the first name and the foliage that is associated with it, the second.

gymnosperms came to dominate plants that dispersed spores for reproduction in late Paleozoic time, but they in turn were dominated in Cretaceous time by the flowering plants. These plants now make up more than 95 percent of the species of vascular plants and include practically all the plants that are useful to humans for food.

Although a strict definition of the flowering plants is difficult, they can be characterized generally as having the following features:

1. The eggs are fertilized by sperm from a pollen tube growing toward them from a collecting organ, the stigma.
2. The ovary is within a flower which evolved as an organ for attracting insects. The success of the flowering plants has depended on insect pollination.
3. Both male and female organs are contained in the flower.
4. Fertilization and the formation of seeds is rapid compared to that in the gymnosperms, usually taking only a few weeks.
5. The pollen tube brings two sperm cells to the

Figure 8-14 Reconstruction of Texas in Permian time. The large tree in the center foreground is *Cordaites*. Three tall *Sigillaria* trees grow nearby. At the extreme right is a *Lepidodendron* and in the distance the early conifers (*Walchia*). The mammal-like reptiles *Dimetrodon* and *Edaphosaurus* lurk in the *Calamites* thickets. (Courtesy of the New York State Museum.)

female gametophyte. One fertilizes an egg cell; the other unites with other nuclei in the gametophyte to produce nutritive tissue with three sets of chromosomes. The seed when released is supplied with a coating of this nourishing tissue that acts as an initial food source as it grows into a new adult plant. This coating is an important component of human diet.

The flowering plants burst upon the world scene in the middle of the Early Cretaceous Epoch. Their

Figure 8-15 Logs of Triassic conifer *Araucarioxylon* in the Chinle Formation, Petrified Forest National Monument, Arizona. (Courtesy of the U.S. Geological Survey.)

Figure 8-16   Reconstruction of an early Mesozoic landscape. In the marshy foreground are several kinds of *Calamites*. Ferns and cycadeoids on bulbous stems are common. The reconstruction of cycadeoids as flowering plants has been questioned. Conifers dominate the background. (*Source:* Reconstruction by J. Augusta and Z. Burian, from Prehistoric Animals, Spring Books, 1956.)

Figure 8-17   Leaf of the angiosperm plant *Liquidambar* from the Cretaceous of the western United States. (Courtesy of the Smithsonian Institution.)

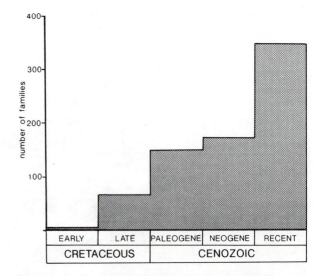

Figure 8-18   Rapid diversification of the anthophytes (angiosperms) in the Late Cretaceous and early Cenozoic. (*Source:* Modified from N. F. Hughes, *Paleobiology of Angiosperm Origins*, 1976.)

appearance without obvious ancestors and their dramatic diversification have posed problems for paleobotanists since the time of Darwin, who called the event "an abominable mystery". The stages of the Lower Cretaceous Series are, from the base upwards, the Neocomian, Barremian, Aptian, and Albian. In Barremian beds a few of the thin flat angiosperm leaves and some pollen believed to be from flowering plants have been found. In Aptian beds several types of angiosperm leaves occur and in the last stage of the Lower Cretaceous, the Albian, these leaves and the characteristic pollen become abundant (Fig. 8–17). During the Late Cretaceous, 67 families of the division Arthrophyta can be recognized in the fossil record (Fig. 8–18).

The identity of the plant group that was ancestral to the flowering plants remains a mystery. Almost certainly the ancestor was a gymnosperm and the seed ferns, the conifers, and the cycadeoids have all been proposed for the role. The flowering plants may not be a monophyletic group, that is, they may have arisen from more than one ancestor. However, no intermediate forms have been discovered to link them with potential ancestors. The sudden success of these advanced plants has been attributed to their more efficient physiology, their more rapid rate of reproduction and greater fertility, their evolution of a partnership with insects in reproduction, their greater adaptability to a range of conditions, and to physical changes in the Early Cretaceous environment. The rise of the angiosperms is likely to have been the result of a combination of these factors.

A relatively late event in angiosperm history had major effects on the evolution of the mammals. In Miocene time the group of flowering plants that we call grasses diversified and rapidly spread across the dry central plains of the continents. In pre-Miocene time most mammals were browsers, that is, they ate soft leaves and flowers and their teeth were not subjected to much abrasion. Grass provided a new source of nourishment to the mammals of the plains but it was nourishment that was hard on the teeth, not only because grass itself is abrasive (it secretes silica), but also because it traps a lot of dust in semiarid climates. The animals that adapted to this new food source are called grazers and they were characterized by high crowned teeth that could resist a lifetime of wear from grass-eating. The story of the grazers, such as the horses and elephants, is continued in Chapter 12.

SUMMARY

The history of plants is marked by four major events. The first was the evolution of the multicelled algae from the unicellular eukaryotes of mid-Proterozoic time. The second was the evolution of vascular tissue that allowed the descendants of algae to live efficiently on land. These first land plants reproduced by means of spores and had two separate generations, one sexual and the other asexual. The third event was the evolution of the seed which was fertilized by pollen. Male and female elements were formed from spores within plants, not in a separate generation. The fourth was the enclosing of the seeds in nutritive tissue and the use of insects attracted by a flower to aid fertilization.

SUGGESTED READINGS

Chuvashov, B., and Riding, R., 1984, Principal floras of Paleozoic marine calcareous algae: Palaeontology, v. 27, p. 487–500.

Doyle, J. A., 1977, Patterns of evolution in early angiosperms. In Hallam, A. (ed.), Patterns of evolution as illustrated in the fossil record: New York, Elsevier, p. 501–546.

Edwards, D., Bassett, M. G., and Rogerson, E. C. W., 1979, The earliest vascular land plants: continuing the search for proof: Lethaia, v. 12, p. 313–324.

Friiss, E. M., Chaloner, W. G., and Crane, P. R. (eds.), 1987, The origin of angiosperms and their biological consequences. Cambridge, Cambridge Univ. Press, 358 p.

Gastaldo, R. A. (ed.), 1986, Land Plants: notes for a short course: University of Tennessee, Studies in Geology, No. 15, 226 p.

Gensel, P. G., and Andrews, H. N., 1987, The evolution of early land plants: American Scientist, v. 75, p. 468–477.

Grey, J., and Boucot, A. J., 1977, Early vascular land plants: proof and conjecture: Lethaia, v. 10, p. 145–174.

Hughes, N. F., 1976, Paleobiology of angiosperm origins: Cambridge, Cambridge Univ. Press, 242 p.

Meyen, S. V., 1987, Fundamentals of Paleobotany: New York, Methuen, 432 p.

Norstog, K., 1987, Cycads and the origin of insect pollination: American Scientist, v. 75, p. 270–279.

Spicer, R. A. and Thomas, B. A. (eds.), 1986, Systematic and Taxonomic Approaches to Paleobotany: Systematic Association Special Volume 31.

Taylor, T. M., 1981: Paleobotany, an introduction to fossil plant biology: New York, McGraw–Hill, 581, p.

Thomas, B. A. and Spicer, R. A., 1987, The evolution and paleobiology of land plants: Portland, Dioscorides Press, 309 p.

Stewart, W. N., 1983, Paleobotany and the evolution of plants: Cambridge, Cambridge Univ. Press, 405 p.

Wray, J. L., 1977, Calcareous algae: Amsterdam, Elsevier, 185 p.

CHAPTER 9

Paleozoic Vertebrates

THE ORIGIN OF VERTEBRATES

The vertebrates were the last major group of metazoans to appear in the fossil record. Isolated fragments of a bonelike material have been described from the Upper Cambrian (l), but the oldest fossils that can be unquestionably identified as fish are of Middle Ordovician age (2).

The ancestors of vertebrates had probably evolved from other metazoans in the Proterozoic, but they had little likelihood of preservation until they evolved the capacity to form bony exoskeletons.

In the absence of a fossil record of the earliest vertebrates and their immediate ancestors, we must look for distinctive features in later fish to establish their relationship to other animals. Among the distinctive features of primitive living vertebrates are (1), a longitudinal rod, the notochord, which occupies the position of the vertebral column in terrestrial vertebrates such as mammals, (2), a dorsal hollow nerve cord, and (3), numerous gill slits in the area of the pharynx (Fig. 9–1). These structures are present in living protochordates, which are clearly more primitive than vertebrates in lacking a brain or paired sense organs of the head. Because they

have a notochord, the protochordates are placed in the phylum Chordata with the vertebrates. Protochordates comprise the subphyla Cephalochordata, Urochordata and Hemichordata.

Phylum—Chordata
    Subphylum—Hemichordata
    Subphylum—Urochordata
    Subphylum—Cephalochordata
    Subphylum—Vertebrata

The protochordates and vertebrates are united by unique features of their early development (discussed in Chapter 5), including the pattern of formation of the coelom, mesoderm, and mouth that distinguish them from the major groups of segmented invertebrates—the molluscs, annelids, and arthropods—and suggest that these two major groups evolved separately from primitive metazoans.

The modern cephalochordates, typified by *Branchiostoma* (amphioxus), provide a good model for the type of animal that may have been immediately ancestral to vertebrates. Amphioxus is locally common in coastal marine waters in many parts of the

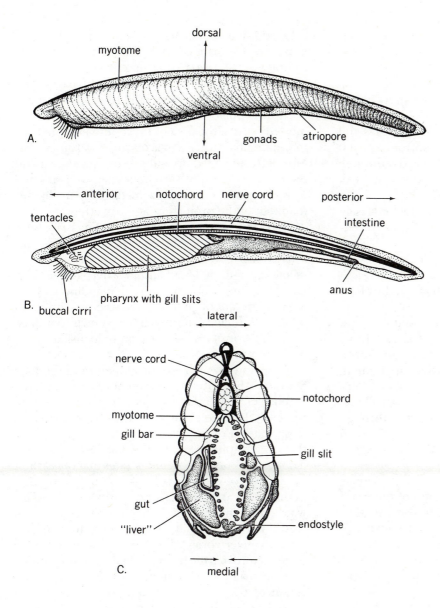

dorsal

myotome

ventral

gonads

atriopore

A.

anterior

tentacles

notochord

nerve cord

posterior

intestine

pharynx with gill slits

anus

B. buccal cirri

lateral

nerve cord

notochord

myotome

gill bar

gill slit

gut

"liver"

endostyle

C.

medial

Figure 9-1 Amphioxus, showing the major features of chordates, and the positional terms used in their description. A. External view of amphioxus. B. Longitudinal section. C. Transverse section.

world. It is bilaterally symmetrical and resembles a small, transparent fish. As an adult, it lives partially buried in the sand and feeds by filtering small particles of food out of the water. The feeding structures of all invertebrate chordates and the larval stage of the lamprey, one of the most primitive living fish, indicate that the ancestors of the vertebrates were filter feeders.

Although amphioxus and the larval stage of the lamprey are sedentary, their body shape and pattern of musculature resemble those of actively swimming fish. Much of the body is formed by a series of muscles that are attached to transversely oriented septae anchored to the notochord. Waves of contraction on alternate sides of the body, running from the head to the tail, produce sinuous lateral undulations that move the animal forward in the water. The arrangement of the muscles and accompanying nerves and blood vessels gives a partially segmented appearance to the body of chordates, but it is not comparable to the metamerism of arthropods and annelids which involves the repetition of internal organs and appendages.

Not only do the muscles, notochord, and nerve cord of amphioxus resemble those of vertebrates, but the pharynx and circulatory system are similar as well. As in fish, the pharynx in amphioxus is pierced by lateral slits. Great amounts of water are drawn in through the mouth by the action of cilia and pass back into the pharynx. Food particles are trapped by mucus secreted from the base of the pharynx. Water is discharged through the slits and enters an external space, the atrium, from which it passes out posteriorly. Food particles pass through the simple gut behind the pharynx, and wastes exit via the anus at the base of the tail.

The geometry of the circulatory system is closely associated with that of the pharynx. The main ventral vessel, the ventral aorta, lies at the base of the pharynx. From the aorta, the branchial arteries extend dorsally between the pharyngeal slits to join the anterior dorsal aortae, through which the blood flows posteriorly. Although the geometry of the vascular system resembles that of primitive vertebrates, it is more primitive in that it lacks a heart and capillaries. The arteries and veins are linked by blood sinuses. Oxygen is carried in solution, without specialized oxygen-carrying pigments, and there are no blood cells. Amphioxus is so small that exchange of respiratory gases can occur through the body surface, and no filamentous gills, such as fish have, are associated with the pharyngeal slits.

Most of the major organ systems in amphioxus are nearly ideal models for the origin of the more complex vertebrate structures. However, neither living animals nor the fossil record give evidence how these basic structures originated. The other groups of protochordates show various chordate traits, but they do not illustrate a simple evolutionary progression.

Cephalochordates are rarely found as fossils, but they have been reported from the Middle Cambrian Burgess shale, and from the Devonian of South Africa (3,4). These fossils superficially resemble amphioxus but provide no additional information as to the origin of chordate characters or vertebrates.

Simple, bilaterally symmetrical animals ancestral to the cephalochordates may have diverged from the early metazoan radiation in the Early Cambrian. These animals evolved an effective method of swimming using segmented muscles anchored to the notochord and a more efficient method of filter feeding by pharyngeal slits. From this lineage diverged the modern cephalochordates and the vertebrates.

In addition to having the anterior end of the nerve cord expanded as a brain associated with complex sense organs for smell (the olfactory or nasal capsules), sight (the eyes), and balance (the inner ear, or otic capsule), vertebrates are advanced over the cephalochordates in the elaboration of the musculature in the area of the mouth and pharynx to supplement the cilia in acquiring food and water. More effective feeding and sensory structures and the complex brain of vertebrates can be associated with the achievement of a metabolic rate higher than that of the protochordates. Invertebrate chordates (protochordates) are not able to increase their metabolic rate to permit short periods of vigorous activity. Vertebrates, in contrast, can increase their metabolic rate at least tenfold when active. This enables them to occupy more demanding ways of life that are not open to the invertebrate chordates (5). Unfortunately, no fossils are known that illustrate the transition between amphioxuslike chordates and vertebrates.

Vertebrates remained uncommon until the Late Silurian. Alfred Romer and others argued that the absence of vertebrates in typical marine deposits in the early Paleozoic was a result of their origin in fresh water. Freshwater deposits are uncommon in the stratigraphic record until Late Silurian time. The fossil record now demonstrates that the earliest members of most groups of primitive fish lived either in shallow nearshore marine deposits, or in brackish water, and only later did they become common in freshwater or normal marine environments. All modern invertebrate chordates are also restricted to nearshore marine habitats. This evidence implies that the origin of vertebrates, involving the elaboration of the brain and cranial sense organs, occurred within the marine environment.

JAWLESS FISH

Vertebrates are divided into two major groups, the Agnatha, or jawless vertebrates including the living hagfish and lamprey (Fig. 9–2), and the jawed vertebrates, or Gnathostomes.

In Upper Silurian and Lower Devonian beds, many distinct groups of early vertebrates are found, but the Ordovician and Lower Silurian fossils are so rare that they do not allow a coherent phylogeny to be formulated. The capacity to form a preservable bony skeleton apparently evolved separately

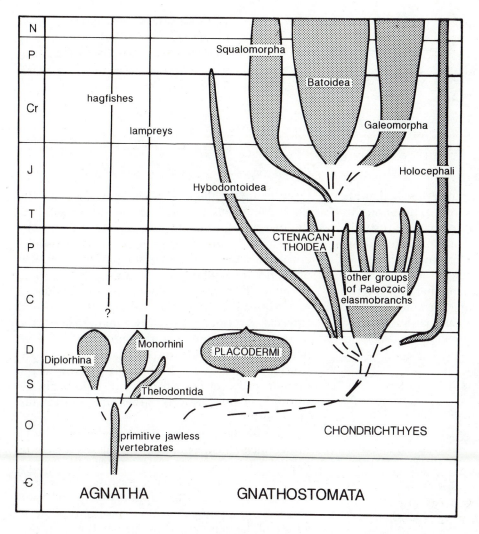

Figure 9-2   Pattern of radiation of fish (excluding Osteichthyes).

in several lineages after the first major radiation of vertebrates. The resulting lineages are all distinct from one another and provide little evidence of their interrelationships.

The first vertebrates were without jaws. Like their ancestors among the invertebrate chordates, they may originally have fed by filtering small prey or particles of food either from the open water or from the substrate. As their size and metabolic requirements increased, they may have become active predators on larger prey, even though they still lacked jaws.

A wide range of jawless vertebrates are known from the Ordovician, Silurian, and Devonian; all are referred to in an informal way as ostracoderms (meaning shell-skinned in Greek) because most are covered with a nearly continuous bony exoskeleton.

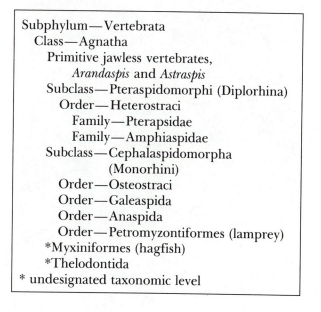

Subphylum—Vertebrata
    Class—Agnatha
        Primitive jawless vertebrates,
            Arandaspis and *Astraspis*
        Subclass—Pteraspidomorphi (Diplorhina)
            Order—Heterostraci
                Family—Pterapsidae
                Family—Amphiaspidae
        Subclass—Cephalaspidomorpha
                (Monorhini)
            Order—Osteostraci
            Order—Galeaspida
            Order—Anaspida
            Order—Petromyzontiformes (lamprey)
        *Myxiniformes (hagfish)
        *Thelodontida
* undesignated taxonomic level

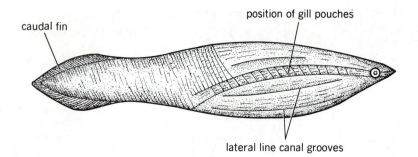

caudal fin

position of gill pouches

lateral line canal grooves

Figure 9-3   The earliest vertebrate, *Arandaspis* from the Middle Ordovician of Australia. (Modified from A. Ritchie & J. Gilbert-Tomlinson, *First Ordovician vertebrates from the Southern Hemisphere.*)

The earliest and most primitive fossils that are clearly those of vertebrates belong to the genus *Arandaspis*, from the Middle Ordovician of Australia. The body reached a length of approximately 14 cm. It was covered anteriorly by a continuous bony carapace and posteriorly by smaller scales (Fig. 9–3). There were no paired, dorsal, or anal fins. A series of gill covers was present along the anterior flank. Remains of a second type of early jawless fish, *Astraspis*, are preserved by the millions in beach de-

posits along the flanks of what were to become the Rocky Mountains, but were, in mid-Ordovician time, the margin of an inland sea. The carapace of *Astraspis* (Fig. 9–4) was made up of numerous polygonal plates that increased in size during the growth of the fish through the addition of smaller marginal elements. This was one solution to a problem faced by early fish: how to grow and yet retain a solid external covering. Many early genera delayed the secretion of the bony covering until the fish were nearly completely grown. Members of more advanced families evolved the ability to grow at the margins of larger bony plates and could begin ossification at progressively earlier stages in the life of the fish.

None of the primitive fish had the capacity to form bone in the deeper layers of this tissue as most modern vertebrates can. Bone clearly served first for protection and only later was used for muscle attachment and internal support of the body and limbs. Bone is composed of calcium phosphate in a form of the mineral hydroxyapatite. Bone is nearly twice as strong as calcium carbonate, which forms the exoskeleton of most invertebrates; it may have been laid down originally as a means of storing phosphates which are vital to many body functions and are in fluctuating supply in nature.

Heterostracans:

Among the most common of the primitive jawless fish were the heterostracans or pteraspidomorphs (Fig. 9–5). Like *Arandaspis*, the anterior portion of the body was covered by a rigid carapace, but the bony plates in most species have growth lines along the margins. Paired openings for the eyes and notches at the front of the head shield show that these animals had sense organs comparable to those of modern vertebrates. The inside surface of the dorsal shield shows impressions of vertical semicircular canals for establishing the direction of move-

Figure 9-4   Dorsal armor of *Astraspis*, showing the mosaic of small bony plates. The entire fish was several cms long. (Photograph courtesy Dr. Tor Ørvig.)

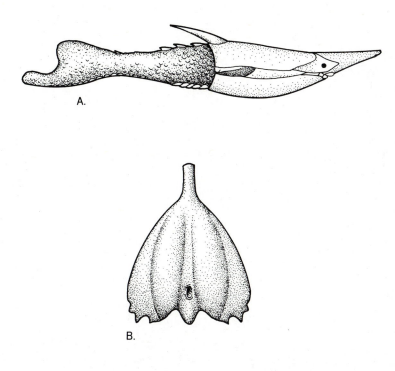

A.

B.

Figure 9-5   Pteraspidomorphs. A. Body of the pteraspid *Pteraspis*. B. Head shield of the amphaspid *Eglonaspis* (note absence of eyes) (J. A. Moy-Thomas & R. S. Miles, *Palaeozoic Fishes*. Courtesy Chapman & Hall, London.).

ments. The semicircular canals, which occupy the inner ear in modern vertebrates, are filled with fluid whose movements displace hairlike structures that send impulses to the brain. Similar structures occur in the lateral line canals of fish which extend along the flanks of the body to the base of the tail and form distinctive patterns along the head and jaws.

Another sensory system that is present in many early fish had the capacity to respond to electrical fields. Flask-shaped structures, revealed by microscopic study of the bones of the head and scales of Paleozoic fish, resemble those of modern species that are used to locate prey and avoid obstacles. These are especially useful in dark or turbid waters. Modern sharks use electrical detectors to locate the electrical impulses emitted during muscle contractions by prey buried in sediments.

The most common heterostracians, the pteraspids, evolved long lateral processes from the carapace that may have helped provide stability in the absence of paired fins. Most of these fish were dorsoventrally flattened, which suggests that they lived on the bottom. Another group, the amphiaspids, are believed to have spent their lives buried in sediments, since the eye openings were covered with bone.

Heterostracan fish have been used by David Dineley (6) to establish correlation between Silurian and Devonian beds in North America and Europe.

The apparently paired nature of the nasal capsules and the absence of internal bone are primitive features of the heterostracians, but they are advanced, relative to other early vertebrates, in having a single external opening for the numerous gill pouches.

Cephalaspidomorphs:

A second major group of primitive armored fish is the Cephalaspidomorpha. This group is also called the Monorhini because the nasal opening is single rather than paired. It opens on top of the head and is connected to the pituitary gland by a common duct. No satisfactory explanation for the evolution of this condition has been proposed, but the presence of a similar pattern in the modern jawless vertebrates, the lamprey and hagfish, suggests that they are closely related to the Cephalaspidomorpha.

Three groups of early armored fish have the monorhine condition: the anaspids, which had a fusiform body for open water swimming, and two groups of flattened bottom dwellers, the osteostracans and the galeaspids (Fig. 9–6). The osteostracans, of which the best known species are referred to as cephalaspids, are the most completely preserved of all the early jawless vertebrates because they had a heavily ossified internal skeleton in the head region. Specimens that have been naturally

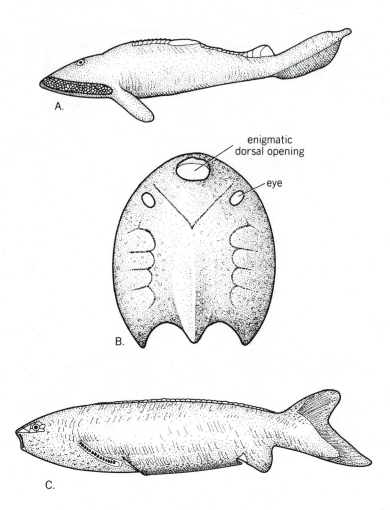

Figure 9-6   Cephalaspidomorphs. A. Osteostracan, B. Head shield of galeaspid, C. Anaspid.

weathered or cut in sections show a multitude of internal structures, including the outline of the brain, the position of the cranial nerves, and two pairs of semicircular canals. Unique to this group are ducts leading from the inner ear to areas on the lateral and dorsal surface of the carapace that may have been sensitive to changes in the pressure of the surrounding water and could therefore warn of the approach of predators (Fig. 9–7).

The surface of the pharynx shows that the many gill pouches opened directly to the exterior rather than via a common duct, as in the Pteraspidomorpha. Some cephalaspids were specialized in the presence of paired anterior (pectoral) limbs just behind the lateral processes of the head shield (Fig. 9–8). No internal skeleton of these appendages has been described and it is uncertain how similar they were to those of more advanced fish. None of the jawless fish had pelvic fins.

The galeaspids resemble the osteostracans in that they have an internal head skeleton and a flattened body, but differ in that they lack cranial sensory fields and have a large opening of unknown function on the upper surface of the skull in front of the eyes. The galeaspids were first recognized as a separate group only a few years ago, but they have since been found to be an extraordinarily diverse assemblage. Galeaspids are restricted to China but range from the Lower Silurian to the Lower Devonian.

The anaspids resemble the galeaspids and osteostracans in the medial position of the nasal opening, but in few other features. Their body was fusiform, suggesting open-water swimming, but the tail, unlike most advanced fish, was bent ventrally. Most anaspids were covered with narrow scales that are presumed to have followed the pattern of underlying muscle segments. Some anaspids, in contrast, have reduced or entirely eliminated the scales. In this they resemble the living jawless fish, the hagfish and lamprey (Fig. 9–9). Primitive members of both of these groups are known from the

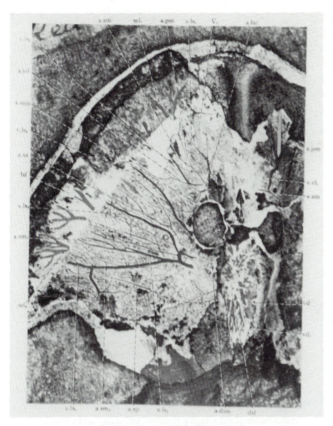

Figure 9-7   Weathered skull of cephalaspid, showing internal anatomical details; lines point to particular nerves and blood vessels. (E. Stensiø, *The Cephalaspids of Great Britain.* By courtesy of the British Museum (Natural History).)

Middle Pennsylvanian of Mazon Creek, Illinois. The living jawless fish differ from their Paleozoic counterparts in that they lack bones and scales; for this reason their fossil record is primarily restricted to this one locality, which also includes a wealth of soft-bodied invertebrates unknown elsewhere (7).

The well-known pteraspidomorphs and cephalaspidomorphs are unlike most modern fish in that they have external armor and peculiar finlike structures. The most enigmatic but possibly most important group of early jawless fish are the coelolepids or thelodonts. Their entire body was covered

Figure 9-8   Progressive evolution of paired fins in osteostracans.

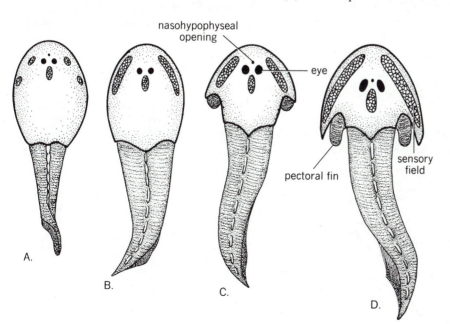

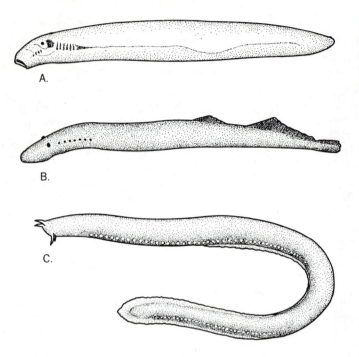

Figure 9-9   Fossil and living agnathans. A. *Mayomazon*, from the Upper Carboniferous, B. Living lamprey *Petromyzon*, C. Living hagfish. (Modified from D. Bardack & R. Zangerl, *First Fossil lamprey: A record from the Pennsylvanian of Illinois*.)

with small scales, but they had no internal skeleton; the fossils are typically so flattened and distorted that few structural details are evident. The body was apparently fusiform, with a slender lateral fold or pectoral fin above a series of gill slits (Fig. 9–10). The thelodonts had a body form that would be appropriate for the ancestry of both jawless and jawed vertebrates. They are known from Lower Silurian to Upper Devonian rocks.

The jawless vertebrates were already diverse when they appeared in the fossil record in the Ordovician. They continued to radiate throughout the Early Devonian, but by the end of that period they had become extinct, except for the lineages that led to the surviving lamprey and hagfish.

JAWED FISH

Although the lamprey and hagfish have a wide distribution today, jawless vertebrates were effectively replaced by jawed vertebrates within the Early Devonian. None of the well-known groups of jawless vertebrates are likely ancestors of the jawed forms. Bobb Schaeffer and Keith Thomson (8) have emphasized that the two groups differ fundamentally in the structure and relation of the gill supports and the filamentous gills. In the modern lamprey and hagfish, the gill supports are lateral to the gill filaments and form a flexible basketlike structure that is compressed by surrounding muscles and expanded by elastic recoil. Jawed vertebrates, in contrast, have the gill supports medial to the gill filaments, and the gill supports are in the form of a series of hingelike elements (Fig. 9–11) whose associated muscles can both contract and expand the pharynx. The different relative position of gill supports and filaments in the two groups suggests that they evolved separately from a common ancestor, like amphioxus, in which the gill slits were not accompanied by filamentous gills.

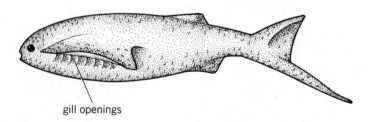

gill openings

Figure 9-10   Thelodont, a primitive vertebrate with paired fins and multiple gill openings. (Courtesy of L. B. Halstead, S. Turner and Elsevier Science Publishers.)

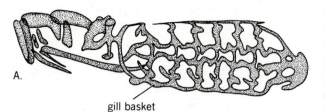

A.

gill basket

Figure 9-11   Nature of gill supports and relative position of gills in jawless and jawed fish. A. Superficial gill basket in the modern jawless lamprey, B. Medially placed hinged gill supports and jaws of a shark, C. Horizontal section through the head of an agnathan to show position of gills, D. Horizontal section through the head of a gnathosome, showing the position of the gills. (J. A. Moy-Thomas & R. S. Miles, *Palaeozoic Fishes*. Courtesy Chapman & Hall, London [C,D].

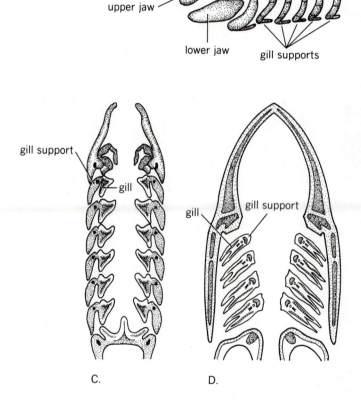

B.

hyoid arch

upper jaw

lower jaw

gill supports

gill support

gill

gill   gill support

C.   D.

The general form of the gill supports in jawed vertebrates resembles the jaws. This is particularly clear in sharks because the primary gill supports and jaws are not covered by superficial bone as they are in bony fish. Primitive vertebrates were thought to have had a series of similar gill supports in the mouth region, whose anterior elements became specialized as jaws. Study of modern jawless fish and Paleozoic fossils shows that the position of the Vth nerve, which innervates the mouth region, is comparable in all vertebrates. Its position suggests that the mouth had a similar position in both jawless and jawed vertebrates and that there never were gill pouches in the mouth area, as would be implied by the theory that the jaws had evolved from gill supports. Malcolm Jollie (9) suggested that the jaws and gill supports evolved as broadly similar structures since they had comparable functional relationships with the muscles that expanded and contracted both the mouth and the pharynx.

The early jawed fish also possessed both pectoral and pelvic fins and an expanded anterior portion of the brain. This suggests that feeding, locomotion, and enhanced neural control were all important aspects in the origin of jawed vertebrates.

Placoderms:

The oldest remains of jawed vertebrates have been collected from Lower Silurian rocks, and several major groups are known from the Upper Silurian and Lower Devonian. The first group of jawed fish to radiate extensively were the placoderms (Fig. 9–12). Unique to this group was the placement of the jaw closing muscles, which were medial rather than lateral to the primary bone of the upper jaw.

Most placoderms were bottom-dwelling fish like the ostracoderms, and the primitive genera were similarly covered with a nearly continuous carapace of bone, with only the tail free. Placoderms are

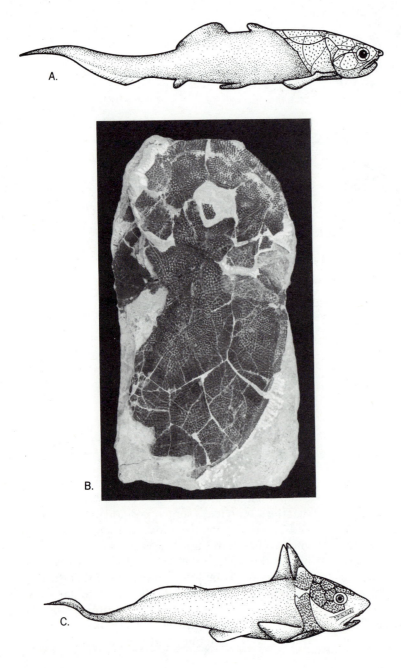

Figure 9-12   Placoderms. A. Reconstruction of *Coccosteus*, a Middle Devonian arthrodire, B. Fossil of the Upper Devonian antiarch *Bothriolepis*, from eastern Canada, dorsal view of head shield and thorax, C. A ptyctodont. (Courtesy of J. Piveteau, *Traité de Paléontologie*, IV, 2 (c) MASSON 1969 [A,C] and Courtesy of Redpath Museum, McGill University [B].)

clearly distinguished from ostracoderms by a joint separating the head from the thorax, which allowed the top of the head to be lifted in opening the mouth. This is presumably an adaptation for bottom-dwelling fish that could not conveniently open the lower jaws.

The largest group of placoderms were the arthrodires. Within this group, numerous lineages show adaption to the role of active, open water predators. The thickness of their armour was reduced, the internal skeleton was replaced with cartilage, and the back of the bony carapace was shortened so that more of the trunk could serve as a flexible swimming organ. Some genera lost the joint between the skull and thorax, and the lower jaw increased in length to provide a wider gape. The largest arthrodires exceeded 3 meters in length and must have been formidable predators.

Other placoderms, such as the antiarchs, accentuated a bottom dwelling way of life. Antiarchs are unique in being very common in fresh water and were the only placoderms to survive into the Carboniferous.

The ptycnodonts (Fig. 9–12C) show the closest resemblance to more advanced jawed fish and have been hypothesized to be closely related to the chimaeroids (Fig. 9–15), which are typically allied with the sharks. Most paleontologists feel that these resemblances are due to convergence, rather than close relationship. Another group of placoderms, the rhenanids, superfically resemble the skates and rays of the Mesozoic and Cenozoic.

The placoderms are almost completely limited to the Devonian Period. The many distinct lineages that are recognized early in that period suggest significant prior evolution. They remained diverse and numerous until near the end of the Devonian before rapidly becoming extinct. The apparent rapidity of their demise may be exaggerated by a change in the pattern of deposition at the end of the Devonian in North America and Europe. Widespread black shales, in which their remains are common in the Late Devonian, are rare in Carboniferous deposits. Only a few antiarchs survive in the Carboniferous. Placoderms are the only extinct class of vertebrates.

The demise of the placoderms may be attributed to the radiation of other, more advanced, jawed fish. All modern jawed fish belong to two large groups, the Chondrichthyes, or cartilagenous fish, including sharks, skates, rays, and chimeroids (or ratfish) and the Osteichthyes, or bony fish, includ-

ing most of the common modern species such as the trout and salmon, swordfish, flatfish, and sea horses, as well as a host of other living and fossil species. The sharks are important among large predators, but the bony fish are far more diverse.

CARTILAGENOUS FISH (CHONDRICHTHYES)

Fish broadly resembling modern sharks were common and diverse in the Late Devonian and Carboniferous. Scattered remains are known from the Lower Devonian, and isolated scales have been reported from the Upper Silurian. There is no evidence that the ancestors of sharks had a solid bony carapace like the ostracoderms and placoderms. Rather, they were characterized by distinctive scales that resemble teeth in that they have a pulp cavity surrounded by dentine and enamel-like material. The internal skeleton is composed of cartilage, which is much lighter than bone but not nearly as strong. Cartilage itself has little chance of being preserved, but in many genera the surface of the cartilage is calcified by a coating of distinctive prismatic granules.

The fossil record of sharks includes isolated teeth, spines, scales, and elements of the skeleton that are calcified. Rainer Zangerl (10) has described the extensive radiation of sharks in the late Paleozoic. Some ten orders are recognized by distinct patterns of the fin spines, dentition, and body form (Fig. 9–13). The xenacanths lived in fresh water, but most sharks, both Paleozoic and modern, are confined to the sea.

Class—Chondrichthyes
  Subclass—Elasmobranchii
      Many Paleozoic orders
    Superorder—Euselachii
      Order—Ctenacanthiformes (ctenacanths and hybodonts)
      Order—Xenacanthida
      Order—Galeomorpha (modern sharks)
      Order—Squalomorpha (modern sharks)
      Order—Batoidea (skates and rays)
  Subclass—Holocephali (modern chimaeroids and many Paleozoic groups)
  Subclass—incertae sedis
    Order—Iniopterygiformes

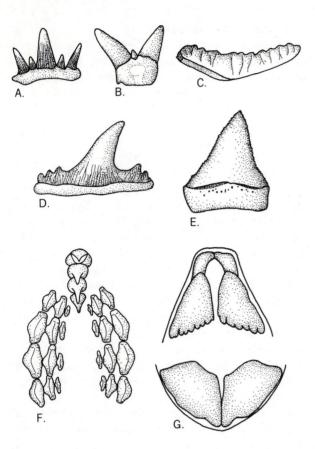

Figure 9-13   Teeth of cartilagenous fish. A. Cladodus tooth, characteristic of *Cladoselache* and other mid-Paleozoic sharks, B. Diplodus tooth, diagnostic of fresh water xenacanth sharks, C. Flat crushing tooth of *Orodus*, D. Hybodont tooth, common to primitive Mesozoic sharks, E. Tooth of modern predaceous galeoid shark *Carcharodon*, F. Tooth plates of Paleozoic petalodont, G. Tooth plates of chimeroid, seen as if looking into an open mouth.

Many groups of Paleozoic chondrichthyians resembled the modern sharks in their general body form, but others were flattened for life on the bottom, like skates and rays. One group, the iniopterygians, had rather short bodies and large skulls. Their dentition is close to the primitive shark pattern, but their limbs were very different (Fig. 9–14).

Most of the shark groups of the Late Devonian and Carboniferous were extinct by the end of the Triassic. One order, the hybodonts, remained dominant into the Mesozoic, and the closely related ctenacanths are close to the ancestry of modern sharks (the neoselachians). The neoselachians are more advanced than early Paleozoic sharks in the development of a tripartite base of the pectoral fin which gave it greater strength and maneuverability, in the evolution of a solid attachment of the left and right halves of the pectoral and pelvic girdles, and in the calcification of the vertebrae. Most importantly, the upper jaws became less solidly attached to the braincase so that they were much more maneuverable. This is very effective in predaceous sharks, where the jaws can be used like saws to cut pieces from their prey.

The neoselachians radiated rapidly in the Late Triassic and Early Jurassic, giving rise to all the modern shark groups, as well as to skates and rays. Leonard Campagno (11) has shown that most of the modern families had become established by the Early Cretaceous, and many modern genera appeared by the early Cenozoic.

The chimaeroids or holocephalians are rare fish in the modern fauna, with large heads, short bodies, and whiplike tails for which they are called ratfish (Fig. 9–15). Chimaeroids are thought to be closely related to sharks because both groups have similar types of calcified cartilage and practice internal fertilization. Chimaeras differ markedly from sharks in having the upper jaw fused solidly to the braincase, in having only a few pairs of large tooth plates rather than rows of rapidly replaced marginal teeth, and in having an operculum covering the gill openings, rather than a series of gill slits opening separately to the outside. The fossil record of modern chimaerid genera goes back to the Jurassic, and

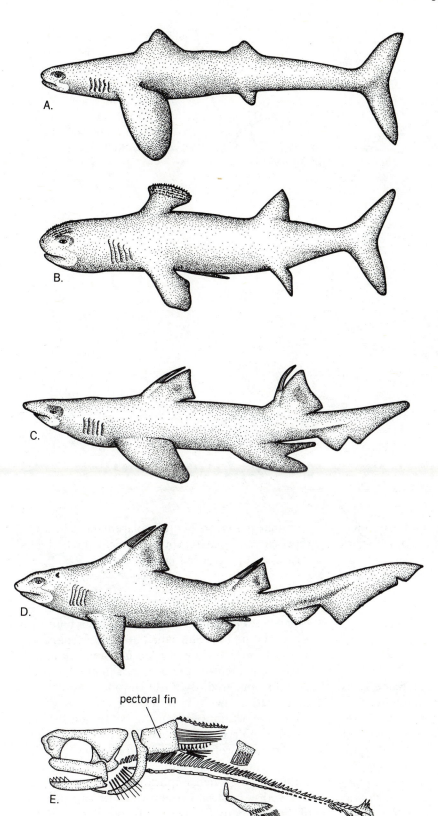

pectoral fin

clasper

Figure 9-14 Body form in a variety of sharks. A. *Cladoselache*, the only Devonian shark in which the body is well known, B. *Stenacanthus*, from the Carboniferous, in which the anterior dorsal fin is covered with large denticles, C. A Lower Carboniferous hybodont shark, D. The common Mesozoic shark *Hybodus*, E. A Carboniferous iniopterygian. Note the bizarre pectoral fin. (E, Courtesy of Dr. R. Zangerl & H. P. Schultze (ed) *Handbook of Paleoichthyology*, vol. 3, Gustav Fischer Verlag.)

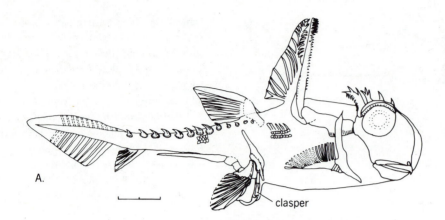

clasper

Figure 9-15   Specimens of the Lower Carboniferous holocephalian *Echinochimaera*. A. Male, showing claspers on pelvic fin, B. Female, lacking claspers. Note that the braincase and upper jaw are fused in contrast with the shark shown in Figure 9-11B. Scale is 1 cm long. (Courtesy Richard Lund.)

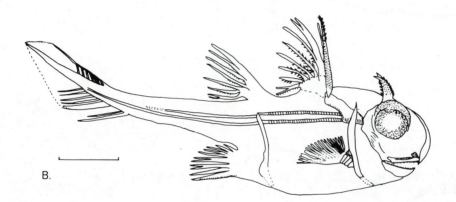

more archaic genera lived during the Carboniferous. A variety of fossils, showing a mosaic of shark and chimaeroid attributes, suggest that the chimaeroids probably evolved from primitive sharks within the Devonian.

OSTEICHTHYES

Acanthodians:

Accompanying the placoderms and archaic sharks in the Devonian and persisting into the Early Permian were an assemblage of bony fish whose relationships remain enigmatic, the acanthodians. Most were small, but some exceeded a meter in length. All are characterized by large spines in front of the paired and dorsal fins; some primitive genera had a series of accessory spines between the girdles as well (Fig. 9–16). Acanthodians resemble the major groups of osteichthyes superficially and in the histology of the scales but differ significantly from both them and the chondrichthyes in that they lack

a regular pattern of tooth replacement. They may represent an isolated group, distinct from both major classes of advanced jawed fish.

Advanced Bony Fish:

Isolated scales similar to those of later bony fish appear in the Late Silurian. They differ from those of sharks in that they overlap one another and in the arrangement of the layers of dentine and enamel-like material. By Early Devonian, several lineages had evolved. Two major groups are recognized; the actinopterygians or ray-finned fish and the sarcopterygians or lobe-finned fish. The sarcopterygians were common and diverse in the Devonian and Carboniferous, but only four genera survive today. Their importance lies in their affinities with terrestrial vertebrates. The vast majority of modern bony fish (some 20,000 species) are actinopterygians, but these were rare and little diversified until near the end of the Paleozoic (Fig. 9–17).

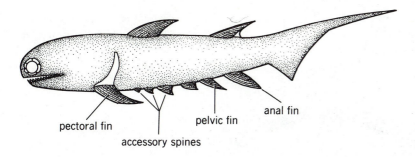

Figure 9-16   A primitive acanthodian fish, showing accessory spines between the pelvic and pectoral fins. Members of this group varied from a few centimeters to a meter in length. (J. A. Moy-Thomas & R. S. Miles, *Palaeozoic Fishes.* Courtesy Chapman & Hall, London.)

Actinopterygians:

Bony fish differ from sharks and their relatives in the presence of an ossified internal skeleton, overlapping scales, and a lung or swim bladder. The swim bladder increases buoyancy and enables advanced bony fish to float at any depth without muscular effort. Primitive bony fish evolved paired pouches from the esophagus that could be filled by swallowing air. In most sarcopterygian fish, these pouches also served as lungs. In advanced actinopterygian fish the pouches became separated from the esophagus and can exchange oxygen with the blood stream. This allows the fish to fill and empty the swim bladder without coming to the surface.

Cartilagenous fish sink unless they are actively swimming, even though their skeleton is lighter than bone. Because they have a lung or swim bladder, bony fish can live in many environments that are not available to sharks, skates, and rays.

Early actinopterygians, the palaeoniscoids, were fusiform fish of generally small size, with triangular fins and a heavy covering of scales (Fig. 9–18). Like sharks, the caudal fin was heterocercal (that is, its main axis tilted upwards). The mouth was small and could not be opened widely, lim-

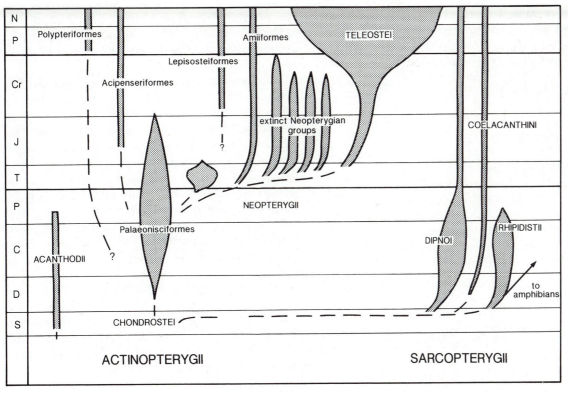

Figure 9-17   Phylogeny of bony fish (Osteichthyes).

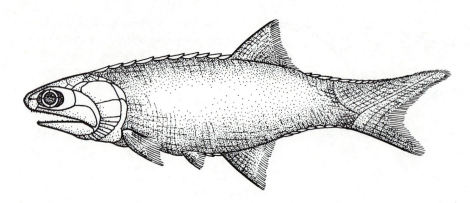

Figure 9-18 The primitive palaeoniscoid fish *Mimia*, from the Upper Devonian, approximately 8 cm long.

iting them to relatively small prey. Unlike sharks, in which the pharynx is long and the gills are well behind the mouth cavity, the pharynx in bony fish is beneath and immediately behind the mouth cavity, and the functions of feeding and respiration are closely linked. As the mouth is opened, the pharynx is expanded laterally as well as dorsally and ventrally, greatly increasing the volume of the mouth and drawing in water. This way of feeding is especially effective for capture of small prey, but was relatively little developed in the palaeoniscoids. Fish of this primitive level of evolution continued into the Cretaceous, but most actinopterygians from the Permian onward had advanced toward the modern bony fish. Several living genera appear to represent little modified descendants of the palaeoniscoids; these include

Polypterus, an elongate fish from the Nile, the sturgeon, and the paddle fish.

A number of distinct orders of bony fish common in the Mesozoic are anatomically intermediate between palaeoniscoids and advanced actinopterygians. They have been called holosteans, but this term no longer has any taxonomic validity because these orders do not have a unique common ancestry separate from that of the higher bony fish (Fig. 9–19). The common bony fish of the Mesozoic had thinner scales than the palaeoniscoids, the number of scale rows covering the fins was reduced, and some genera had ossified vertebral centra. The tail was more symmetrical, indicating the more effective function of the swim bladder. This level of actinopterygians evolution is represented by two living genera, *Amia* and *Lepisosteus* (the gar).

The exact ancestry of the higher, ray-finned fish, the teleosts, is still obscure, but they appear to have diverged from the palaeoniscoids by the end of the Permian. In the Upper Triassic and Jurassic, the teleosts are represented by two groups of small, fusiform fish, the pholidophoroids and the leptolepids. Within these groups, the scales are further reduced, until they retain only a flexible bony layer, the caudal fin becomes superfically symmetrical, and the vertebrae become fully ossified. The caudal vertebrae have a particular shape that could be ancestral to that of all more advanced teleosts of the late Mesozoic and Cenozoic.

The teleosts remained relatively rare until the Late Cretaceous, during which time they diversified explosively and approached the structure of modern groups (Fig. 9–20). Members of families with an adequate fossil record had reached an essentially modern form by the end of the Cretaceous, and modern genera are known from the Early Cenozoic. Today, teleosts are by far the most diverse vertebrate group.

Class—Osteichthyes
    Subclass—Acanthodii
    Subclass—Actinopterygii
        Infraclass—Chondrostei
            Order—Palaeonisciformes
            Order—Polypteriformes
                (*Polypterus*)
            Order—Acipenseriformes
                (sturgeon)
        Infraclass—Neopterygii
            Order—Lepisosteiformes (gars)
            Order—Amiiformes (*Amia*)
        Division—Teleostei
    Subclass—Sarcopterygii
        Order—Dipnoi
        Order—Crossopterygii
            Suborder—Rhipidistii
            Suborder—Coelacanthini

Figure 9-19 The "Holostean" fish *Lepidotes*, from the Lower Jurassic of Germany, 15 cm long. (Photograph courtesy Rupert Wild.)

Sarcopterygians:

The lobe-finned fish are today represented by only four genera, three lungfish, one each in Australia, South America, and Africa, and the coelacanth *Latimeria*, restricted to moderately deep water off the north coast of Madagascar. In the Devonian and Carboniferous, by contrast, they were the dominant freshwater predators. Sarcoptgerygians appeared at the very beginning of the Devonian. Early fossils come from both freshwater and marine deposits, but most later members of the group lived in fresh water. They were generally of larger size than the early actinotperygians. The olfactory capsules were more highly developed and the orbits were smaller than in actinopterygians, suggesting that the sense of smell was more important than vision. Early sarcopterygians show evidence of an elaborate electrosensory system, suggesting life in deep or turbid water. This may be associated with their particular limb structure.

The fins of actinopterygians are supported by a

Figure 9-20 Eocene teleost fish. A. *Pristacara* from the Green River Formation of Wyoming. Note short body and spiny dorsal fin, typical of advanced spiny teleosts. B. *Caratoichthys* from Monte Bolca, northern Italy. (Courtesy Redpath Museum, McGill University [A] and J. Blot, *Les poissons fossiles du Monte Bolca*. Courtesy Museo di Scienze Naturale, Verona [B].)

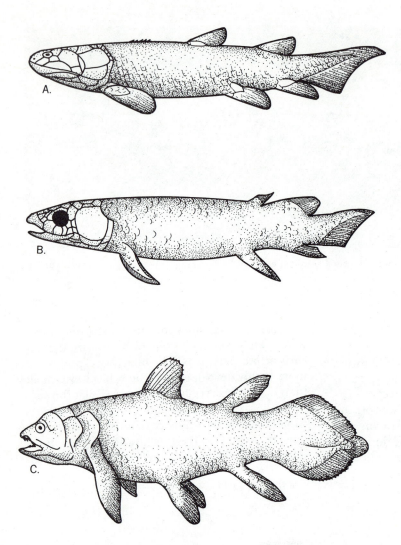

Figure 9-21   Sarcopterygian fish. A. A primitive rhipidistian, *Osteolepis* × ½, B. A primitive lungfish, *Dipterus* × ½, C. The modern coelacanth, *Latimeria* × ⅒.

number of slender, parallel fin rays. There are no muscles in the fins, but they can be controlled by muscles in the body wall. In sarcopterygians, in contrast, the paired fins are supported by a strong, bony axis that had its own musculature, giving the fins a thickened, lobed appearance. The function of such fins may have been to push the fish along the bottom. The main force of swimming, as in other fish, was the axial musculature.

Two major groups of sarcopterygians are known at the base of the Devonian, the lungfish and the crossopterygians. The postcranial skeleton is similar, with heavy rectangular scales, two dorsal fins (rather than one as in actinopterygians), and a heterocercal tail with a small dorsal lobe (Fig. 9–21).

In palaeoniscoids, the chamber for jaw muscles was small, and the force of jaw closure would have been limited. This chamber was gradually modified

throughout the Late Paleozoic and Mesozoic by the reduction of the cheek and the lateral expansion of the muscles. Both crossopterygians and lungfish evolved very different patterns of jaw musculature by the Early Devonian (Fig. 9–22). Crossopterygians retained a relatively small chamber for the jaw muscles but evolved large muscles that ran beneath the braincase. The braincase itself ossified in two sections that hinged on one another. As the jaws were closed, the muscles beneath the braincase pulled down on the front segment, which was attached to the palate and upper jaw, adding force to the bite. The front portion of the braincase moved through only a small arc, but this movement was enough to force the very large teeth on the palate through the heavy scales of the fish on which they fed.

The changes in lungfish involved the fusion of

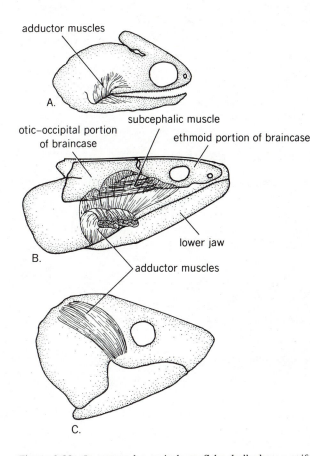

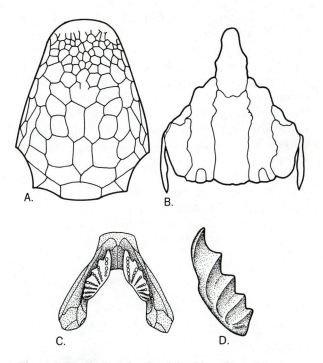

Figure 9-23 Skull roof and and dentition of lungfish. A. Skull roof of the Lower Devonian *Dipnorhynchus*. B. Ossified portion of the skull roof of the living genus *Epiceratodus*. C. Lower jaw of *Dipterus* from the Upper Devonian. D. Individual tooth plate of *Sagenodus*.

Figure 9-22 Jaw musculature in bony fish; skulls drawn as if transparent. A. A primitive palaeoniscoid, in which the adductor chamber is small and entirely enclosed by bone, B. A rhipidistian fish, in which the adductor musculature is supplemented by the subcephalic muscle which pulls the anterior portion of the braincase and attached palate down on the prey, C. A lungfish, in which the adductor musculature has expanded out of the adductor chamber and over the braincase.

the palate to the braincase, the forward movement of the jaw joint, and the opening of the adductor chamber dorsally and posteriorly so that the jaw musculature spread out over the top of the braincase and eventually out the back of the skull. By the Middle Devonian, the pattern of the jaw musculature was already essentially like that of modern lungfish, to judge by the similarity of the skull. Because of the clear distinction between their jaw structures, crossopterygians and lungfish have been placed in separate orders, although the postcranial anatomy suggests an ultimate common ancestry.

Fossils from the very base of the Devonian in China (12) suggest a close link between the two groups. The skull of *Diabolichthys* is like that of later lungfish (Fig. 9–23), and the distribution of the teeth on the palate and inside surface of the lower jaw closely resembles the arrangement retained in the toothplates of later lungfish. The palate is not fused to the braincase, however, and the braincase itself is ossified in two segments, as in the crossopterygians. In fact, the braincase in all descendants of the primitive bony fish shows some degree of separation into two units. This does not indicate that lungfish evolved from the crossopterygian pattern, but rather that they diverged from a common stock that had the potential for specialization in either direction.

Lungfish:

Lungfish show very rapid evolution in the Devonian, and by the end of that period they had achieved many of the features of their living descendants. The large size of the jaw muscles and the solid attachment of the palate to the braincase made them effective in feeding on a wide range of prey, as is further suggested by the diversity of patterns of their tooth plates and different configurations of the body. Lungfish remained common and diverse into the Triassic, but only two or three lineages are known in the later Mesozoic and Cenozoic. Two major groups survive today. *Neoceratodus* from Australia has a heavy body that most closely resembles the Paleozoic forms, except for the migration of the dorsal and anal fins so that they are confluent with the tail. The ossification of the skull roof, braincase, and scales is much reduced.

Protopterus from Africa and *Lepidosiren* from South America have much more slender bodies, and the fins are reduced to narrow filiments. Both can aestivate; that is, during dry seasons, they can coil up in the mud surrounded by a water-tight covering

of mucus and remain for more than a year in a state of suspended animation waiting for the next rainy season. They make use of atmospheric oxygen during this period and will drown if they are not able to breath air. Aestivation is an ancient habit; lungfish burrows are known from as early as the Carboniferous.

Crossopterygians:

Several major groups of crossopterygians are known from the Devonian, and one genus survives to the present. The living genus *Latimeria* provides evidence of the soft anatomy of this primarily Paleozoic group, including the presence of muscles under

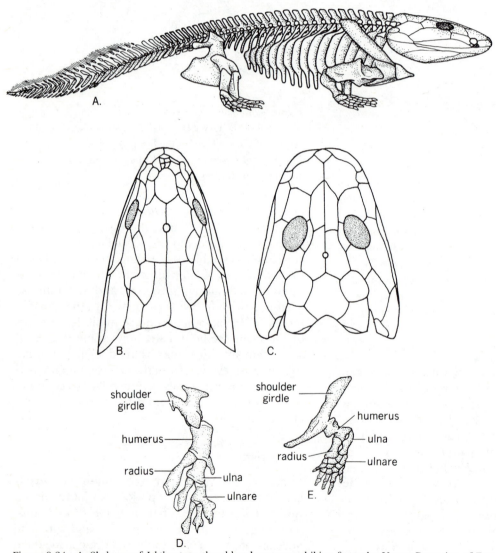

Figure 9-24   A. Skeleton of *Ichthyostega*, the oldest known amphibian from the Upper Devonian of East Greenland, B,C. Skulls of rhipidistian fish, Eusthenopteron and *Ichthyostega*, D. Limbs of rhipidistian and early tetrapod. (E. Jarvik, *Scientific Monthly*, 80: 141–154. Copyright 1955 by the AAAS [A], E. Jarvik, *On the fishlike tail in the ichthyostegid stegocephalians* [C] and H. P. Schultze, 1986, *Journal of Morphology, Suppl.* 1 [D,E].)

the braincase that had been hypothesized as being present in the Devonian crossopterygians. *Latimeria* belongs to the suborder Coelacanthina, whose oldest representatives are Late Devonian in age. The modern genus is marine, but in the Carboniferous the group was common and diverse in both fresh and salt water. Fossil coelacanths were common in nearshore marine deposits in the Cretaceous but disappeared from the fossil record in the Cenozoic, presumably because the only surviving lineage had adapted to a deep-sea habitat where the chances of preservation were limited.

The remaining crossopterygians are included in the suborder Rhipidistia. Two major groups lived during the Devonian, the porolepiforms, which are closely related to the lungfish and may be the stem group of all sarcopterygians, and the osteolepiforms. The osteolepiforms appeared first in the Middle Devonian and are specialized in the presence of a third opening in the nasal capsule within the mouth cavity. This became the internal nostril of terrestrial vertebrates. Most fish have only two openings in the nasal capsule, an anterior, incurrent opening, and a posterior, excurrent opening. All early bony fish had lungs and could breath air was well as making use of oxygen from the water. The internal nasal opening in rhipidistians prob-

ably evolved initially to facilitate smelling within the mouth cavity, since they could have breathed through the mouth, whether or not the nasal passage was used.

Rhipidistians were common in shallow, freshwater environments where the presence of lungs would have been of selective advantage during times of oxygen depletion and where the stout limbs would have been useful in propelling the body along the bottom, even when the water was very shallow. Alfred Romer (13) suggested that the evolution of limbs cabable of supporting the body on land may have been selected to enable rhipidistians to return to deep water during periods of drought.

AMPHIBIANS

Amphibians appeared in the fossil record at the end of the Devonian Period. The earliest known amphibians, from East Greenland, were about a meter long (Fig. 9–24). They had well-developed feet for walking on land, but retained a fishlike tail. Despite the changes in the limbs and limb girdles, the rest of the skeleton, especially the skull and vertebrae, is very similar to that of osteolepiform rhipidistians. One pecularity they share is the en-

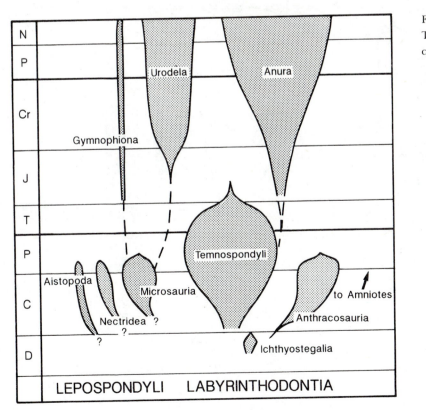

Figure 9-25   Radiation of amphibians. The Gymnophiona are also referred to as caecilians.

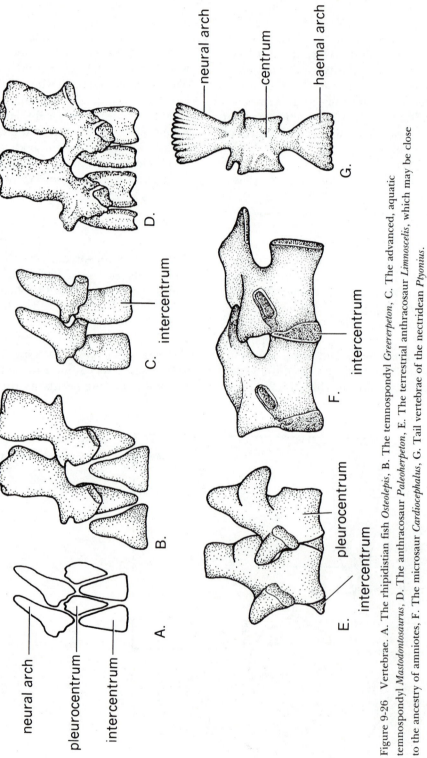

Figure 9-26   Vertebrae. A. The rhipidistian fish *Osteolepis*, B. The temnospondyl *Greererpeton*, C. The advanced, aquatic temnospondyl *Mastodontosaurus*, D. The anthracosaur *Paleoherpeton*, E. The terrestrial anthracosaur *Limnoscelis*, which may be close to the ancestry of amniotes, F. The microsaur *Cardiocephalus*, G. Tail vertebrae of the nectridean *Ptyonius*.

folding of the dentine of the teeth, which in cross section resembles the paths of a labyrinth. For this reason, the early amphibians are referred to as labyrinthodonts. The earliest labyrinthodonts, the ichthyostegids, retained the break between the front and back of the braincase common to rhipidistians, but the two sections were firmly attached in other amphibians.

More advanced labyrinthodonts are divided into two large groups, the temnospondyls and the anthracosaurs (Fig. 9–25). They differ in the structure of the vertebrae. Many rhipidistians and all primitive amphibians had vertebrae made of several pieces, in contrast with the cylindrical vertebrae of modern terrestrial vertebrates. The intercentrum forms the base of the anterior portion of each segment, and the pleurocentra (primitively paired) form the posterior portion and support the neural arch (Fig. 9–26). The intercentrum becomes the largest element in temnospondyls, and in some advanced groups assumes the shape of a spool. The pleurocentra become spoolshaped in the anthracosaurs. The structure of the vertebrae does not appear to have affected the general pattern of locomotion or the way of life of these groups, both of which radiated into a broad range of habitats from semiaquatic to terrestrial.

```
Class—Amphibia
    Subclass—Labyrinthodontia
        Order—Ichthyostegalia
        Order—Temnospondyli
        Order—Anthracosauria
    Subclass—Lepospondyli
        Order—Aistopoda
        Order—Nectridea
        Order—Microsauria
    Subclass—Lissamphibia
        Order—Anura (frogs)
        Order—Urodela (salamanders)
        Order—Gymnophonia (caecilians)
```

Many lineages of early amphibians became secondarily aquatic, going back to the environment of their rhipidistian ancestors. Unlike the fish, the trunk region of these amphibians was usually elongated. The limbs retained the structure of their terrestrial antecedents but were relatively smaller. Most secondarily aquatic amphibians had flat heads giving them the appearance of gigantic aquatic salamanders. Others may have resembled crocodiles in their habits.

Labyrinthodonts were the dominant terrestrial vertebrates during the Carboniferous, but competition and predation from their own descendants, the reptiles, resulted in the demise of most of the terrestrial families by the end of the Early Permian. During the Late Permian and Triassic, a number of labyrinthodont lineages persisted, but most were obligatorily aquatic, which presumably allowed them to survive competition and predation from the primarily terrestrial reptiles. The last of the labyrinthodonts are known from the Middle Jurassic of China.

Accompanying the labyrinthodonts, which were primarily of moderate to large size (up to 3 m long), were many smaller species, collectively termed lepospondyls. All had spool-shaped vertebrae and had lost the labyrinthine enfolding of the teeth and large fangs on the palate that characterized the labyrinthodonts. Despite these similarities, there is no evidence that the lepospondyls had a common ancestry; in fact, the fossil record provides no evidence of the ancestry of any of the individual groups.

There were three major groups of lepospondyls. The most highly specialized were the aistopods, which had no trace of either girdles or limbs, but were elongate, snakelike species with up to 200 vertebrae (Fig. 9–27). The nectrideans include newtlike forms with laterally compressed tails. The tail vertebrae were expanded vertically to form an effective organ of aquatic propulsion. The microsaurs, the most diverse of the lepospondyl groups, included some aquatic genera and others that resembled lizards. Some very elongate microsaurs with reduced limbs may have been burrowers. Lepospondyls are not known after the Early Permian and, except for one genus from northern Africa, have been found only in North America and Europe.

Almost no small amphibians are known between the Lower Permian and the Lower Jurassic, when the first frogs, salamanders and caecilians (a group of poorly known limbless amphibians, currently restricted to the damp tropics) appear in the fossil record. The modern amphibian orders are very different from both the labyrinthodonts and lepospondyls in their anatomy and the interrelationships of these groups have not yet been demonstrated (Fig. 9–28).

The only affiliation that is well established is be-

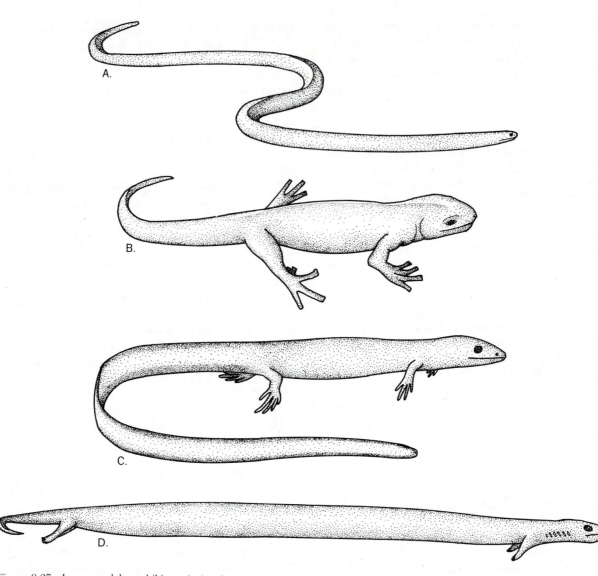

Figure 9-27   Lepospondyl amphibians. A. An aistopod, B. A microsaur, C. A nectridean, D. A lysorophid.

tween frogs and labyrinthodonts. One group of small temnospondyls, the family Dissorophidae, has a vaguely froglike skull and some genera have nearly spool-shaped vertebrae. A deep notch at the back of the skull, as in other terrestrial temnospondyls, is thought to have supported an eardrum similar to that of frogs. A narrow stapes, which would have transmitted airborne vibrations to the inner ear, extends medially from the eardrum. Frogs are the only group of modern amphibians that have an ear structure like that of the dissorophids.

Unlike frogs and some temnospondyls, neither salamanders nor caecilians have an eardrum, and the stapes is a heavy structure linking the braincase with the cheek. This arrangement is not unlike that present in microsaurs, among the lepospondyls. Robert Carroll, Philip Currie, and Robert Holmes (14, 15) have suggested that these groups may be related. Thomas Parsons and Ernest Williams (16) have argued, in contrast, that frogs, salamanders, and caecilians should be included in a single group, the Lissamphibia, because of the similarity of the skull and the common presence of a specialized tooth structure, termed pedicellate, in which the base and crown of the teeth are divided by fibrous tissue and can bend to facilitate feeding on large

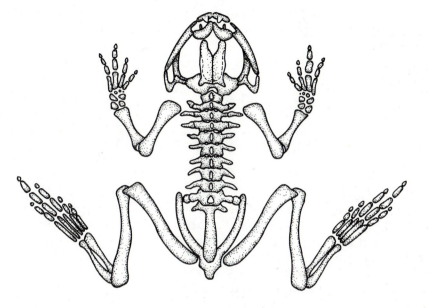

Figure 9-28 The Lower Jurassic frog *Vieraella*. It closely resembles modern frogs except for the retention of vestigial ribs.

prey. There are no described fossils linking either salamanders or caecilians with any of the Paleozoic amphibians.

One fossil, the only specimen of the genus *Triadobatrachus*, appears to link frogs with small labyrinthodonts. This genus, from the Lower Triassic of Madagascar, has a froglike skull, a relatively short vertebral column, and long pelvic bones as do frogs.

Fossils of frogs and salamanders are rare in Mesozoic rocks, but the major groups had apparently all differentiated by the end of the Cretaceous (17). We have only a very sketchy notion of the interrelationships of the major families. Only about 2 percent of the living species are known as fossils, but several modern genera are known from rocks as old as Miocene age. Vertebrae similar to those of modern caecilians have been described from the Upper Cretaceous and Lower Cenozoic.

REPTILES

Amphibians are primitive in being tied to the water for reproduction. Most lay their eggs in the water, and many have a distinct aquatic larval stage. Reptiles, birds, and mammals have evolved special membranes associated with the egg that provide protection, support, and gas exchange so that the eggs can be laid on land. The young are typically hatched or born as small replicas of the adults. Together, these groups are termed amniotes, because

they possess an amnion, one of the extraembryonic membranes that provides support and protection.

```
Amniota
    Class—Reptilia
    Class—Aves
    Class—Mammalia
```

Reptiles appear suddenly in the fossil record in the Upper Carboniferous of Nova Scotia. Their remains are found in the upright stumps of lycopods of the genus *Sigillaria*. According to Sir William Dawson, who discovered these fossils in the mid-nineteenth century, this peculiar mode of preservation was the result of fluctuations in the water level. The lycopod forests were periodically flooded and partially buried, killing the trees, but leaving their stumps in a vertical position. The centers of the stumps rotted out, but the strong outside tissue remained. When the water subsided, a new land surface was developed with the stumps still in place, but now appearing as gigantic traps, open to unwary animals walking on the new land surface. They fell into the openings in the stumps, some of which were 3 to 5 meters deep, and could not crawl out. They were eventually buried and fossilized in place. The importance of this form of burial is that it preferentially selected such truly terrestrial animals as the early reptiles.

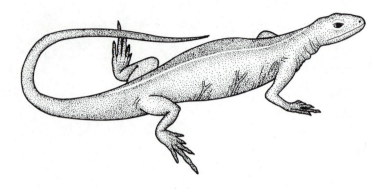

Figure 9-29   The oldest known reptile: *Hylonomus* from the Upper Carboniferous of Nova Scotia.

Reptiles, such as *Hylonomus* (Fig. 9–29), that were preserved in the stumps were small and superficially resembled primitive modern lizards and the primitive living reptile *Sphenodon*. Their palates are distinguished from those of more primitive tetrapods by a transverse flange on the pterygoid, a process from which the pterygoid muscle in modern reptiles originates. This muscle contributes to producing a strong force on the lower jaw through the entire arc of its closure (Fig. 9–30).

The structure of the vertebrae in early reptiles indicates that they evolved from primitive anthracosaurian amphibians. Within that group, the limnosceloids resemble reptiles most closely in the presence of a transverse flange of the pterygoid and

the loss of large palatal teeth and the posterior embayment of the cheek. Limnosceloids remain primitive in retaining labyrinthine enfolding of the teeth that is lost in reptiles. Early reptiles were primitive in lacking space for the attachment of the eardrum. They had a very heavy stapes, indicating that their ear was not highly sensitive to airborne vibrations.

The initial radiation of amniotes occurred within the Late Carboniferous (18). The major divisions of this groups are based on the number and distribution of openings in the skull over the area of the jaw muscles. In advanced members of each group the pattern of the jaw muscles is much altered in relationship to these openings, but the initial ap-

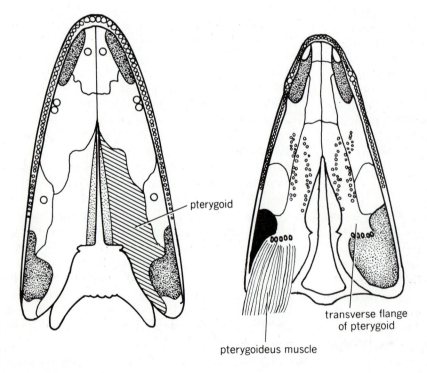

pterygoid

pterygoideus muscle

transverse flange
of pterygoid

Figure 9-30   Palates of a Carboniferous amphibian, and the early reptile *Paleothyris*, to show the position of the transverse flange of the pterygoid.

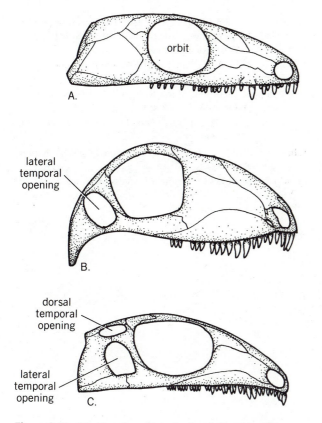

Figure 9-31   Temporal openings in the skulls of reptiles which are used as a basis for their classification. A. The anapsid condition in the earliest reptiles, B. The synapsid condition of mammal-like reptiles, C. The diapsid condition of the ancestors of lizards.

pearance of the openings may have lightened the skull and redistributed stress. The primitive condition, inherited from amphibians, is a solid bony covering over the cheek region. This pattern is termed anapsid. Turtles have openings in the back of the skull but they are usually classified within the anapsids. The group of primitive amniotes that was ancestral to mammals has a single opening low in the cheek, the synapsid condition. The remaining reptiles, represented today by *Sphenodon*, lizards, snakes, and crocodiles, have two openings, or evolved from genera with two openings, the diapsid condition (Fig. 9–31).

The synapsids (mammal-like reptiles) were the first group to diverge from the ancestral amniotes, appearing as contemporaries of *Hylonomus* in the tree stump fauna of Joggins, Nova Scotia. The diapsids appeared by the end of the Late Carboniferous. The first turtles appeared in the the Late Triassic,

but a possible ancestral group, the family Captorhinidae, is known from Lower Permian rocks.

REFERENCES

1. Repetski, J. E., 1978, A fish from the Upper Cambrian of North America: Science v. 200, p. 529–531.

2. Ritchie, A., and Gilbert-Tomlinson, J., 1977, First Ordovician vertebrates from the Southern Hemisphere: Alcheringa v. 1, p. 351–368.

3. Conway-Morris, S., 1979, The Burgess Shale (Middle Cambrian) fauna: Annual Reviews of Ecology and Systematics, v. 10, p. 327–349.

4. Oelofsen, B. W., and Loock, K., 1981, A fossil cephalochordate from the early Permian Whitehill Formation of South Africa: South African Journal of Science, v. 77, p. 178–180.

5. Ruben, J. A., and Bennett, A., 1980, Antiquity of the vertebrate pattern of activity metabolism and its possible relation to vertebrate origin: Nature, v. 286, p. 886–888.

6. Dineley, D. L., 1964, New specimens of *Traquairaspis* from Canada: Palaeontology, v. 7, p. 210–219.

7. Nitecki, M. H. (ed.), 1979, Mazon creek fossils: New York, Academic Press, 581 p.

8. Schaeffer, B. and Thomson, K. S., 1980, Reflections on agnathan-gnathostome relationships. In : Jacobs, L. L. (ed.), Aspects of vertebrate history; essays in honor of Edwin Harris Colbert, Flagstaff, Museum of Northern Arizona Press, p. 19–33.

9. Jollie, M., 1971, A theory concerning the early evolution of the visceral arches: Acta Zoologica, v. 52, p. 85–96.

10. Zangerl, R., 1981, Chondrichthyes I. Paleozoic elasmobranchs. In: Schultze, H. P. (ed.), Handbook of Paleoichthyology, v. 3: Stuttgart, Gustav Fischer Verlag.

11. Compagno, L. J. V., 1977, Phyletic relationships of living sharks and rays: American Zoologist, v. 17, p. 303–322.

12. Chang Mee-Mann, and Yu Xiaobo, 1984, Structure and phylogenetic significance of *Diabolichthys speratus* gen. et sp. nov., a new dipnoan-like form from the Lower Devonian of Eastern Yunnan, China: Proceedings of the Linnean Society of New South Wales, v. 107, p. 171–184.

13. Romer, A. S., 1957, Origin of the amniote egg: Scientific Monthly, v. 85, p. 57–63.

14. Carroll, R. L. and Currie, P. J., 1975, Microsaurs as possible apodan ancestors: Zoological Journal of the Linnean Society, v. 57, p. 229–247.

15. Carroll, R. L. and Holmes, R., 1980, The skull and jaw musculature as guides to the ancestry of salamanders: Zoological Journal of the Linnean Society, v. 68, p. 1–40.

16. Parsons, T. and Williams, E., 1963, The relationship of modern Amphibia: a reexamination. Quarterly Review of Biology, v. 38, p. 26–53.

17. Estes. R., 1981, Gymnophiona, Caudata. In: O. Kuhn (ed.), Handbuch fur Palaeoherpetologie. 2: Stuttgart, Gustav Fischer Verlag.

18. Carroll, R. L., 1982, Early evolution of reptiles. Annual Reviews of Ecology and Systematics, v. 13, p. 87–109.

SUGGESTED READINGS

Cappetta, H., 1987, Chondrichthyes II. Mesozoic and Cenozoic Elasmobranchii. In: Schultze, H. P. (ed), Handbook of Paleoichthyology, v. 3B: Stuttgart, Gustav Fischer Verlag.

Carroll, R. L., 1987, Vertebrate Paleontology and Evolution: New York, Freeman, 698 pp.

Denison, R. H., 1978, Placoderms. Handbook of Paleoichthyology. 2: Stuttgart, Gustav Fischer Verlag.

Elliott, D. K., 1987, A reassessment of *Astraspis desiderata*, the oldest North American vertebrate: Science v. 237: 190–192.

Gardiner, B. G., 1984, The relationships of the palaeoniscid fishes, a review based on new specimens of *Mimia* and *Moythomasia* from the Upper Devonian of Western Australia: Bull. Brit. Mus. (Nat. History) v. 37: 173–428.

Greenwood, P. H., Miles, R. S., and Patterson, B., 1973, Interrelationships of fishes: Zoological Journal of the Linnean Society, supplementary series v. 53, 536 p.

Lauder, G. V., and Liem, K. F., 1983, The evolution and interrelationships of the actinopterygian fishes: Bulletin of the Museum of Comparative Zoology, v. l50, p. 95–197.

Hallam, A., (ed.), 1977, Patterns of Evolution: Amsterdam, Elsevier, 591 pp.

Moy-Thomas, J. A., and Miles, R. S., 1971, Palaeozoic fishes: London, Chapman & Hall, 259 p.

Romer, A. S., and Parsons, T., 1986, The vertebrate body, 6th ed: Philadelphia, Sanders, 679 p.

Schaeffer, B., 1987, Deuterostome monophyly and phylogeny: Evolutionary Biology, v. 21, p. 179–235.

Stensio, E., 1964, Les Cyclostomes fossiles ou Ostracoderms. In: Piveteau, J. (ed.), Traité de Paléontologie, Tome IV, v. 1: Paris, Masson et Cie., p. 96–382.

Thomson, K. S., 1971, Adaptation and early evolution of early fishes: Quarterly Review of Biology, v. 46, p. 139–166.

Chapter 10

The Second Wave:

INVERTEBRATES OF THE MESOZOIC AND CENOZOIC ERAS

INTRODUCTION

At the close of the Paleozoic Era the seas were withdrawn from the continental platforms into the ocean basins, the continents were gathered into a supercontinental mass called Pangaea, the continental shelves where marine invertebrates thrived were highly restricted, and the coastlines of the world were shorter than they had been since late in the Proterozoic, shorter than they would ever be again. Much of the land was arid and where sections of the sea were partially isolated from the oceans many of the world's evaporite deposits were laid down. Marine deposits of Late Permian and Early Triassic age are therefore relatively thin and cover small areas. The effect of these inhospitable conditions on marine life was reviewed at the end of Chapter 7. The marine invertebrate fauna was restricted as never before or since. Similar changes did not occur in land faunas or floras.

Recovery was not immediate. New invertebrate groups began to appear in Middle and Late Triassic time but rapid diversification did not take place until Jurassic time.

FORAMINIFERIDA

The foraminiferans that live on the sediment surface and within it are called benthonic; those that float in the water column are planktonic.

Benthonic Foraminiferida:

The Permian crisis completely wiped out the fusulines and almost eliminated the endothyrids but several families of agglutinated forms (Textulariina) were little affected by its rigors. Recovery from the late Paleozoic crisis was slow, and Triassic foraminiferan faunas are neither diverse nor abundant. The benthonic forams rapidly diversified in Jurassic time both in shallow and deep-water environments; members of the suborders Miliolina and Rotaliina led the way. By mid-Jurassic time the benthonic foraminiferans were again making important contributions to the formation of shallow-water limestones. The Miliolina appear to have arisen from such agglutinated foraminiferans as *Ammodiscus* (Fig. 10–1C) in Carboniferous time, and the Rotaliina, from the endothyrids in the Triassic

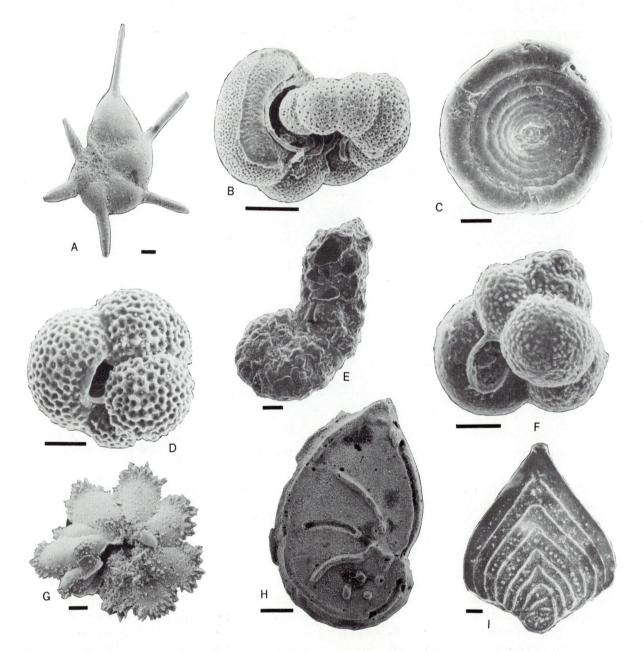

Figure 10-1   Scanning electron micrographs of fossil foraminiferans. The scale bars are 0.1 mm. A. *Hantkenina* (planktonic, Rotaliina), Middle Eocene, Blake Plateau. B. *Ticinella* (planktonic, Rotaliina), Lower Cretaceous, Blake Plateau, C. *Ammodiscus* (agglutinated, Textulariina), Eocene, North Sea. D. *Globigerina* (planktonic, Rotaliina), Miocene, Labrador shelf. E. *Ammobaculites* (coarsely agglutinated, Textulariina), Eocene, Labrador shelf. F. *"Globigerina"* (oldest planktonic foram, Rotaliina), Middle Jurassic, Atlantic shelf. G. *Morosovella* (planktonic, Rotaliina), Middle Eocene, Blake plateau. H. *Lenticulina* (benthonic, Rotaliina), Lower Cretaceous, Blake plateau. I. *Neoflabellina* (benthonic, *Rotaliina*), Upper Cretaceous, Atlantic shelf. (Courtesy of Felix Gradstein and the Geological Survey of Canada.)

Period. The Rotaliina achieved their dominant position in modern foraminiferan faunas by rapid diversification in the middle of Cretaceous time. In some brackish and deep-water environments, the agglutinated Textulariina became dominant (Fig. 10–1).

In late Mesozoic time several groups evolved into giants comparable in size to the fusulines. They are

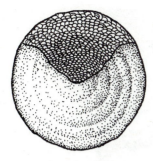

A. *Orbitolites*

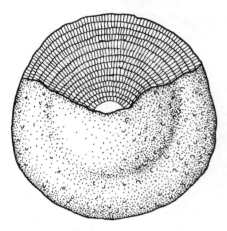

B. *Discocyclina*

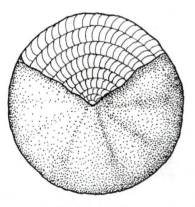

C. *Nummulites*

Figure 10-2   Larger benthonic foraminiferans cut away in one sector to show the internal structure. All can be as large as several centimeters across. A. *Orbitolites*. B. *Discocyclina* (an orbitoid). C. *Nummulites*.

giants only relative to their miniscule contempories, but many reached the size of a small coin. In Jurassic time giants developed in the agglutinated foraminiferans, and in Cretaceous time in the Miliolina and the Rotaliina. The milioline giants include disk-shaped genera, like the common early Cenozoic genus *Orbitolites* (Fig. 10–2A). One family of the miliolines, the Alveolinidae, which were particularly abundant in Late Cretaceous and Eocene time, evolved forms that are remarkably similar to the late Paleozoic fusulines. The giant rotaliines include the nummulites and orbitoids. The most common genus of the first is *Nummulites* (Fig. 10–2C). These coin-shaped fossils are so common in the Eocene limestones of the Mediterranean region that in Europe this interval has been called the Nummulitic Epoch. Because the limestones were used to build the pyramids of Egypt, these fossils came to the attention of the ancient Greeks and Romans. Heroditus, (about 500 B.C.) described them as the petrified remains of lentiles that the pyramid builders had for lunch. The test of the nummulites is basically a planispiral coil like that of an ammonoid in which the last chambers cover the inner ones. However, the test is complicated by the subdivision of the chambers by internal partitions. The test of orbitoids is thinner and the chambers are added initially in a spiral pattern and in later stages in rings around the periphery of the disk (Fig. 10–2B). Many of the larger foraminiferans, and also the planktonic ones, have developed a symbiotic relationship with unicellular algae that supply them with oxygen through photosynthesis. The relationship is similar to that of molluscs and corals (see following discussion).

Some of the large rotaliines attached themselves to hard surfaces and became encrusting in habit. Two genera are presently common in tropical seas: the red *Homotrema* whose color adds a pink cast to Bermuda beaches, and the brown *Gypsina* which is an important binding agent in Caribbean reefs.

Planktonic Foraminiferans:

In Middle Jurassic time one group of the order Rotalliina adapted to a planktonic existence, floating near the surface of the ocean. These forams have a simple test composed of chambers coiled in a trochoid spiral like that of a gastropod. Many are equipped with spines that may be several times the diameter of the test in length (Fig. 10–3B). The two families, Globigerinidae and Globorotaliidae

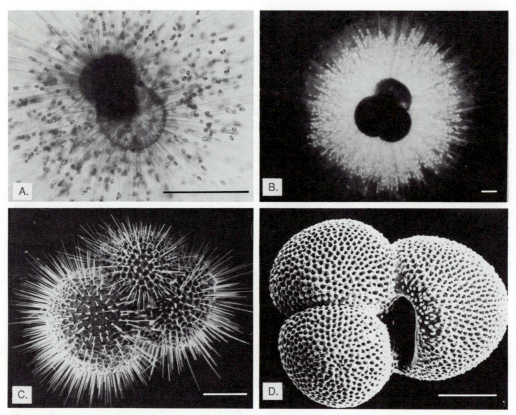

Figure 10-3　Recent planktonic foraminiferans. The scale bars are 0.1 mm. A. Juvenile spinose form showing the symbiotic algae in the pseudopods extended widely from the test. B. *Globigerinoides ruber*, a living specimen. C. Dead test of *Globigerinoides sacculifera* with spines. D. Test of *G. sacculifera* at stage of gametogenesis when the test is thickened. (Courtesy of Roger Anderson, Lamont-Doherty Geological Laboratory and *Micropaleontology*, v. 23(2), pp. 155–179.)

became very abundant in mid-Cretaceous time and were important contributors to the formation of the typical Late Cretaceous carbonate sediment, chalk. Although 87 percent of the genera of planktonic forams became extinct in the Late Cretaceous crisis, the rest made a rapid recovery in Early Cenozoic time and continue to thrive and contribute their tests to the layers of *Globigerina* ooze that now cover 127 million square km of the deep ocean floor. However, both benthonic and planktonic foraminifera were apparently more diverse and abundant at the end of the Mesozoic than they are today. The ranges in time of the species of planktonic foraminifera and their evolution have been studied in great detail because they provide a method of accurately dating the sedimentary layers that are penetrated by drilling in the floor of the oceans.

OTHER PLANKTONIC MICROORGANISMS

Only a few plants and animals that have left a paleontological record lived a planktonic existence in the surface waters of the sea in the Paleozoic Era, but this habitat became much more densely populated in Mesozoic and Cenozoic time. Marine plankton is at the base of the food pyramid. Planktonic plants are the major primary producers of the sea, as they convert inorganic nutrients to food usable by animals. The increasing availability of such microscopic food in Mesozoic time must have influenced the diversity and abundance of marine life. The planktonic foraminiferans and other animals that share this habitat comprise the zooplankton; plants, such as the coccoliths, diatoms, and dinoflagellates, described in the following discussion, constitute the phytoplankton.

Coccoliths:

If Cretaceous chalk is examined under the high powers of the scanning electron microscope, the grains are revealed to be composed of subangular calcite particles which are locally united into circlets of overlapping platelets (Fig. 10–4A). The circlets, called coccoliths, are secreted by single cells of yellow-green algae (Chrysophyta) which live today for part of their life cycle in great abundance in the surface waters of the sea and are known from rocks as old as Triassic. During life, these coccoliths, a

Figure 10-4   Coccoliths and coccosphere. The scale bars are 1 micrometer. A. Coccoliths in Cretaceous chalk. (Courtesy of C. W. Stearn & A. J. Mah.) B. Coccosphere of the living species *Emiliana huxleyi*. (Courtesy of Margaret Goreau.)

few micrometers across, are arranged around the algal cell to form a coccosphere (Fig. 10–4B). They fall off the sphere when the cell dies, and tend to disintegrate to their constituent platelets which are the grains of chalk. The great chalk cliffs of the English channel are evidence of the abundance of these plants in Cretaceous seas.

Diatoms:

Other members of the chrysophyte algae that appeared in Early Triassic time and became abundant in the plankton of Cretaceous time are the diatoms. These algae secrete a minute skeleton (commonly 10 to 100 micrometers across) whose preservable part is composed of opal—hydrous and noncrystalline silica. The tests are commonly discoidal or oblong in shape, ornamented with knobs and ridges, and consist of two valves that fit into each other like the lid and base of a pill box (Fig. 10–5). Diatoms have, since the Cretaceous, thrived in fresh (Fig. 10–6) and marine water and left their tests behind in the form of a light and minutely porous rock called diatomite. Deposits in the Miocene rocks of California (Monterey Formation) and the Pleistocene of Germany have been exploited for use as fillers, abrasives, filtering media, and originally as the medium which Alfred Nobel used to absorb nitroglycerine to make dynamite.

Dinoflagellates and Acritarchs:

The dinoflagellates are a group of algae some of which secrete cellulose cysts that can be preserved (Fig. 10–7D,E). As the name implies, they are equipped with flagellae. At times some of these plants multiply rapidly in modern oceans and secrete a toxin that causes mass mortality in fish and makes filter feeding clams that are normally edible, poisonous. As the organisms causing this phenomenon, which have reddish pigments, become abundant enough to color the water, these mass mortalities are referred to as red tides. The cysts can be extracted from sedimentary rocks by dissolving the rock in hydrofluoric acid and can be used for dating rocks in the relative time scale. Dinoflagellate cysts first become common in rocks of Late Triassic age but a single genus, regarded by some as the oldest dinoflagellate, is known from Upper Silurian rocks. Carbonized organic capsules commonly ornamented with fine spines have been found

Figure 10-5   The marine diatom *Odontella*. Scanning electron micrograph. The scale bar is 10 micrometers. (C. W. Stearn.)

in insoluble residues of Paleozoic and late Proterozoic rocks (Fig. 10–7F,G). Some paleontologists believe these are related to the dinoflagellates but, since their affinity to a living group is unknown, they are referred to by the nonspecific name, acritarchs. They have also been used in correlation of Paleozoic rocks and have the potential for use as correlators in late Proterozoic rocks.

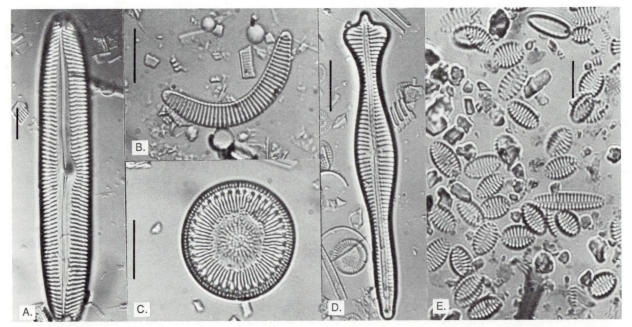

Figure 10-6   Freshwater fossil diatoms from Pleistocene sediments. The scale bars are 10 micrometers. A. *Pinnularia viridis*, B. *Semiorbis hemicyclis*, C. *Cyclotella comta*, D. *Gomphonema acuminatum*, E. *Frigilaria* sp. (Courtesy of John Smol, reproduced with the permission of the Geological Association of Canada, *Geoscience Canada*, v. 14, p. 210.)

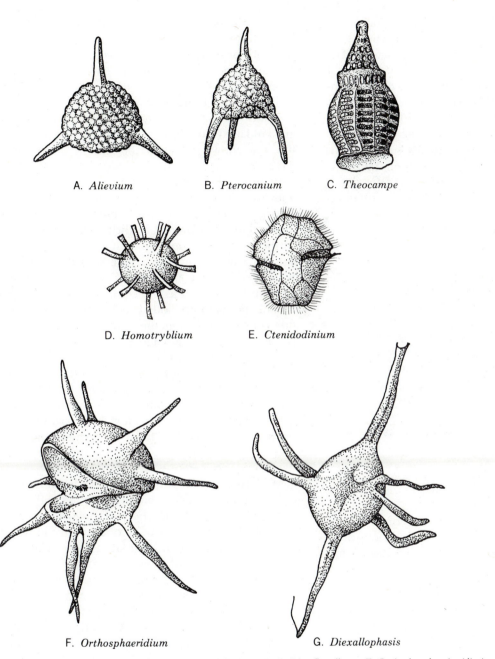

A. *Alievium*   B. *Pterocanium*   C. *Theocampe*

D. *Homotryblium*   E. *Ctenidodinium*

F. *Orthosphaeridium*   G. *Diexallophasis*

Figure 10-7   Planktonic organisms. A-C. Radiolarians, D, E. Dinoflagellates. F, G. Acritarchs. A. *Alievium* (Cretaceous) (0.2 mm). B. *Pterocanium* (Miocene) (0.2 mm). C. *Theocampe* (Cretaceous) (0.2 mm). D. *Homotryblium* (Cenozoic) (0.05 mm). E. *Ctenidodinium* (Jurassic–Cretaceous) (0.075 mm). F. *Orthosphaeridium* (Ordovician) (0.07 mm). G. *Diexallophasis* (Silurian) (length 0.1 mm).

Radiolaria:

Like the dinoflagellates, the radiolarians secrete a skeleton of opaline silica. Their siliceous tests are spherical with radial spines of great delicacy, or they are helmet-shaped (Fig. 10–7). Although radiolar-ians are known from Paleozoic rocks (and were locally important contributers to sediments), they did not become abundant until the Mesozoic and are common constituents of modern plankton. Radiolarian oozes are accumulating now on the deep ocean floor in an equatorial belt west of Ecuador.

Older deposits to which radiolarians have made significant contributions include the siliceous sediments and cherts of the Arkansas novaculite (Devonian), the cherts of the Cache Creek Group of British Columbia (Permian), the cherts of the Franciscan Formation of California (Jurassic), and the Oceanic Formation of Barbados (middle Cenozoic). Radiolarians are believed to have been contributors to the formation of silceous sediments, such as chert, throughout Phanerozoic time but their skeletons are easily dissolved and reprecipitated after deposition leaving little trace. About one-third of all radiolarians contain algae living in their tissues which supply them with oxygen.

Chitinozoa:

Chitinozoa are minute carbonized flasks found when Paleozoic rocks are dissolved in acid or disaggregated into particles (Fig. 10–8). As they are composed of chitin, they are assigned to the zooplankton, but their affinity is unknown. Some paleontologists have suggested that they are the egg capsules of invertebrates, such as graptolites. They are most abundant in rocks of Ordovician to Devonian age (the time when the graptolites were most abundant) but have been found in rocks as young as Permian in age.

SCLERACTINIAN CORALS

The second wave of corals, the Scleractinia, did not appear in the paleontological record until long after their relatives, the Rugosa, had ceased to be important as reef builders. Although rugose corals lasted until the end of the Permian Period, they were not prominent contributors to the late Paleozoic reefal facies which was formed largely by algae, sponges, and bryozoans. They steadily decreased to extinction at the end of the Paleozoic Era as conditions for invertebrates worsened. No corals of any kind are found in the Lower Triassic Series and the new corals appear at the base of the Middle Triassic.

In septa, dissepiments, tabulae, corallite shape, and growth form the scleractinians do not differ greatly from the rugose corals but their septa were inserted in a fundamentally different way and they formed their skeleton of aragonite, rather than calcite. In rugose corals after the six original septa have been formed, additional septa are inserted serially in four places as the coral grows. In scleractinians the septa are added six at a time or in multiples of six. In addition, the exterior wall of corallites is not like that of the rugose corals, where it is an independent structure, but is formed by the joining of the ends of the septa. As no intermediates have been found between the two types of corals and their time ranges are separated by the Early Triassic gap, the scleractinians do not appear to have evolved from rugose stock but from uncalcified and unpreserved cnidarians like modern anemones that developed a calcareous skeleton (1).

Modern scleractinians are among the most successful of limestone-forming invertebrates and build atolls, barrier, and fringing reefs in tropical seas (Fig. 10–9). However, not all of them live in tropical reefs; many live at high latitudes in relatively cold and deep water. The reef corals are called hermatypic and the nonreef corals, ahermatypic. The hermatypic corals have developed a symbiotic relationship with unicellular dinoflagellate algae that live within their soft tissue. A symbiotic relationship is one that benefits both organisms—in this particular relationship, the coral obtains oxygen for respiration from the alga, and the alga obtains the carbon dioxide that it needs from the respiration of the waste gas by the coral. The algae make oxygen by photosynthesis and for photosynthesis they need light. Since hermatypic corals have given up other methods of obtaining oxygen, they have become indirectly dependent on light and can only live at depths where sufficient light penetrates to satisfy photosynthetic requirements of their symbiotic algae. The depth below which the corals cannot grow is called the oxygen compensation level. Below this depth they can be said to drown for they

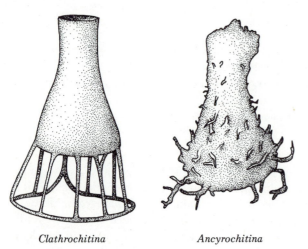

Clathrochitina   *Ancyrochitina*

Figure 10-8   Chitinozoans. A. *Clathrochitina* (Ordovician) (0.2 mm). B. *Ancyrochitina* (Ordovician) (0.14 mm).

Figure 10-9   Assemblage of scleractinians on a living fringing reef, Discovery Bay, Jamaica. (C. W. Stearn.)

do not get enough oxygen for respiration. The depth varies with all the factors that affect penetration of light into the ocean such as turbidity, turbulence, cloud cover, and latitude, but rarely exceeds about 90 meters. Prolific growth of corals is confined to the upper 20 meters of water. Coral reefs cannot grow to the surface from water that is deeper than 90 meters, and thicker accumulations of corals must have been constructed as the growing corals kept pace with subsidence of the seafloor on which they were established.

The symbiotic algae (a type of dinoflagellate) are called zooxanthellae. They have been suspected of helping the coral to calcify and to eliminate nitrogenous wastes as well as to respire, but the intracellular chemistry by which they operate is still not clear in detail. Corals can live in water that is poor in nutrients and planktonic food because their symbiotic algae supply them with many of the essentials of life. George Stanley (5) believed that the symbiotic relationship was not established until late in the Triassic Period because not until then did the scleractinians become important frame builders. Although the aragonite of modern corals with zooxanthellae may have a different isotopic signature from others, no method has yet been found of determining whether fossil scleractinians or rugose corals had these symbiotes because in their skeletons the isotope ratios have been disturbed.

The growth forms of the Scleractinia duplicate those of Paleozoic colonial invertebrates (Chapter 6). In the brain corals, such as *Diploria* (Fig. 10–10), they evolved a new form. The corallites elongated, become confluent with adjacent ones, and entwined sinuously to produce a colony whose convoluted surface resembles a brain. A few hermatypic and ahermatypic corals are solitary, but most are colonial. The rate of growth and the amount of calcium carbonate extracted from the sea by these animals is remarkable. Hemispherical corals in the Caribbean add 1 cm of skeleton per year, and branching corals may extend by 25 cm per year. The corals on a small, not particularly healthy, Barbados reef extract 2 kg per square meter per year of calcium carbonate from seawater. Modern reefs where corals and algae thrive are the carbonate factories of the sea.

The scleractinians increased in abundance and diversity from Late Triassic to Late Cretaceous time when their ascendancy in the reef community was threatened briefly by the extraordinary success of the rudist bivalves. Since the beginning of the Cenozoic Era the scleractinians have consolidated their dominance of the reef environment.

BIVALVIA

Brachiopods and bivalves are both largely sedentary, small invertebrates enclosed in two calcareous

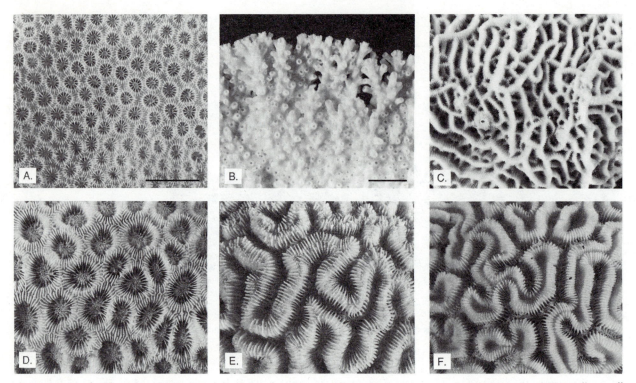

Figure 10-10   Surfaces of modern scleractinian corals from the Caribbean. The scale bars are 1 cm and the bar in B applies to all photographs except A. A. *Montastrea annularis*. B. *Acropora palmata*. C. *Agaricia agaricites*. D. *Montastrea cavernosa*. E. *Diploria strigosa*. F. *Diploria labryrinthiformis*. (C. W. Stearn.)

shells. One dominated shallow marine environments of Paleozoic time and the other has dominated the same environments since the end of that era. We have seen that the brachiopods yielded the inshore and estuarine environments to the bivalves progressively as the Paleozoic passed, but this slow change was interrupted by the Permian invertebrate crisis. David Raup (2) has estimated that 96 percent of brachiopod species became extinct at the end of the Permian Period, yet the major divisions of the bivalves seemed to have been little affected. Some brachiopods survived and continued to live as minor elements of the fauna in Mesozoic and Cenozoic seas but whereas these brachiopods merely survived, the bivalves continued to expand in diversity and number.

In the muscular, digging foot the bivalves had an organ that allowed them literally to penetrate environments that were unavailable to the brachiopods. Most Paleozoic bivalves lived on or near the sediment surface, but a few were burrowers. In the Permian Period many more turned to an existence within the sediment using the foot for burrowing. This movement into the subsurface may have been made possible by the fusion of the two lobes of the mantle to keep sediment grains out of the interior and to form siphons by which the buried animal could stay in contact with the water above. Why did many bivalves move into the sediment at the close of the Paleozoic Era? Geerat Vermeij (3) has suggested that they were escaping recently evolved predators capable of crushing hard shells. Bivalves evolved forms that could live as far as a meter beneath the surface.

Not all post-Paleozoic bivalves were burrowers; they diversified into many other aquatic environments. Some adapted to boring into hard skeletal tissue of other invertebrates or into rock or wood; others retained into adulthood the larval stage of attachment by byssal threads; still others cemented themselves to hard bottoms or other organisms, or reclined on soft sediment. The form of the shells of living bivalves has been accurately correlated with their habitats. With this knowledge paleontologists can deduce the habitat of fossil bivalves.

Classification:

Unfortunately, shell form is not a useful character for establishing relationships because it is highly

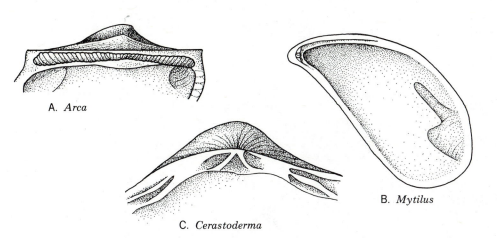

A. *Arca*

B. *Mytilus*

C. *Cerastoderma*

Figure 10-11   Patterns of bivalve dentition. A. Taxodont (*Arca*). B. Dysodont (*Mytilus*). C. Heterodont with central cardinal and lateral teeth (*Cerastoderma*).

correlated with current way of life rather than ancestry. Fossil bivalves, therefore, may be difficult to place in classifications of living bivalves that are based mostly on soft part anatomy. The combination of laminar, cross-laminar, and prismatic layers of calcite and aragonite in mollusc shells produce a bewildering number of microstructures (4) which are considered further in Chapter 19. The teeth are either long lateral ridges placed on either side of the umbo parallel to the hinge line or short cardinal teeth located below the umbo at high angles to the hinge line. The most common pattern, called heterodont, consists of a combination of these two types of teeth (Fig. 10–11C). In taxodont dentition the teeth are all short, like cardinal teeth, but are arranged in a row along the hinge with axes forming a fan pattern. This type of dentition occurs early and persistently in such primitive bivalves as the nuculoids and in modern ark shells (genus *Arca*, Fig. 10–11A). The dysodont type of dentition, found in *Mytilus* and its relatives, consists of single lateral teeth along the edge of the hinge (Fig. 10–11B). Several other patterns of teeth are found in specialized families.

Phylum—Mollusca
　　Class—Bivalvia
　　　　Subclass—Paleotaxodonta
　　　　Subclass—Isofilibranchia
　　　　Subclass—Heteroconchia
　　　　Subclass—Pteriomorphia
　　　　Subclass—Anomalodesmata

Burrowing Bivalves:

Bivalves that live unattached within sediments tend to be bilaterally symmetrical and to have two subequal adductor muscles. Their shape is adapted to the depth to which they burrow and the nature of the sediment, whether sand, mud, or gravel. Because these bivalves rock their shells forwards and backwards as they move into the substrate, any surface ornament, such as flanges, that interlocks like a rachet with the sediment grains to impede slipping is advantageous. Once in place, most burrowing bivalves move little and only when their position with respect to the surface is altered by scouring of the sediment in storms or by burial by sudden deposition. The lack of a mechanism for any but the most simple burrowing movements within the sediment has limited the environments that this class of molluscs has been able to exploit to a relatively narrow range compared to those exploited by the gastropods. Few Paleozoic bivalves show the pallial sinus that is evidence of the retractable siphons of deep-burrowing clams.

Many familiar and edible clams, such as the quahog (*Mercenaria*) (Fig. 7–1) and the cockle (*Cerastoderma*), today occupy the burrowing habitat. Shallowly burrowing clams generally have thick, highly ribbed shells to resist predator attacks but deeper burrowers have thinner, smoother shells. In such deeply burrowing forms as *Mya* (Fig. 10–12E), the siphons are so large that they cannot be retracted and the valves gape at the back end. Among the most abundant Mesozoic bivalves are a group typified by the genus *Trigonia* (Fig. 10–12D), that de-

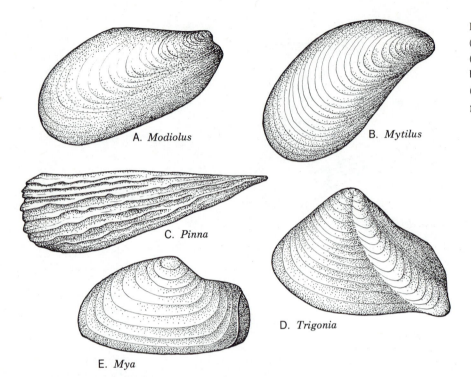

Figure 10-12   Bivalves. *Modiolus* (Carboniferous–recent). B. *Mytilus* (Triassic–recent). C. *Pinna* (Carboniferous–recent). D. *Trigonia* (Triassic–Cretaceous). E. *Mya* (Oligocene–recent).

A. *Modiolus*

B. *Mytilus*

C. *Pinna*

D. *Trigonia*

E. *Mya*

veloped heavy shells with a distinctive ornament of flanges and knobs that meet along a prominent ridge from the umbo to the lower edge. Freshwater clams first appear in Devonian rocks but are not abundant in the paleontological record at any time. The largest group belongs in the order Unionoida, but some other bivalves also reached this environment. Their shells are characterized by a thick organic coating that protects the carbonate from dissolution. This coating is insubstantial in most marine bivalves because they live in a solution saturated with calcium carbonate.

Byssally Attached Bivalves:

When the swimming larva of most bivalves settles to the bottom, it attaches itself to grains of sediment by fibers that together form the byssus. The byssus is secreted by a gland in the foot. The byssus keeps the minute shell from being swept from its resting place. Most bivalves abandon this attachment as adults to become burrowers, borers, and swimmers, but many retain it. Several lineages remained partly or completely embedded in sediment and anchored by the byssus to sediment grains, but other byssate groups evolved to live on the surface attached to rocks or the shells of other organisms. Those that remained within the sediment have valves that are mirror images, like those of burrowers, but many

are highly asymmetric (Fig. 10–12A,B,C). Modern bivalves that live a half-buried life style, typified by the modern genus *Modiolus* (Figs. 10–12A, 10–13B,C), are shaped like some Paleozoic genera that are thought to have lived the same way. The posterior part that stuck out of the sediment surface is elongated and expanded and the anterior part reduced. Another common Jurassic to recent byssally attached bivalve that lives half-submerged in sediment is the fanlike *Pinna* (Fig. 10–12C). Partly buried byssate bivalves develop inequality in the size of the two adductor muscles; the anterior one, submerged in sediment, becomes small, and the posterior one, large (Fig. 10–13). This inequality increased in those bivalves that evolved to live on the sediment surface.

A diverse group of byssate bivalves abandoned life within the sediment to live on the surface. In those that maintained the plane between the valves in a vertical orientation, the valves remained mirror images, but in those that reclined on one side, the valves became not only asymmetrical but also markedly unequal in size. Of the first group *Mytilus* (Triassic to recent), the mussel has a shell flattened at the bottom where it is attached by the byssus, and the muscles are unequal in size (Figs. 10–12B, 10–13D). *Mytilus* grows in great numbers over rocks and its own shells in the tidal zone where it must resist daily drying cycles and freshwater influx. The

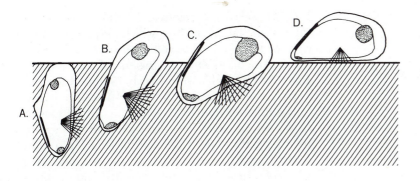

Figure 10-13   Transition in modern mytilid bivalves from completely buried to attached to the sediment surface. Muscles are dotted. A. *Brachidontes*. B, C. *Modiolus*. D. *Mytilus*. (*Source:* Modified from S. Stanley, *Journal of Paleontology,* v. 46, pp. 361–385.)

epifaunal byssate group also includes shells of the *Arca* type (Fig. 10–11A).

The region where the byssus is attached is commonly marked by a re-entrant in the shell known as the byssal notch. The bivalves of the superfamily Pteriacea reclined on one valve attached by a byssus in a notch located below an extension of the hinge line in front of the umbo called an auricle (Fig. 10–14). Similarly shaped valves are common from Ordovician time onward.

A byssate bivalve group in which the valves remain mirror images is represented by the common Cretaceous genus *Inoceramus* (Fig. 10–15F). This thick-shelled bivalve is marked by concentric coarse annulations. The disintegration of its shell after death contributed large quantities of calcite prisms as sediment to the accumulation of chalks of this period.

Swimming Bivalves:

Scallops are characterized by a symmetrical shell that is usually radially ribbed, has the hinge line extended into auricles on either side of the umbo, and has weak teeth (Fig. 10–15E). The typical genus is *Pecten*, which is widely displayed as the symbol of the Shell Oil company. Although *Pecten* and its relatives are byssally attached in the early stages of life (Fig. 10–14), they recline on one valve as adults and are capable of moving through the water by jetting water from the mantle around either side of the auricles. The jets are produced when the valves are snapped together by the contraction of a single drum-shaped adductor muscle that we eat when served a dish of scallops. The snapping motion evolved in most bivalves to expel sediment from between the valves but the scallops have adapted it for propulsion. Fossil bivalves of this shape, and presumably of this life-style, occur in rocks as old as Carboniferous in age. Many members of the superfamily Pectinacea do not swim but retain the byssal connection thoughout life and some

can detach the byssus and swim if motivated by predators to flight. The swimming bivalve *Lima* aids the jet propulsion by rowing with tentacles that are extensions of the mantle around the edges of the valves.

Cemented Bivalves:

In late Paleozoic time a group of pteriomorph bivalves, instead of attaching themselves in a reclining position by a byssus, cemented the surface of one of the valves to rocks or other shells. These attached bivalves are commonly called oysters. The lower valve grew larger than the caplike top valve and may conform to the shape of the object to which it was cemented. Only one muscle is present. The oysters were, and are, a highly successful group and encrust hard surfaces in shallow water. Even where the seafloor is soft, once a few shells have made a foothold, oysters will grow on the dead shells of their ancestors to build large carbonate banks. Some oyster banks may reach the dimensions of reefs and are important traps for petroleum in Mesozoic and Cenozoic successions.

Some of the oyster family did not remain cemented throughout their lives but as adults broke free and reclined on the sediment. *Gryphaea* was a common Jurassic oyster with a very thick lower valve coiled like a ram's horn in a vertical plane (Fig. 10–15B). The evolution of *Gryphaea* in the British Jurassic rocks has been studied intensely because its progressive coiling was cited as an example of an evolutionary trend of long duration. *Exogyra*, a common Cretaceous relative, developed an equally thick shell but the umbo coiled in the dorso-ventral plane (Fig. 10–15D).

Rudists:

Descendants of the megalodont bivalves of the Devonian reef facies thrived in Late Jurassic and Cre-

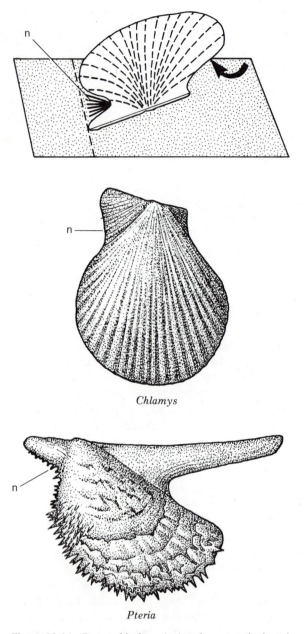

Chlamys

Pteria

Figure 10-14   Byssate bivalves. A. Attachment at the byssal notch (n) and function of the auricle to prevent overturning by wave action (curved arrow). B. *Chalmys* (Triassic–recent), a pectinacean with a byssal notch. C. *Pteria* (Cretaceous–recent), a pteriomorph with a byssal notch.

taceous time. Nearly all the advanced forms of this group, referred to as rudists, became extinct at the end of the Cretaceous after phenomenal success. A few persisted into the Paleocene of Russia and Cuba, and the thick-shelled reef oyster *Chama* (Fig. 10–

15C) of today is a relative. The lower cemented valve of the largest rudists, which may be either the right or left, evolved to conical form and the other valve formed a cap to close the end of the cone (Fig. 10–16). The teeth were highly modified and the upper valve, instead of hinging open, lifted vertically guided by long teeth sliding in slotlike sockets. *Hippurites* and *Radiolites* are typical of this group. In other rudists both valves are thickened into coiled forms (*Diceras*) and reclined on the sediment (Fig. 10–16A). In still others one is coiled and the other is conical (*Caprina*, Fig. 10–16B). During the Cretaceous, caprinids, hippuritids, and radiolitids displaced the corals and sponges as the principal reef builders and built carbonate reefs of regional extent that are important oil exploration targets in the Mediterranean and Caribbean areas. Not only did these bivalves mimic solitary rugose corals in their external form but they also divided the interiors of their valves with horizontal partitions like tabulae and dissepiments, and radial ones like septa. The walls were not solid but were composed of a network of cells. The largest hippuritids are 2 meters in height and grew in thickets so that they supported one another in upright positions. They are widely believed to have achieved such massive skeletons through a symbiotic relationship with photosynthetic algae like the scleractinian corals. The upper valve in some appears to have allowed the mantle to remain exposed when the valves closed to cover the viscera, and in others the valve is so thin that it may have been transparent to allow the algae to photosynthesize when it was closed. The South Pacific genus *Tridacna* is the best-known living bivalve with this relationship to algae. The genus includes the largest known living bivalves but is not related to the rudists. When its valves are open, the edges of the mantle exposed to the light are vividly colored by various strains of symbiotic algae.

The first evolutionary expansion of the rudists in Early Cretaceous time resulted in the deposition of the thick-bedded, cliff-forming limestones around the Mediterranean. Similar rocks of the El Abra Formation are major producers in the Golden Lane and more recently discovered oil fields of southern Mexico. In the second major expansion of Late Cretaceous time the radiolitids spread from the tropics as far north as England and southern Canada. The catastrophic decline of the rudists at the end of the Cretaceous took place at the height of their abundance and diversity.

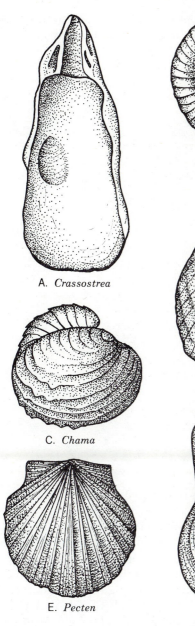

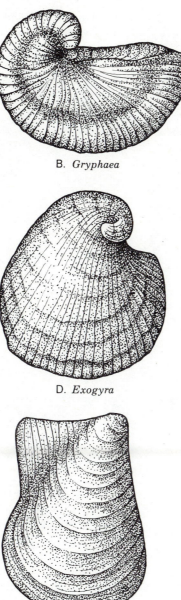

A. *Crassostrea*

B. *Gryphaea*

C. *Chama*

D. *Exogyra*

E. *Pecten*

F. *Inoceramus*

Figure 10-15 Bivalves. A. *Crassostrea* (left, interior) (Cenozoic). B. *Gryphaea* (Triassic–Jurassic). C. *Chama* (Cenozoic). D. *Exogyra* (Cretaceous). E. *Pecten* (Cretaceous–recent). F. *Inoceramus* (Jurassic–Cretaceous).

Borers:

A small number of bivalves of the subclasses Heteroconchia and Pteriomorpha developed the ability to bore into hard substrates, an adaptation that allowed them to escape predators. These borers chip, grind, and file with the rough front edges of the valves against rock, wood, or skeletal tissue of other animals. Some secrete chemicals that dissolve the substrates they attack. Most, such as *Lithophaga* and

Pholas, are round in cross section and elongate in form (Fig. 10–17). These bivalves are major agents in the destruction of reef limestones in the tropics today and are therefore important in the study of bioerosion. The shipworm, *Teredo*, is typical of bivalves that have developed the ability to penetrate and digest wood. Wooden ships and piers in modern oceans soon fall apart under the attack of these bivalves unless protected by antifouling paints. The shell of many members of the family covers only a

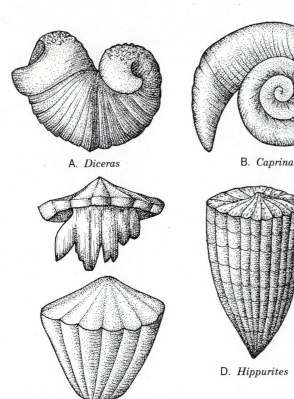

A. *Diceras*

B. *Caprina*

D. *Hippurites*

C. *Radiolites*

Figure 10-16   Rudist bivalves. The specimens are several centimeters to tens of centimeters in length. A. *Diceras* (Jurassic). B. *Caprina* (Cretaceous). C, D. *Radiolites* (Cretaceous), right valve showing the teeth and complete specimen. E. *Hippurites* (Cretaceous).

small area of the front of the animal and the elongate body resembles a worm more closely than a mollusc because it is mostly naked.

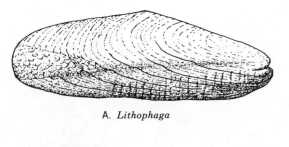

A. *Lithophaga*

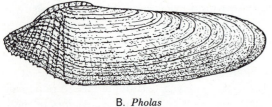

B. *Pholas*

Figure 10-17   Boring bivalves. A. *Lithophaga* (Cenozoic). The back of the shell is commonly covered with a calcareous encrustation. B. *Pholas* (Cretaceous–recent).

GASTROPODA

Gastropod faunas of the early Mesozoic were dominated by orders common in Paleozoic rocks, the Archaeogastropoda and Mesogastropoda. The major event for the snails in the Mesozoic was the evolution from mesogastropod stock of the neogastropods in the middle of the Cretaceous Period. The Neogastropoda are distinguished from more primitive gastropods by details of soft part morphology related to the radula and the gill. Most are predatory gastropods with prominent siphons and modification of the aperture margin to accomodate them. The shapes of their shells are so close to those of the mesogastropods that the two orders are difficult to distinguish.

Geerat Vermeij (3) has noted that many late Mesozoic gastropods are ornamented with conspicuous nodes, spines, and ridges that strengthen the shell against crushing forces. This is particularly prominent in the muricids, illustrated by the genera *Murex* and *Ecphora* (Fig. 10–18C,D). Others protected their apertures by narrowing them to slits as in the neogastropod *Conus* (Fig. 10–18A), and the

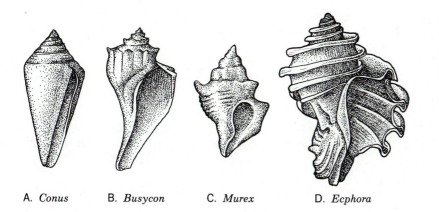

Figure 10-18   Neogastropods from Mesozoic and Cenozoic. About natural size. A. *Conus* (Eocene–recent). B. *Busycon* (Oligocene–recent). C. *Murex* (Eocene–recent). D. *Ecophora* (Cretaceous–Miocene).

A. *Conus*   B. *Busycon*   C. *Murex*   D. *Ecphora*

mesogastropod cowrie shells (family Cypraeidae). The adoption of these protective devices appears to have coincided with the rise of fish and arthropods that fed by crushing molluscan shells. The drilling gastropods that make neat circular holes in the shells of other molluscs became common in the late Mesozoic. Similar holes found in Paleozoic brachiopods were made by an unknown organism.

The three orders of gastropods were all abundant in the Cenozoic Era, but only the mesogastropods and neogastropods were rapidly increasing in diversity.

AMMONOIDS

By late Paleozoic time the basic architecture of the ammonoid cephalopods had been established. They were swimmers suspended in the water column by buoyant gas-filled chambers located in a coiled shell above the body, and the partitions between the chambers (the septa) were fluted or folded along their periphery where they joined the wall of the shell. The siphuncle followed the outside margin of the coil, the venter, and was used to empty water from recently formed chambers to keep the animal near neutral buoyancy. The evolution of the ammonoids has been described from studies of the folding of the septal edges that is illustrated by the form of the septal suture line.

The Ammonoid Suture:

The septal suture is evident on most fossil ammonoids because the wall of the shell is usually broken away exposing the folded edges of the septa. When the suture line is illustrated, the venter is labelled

on the suture line by an arrow pointing toward the aperture, and the edge where the suture is covered on the dorsum by the overlap of the former whorl is marked by a vertical line (Fig. 10–19B). The upward folds of the suture illustrated in this way have the shape of a horse's saddle seen from the front and are called saddles. The folds between them, which are concave toward the aperture, are called lobes. Specialists have developed complicated systems of naming the lobes and expressing the form of the whole suture as a formula.

In certain forms of preservation the successive chambers of the ammonoid can be separated back to the tiny first chamber secreted in the egg. From these isolated chambers the changes in the form of the suture from the ammonoid's infancy to its maturity can be plotted. The first sutures were simple, like those of the oldest ammonoids, and more lobes and saddles of greater complexity developed as the animal matured.

The three major suture types in the ammonoid cephalopods are almost always referred to as goniatitic, ceratitic, and ammonitic. Goniatitic sutures, which are characteristic of most late Paleozoic shells, have simple lobes and saddles, and most have only a few of them (Fig. 7–12). Ceratitic sutures are particularly characteristic of Triassic ammonoids but appeared in Early Carboniferous and Cretaceous shells and have smooth saddles and lobes that are finely frilled (Fig. 10–19A). In ammonoid sutures both lobes and saddles are refolded and highly complex (Fig. 10–19B). Although such sutures are characteristic of Jurassic and Cretaceous ammonoids, they are also found in some Permian and Triassic shells. Although knowledge of the sutures is essential in the identification and classification of ammonoids, the shape and ornamentation of the surface of the shell are also important.

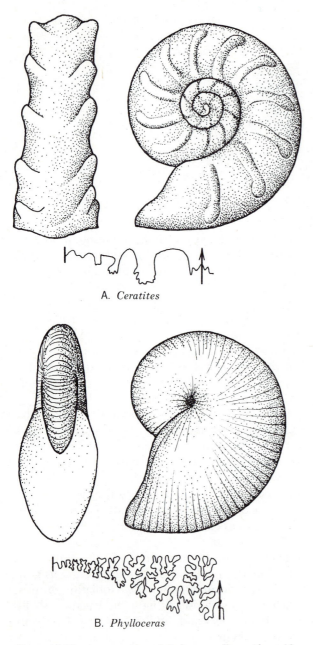

A. *Ceratites*

B. *Phylloceras*

Figure 10-19   Ammonoids and their suture lines. About 10 cm across. A. *Ceratites* (Triassic). B. *Phylloceras* (Jurassic–Cretaceous).

The Ammonoid Shell:

Shells range in diameter from a few millimeters to about 2 meters. Ammonoid giants are characteristic of Late Jurassic and Cretaceous time in several parts of the world (Fig. 10–20). Most Mesozoic ammonoids are about 10 cm across. The cross section of the coiling cone ranges from fat and rounded to

thin and streamlined (Fig. 10–21). It may be ornamented by keels and ridges that wind around the shell, but most surface ornamentation is transverse to the length of the cone. Nodes, knobs, spines, ridges, and folds interrupt the surface of many post-Paleozoic ammonoids but are uncommon on Paleozoic shells. They have been interpreted as having both a protective and strengthening function. These transverse features are independent of the fine growth lines that show former positions of the aperture margin.

When ammonoids reached maturity they stopped secreting their shells. Mature shells can be distinguished from immature ones by the closer spacing and increased thickness of the last few septa and, in some genera, by forward extensions of the sides of the aperture. The most distinctive of these are tonguelike extensions called lappets (Fig. 10–21C). They may be a dimorphic structure secreted by only one sex, but their function is obscure.

In many deposits of ammonoids, morphologically similar large shells (macroconchs) and smaller shells (microconchs) are found associated. The macroconchs usually have several smooth, inflated, extra whorls but their interior whorls are like those of the microconchs. The microconchs may have lappets and the macroconchs none. Paleontologists have identified such pairs of ammonoids as sexual dimorphs but do not agree on which is the male (Fig. 3–10).

Associated with ammonoid shells in fossil depos-

Figure 10-20   Giant Late Jurassic ammonite *Parapuzosia*, 1.7 m across. (Courtesy of Ulrich Lehmann, *The Ammonites and Their World*, Cambridge University Press, 1981.)

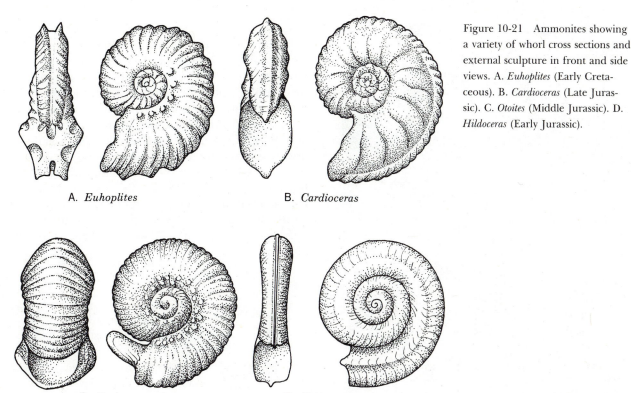

Figure 10-21 Ammonites showing a variety of whorl cross sections and external sculpture in front and side views. A. *Euhoplites* (Early Cretaceous). B. *Cardioceras* (Late Jurassic). C. *Otoites* (Middle Jurassic). D. *Hildoceras* (Early Jurassic).

A. *Euhoplites*

B. *Cardioceras*

C. *Otoites*

D. *Hildoceras*

its are pairs of hinged calcite plates called aptychi (single, aptychus) or single, phosphatized, proteinaceous plates, called anaptychi. Although originally thought to be plates that closed the aperture when the animal was withdrawn into its shell, anaptychi are now identified as lower jaw parts. Aptychi may have served as both jaws and closures. Radulas have been found in exceptionally well-preserved body chambers.

Heteromorphs and Habitats:

Several times in the history of the ammonoids a group abandoned the tight planispiral coil and secreted shells that were loosely coiled, planispirally coiled at first and then straight, hooked at the end, trochoidally coiled like a gastropod, completely irregular, and various combinations of these (Fig. 10–22). These shells are called heteromorphs. A small group developed heteromorph shells in Late Triassic time. Heteromorphs reappear beginning in Middle Jurassic time but the time of their greatest abundance was the Late Cretaceous. Although their shapes may appear bizarre and aberrant, they must have been well adapted to their habitats, for many, such as the gastropodlike *Turrilites* (Fig. 10–22C),

achieved worldwide distribution and a long range. Unlike the gastropods that some of them resemble, the heteromorph ammonoids have a chambered shell and a siphuncle. The position of the siphuncle, the organ that controlled the water level in the chambers, has led some paleontologists to speculate that the heteromorphs had a higher water level and therefore less buoyancy than normal cephalopods and that they foraged for food near the bottom. The large Cretaceous ammonoid, *Baculites*, made long, straight shells after an immature coiled stage. This ammonoid was not weighted by calcareous deposits and probably floated head down. In the suborder Ancyloceratina (Late Cretaceous), typified by the genus *Scaphites* and its relatives (Fig. 10–22B), the living chamber grows away from the coiled shell then turns back toward it. Such shells must have floated in the water column feeding on small organisms.

All living cephalopods are carnivores or scavengers, as were, presumably, the ammonoids. Like *Nautilus* they could swim backwards or forwards by the jet action of the funnel but only a few seem to have evolved a streamlined shape for speed. The shell form of most ammonoids suggests that they were not fast swimmers but moved slowly near the

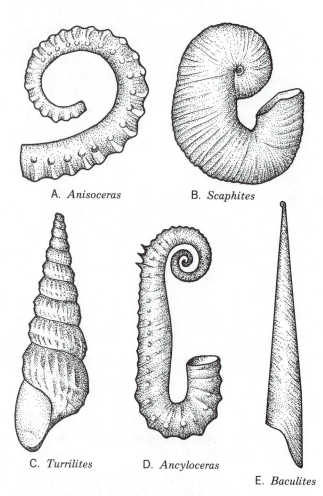

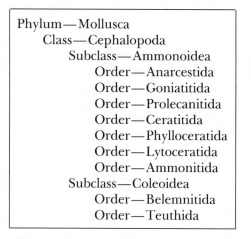

A. *Anisoceras*   B. *Scaphites*

C. *Turrilites*   D. *Ancyloceras*

E. *Baculites*

Figure 10-22   Heteromorph ammonites. A. *Anisoceras* (Cretaceous). B. *Scaphites* (Cretaceous). C. *Turrilites* (Late Cretaceous). D. *Ancyloceras* (Early Cretaceous). E. *Baculites* (Late Cretaceous).

idly. As a result they have been extensively used for dating and correlating the sedimentary rocks in which they are preserved. The basic principles of biostratigraphy were formulated by paleontologists determining the stratigraphic distribution of ammonoids. Late Paleozoic and Mesozoic rocks have been divided into 230 zones on the basis of ammonoid stratigraphic ranges. Within a basin of deposition, such as the western States, time resolution on the basis of ammonoid ranges may approach a few hundred thousand years. Because almost identical adaptations recur in ammonoid history, even a specialist may need to know the approximate stratigraphic position of a shell before making a specific identification.

Phylum—Mollusca
   Class—Cephalopoda
      Subclass—Ammonoidea
         Order—Anarcestida
         Order—Goniatitida
         Order—Prolecanitida
         Order—Ceratitida
         Order—Phylloceratida
         Order—Lytoceratida
         Order—Ammonitida
      Subclass—Coleoidea
         Order—Belemnitida
         Order—Teuthida

Ammonoid history has been used to illustrate mechanisms of evolution, but most of the older studies have recently been discredited. Recognition of the pervasiveness of sexual dimorphism and the extent of variation within species has resulted in major revisions in the taxonomy of the ammonoids and reassessment of most evolutionary successions described in the first half of this century.

The largest group of late Paleozoic ammonoids were the goniatites but accompanying them were another order, the Prolecanitida, that survived when all but one line of the goniatites became extinct at the end of the era. Ammonoids with ceratite sutures appeared in Early Carboniferous time and also survived the invertebrate crisis at the end of the Paleozoic. Ammonoid history records other close brushes with extinction in Late Devonian, Late Triassic, and Late Permian time before they did become extinct at the end of the Mesozoic (Fig. 10–23). Expansion of the order Ceratitida in the Early Triassic was slow as the seaways spread slowly across

bottom where they are thought to have fed on small animals in the water. As the distribution of many ammonoids is affected by the depositional environments in which they lived and groups are restricted to specific regions and faunal provinces, most are unlikely to have floated in the open ocean, but must have been shallow-water predators influenced by the bottom conditions. However, some ammonites inhabited open oceans and deeper water. The shells of a few large species bear scars of attacks by large marine reptiles of the Mesozoic.

History:

During their long history from the Devonian to the end of the Cretaceous the ammonoids evolved rap-

into Cretaceous time giving rise to many of the heteromorph ammonites (Fig. 10–23).

The shallow-water niches abandoned by the Ceratitida were rapidly occupied in Jurassic and Cretaceous seas by the Ammonitida. These are the ammonoids whose sutures are characterized by intricate folding of both lobes and saddles. They diversified to become some of the most abundant invertebrate fossils in late Mesozoic rocks and the inspiration of many amateur and professional collectors. They thrived in the extensive shelf seas of middle Cretaceous time that spread across western North America and Europe in the last of the great transgressions and began to decline when the seas ebbed as the period drew to a close. By the last age in the Cretaceous, the Maastrichtian (also spelled Maestrichtian) ammonite diversity was only 40 percent of what is was in the middle of the period (Fig. 16–8).

The cause of the eventual demise of the ammonites has puzzled paleontologists for over a century and its coincidence with the extinction of the great reptiles of land, sea, and air and the tiny planktonic unicells of the oceans has stimulated endless speculation. Some of the mechanisms that have been proposed to have caused this and similar catastrophic extinctions are reviewed in Chapter 16.

BELEMNITES

Cigar-shaped fossils of the relatives of the squid are common in Jurassic and Cretaceous rocks. Their name, belemnites, is derived from the Greek word for dart and, before their molluscan origin was first suggested in the 17th century, they were thought to be thunderbolts. The similarity of belemnites to squid has been demonstrated by the outlines of their soft tissues preserved in specimens in the Jurassic Holzmaden Shale of Germany. In contrast to the belemnites that formed a solid, often preserved skeleton, modern squid have practically no hard parts—only a thin chitinous blade called a pen that strengthens their back. The squid, octopus, and their relatives are placed in the the order Teuthida of the subclass Coeleoidea.

The cigar-shaped fossils are the most resistant and most commonly collected of the three parts of the belemnite shell (Fig. 10–24). This guard, or rostrum, is smoothly cylindrical, narrows to a point at what was the rear of the animal, and is composed

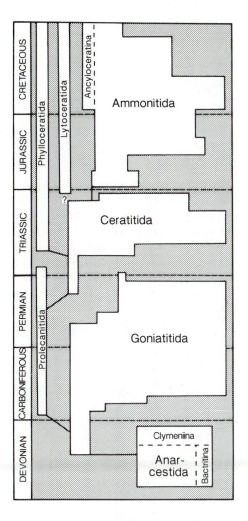

Figure 10-23 Relationships of the orders of the Ammonoidea. Suborders mentioned in the text are also plotted. The width of the boxes is approximately proportional to the number of families. (*Source:* Modified from M. R. House and J. R. Senior, *The Ammonoidea,* Systematics Assoc. Sp., v. 18.)

the emergent continental platforms, but by Middle Triassic time almost 40 families can be recognized in the order. One of the early divergent branches from ceratite stock, the order Phylloceratida, became the only survivors of the end-of-Triassic extinctions that swept the Ceratitida away. Paleontologists have postulated that these survivors inhabited deeper waters of the open oceans and survived in this stable environment when the habitats for shallow-water species were restricted by widespread regressions. Another open ocean order, the Lytoceratida, arose in Jurassic and continued

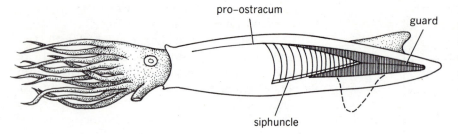

Figure 10-24 Diagrammatic restoration of the soft parts of a belemnite, showing the relationship of
these to the three parts of the skeleton.

of calcite crystals radiating from a central axis. A
conical depression at the front of well-preserved
specimens is occupied by a series of chambers separated by plain, saucer-shaped septa, and joined by
a small, cylindrical siphuncle along the ventral margin. This chambered part of the shell is much like
that of a short, slightly curved Paleozoic nautiloid.
The part of the chambered shell preserved within
the guard is only the rear end of a larger chambered
shell approximately the same length as the guard
whose front end is rarely preserved. Rarely the third
element of the shell is preserved—a thin and narrow plate, called the pro-ostracum, that extended
forward from the top of the chambered shell.

These three skeletal units were enclosed within
the body of a squidlike animal. On the head it had
ten tentacles equipped with small hooks for trapping prey. The funnel beneath the head was capable of driving the animal backwards by jet propulsion but its swimming speed is unlikely to have
equalled that of squid because the belemnite shell
was much heavier. Like a squid, the back, pointed
part of the belemnite had lateral fins to control the
backward movement. The body length was probably six times the length of the guard and a total
length of a meter is possible. Like the ammonoids,
the belemnites appear to have swum in shallow shelf
seas rather than in the open ocean. That the shell
was internal is shown by its smoothness and markings of canals in the soft tissue that are impressed
on the guard (Fig. 10–25) but its functions are not
clear. Presumably, the chambered part was buoyant
and the guard acted as a counterweight to keep the
animal horizontal in much the same way as did the
internal deposits of the early Paleozoic straight nautiloid cephalopods. The preservation of masses of
belemnites on bedding planes suggests that they
travelled in schools and were subject to mass mortality, possibly, like modern squid, after mass mating.

The earliest belemnites have been found in Early
Devonian rocks. A few survived the end of the Cre-

taceous and in the earliest Cenozoic were replaced
by the octopus and cuttlefish. Fossils of squidlike
animals that are not belemnites have been found
in rocks as old as Devonian in age. Belemnites are
believed to have arisen from the bactritids in the
Devonian by the growth of soft tissue over the shell

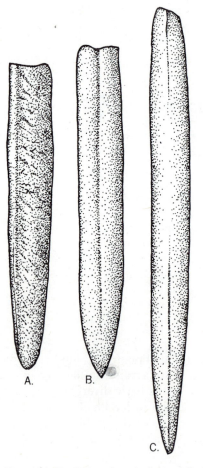

Figure 10-25 Belemnites. A. *Belemnitella* (Cretaceous) (12 cm
in length), lateral view showing the vascular marking. B.
Belemnopsis (Jurassic) (7 cm in length), ventral view showing
groove. C. *Cylindroteuthis* (Jurassic) (18 cm in length), showing
the ventral groove.

surface and the secretion of the guard as an external counterweight. This adaptation was not immediately successful and Paleozoic belemnites are rare fossils. Although most belemnite guards are superficially similar, variations in markings, grooves, and cross section permit experts to distinguish many different species and to use their ranges in Cretaceous rocks for paleontological correlation and zonation.

ECHINOIDS

Although echinoids (sea urchins) appear in rocks as old as Ordovician in age, they did not rise to prominence in the fossil record until Mesozoic time. Paleozoic echinoids are rare enough to qualify as museum specimens. The urchins are divided here into three informal groups: regular, irregular, and Paleozoic. A discussion of the last is postponed until the relation between the skeleton and life habits of living forms has been described.

Phylum—Echinodermata
  Class—Echinoidea
    Subclass—Perischoechinoidea
      Order—Bothriocidaroidea
      Order—Echinocystitoidea
      Order—Cidaroidea
    Subclass—Euechinoidea
      Superorder—Diadematacea
      Superorder—Echinacea
      Superorder—Gnathostomata
        Order—Holectypoidea
        Order—Clypeasteroidea
      Superorder—Atelostomata
        Order—Holasteroidea
        Order—Cassiduloidea
        Order—Spatangoidea

Regular Echinoids:

At one time the class was divided into the subclasses Regulares and Irregulares, but now most paleontologists agree that this classification does not reflect their history because different members of the Irregulares were decended from different stocks of the regular echinoids, that is, the group Irregulares is polyphyletic. Modern regular echinoids move about on the seafloor and are the most conspicuous group; the irregular forms generally live buried in the sediment. The regulars have a conspicuous radial pentameral symmetry which reflects their behavior. They show no preference for moving with a particular part of their body forward but move backward, forward, or sideways across the sediment surface. Because they meet their environment on all sides, they develop radial symmetry. The test of the echinoid is composed of rows of plates whose structure is porous, like that of all echinoderms, and was described in Chapter 7. The plates are arranged in 20 vertical rows defining ten sectors of two rows each. Ten (five pairs) rows are ambulacral areas and ten (five pairs), interambulacral areas (Figs. 10–26, 10–27). The anus is located in the center

Figure 10-26 Top and side views of a cidaroid echinoid. Two stout spines detached from their sockets are preserved on the top surface. (C. W. Stearn.)

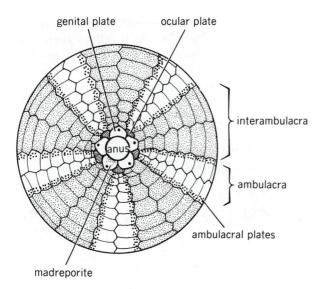

Figure 10-27 Dorsal view of the echinoid *Echinus*, showing the names of the plates around the anus.

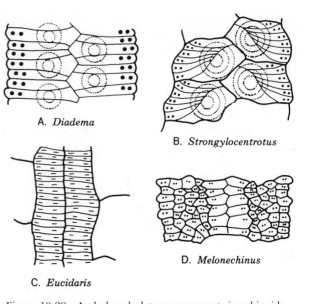

Figure 10-29 Ambulacral plate arrangements in echinoids. A. *Diadema* (Recent), each ambulacral plate is divided into three plates each with a double pore. B. *Strongylocentrotus* (Recent), each plate is divided into many demiplates. C. *Eucidaris* (Recent), many small plates form a narrow ambulacral area. D. *Melonechinus* (Carboniferous), the number of rows of plates in the ambulacral area is increased.

of the top surface and the mouth at the center of the bottom surface. Each is surrounded by a leathery membrane studded with irregular plates that are rarely preserved in fossil echinoids.

The radial water canals run from a ring canal around the mouth, beneath the ambulacral plates to single terminal pores in five ocular plates around the anus (Fig. 10–27). Each ambulacral plate is pierced by two pores, which allow a single tube foot to connect to its ampulla inside and to the radial canal (Fig. 10–28). The tube feet are used mostly for locomotion and food gathering in the regular echinoids. They attach themselves to objects and,

by their coordinated motion, pull the animal along behind them. The number of tube feet is increased by division of simple ambulacral plates into compound units of up to 16 demiplates, each with its own pore pair (Figs. 10–29, 10–30). The echinoid test grows both by the addition of new plates below the ocular plates and by growth of the plates themselves.

Both ambulacral and interambulacral plates are the bases for spines of a range of sizes. The largest may be the size of pencils, and spines of one genus were used as slate pencils in the last century. The smallest are the size of hairs. Each is attached to a plate by a ball-and-socket joint and muscles that can move it in all directions (Fig. 10–30). The spines are not only a deterrent to attackers but are used for walking across the sediment surface and handling food. In a few echinoids they are equipped with poison glands. However, the pain of most spine wounds to swimmers is not caused by poison but by the rapid decay of, and subsequent infection by, the organic matter pervading the spine tip broken off in the flesh. Spines commonly fall off the test after death and are separated from it, but their size and arrangement are indicated by the bases that

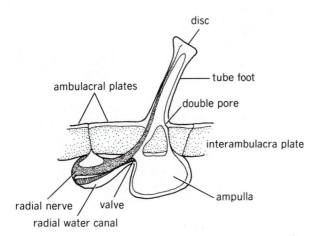

Figure 10-28 Cross section of an echinoid tube foot, ampulla and radial water canal.

Figure 10-30 Ambulacral area and adjacent interambulacral plates of the recent echinoid *Strongylocentrotus*, showing the multiplication of the double pores through the formation of many demiplates, the large primary tubercules on which spines were attached, and two orders of spine bases between them. Some of the spine bases have been abraded on the specimen. Width of field of view is about 1 cm. (C. W. Stearn.)

cover the plates. The surface of living echinoids is also covered with many tiny, snapping, beaklike appendages that are equipped with poison glands. They repel and remove the larvae of marine organisms that settle on the echinoid test.

Just inside the mouth is an inverted pyramid of braces and struts which is called Aristotle's lantern because it was first described by Aristotle and bears a slight resemblance to a lantern (Fig. 10–31). The skeletal elements are controlled by 60 muscles that

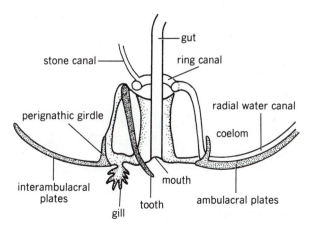

Figure 10-31 Diagrammatic cross section of Aristotle's lantern, the perignathic girdle and adjacent plates of a regular echinoid.

cause five teeth to converge at the mouth. The teeth are capable of disintegrating organic matter on which the echinoid is feeding but also can excavate hard substrates on which the echinoid is grazing. In a reef in Barbados, 5 kg per square meter of limestone was eroded yearly by the echinoid *Diadema* as it scraped algae off the reef surface. The muscles that work the lantern are attached to extensions of the plates around the mouth upward into the coelom. Between the mouth and the anus the intestine is looped and coiled in the water-filled coelom. During much of the year the coelom space is filled only with fluid but at breeding times it is filled with eggs that are eaten as a delicacy in certain parts of the world. At the edges of the membrane around the mouth, the gills which aid the tube feet in respiration may be situated (Fig. 10–31). Fossil echinoids that had gills can be identified by the slots that the gills occupy in the edges of the test.

Five or ten plates may border the flexible area adjacent to the anus. If ten plates form the circle, five of them are ocular plates at the head of the ambulacral areas, and the other five are called genital plates because they have a single pore through which the eggs or sperm are released to unite in the water. One of the genital plates (the madreporite) is minutely porous and filters water entering the water vascular system. In some echinoids the five genital plates are the only ones that border the anal area directly and the ocular plates do not reach the inner margin (Fig. 10–32).

Irregular Echinoids:

In Early Jurassic time regular echinoids gave rise to several groups that lived in burrows. This move into a new habitat was accompanied by a great increase in echinoid diversity and at present about half of echinoid species live within marine sediment. The move to subsurface living involved drastic changes. The symmetry of the test changed from radial to bilateral. Specific front and back parts developed because the animal moved in only one direction through the sediment. As seen from above many of these burrowing urchins have an oval or heart-shaped outline. The anus moved backward to a position near the lower surface, and the mouth moved forward. The space left by the migration of the anus is commonly filled by the expansion of the madreporite (Fig. 10–32). In some groups, one of the genital plates disappeared following the change of postion of the anus. In urchins that feed

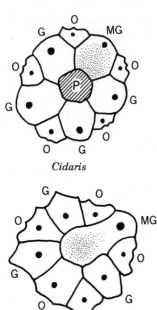

Cidaris

Micraster

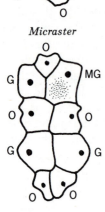

Echinocorys

Figure 10-32   Modification of plates on the dorsal surface of echinoids with the migration of the anus to the posterior. G = genital plate, O = ocular plate, MG = madreporite, P = area around anus. A. *Cidaris*: a regular echinoid in which five genital plates (one of which acts as the madreporite) surround the anus. B. *Micraster*: in this irregular echinoid, the anus has moved off the dorsal surface and its space is filled by the expansion of the madreporite. One genital plate is lost. C. *Echinocorys*: The space left by the migration of the anus is filled by the closing inwards of both ocular and genital plates. The posterior genital is lost, as in *Micraster*.

on small organic particles in the sediment that can be handled by tube feet around the mouth, the strong grinding action of the teeth is superfluous. The whole lantern apparatus disappeared in such advanced irregular echinoids.

An animal enclosed in sediment has respiratory and sanitary problems not faced by one living on the surface. Shallow burrowers in coarse sediment can obtain oxygenated seawater through the interstices between the sediment grains. Deeply burrowing echinoids get oxygen from the overlying seawater through a channel lined with mucus in the sediment formed and maintained by tube feet that may be several times the length of the test (Fig. 10–33). Water is drawn down this channel by specially modified ciliated spines beating in unison. The granular bands of fine tubercles to which these current-forming spines were attached form distinctive bands on the test. To extract oxygen from this downward current, which impinges on the top of the test, the ambulacral areas are modified to the shape of the petals of a flower (Fig. 10–34) and set in grooves to channel the water current. The tube feet in these petalloid areas become straplike

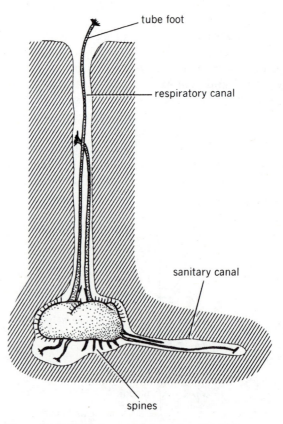

Figure 10-33   The burrowing irregular echinoid *Echinocardium* (recent) in its burrow with tube feet building respiratory and sanitary canals. (*Source:* Modified from D. Nichols, *Transactions Royal Society*, B 242.)

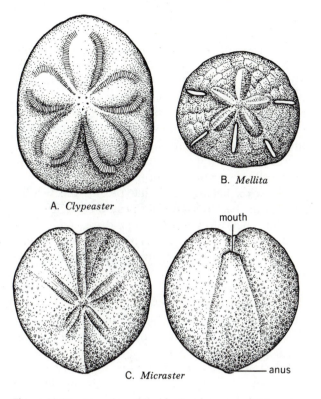

A. *Clypeaster*

B. *Mellita*

mouth

C. *Micraster*

anus

Figure 10-34   Irregular echinoids. Specimens are 7-15 cm in diameter. A. *Clypeaster* (Eocene–recent), dorsal view. B. *Mellita* (Miocene–recent), dorsal view. C, D. *Micraster* (Cretaceous–Paleocene), dorsal and ventral views.

to expose more area to gas exchange, and the pores that they pass through are elongate. Exchange of gases takes place largely through these tube feet, which assume the function of the gills. In advanced irregular echinoids the gills are absent. The double pore connecting each tube foot to its ampulla allows a one-way current carrying respiratory gases to circulate in and out of the tube foot. The granular bands also produce mucus that covers the test and prevents sediment from coming into contact with it.

The tube feet at the back of the urchin solve the sanitary problem by building a channel behind the animal into which the waste products are flushed by currents of water produced by the ciliated spines near the anus (Fig. 10–33). As the animal moves forward through the sediment to feed, the respiratory channel collapses and the tube feet form another channel above the new postion of the body. In adapting to a burrowing habit, the irregular echinoids reduced most of the spines on the test to the size of coarse hairs. Some of the spines are specialized to assist the animal in burrowing and moving forward in the sediment.

Suspended food is brought by water currents descending the respiratory channel and waste organic matter in the sediment falls on to the top of the test. These particles are channeled forward over the edge of the test to the mouth on a string of mucus or on fine cilia in a groove on the upper surface (Fig. 10–34C,D).

A group of irregular echinoids known as sand dollars live lightly covered just below the sediment surface. These urchins, which belong to the order Clypeasteroidea, have depressed tests some of which have holes through them (Fig. 10–34). The mouth remains on the center of the base and is commonly surrounded by radial food grooves in branching depressions. Fine cilia in these grooves bring food to the mouth. The anus is on the back edge of the discoidal test. Clypeasteroids are characterized by internal partitions supporting the thin test and many accessory tube feet that penetrate interambulacral areas. These additional tube feet allow the clypeasteroids to feed on fine organic particles trapped in fine grained sediment. The ambulacral areas, like those of many other irregular echinoids, are conspicuously petalloid. The clypeasteroids rose to prominence at the beginning of Cenozoic time and are common in modern seas.

Paleozoic Echinoids:

These can be divided into two informal groups on the basis of the nature of the sutures between the plates. In one group (Echinocystitoidea) the plates overlap like shingles on a roof and the test is flexible; in the other they abut one another squarely, and the test is rigid. Some post-Paleozoic echinoids have flexible tests of overlapping plates, but they are not believed to be directly descended from the Paleozoic forms. Paleontologists do not agree on whether rigid or flexible echinoids are most primitive as both first appear in Ordovician rocks. The rigid type is represented by the genus *Bothriocidaris*, which is unique among the echinoids in having only a single row of plates in the interambulacral areas. The genus *Aulechinus* (Fig. 10–35) is representative of the flexible forms. In this genus the radial water canal occupies a groove on the outside of the test. In later Paleozoic echinoids it moved inward, first becoming enclosed within the ambulacral plates, then totally inside the test. Both flexible and rigid echinoids multiplied the rows of tube feet, not by dividing the ambulacral plates into composites as did post-Paleozoic echinoids, but by adding addi-

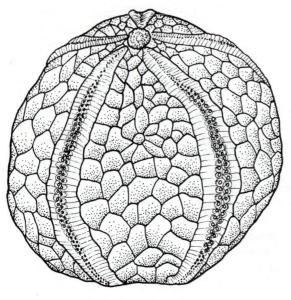

Aulechinus

Figure 10-35   The Ordovician echinoid *Aulechinus* showing the flexible overlapping plates. About 4 cm in width.

tional rows of ambulacral plates. In *Melonechinus*, ten rows of plates occur in each ambulacral sector (Fig. 10–29D).

In Early Carboniferous time a group of echinoids called cidaroids appeared. The cidaroids typically have good radial symmetry, narrow sinuous ambulacra composed of small plates, and large interambulacral plates bearing large spines (Fig. 10–26). When all other groups of echinoids died out at the end of the Paleozoic Era, the single genus of cidaroids, *Miocidaris*, survived and gave rise to the resurgence of the echinoids of Triassic time.

BRYOZOANS

The Permian invertebrate crisis devastated the bryozoans, eliminating the fenestrates, trepostomes, and cryptostomes and greatly reducing the tubuliporates (cyclostomes). Few bryozoans of any sort are known from Triassic rocks, but in the Jurassic the surviving tubuliporates again became abundant. A new order of bryozoans appeared from unknown ancestors in mid-Jurassic time and had soon expanded to eclipse the tubuliporates. These are the order Cheilostomata. They are distinguished by the closure of the aperture of the zooecium by a trapdoor.

Most cheilostome bryozoans are encrusting, sheetlike colonies composed of low, oblong, boxlike or coffin-shaped zooecia fitted closely together and communicating by microscopic tissue strands (Fig 10–36). In many the front of the box is a flexible noncalcified membrane that is not preserved in fossils. The membrane can be drawn inward by muscles attached to the base of the box and as it moves inward the pressure in the fluid inside the box is increased, pushing the animal out through the aperture. Retractor muscles bring the animal back into the box, displacing fluid that pushes the flexible membrane back to its former position (Fig. 10–37). In some cheilostomes the zooid is displaced by muscular expansion of a sack that draws water from a pore near the main aperture. The trapdoor operculum is arranged so that when it opens outward to let the individual out, the section beyond the hinge opens inward to let water into the sack. Some of the zooecia in most cheilostomes are modified by caplike extensions in which the fertilized cells are kept until the larvae develop enough to be released into the water (Fig. 10–36). The form of these brood pouches may be important in the identification of species of bryozoans. Fossil cheilostomes are also characterized by small sockets that housed modified individuals that cleaned the surface of the colony. The cleaning individuals are not calcified.

Although cheilostomes are abundant in modern seas, they are rarely conspicuous as they tend to live in crevices and cavities. Perhaps they have not been able to compete with modern corals for living space on reefs and have been forced into habitats where the corals cannot go for lack of light. Locally bryozoans form small reeflike bodies but cheilostomes have not regained the prominence the trepostomes had in early Paleozoic reefs.

THE LATE CRETACEOUS INVERTEBRATE CRISIS

In post-Paleozoic time the second wave of invertebrates increased steadily in abundance while invertebrates typical of the Paleozoic Era declined (Fig. 7–60). The second wave was characterized by new calcareous forams (benthonic and later planktonic), scleractinian corals, cheilostome bryozoans, neogastropods, many new types of bivalves, new ammonoid orders, and burrowing echinoids. This rise in diversity was interrupted by an extinction

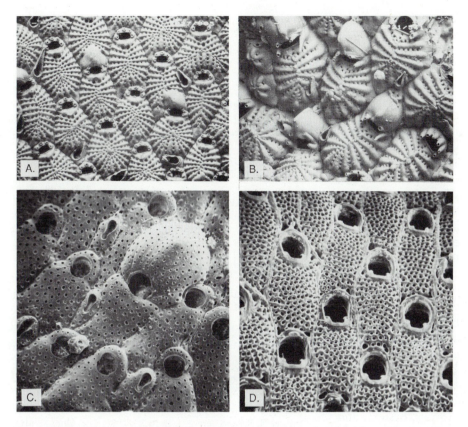

Figure 10-36 Surfaces of recent cheilostome bryozoans. Most show brood pouches on some zooecia and the bases of zooecia modified for cleaning and protective functions. The zooecia are approximately 0.5 mm in width. A. *Cribrilaria venusta*. B. *Cribrilina pedunculata*. C. *Trypostega claviculata*. D. *Uscia mexicana*. (Courtesy of J. D. D. Bishop, P. J. Chimonides, and Patricia Cook, and the British Natural History Museum.)

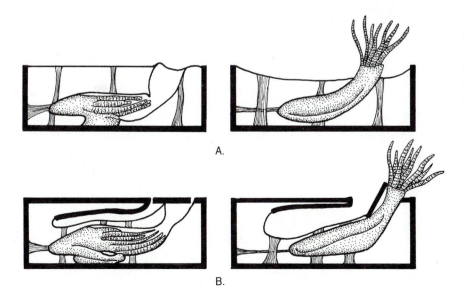

Figure 10-37 Extrusion apparatus in cheilostome bryozoans. Extruded position is on the right. A. A simple flexible frontal membrane is drawn in by muscles. B. A compensating sack is inflated by muscles displacing the zooid. The trapdoor opens both sack and opening for the zooid at the same time.

A.

B.

event at the end of the Triassic Period and a major one at the end of the Mesozoic Era. The largest group to suffer in the first of these was the cephalopods (Ceratitida), of which 35 families became extinct. Other groups affected by the crisis were the gastropods (six families), bivalves (42 percent of genera), brachiopods, and conodonts.

Only the Late Permian event was more traumatic to invertebrate life than that which closed the Mesozoic Era. Certainly no event has stimulated as much discussion as the late Cretaceous crisis for, unlike the Permian event, it affected not only marine life but also the reptiles of the land, sea, and air. Dale Russell (6, in Emiliani, and others) has estimated that half of the genera of the world biota became extinct. Like all similar events that have been studied, it did not affect all life equally; some animals and plants seem to have been immune. When about 85 percent of the genera of coccoliths and planktonic foraminiferans died out, the dinoflagellates were little affected. Although the ammonoids and belemnites became extinct (except for some dubious early Cenozoic belemnites), the nautiloids were relatively immune and these survivors flourished in the early Cenozoic environments abandoned by their relatives. Corals, oysters, and particularly rudists were reduced or eliminated but the diversification of benthonic forams continued unabated. Any mechanism proposed to explain the crisis must account for significant changes in the histories of the coccoliths, planktonic foraminiferans, ammonoids, belemnoids, corals, and rudists—organisms that do not have a great deal in common in their habitats and lifestyles. For some of these groups, such as the ammonites, troubles started long before the period closed; for others, such as the rudists, the end appears to have been sudden.

REFERENCES

1. Oliver, W. A., 1980, The relationship of the scleractinian corals to the rugose corals: Paleobiology, v. 6, p. 146–160.

2. Raup, D. M., 1980, Size of the Permo–Triassic bottleneck: Science, v. 206, p. 217–218.

3. Vermeij, G. J., 1977, The Mesozoic marine revolution: Evidence from snails, predators, and grazers: Paleobiology, v. 3, p. 245–258.

4. Carter, J. G., and Clark, G. R., 1985, Classification and phylogenetic significance of molluscan shell microstructure. In Bottjer, D. J., Hickman, C. S., and Ward, P. D. (eds.), Mollusks: University of Tennessee, Studies in Geology, v. 13, p. 150–72.

5. Stanley, G. D., 1981, Early history of scleractinian corals and its geological consequences: Geology, v. 9, p. 507–511.

6. Emiliani, C., Kraus, E. B., and Shoemaker, E. M., 1981, Sudden death at the end of the Mesozoic: Earth and Planetary Sciences Letters, v. 55, p. 317–334.

SUGGESTED READINGS

Buzas, M. A., and Sen Gupta, B. K. (eds.), 1982, Foraminifera: University of Tennessee Studies in Geology, No. 6, 219 p.

Kennedy, W. J., 1977, Ammonite Evolution. In Hallam, A. (ed.), Patterns of Evolution as Illustrated by the Fossil Record: Amsterdam/New York, Elsevier, p. 251–304.

Kennedy, W. J., and Cobban, W. A., 1976, Aspects of Ammonite Biology, Biogeography, and Biostratigraphy: Palaeontological Association Special Papers in Palaeontology No. 17, 93 p.

Haynes, J. R., 1981, Foraminifera. New York, Wiley, 433 p.

House, M. R., and Senior, J. R. (eds.), 1981, The Ammonoidea. Systematics Association Special Volume 18: New York, Academic Press, 593 p.

Lehmann, U. 1976, The Ammonites: their life and their world: London/New York, Cambridge Univ. Press, 246 p.

Smith, A., 1984, Echinoid Paleobiology: London, Allen & Unwin, 202 p.

Stanley, S. M., 1975, Adaptive themes in the evolution of the Bivalvia (Mollusca): Annual Review of Earth and Planetary Sciences, v. 3, p. 361–385.

Stanley, S. M., 1972, Functional morphology and evolution of byssally attached bivalve mollusks: Journal of Paleontology, v. 46, p. 185–212.

Tappan, H., and Loeblich, A., 1973, Evolution of ocean plankton: Earth Science Reviews, v.9, p. 207–240.

Tappan, H., 1986, Phytoplankton: below the salt at the global table: Journal of Paleontology, v. 60, p. 545–554.

Tozer, E. T., 1984, The Trias and its ammonites: the evolution of a time scale: Geological Survey of Canada, Miscellaneous Report No. 35, 171 p.

Chapter 11

Evolution of Vertebrates During the Mesozoic

MESOZOIC TERRESTRIAL REPTILES

During the Carboniferous and Early Permian, the terrestrial environment was dominated by archaic amphibians. Reptiles originated in the Late Carboniferous, but it was only in the Late Permian and Mesozoic that they became conspicuous elements of the fauna. During the Mesozoic, the dinosaurs dominated the land while aquatic reptiles abounded in the seas and the pterosaurs and birds took to the air. The Mesozoic also witnessed the emergence of the turtles, lizards, snakes, and crocodiles.

Turtles:

Because they are common members of the modern fauna, turtles do not appear exotic, but their extensive bony armor makes them among the most specialized of all vertebrates. Turtles, already clearly recognizable and providing no evidence of their immediate ancestry, appear in the fossil record in the Upper Triassic (1) (Fig. 11–1). Their lack of lateral or dorsal temporal openings suggests they are not closely related to either diapsids or mammal-like reptiles but evolved from the ancestral or stem reptiles. Turtle skulls are distinguished by

openings on the posterior surface through which the jaw muscles extend. Even the earliest fossil turtles had completely lost their marginal teeth and had evolved a horny beak.

No fossils are known that show early stages in the evolution of the turtle shell, although elements of the shell fossilize readily and are easily recognized. The ventral part of the shell, the plastron, incorporates into its anterior margin the ventral portions of the dermal shoulder girdle of primitive reptiles. The remainder of the plastron and the entire dorsal portion of the shell (the carapace) evolved *de novo*. The shell of early turtles not only included all the elements present in modern turtles but also incorporated a fringe of extra marginal elements in the carapace and extra bones in the plastron.

By the Early Jurassic, turtles had split into the two major groups recognized in the modern fauna, the cryptodires and the pleurodires, which are distinguished by the way the head is retracted into the shell (Fig. 11–2). The specialization of the neck vertebrae for vertical or horizontal retraction was not fully evident until the Cenozoic, but early members of both groups can be recognized by the different ways that the palate and braincase are attached and by the modification of different bones

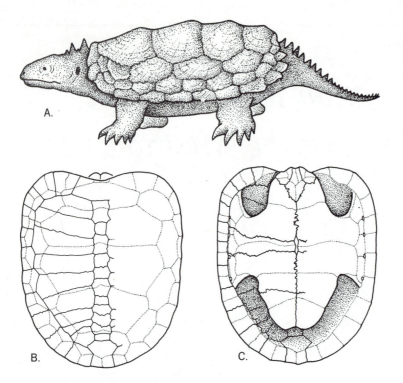

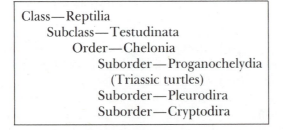

Figure 11-1 Mesozoic turtles. A. Restoration of *Proganochelys*, the oldest known turtle from the Upper Triassic of Germany. B, C. Carapace and plastron of the Lower Jurassic cryptodire turtle, *Kayentachelys*. The solid lines on the left show the division between the bony plates. The dotted lines show the outline of the overlying scutes. *Kayentachelys* has lost the extra marginal plates present in *Proganochelys*. (B & C: Courtesy of E. S. Gaffney and *Science*, v. 237: 289–291, 17 July 1987. Copyright 1987 by the AAAS.)

to form a pulleylike structure for the passage of the major jaw-closing muscles (Fig. 11–3).

Class—Reptilia
    Subclass—Testudinata
        Order—Chelonia
            Suborder—Proganochelydia
            (Triassic turtles)
            Suborder—Pleurodira
            Suborder—Cryptodira

Pleurodires appeared in the Late Triassic and were common in the late Mesozoic and early Cenozoic but are now largely limited to the southern continents. The cryptodires apparently achieved a position of dominance at a later time and are now common throughout the world. Of the three modern superfamilies, the sea turtles (Chelonioidea) and the soft-shelled turtles (Trionychoidea) both appeared in the Late Jurassic, while the tortoises and freshwater turtles (Testudinoidea) were first recognized in Lower Cenozoic rocks.

DIAPSID REPTILES

Aside from the stem reptiles of the late Paleozoic, the turtles, and the ancestors of mammals, most

reptiles belong to a single extremely diverse assemblage, the Diapsida. In the modern fauna, diapsids are represented by crocodiles, snakes, lizards, and the superficially lizardlike genus, *Sphenodon*, now

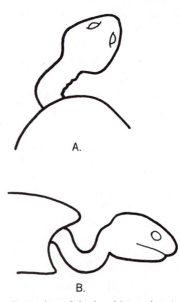

Figure 11-2 Retraction of the head in turtles. A. The horizontal curvature of the neck typical of advanced pleurodires (side necked turtles). B. The vertical curvature of the neck characteristic of advanced cryptodires.

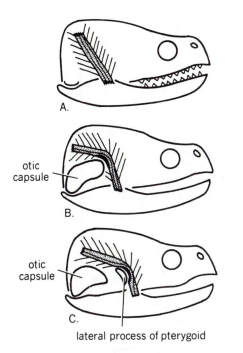

Figure 11-3 Specializations of the jaw closing musculature in turtles. A. Schematic diagram of the major jaw closing muscles in primitive reptiles including *Proganochelys*. The muscles pass in a straight line from the top of the jaw to the back of the skull. B. In cryptodires, the muscles pass over the capsule of the inner ear. C. In pleurodires, the muscles pass over a lateral process of the pterygoid bone. In all turtles the muscle extends posteriorly through openings in the back of the skull. The greater length of the muscle allows the jaw to be opened more widely. (From E. S. Gaffney, *Bulletin of the American Museum of Natural History*, vol. 155.)

limited to a few islands off the coast of New Zealand.

Diapsids were much more diverse in the Mesozoic Era. In addition to early representatives of all the major modern groups, they included two orders of dinosaurs, the flying pterosaurs, and several groups of secondarily aquatic reptiles. Birds were derived from Mesozoic diapsids and should be considered part of this reptilian radiation. Diapsids are the most diverse of all groups classified as reptiles and were the dominant vertebrates on the land, in the sea, and in the air throughout the Mesozoic (Fig. 11–4).

Early Diapsids:

The oldest known diapsid is *Petrolacosaurus* from the Upper Carboniferous of Kansas (Fig. 11–5). It was less than a meter long, including a very long tail. Relative to the trunk, the limbs and neck were both longer than those of such Early Carboniferous reptiles as *Hylonomus*. The most significant differences were in the skull, which had two pairs of clearly defined temporal openings and a new opening in the palate, the suborbital fenestra. These openings may have served to lighten the skull and distribute mechanical forces more effectively.

The fossil record of diapsids is very incomplete during the Permian, but toward the end of that period and at the beginning of the Triassic many lineages appeared, including the ancestors of all the major Mesozoic and Cenozoic groups. A divergent and specialized family, the Coelurosauravidae, appears in the Upper Permian rocks of Madagascar and Europe. The body form of the coelurosauravids was lizardlike, except for the ribs, which were greatly elongated and almost certainly supported a gliding membrane as in the modern lizard *Draco* (Fig. 11–6). Most other diapsids can be grouped into one or the other of two large assemblages, the Lepidosauromorpha and the Archosauromorpha. Lepidosauromorphs include the modern lizards, snakes, and *Sphenodon*, and several broadly similar Mesozoic families. Most lepidosauromorphs are relatively small and, like primitive reptiles, have a sprawling posture and use lateral undulation of the trunk for locomotion. These habits are carried to the extreme by snakes. The archosauromorphs include the living crocodiles and all the large terrestrial reptiles of the Mesozoic. In this group, the limbs are drawn to some degree under the body, and move in a fore and aft direction. The patterns of posture and locomotion are reflected in the bones of the girdles and limbs (Fig. 11–7).

Class—Reptilia
  Subclass—Diapsida
        Order—Araeoscelida (including
          Petrolacosaurus)
    Infraclass—Lepidosauromorpha
      Order—Eosuchia
      Superorder—Lepidosauria
      Order—Sphenodontida
      Order—Squamata
          Suborder—Lacertilia (lizards)
          Suborder—Serpentes (snakes)
      Superorder—Sauropterygia
      Order—Nothosauria
      Order—Plesiosauria
          Superfamily—Plesiosauroidea
          Superfamily—Pliosauroidea

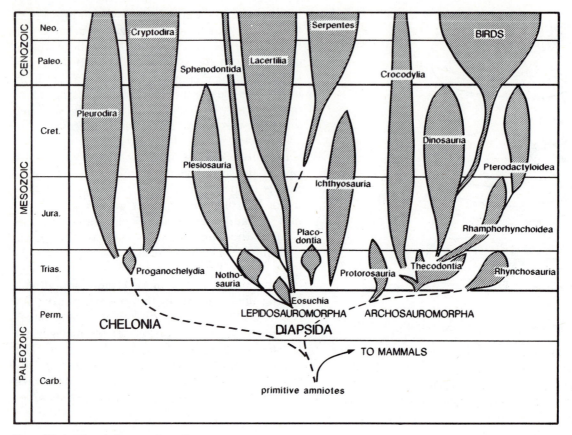

Figure 11-4   The phylogeny of reptiles.

Primitive lepidosauromorphs from Upper Permian and Lower Triassic beds are termed eosuchians. The eosuchians resembled lizards in their body form. They were more advanced than primitive diapsids, having evolved a large sternum (or breast bone) which served as a basis for the rotation of the shoulder girdle to extend the stride of the front limbs. From animals of this general pattern evolved the modern lepidosaur groups, the Squamata, including lizards and snakes, and the Sphenodontida. Both of these groups differ from the eosuchians in that there are separate areas of ossification at the ends of the limb bones, the epiphyses. Epiphyses serve to limit growth to a specific size. Growth occurs in a narrow band of cartilage between the shaft of the limb bone and the epiphyses. At a specific age, the cartilage becomes ossified and growth ceases. The bony epiphyses also form joint surfaces that are better defined that those of other reptiles.

Like frogs, lepidosaurs evolved an ear structure that was sensitive to airborne sounds. Unlike frogs, a different bone in the cheek, the quadrate, became specialized to support the tympanum (ear drum). In both squamates and sphenodontids, the stapes is a long, narrow rod. The living genus *Sphenodon* lacks a tympanum and a middle ear cavity, as do many burrowing lizards and snakes. Early sphenodontids, from Upper Triassic and Lower Jurassic rocks, more closely resemble typical lizards in the bony structure of the ear region and probably had a large tympanum.

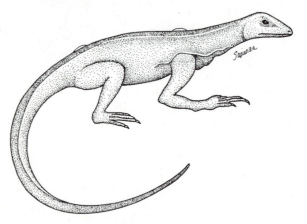

Figure 11-5   Restoration of the oldest known diapsid reptile *Petrolacosaurus*, from the Upper Carboniferous. The skull is illustrated in Figure 9-31.

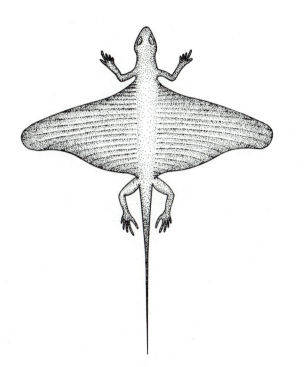

Figure 11-6   Restoration of *Coelurosauravus*, a primitive diapsid reptile from the Upper Permian of Madagascar in which the ribs are greatly elongated to form a gliding surface like that of the modern lizard *Draco*.

Sphenodontida:

Both sphenodontids and lizards appeared in the Triassic Period. Sphenodontids are lizardlike in their general appearance, but most species differ from squamates in having a solid bar below the lower temporal opening, as in primitive eosuchians. Sphenodontids also differ from most other reptiles in having their teeth fused to the margin of the jaw. In most lizards, the teeth are loosely attached to the side of the jaw. Unlike most other reptiles, the teeth of sphenodontids are not replaced, but are subject to wear throughout their life. In some sphenodontids, the edges of the teeth in the upper and lower jaws shear past one another vertically. In the living genus *Sphenodon*, the lower jaw moves anteriorly and posteriorly so that the teeth move across their prey like the teeth of a saw.

Sphenodontids were common and diverse from the Late Triassic through the Jurassic but their fossils are rare in Cretaceous deposits. Sphenodontids have no Cenozoic fossil record. Presumably they reached New Zealand during the early Mesozoic, before it became isolated from Australia. The survival of *Sphenodon* on these isolated islands may be explained by the rarity of competing lizards. *Sphen-*

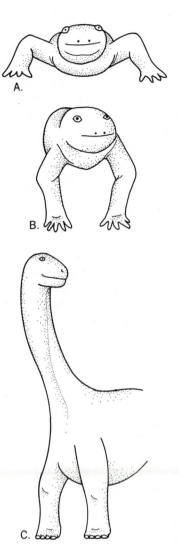

Figure 11-7   A. The sprawling posture of primitive amphibians and reptiles, B. The semi-upright posture of crocodiles and advanced mammal-like reptiles, C. The upright posture of a dinosaur.

odon is typically cited as an example of a living fossil (12) since it closely resembles its Triassic ancestors. On the other hand, a group of marine reptiles, the pleurosaurs, diverged from the same ancestral stock and evolved rapidly during the Jurassic and Early Cretaceous to become highly adapted for aquatic locomotion.

Squamata:

A skull from the Upper Permian or Lower Triassic rocks of southern Africa is the oldest fossil that may belong to a lizard (Fig. 11–8). The quadrate, unlike that of eosuchians and sphenodontids, is not firmly attached to the palate and the lower temporal bar is lost. The upper surface of the quadrate is articulated with the skull so that the lower end of this bone could move forward and backward. Mobility of the quadrate, termed streptostyly, is characteristic of most squamates (Fig. 11-9).

Primitive members of most of the modern lizard infraorders appeared by the Late Jurassic, and the modern families by the Late Cretaceous (3). Nearly 4000 species are present in the modern fauna. Lizards have a sparse record in the late Mesozoic and Cenozoic, but many modern genera are known from the Miocene.

Most fossil lizards broadly resemble the modern forms, with one striking exception. During the Late Cretaceous, one family related to the modern Komoto dragon *Varanus* became highly adapted to an aquatic existence. This family, the Mosasauridae, were among the largest marine reptiles in the latest Cretaceous seas and achieved a worldwide distribution in shallow, nearshore marine environments.

Figure 11-8   A. The skull of the oldest known lizard, *Paliguana* from the Upper Permian or Lower Triassic of southern Africa. The quadrate is embayed posteriorly for support of a tympanum, and the lower temporal bar is lost. B. Restoration of a paliguanid lizard, slightly less than life size.

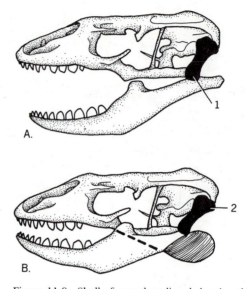

Figure 11-9   Skull of a modern lizard showing the function of quadrate mobility (streptostyly). A. Skull with jaw open and jaw articulation (1) in posterior position, B. Skull with jaw nearly closed. Jaw articulation has moved forward as the joint between the quadrate (black) and the skull roof (2) becomes functional. Dashed line shows the course of the pterygoideus muscle. Muscles can exert their greatest force when they act at right angles to the bone being moved. The pterygoideus is nearly horizontal so it can exert a greater force in moving the nearly vertical quadrate than it can on the lower jaw when it is nearly closed. (Courtesy of K. K. Smith. Reprinted by permission from *Nature* Vol. 283, pp. 778–779. Copyright (©) 1980 Macmillan Magazines Ltd.)

Mosasaurs reached a length of up to 10 meters and had a fusiform body, like moderately fast-swimming fish (Fig. 11-10). The fore and hind limbs were in the shape of paddles and the tail evolved a caudal fin. Tooth marks of mosasaurs are found in many large ammonites. The aigialosaurs from the mid-Cretaceous are nearly ideal intermediates between terrestrial varanoid lizards and mosasaurs.

Varanoid lizards may also include the ancestors of snakes. Snakes are represented by isolated vertebrae in deposits as old as Early Cretaceous (4). There are approximately 3000 species of snakes in the modern fauna, but their fossil record is difficult to interpret because most specimens are nothing but vertebrae. The skulls of most snakes have a very flexible, latticelike construction which allows them to swallow very large prey. Elongation of the body with reduction of the limbs has evolved among many lizard groups. The closest intermediates between lizards and snakes appear to be the genera *Pachyrhachis* and *Ophiomorphus* from the Lower Cretaceous of Israel (5).

Figure 11-10   A mosasaur, a giant marine lizard from the Upper Cretaceous. Some species were as long as 10 meters.

Unlike other reptilian groups, the snakes are still a rapidly evolving assemblage. Most modern genera belong to families whose major radiation has occurred in the late Cenozoic.

ARCHOSAUROMORPHA

Primitive Archosauromorphs:

The archosauromorphs are first known in Upper Permian rocks. By the Early Triassic, several major orders had evolved, including the ancestors of the dinosaurs, pterosaurs, and crocodiles. Divergent groups included the protorosaurs and the rhynchosaurs. All archosauromorphs are characterized by advances in the structure of the rear limbs to facilitate a more upright posture. In primitive archosauromorphs, the ankle joint allowed rotation so that the foot faced more or less forward throughout the stride, although the knee still swept through a lateral arc during retraction. Much of the rest of the skeleton remained primitive. A sternum of the type seen in lepidosauromorphs never evolved in this group, and the bones of the wrist and hand are reduced.

Class—Reptilia
  Infraclass—Archosauromorpha
      Order—Protorosauria
      Order—Rhynchosauria
     Superorder—Archosauria
      Order—Thecodontia
        Suborder—Proterosuchia
        Suborder—Ornithosuchia
        Suborder—Aetosauria
        Suborder—Phytosauria
      Order—Crocodylia
      Order—Pterosauria
        Suborder—Rhamphorhynchoidea
        Suborder—Pterodactyloidea
      Order—Saurischia
      Order—Ornithischia

The protorosaurs appeared in the Late Permian and continued into the Early Jurassic. They are specialized in the great elongation of the neck and the reduction of the lower temporal bar (Fig. 11–11). Rhynchosaurs were large-headed herbivores common for most of the Triassic.

The most important of the early archosauromorphs are the thecodonts, a group of primarily carnivorous families that include the ancestors of all the large terrestrial reptiles of the later Mesozoic as well as the crocodiles and birds. Thecodonts can be distinguished from other primitive archosauromorphs by the presence of an opening in front of the eye, the antorbital fenestra, of unknown function (Fig. 11–12).

Thecodonts appeared at the end of the Permian and diversified throughout the Triassic Period. The most primitive of the several groups are the protosuchians such as *Chasmatosaurus*, which were sprawling animals resembling crocodiles. Other groups, including the heavily armored, herbivorous aetosaurs and the long-snouted aquatic phytosaurs, were limited to Late Triassic time.

Several thecodont groups were more advanced than these in their limb structure and may have been at least facultatively bipedal. Included in these advanced groups are *Gracilosuchus*, which may be close to the ancestry of crocodiles, and the ornithosuchids, which may be close to the ancestry of dinosaurs (Fig. 11–13).

The most dinosaurlike of all the thecodonts are the lagosuchids (Fig. 11–14A), in which the head of the femur is inflected so that the leg can be held nearly vertical. The knee and ankle joints were modified so that the rear limb could be moved directly fore and aft.

Crocodiles:

Crocodiles appeared in the Late Triassic and continue to be important elements in the world fauna. The skulls of even the earliest crocodiles resemble their modern counterparts in having a very low profile and a solidly attached jaw suspension to re-

Figure 11-11   Primitive archosauromorph reptiles. A. The extremely long-necked genus *Tanystropheus* from the Middle Triassic of Europe, B. A Triassic rhynchosaur. Rhynchosaurs were herbivores characterized by the presence of many rows of teeth set in deep sockets. (A: Modified from R. Wild, *Schweiz. Palaeontol. Abh.*, v. 95. B: From S. Chatterjee. Courtesy of *Mém. Soc. géol. France, N. S.* 139, p. 64, fig. 4.)

sist the massive jaw musculature. The long limbs of early crocodiles suggest that they were fast-running, terrestrial animals, without skeletal evidence of the semi-aquatic or amphibious habits of most modern crocodiles and many of the Jurassic and Cretaceous genera (Fig. 11–14B). Four suborders of crocodiles can be recognized. The sphenosuchids, restricted to the Late Triassic, are the most primitive and show a close link with the thecodonts. The proterosuchians (Upper Triassic and Lower Jurassic) retained the primitive terrestrial life-style, but the skull is more similar to that of modern crocodiles. Most of the Jurassic and Cretaceous crocodiles are included in the Mesosuchia, which consists primarily of semi-aquatic and marine forms. Several families were highly adapted to an aquatic way of life, with a greatly elongated snout and a variable degree of modification of the limbs into paddles.

All modern crocodiles and most of the Late Cretaceous and early Cenozoic genera are included in the suborder Eusuchia, characterized by the great posterior extent of the secondary palate and the specialized articulating surfaces between the ver-

tebrae. The range and diversity of crocodiles has been steadily reduced throughout the Cenozoic because of the cooling of the climate and the gradual geographical restriction of the tropics.

DINOSAURS

Although many dinosaurs have been described, knowledge of their fossil record is still very incomplete, particularly in the Middle to Late Triassic when the group began its major radiation. The earliest dinosaur, *Staurikosaurus*, is known from upper Middle Triassic beds of South America (6). Its remains are incomplete, but suggest that it may have reached a length of approximately 2 m. The large skull and piercing teeth indicate a carnivorous habit, and the great length of the rear limbs suggests that it was a fast-running animal. Features other than the skull and teeth, which are like later theropods, are primitive for dinosaurs, and this genus may be close to the ancestry of the entire group. In the Upper Triassic, representatives of a least six major

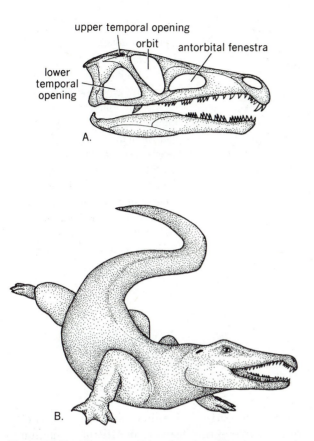

Figure 11-12 The protosuchian thecodont *Chasmatosaurus*. A. Skull showing antorbital fenestra, B. Reconstruction. This genus was approximately 1.5 meters long. (Courtesy of A. J. Charig and H.-D. Sues. From Kuhn/Wellnhofer, *Handbuch der Paläoherpetologie*, Part 13 (Charig/Krebs/Sues, Thecodontia). Gustav Fischer Verlag Stuttgart, 1976.)

lineages are known, but their specific interrelationships have not been established.

Dinosaurs have long been separated into two distinct orders, the Saurischia and Ornithischia, based on different patterns of the pelvis (Fig. 11–15). It is now recognized that all dinosaurs show similar advanced features of the pelvis and rear limbs associated with a more upright posture that indicate an immediate common ancestry. The triradiate pattern of the ilium, ischium, and pubis in saurischians is primitive for all dinosaurs and so does not demonstrate that the saurischians had a separate ancestry from that of ornithischians. The ornithischian pelvis, in which the pubis is directed posteriorly, does demonstrate the common origin of this group, since it is a unique specialization.

Two major groups have been classified among the saurischians, the carnivorous theropods and the herbivorous sauropodomorphs (Fig. 11–16).

```
Order—Saurischia
    Suborder—Staurikosauria
    Suborder—Theropoda
    Suborder—Sauropodomorpha
        Infraorder—Prosauropoda (Plateosauria)
        Infraorder—Sauropoda
Order—Ornithischia
    Suborder—Ornithopoda
    Suborder—Pachycephalosauria
    Suborder—Stegosauria
    Suborder—Ankylosauria
    Suborder—Ceratopsia
```

Carnivorous Dinosaurs:

The theropod dinosaurs were the largest terrestrial carnivores that ever existed, weighing as much as 7000 kg with lengths of 15 m. All were obligatorily bipedal. The skull of the Late Cretaceous genus *Tyrannosaurus* was more than a meter in length, with individual teeth 15 cm long. The largest and latest genera, in the family Tyrannosauridae, have only two fingers in the hand and such a short front limb that it is difficult to understand how it was used (Fig. 11–17).

Figure 11-13 The ornithosuchian thecodont *Riojasuchus* from the Upper Triassic of South Africa. The limbs were held nearly vertically but the gait remained quadrupedal.

Figure 11-14   A. The faculatitively bipedal Middle Triassic thecodont *Lagosuchus* that may be close to the ancestry of dinosaurs and pterosaurs. This genus walked on its toes, as did primitive dinosaurs, rather than flat on its feet, as did more primitive reptiles. B. The primitive crocodile *Terrestrisuchus* from the Upper Triassic of Europe. The limbs were very long, in contrast to those of most later crocodiles. (A: Modified from J. F. Bonaparte, *Opera Lilloana*, 26. B: Modified from P. J. Crush, *Palaeontology*, v. 27.)

For a long time, theropods were divided into two subgroups, the carnosaurs, generally with heavy bodies, short forelimbs, and large skulls, and the smaller, more lightly built coelurosaurs, most of which had larger forelimbs, but smaller skulls. Discoveries during the past ten years of several large genera with long arms and large skulls, which can-not be categorized as either carnosaurs or coelurosaurs, makes this distinction difficult to maintain. Approximately 17 families of theropods have been recognized.

The early theropods included such small and agile genera as *Coelophysis* of the Late Triassic, which reached a length of approximately 3 meters. From

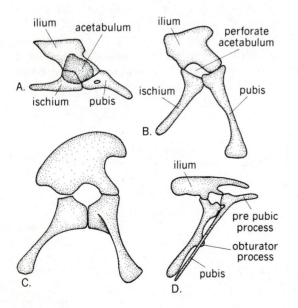

Figure 11-15   The pelves of archosaurs. A. The Lower Triassic thecodont *Chasmatosaurus*. The pubis and ischium form a flat plate for the origin of the large muscles to lift the body on the rear limbs. B. The primitive dinosaur *Staurikosaurus* from the late Middle Triassic of South America. The pubis and ischium are elongated. They now serve for the origin of muscles that swing the rear limbs anteriorly and posteriorly. The area of the acetabulum is not ossified because the main force of the rear limbs is directed dorsally towards the ilium, rather than medially, as in primitive reptiles with a more sprawling posture. C. A sauropod dinosaur. The ilium is expanded anteriorly and posteriorly from the attachment of up to six pairs of sacral ribs. D. The ornithischian dinosaur *Hypsilophodon*. Typical of this group, the pubis is oriented posteriorly and lies alongside the ischium. A new structure, the prepubic process, extends anteriorly. The obturator process is a characteristic of ornithopod dinosaurs that is not present in other ornithischians.

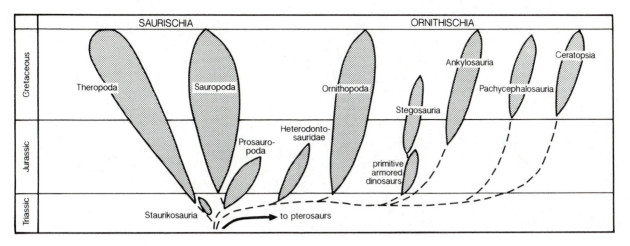

Figure 11-16   The phylogeny of dinosaurs.

such animals, several lines of small to moderate size genera evolved. *Deinonychus*, described in detail by John Ostrom (7), appears to have been a particularly effective predator with an extremely large claw on the second digit of the rear foot that was probably used to disembowel prey. *Deinonychus* and members of the related family Saurornithoididae had very large braincases, suggesting a brain of comparable size to that of advanced mammals. A divergent lineage, the ornithomimids or ostrich dinosaurs, lost their teeth and had birdlike skulls.

Larger and more massively built theropods, assigned to the family Megalosauridae, are known from the Late Triassic into the Late Cretaceous. It is not known whether the well-known genera *Allosaurus* and *Tyrannosaurus* shared a close common ancestry with them or evolved separately from smaller theropods.

Herbivorous Dinosaurs:

All other dinosaur groups were herbivores. The largest of the herbivores were the sauropods, which were common throughout much of the world in the Jurassic and Cretaceous. Edwin Colbert estimated that some reached sizes in excess of 80,000 kg (8). Incomplete remains are known of individuals that may have been considerably larger. The well-known genus *Diplodocus*, with its small head, long neck, stocky body supported on elephantine legs, and long tail, is characteristic of sauropods. Six families of sauropods are recognized on the basis of consistent differences in the number of trunk and neck vertebrae, limb proportions, and

details of skull anatomy. To reduce their weight, the vertebrae of all sauropods have large openings that give them the appearance of lattice work. The limb bones were solid and held in a nearly vertical position. The joint surfaces of the ankles and wrists were much less well ossifed than in smaller dinosaurs. The presence of large amounts of cartilage distributed the weight more uniformly, since cartilage is more capable of deformation without breakage than is bone. The great size of sauropods has led to speculations regarding the ability of the bones to support the body. It was once thought that they spent their lives in the water, since the buoyancy of water would assist in their support. Foot prints demonstrate that sauropods were capable of supporting themselves on land, and recent reconstructions imply considerable agility (Fig. 11–18). The skulls of sauropods are relatively much smaller than those of theropods, and the relative brain size was probably smaller than that of modern crocodiles.

Fossils from the Upper Triassic and Lower Jurassic are transitional between typical, obligatorily quadrupedal sauropods and a group of smaller facultatively bipedal herbivores from the Upper Triassic, the prosauropods. Prosauropods were a common and diverse assemblage on most continents. Their skeleton resembled that of the most primitive dinosaurs, but the laterally compressed and crenulated teeth are like those of some modern lizards, indicating a herbivorous diet. The hind limbs of primitive prosauropods were considerably larger than the front limbs, indicating at least facultative bipedality and close affinities with the lagosuchids and staurikosaurids.

Figure 11-17   Theropod dinosaurs. A. The Upper Triassic genus *Coelophysis*, B. The Lower Cretaceous genus *Deinonychus*, C. *Tyrannosaurus*, from the Upper Cretaceous. This genus is known from both North America and Asia.

The ornithischians constitute a second, much more diverse assemblage of herbivorous dinosaurs, common from the end of the Triassic until the close of the Mesozoic. In addition to the structure of their pelvis, they are distinguished from saurischians by the presence of a medial bone at the front of the lower jaws, the predentary.

The Early Jurassic genus *Fabrosaurus* was a small, bipedal dinosaur that may be close to the ancestry of all other ornithischians (Fig. 11–19). Its forelimbs were much smaller than the hind. The distal bones of the rear limbs were longer than the femur, as in modern fast-running mammals. The skull is characterized by a small bone, the supraorbital,

common to most late ornithischians, which extends across the opening for the eye. The teeth are small, with flattened, crenulated crowns, similar to those of modern herbivorous lizards.

The ornithischians underwent a major radiation toward the end of the Triassic, leading to several distinct lineages: the ornithopods, the armoured dinosaurs, the ceratopsians, and the pachycephalosaurids. *Fabrosaurus* is typical of primitive ornithopods. Most members of this group were more advanced than other herbivorous dinosaurs in having the tooth rows inset from the margin of the jaw. The bones of the upper and lower jaw are ridged above and below the tooth row, suggesting the pres-

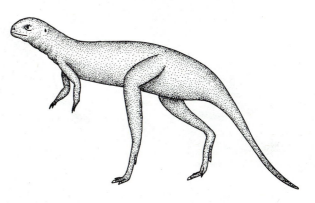

Figure 11-19   Restoration of the Lower Jurassic ornithopod ornithischian *Fabrosaurus*. This animal was approximately .5 meters tall. The short front limbs and long hind limbs indicate that it must have been bipedal. In common with *Lagosuchus* (Fig. 11-14a) the lower segment of the rear limb is longer than the femur. Mammals with similar proportions are rapid runners. (Modified from R. A. Thulborn, *Palaeontology*, v. 15.)

Figure 11-18   Sauropodomorpha. A. A relatively small Upper Triassic prosauropod. The great length of the rear limbs is a heritage of its bipedal ancestry. B. The gigantic sauropod *Apatosaurus* (*Brontosaurus*). Bakker's restoration suggests a very athletic posture. (A: Modified from P. M. Galton, *Postilla, Yale Peabody Museum*, v. 169. B: Courtesy of R. T. Bakker.)

ence of a fleshy cheek, as in mammals, to retain the food while it was chewed. The upper and lower teeth are worn where they moved across one another in a particular direction to insure that the food was broken into small pieces.

The iguanodontids were larger ornithopods, common in Europe and North America from the Middle Jurassic to the end of the Cretaceous. The hadrosaurs were the most common and varied of dinosaurs in the Late Cretaceous of North America and Eurasia. They were heavier than the iguanodontids, and hoofs on the forelimbs indicated that they were at least facultatively quadrupedal. Hadrosaurs had multiple tooth rows so that hundreds of shearing surfaces could be used to break up the food.

Many hadrosaurs had crests of various shapes and sizes on the skull, which were occupied by extensions of the nasal passages (Fig. 11–20). They may have been used to produce different call notes that would have been useful in species recognition (9). Jack Horner (10) has discovered the nests of many hadrosaurs that imply different patterns of parental care.

The remaining ornithischian groups appear to have diverged in the Late Triassic from a common ancestry with the ornithopods.

The pachycephalosaurids retained the bipedal stance of early ornithischians, but did not develop the complex dentition of the ornithopods. They are notable for having dome-shaped heads resulting from a greatly thickened skull. Peter Galton (11) suggested that the top of the head may have been used as a battering ram in intraspecific combat, like the horns of sheep and goats (Fig. 11–21). Pachycephalosaurids are limited to the Cretaceous but ranged through Asia, North America, and Madagascar (11).

The earliest of the armored dinosaurs appeared in Lower Jurassic rocks. *Scutellosaurus* broadly resembles *Fabrosaurus* except for the presence of bony plates along the backbone and extending over the flanks (Fig. 11–22). The stegosaurs, characterized by large, bony plates along the dorsal midline and bony spikes on the tail, lived from the Middle Jurassic into the Late Cretaceous. *Stegosaurus* is usu-

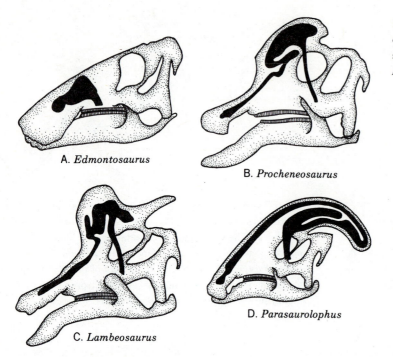

A. *Edmontosaurus*

B. *Procheneosaurus*

C. *Lambeosaurus*

D. *Parasaurolophus*

Figure 11-20 Skulls of hadrosaurs showing a variety of crests and the configuration of the nasal passages. (Modified from J. H. Ostrom, *Postilla, Yale Peabody Museum*, v. 62.)

ally restored in a quadrupedal pose, but the rear limbs were much longer than the front, reflecting their bipedal ancestry.

The most heavily armoured of all dinosaurs were the anklyosaurs, which appeared in the Middle Jurassic in Europe and were widespread during the Late Cretacous in North America and Eurasia. In addition to armor plating over the entire body, the skull was covered with dermal plates, and some genera had a massive club at the end of the tail.

Ceratopsians were among the most spectacular dinosaurs of the Late Cretaceous in North America, distinguished by their massive, commonly horned frill, extending from the skull over the neck. The frill would have offered a measure of protection but was also used for the attachment of long muscles to close the jaws and may have served for sexual display as well. The advanced ceratopsians of the Late Cretaceous were fully quadrupedal, but more primitive genera from the Early Cretaceous had long rear limbs and were probably capable of bipedal locomotion (Fig. 11–23).

Aspects of the biology and extinction of dinosaurs are discussed in Chapters 14 and 16.

Figure 11-21 Intraspecific combat between pachycephalosaurids. Their heads are greatly thickened. This may have provided protection, as do the horns of sheep and goats in which males fight for the attention of females, about 1 m long. (Modified from P. M. Galton, *Discovery*, v. 6(1).)

Figure 11-22   Armored dinosaurs. A. The small Lower Jurassic genus *Scutellosaurus* from North America. As in many primitive ornithopods, the rear limbs are much longer than the front limbs. Small armor plates cover the trunk region and extend down the tail. Some thecodonts and crocodiles show a similar pattern of armor. B. *Huayangosaurus*, a stegosaur from the Middle Jurassic of China. C. *Polacanthus*, a primitive ankylosaur from the Lower Cretaceous of England.

FLYING REPTILES— THE PTEROSAURS

During the Mesozoic, archosaurs were dominant not only on land but also in the air. Two groups of archosaurs became highly adapted for flight: the pterosaurs, and the ancestors of birds. These two groups evolved from different ancestors, at different times, and show different flight apparatus, but both share with dinosaurs a fully upright posture of the rear limbs, and required a high metabolic rate for active flight.

Pterosaurs differed from birds in having membraneous wings like bats rather than feathers, but unlike bats, only a single finger, the fourth, supports the membrane (13). The first three fingers retained their primitive proportions and claws. The pterosaur wing is narrow and attached to the trunk at the base of the rear limb. Early reconstructions suggested a sprawling, awkward posture, but Kevin Padian argued that the rear limbs were held vertically and moved in a fore and aft direction, as in birds (14).

Pterosaurs evolved from advanced bipedal thecodonts, near the base of the dinosaur radiation. Kevin Padian suggested that their closest relationship may be with *Scleromochlus*, a tiny thecodont of the Late Triassic. Pterosaurs were common throughout the Jurassic and Cretaceous, and recently discovered fossils show that they were al-

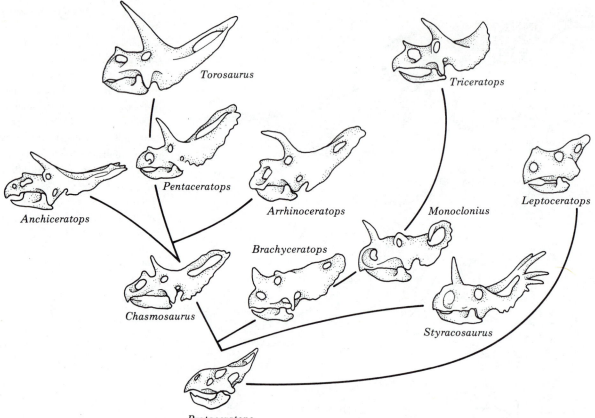

Figure 11-23 Skulls of ceratopsian dinosaurs showing a variety of patterns of horns. The bony frill extending over the neck served for the origin of jaw muscles, provided protection, and perhaps served for sexual display. (From J. H. Ostrom. Courtesy of *Discovery*, (Yale Peabody Mus. Nat. Hist.) 1(2).)

ready highly evolved in the Late Triassic (Fig. 11–24). Triassic pterosaurs were primitive in the proportionately short wings, but all the basic characters of the order were fully evident.

Pterosaurs are divided into two major groups. The primitive rhamphorhynchoids, restricted to the Late Triassic and Jurassic, and the pterodactyloids, from the Late Jurassic and Cretaceous. The rhamphorhynchoids had long tails, stiffened by ossified tendons. The skull was large, but had an open structure, with thin struts of bone between the openings for the eyes, nose, and jaw muscles. Elongate teeth were retained from the thecodonts.

Pterodactyloids had much shorter tails, as do modern birds, which provided for greater maneuverability. The teeth were reduced or lost, and the girdles were rigidly attached to the trunk.

Pterosaurs ranged in size from that of a sparrow to *Quetzalcoatlus*, with a wing span of 11 to 12 meters and an estimated weight of 65 kg (15). The remains of most pterosaurs are found in marine deposits.

At Solnhofen, in the Upper Jurassic of southern Germany, the fine-grained limestone preserves impressions of the wing membranes. Because the pterosaurs are classified as reptiles, most of which have a low metabolic rate, they were thought to have been poor flyers, relying on gliding rather than active flapping flight. However, the structure of the skeleton is very similar to that of birds in many features associated with active flight including the relatively large size of the brain (indicated by endocasts), the presence of a sternum (or breastbone) for attachment of flight muscles, pneumatic foramina for the extension of air sacs into the bones, and the strong integration of the limb girdles and vertebrae. These structures would not be necessary if pterosaurs did not have flight capabilities comparable to those of birds.

Five families of rhamphorhynchoids and five families of pterodactyloids are recognized. The rhamphorhynchoids became extinct during a short period of time during the Late Jurassic, possibly as

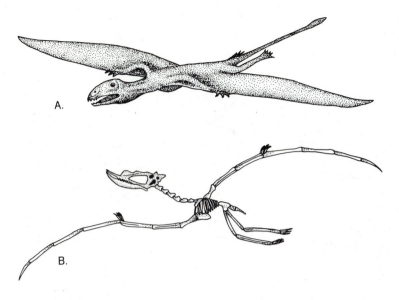

Figure 11-24 Pterosaurs. A. The Upper Triassis rhamphorhynchoid *Eudimorphodon*. B. The pterodactyloid *Dsungaripterus*, which had a wing span of 3 m. The long tail, characteristic of rhamphorhynchoids has been much reduced. (A: Courtesy of Rupert Wild. B: Modified from Dong Zhiming, *Dinosaurs from China*.)

a result of competition with their own descendants, the pterodactyloids. The pterodactyloid families became extinct gradually during the Cretaceous, but three genera, the widespread *Pteranodon* and *Nyctosaurus* and the rare *Quetzalcoatlus*, continued to the end of the Mesozoic.

AQUATIC REPTILES OF THE MESOZOIC

During the time of the dinosaurs, other groups of reptiles dominated the sea. Only a single group of amphibians, the Lower Triassic trematosaurs, managed to adapt to the marine environment. Most amphibians have such permeable skin that they would immediately lose their body fluids by osmosis if they lived in salt water. As a result of evolving a thick, more nearly impermeable skin, the reptiles were able to adapt to life in the sea. From the Late Permian on, group after group of reptiles adapted to life in an aquatic environment. The first known marine reptiles were the mesosaurs of the Middle to Late Permian (Fig. 11–25). Their remains are known from shallow marine deposits in what is now western Africa and eastern South America. Long before the idea of continental drift was accepted by geologists, this distribution suggested that these land masses might have been juxtaposed in the Late Paleozoic. Mesosaurs lack any temporal openings and apparently evolved from the ancestral reptilian stock. They are not related to any other groups of marine reptiles.

Ichthyosaurs:

The most highly specialized of all aquatic reptiles were the ichthyosaurs. They first appeared in Lower Triassic beds of China, Japan, North America, and Spitzbergen. The body was already fusiform to facilitate aquatic locomotion, but the limbs retained the elements of primitive terrestrial reptiles (Fig. 11–26). The skulls of all ichthyosaurs differ from those of other early reptiles in the great size of the quadratojugal (a cheek bone that is much reduced in other groups) and the elongation of the snout. They have an upper temporal opening like those of diapsids but lack a lateral opening. Both their origin and the length of time required for the early stages of aquatic adaptation remain unknown. During the Triassic, ichthyosaurs underwent an adaptative radiation characterized by the evolution of a variety of patterns of the limbs and body proportions. The anatomy and interrelationships of the Triassic families remain poorly known.

Fossils from the Early Jurassic show the beginning of a second radiation based on a more sophisticated pattern of locomotion. The Triassic ichthyosaurs typically had long fusiform bodies, as do moderately fast-swimming modern fish in which the propulsive force is produced by undulation of both the trunk and tail. Jurassic and Cretaceous ichthyosaurs, in contrast, have shorter, spindle-shaped bodies, and symmetrical, lunate tails similar to those of the fastest swimming of modern bony fish, including the tuna, swordfish, and marlin. This body form indicates that Jurassic and Cretaceous

Figure 11-25 *Mesosaurus*, an aquatic reptile from the Upper Permian of western Africa and eastern South America. The numerous, extremely long and slender teeth may have been used to strain prey such as small crustaceans that are common in the same deposits. The limbs are only slightly modified as paddles, but the tail is laterally compressed, about 1 m long. (Modified from J. Augusta & Z. Burian, *Prehistoric Animals*.)

ichthyosaurs were capable of very rapid locomotion.

In numerous fossils from the Lower Jurassic shales of Holzmaden, in southern Germany, the body is preserved as a coalified impression that shows the outline of a symmetrical tail fin and a triangular dorsal fin. These fossils also show that ichthyosaurs gave birth to live young. The advanced ichthyosaurs underwent a period of adaptive radiation in the Early Jurassic, diversifying into several families that show various patterns of skull proportions and fin structure, but all retained a constant body form until their extinction in the Cretaceous. Ichthyosaurs gradually declined in number and diversity after the Early Jurassic and became extinct long before the end of the Mesozoic.

Nothosaurs and Plesiosaurs:

In contrast with the ichthyosaurs, a second major assemblage of Mesozoic reptiles, the sauropterygians, show a more gradual, although stepwise adapation toward an aquatic way of life during the Mesozoic. The sauropterygians are divided into two large groups: the entirely Triassic nothosaurs and the primarily Jurassic and Cretaceous plesiosaurs. Plesiosaurs include gigantic forms that fit the modern concept of sea serpents, with rounded bodies, generally long necks, and paddle-shaped limbs (Fig. 11–27). Two superfamilies are recognized. The plesiosauroids include the more primitive members of the suborder and the extremely long-necked elasmosaurids. The pliosauroids had shorter necks but enormously elongated skulls; the skull of *Kronosaurus* reached a length of more than 3 meters. Plesiosaurs remained diverse until the very end of the Cretaceous.

The Triassic nothosaurs were smaller than the plesiosaurs, with the limbs less specialized as paddles. Structures of the skull link nothosaurs closely with primitive diapsids of the Late Permian. The lower temporal bar has been lost, as in lizards, but unlike lizards the quadrate of these aquatic reptiles is solidly attached to the palate medially and is not movable. The ribs of nothosaurs are thickened, as in mesosaurs to increase their body weight.

Nothosaurs differ from plesiosaurs in the more complete reduction of the openings in the palate that were present in primitive diapsids. *Pistosaurus* from the Middle Triassic differs from primitive

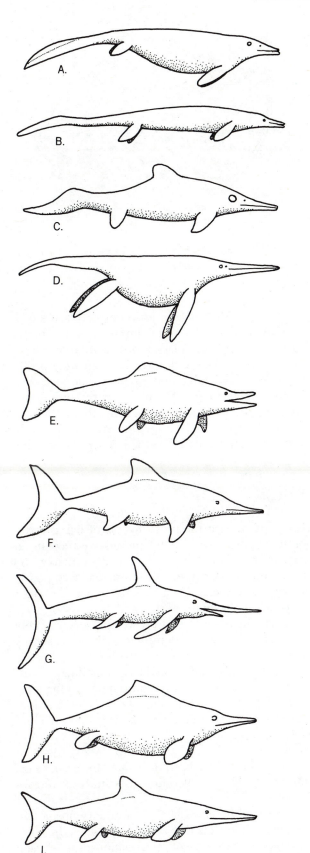

Figure 11-26   a–d. Various shapes of ichthyosaurs from the Triassic. A. *Utatsusaurus* from the Lower Triassic of Japan. B. *Cymbospondylus* from the Middle Triassic of North America. C. *Mixosaurus*, from the Middle Triassic of Spitzbergen, central Europe, Canada, and East Indies. D. *Shonisaurus* from the Upper Triassic of western North America. This is the largest known ichthyosaur, 15 meters long. E–I. Jurassic and Cretaceous ichthyosaurs, all of which have similar body form. The trunk is spindle-shaped and the tail is a high, lunate structure as in the fastest swimming modern fish. The body outline is preserved in specimens from Holzmanden, southern Germany.

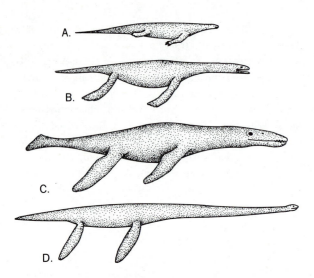

Figure 11-27   Sauropterygians. A. The primitive mid-Triassic nothosaur *Pachypleurosaurus*, about 1 meter long. Much of the skeleton resembles that of primitive eosuchian reptiles. B. *Pistosaurus*, from the Middle Triassic that may link nothosaurs and plesiosaurs, C. The pliosaur *Liopleurodon* from the Upper Jurassic of Europe, D. An elasmosaur from the Upper Cretaceous of North America that was more than 12 meters long.

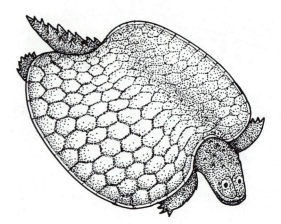

Figure 11-28   The armored placodont *Henodus* from the Upper Triassic of Germany.

sauropterygians in having a palatal structure like that of plesiosaurs, which suggests that it may form a link between these groups, although the relatively short and unspecialized limbs have led in the past to its classification with nothosaurs (16).

Plesiosaurs, in contrast with most marine reptiles other than turtles, must have relied primarily on their limbs for aquatic locomotion. Other reptiles depend primarily on the trunk and tail for propulsion in the manner of fish. Relative to the length of the trunk, the limbs are much larger in plesiosaurs than in nothosaurs, and no fossils are known that show intermediate proportions. Differing body proportions among the nothosaurs suggest that they may have experimented with different modes of propulsion. Larger and more specialized genera developed a more rigid trunk region, while reducing the size of the tail.

Another group of aquatic reptiles, the placodonts, were common and diverse in shallow coastal waters in Europe during the Middle Triassic. They have long been classified with the nothosaurs on the basis of the limited specialization of the limbs for aquatic locomotion, but placodonts differ in having a solidly ossified cheek and very long transverse processes on the vertebrae. There is no strong

evidence for the specific affinities of placodonts, but they probably evolved from the primitive diapsid stock (17). Most placodonts have large, blunt teeth on the palate and lower jaw that were probably used for crushing molluscs or crustaceans. One family, the Heleodontidae, had a nearly continuous covering of dermal armor, giving them the appearance of turtles. Within this group, the teeth were lost and replaced with a horny beak (Fig. 11–28).

BIRDS

Birds are among the most diverse of living vertebrate groups. Nearly 9000 species are recognized among the 27 modern orders. Unfortunately, the evolutionary history of birds remains very poorly known. The skeletons of most birds are very light in order to facilitate flight, and because of their fragility, the bones are rarely well preserved. Many fossil birds have been identified on the basis of only a few bones, such as those in the lower leg that are relatively robust and so resist crushing during preservation. The mechanical requirements of flight restrict the extent of variability of the skeleton so that comparable bones of different avian orders are more similar to one another than are bones of different orders of other vertebrate groups. Many fossils of Cenozoic birds have been discovered in recent years, but relatively few paleontologists study avian evolution.

Nearly all fossils of birds from the Cenozoic can be placed in orders that are living today (18). Very few provide evidence of interrelationships between the orders. Study of the biochemistry and molecular genetics of living birds appears to hold greater promise for establishing specific relationships among and between the families and orders. As more fossils of early Cenozoic and Late Cretaceous birds are described, they can provide a test for hypotheses of relationship based on modern species.

The most interesting episode in the history of birds is the time of their origin (19). Fortunately, a very early stage in this process is well documented in the fossil record. The earliest known bird, *Archaeopteryx*, is known from four nearly complete skeletons from the Upper Jurassic of southern Germany (Fig. 11–29). They are preserved in extremely fine-grained limestone, deposited by precipitation in a quiet lagoonal environment. These fossils show impressions of the feathers that were attached to the forelimb and tail. Both the individual structure and the arrangement of the feathers on the wings are very similar to those of modern flying birds. The main axis of the feather is not symmetrically placed, but is closer to the leading edge to give the shape of an airfoil to provide lift.

Figure 11-29   Comparison of the skeleton of *Archaeopteryx* (A), and a modern bird (B). Bones indicated in black have changed most radically. (From E. H. Colbert, *Evolution of the Vertebrates*. John Wiley, 1980.)

Although the configuration of the feathers is essentially modern, the skeleton of *Archaeopteryx* is much more primitive than that of any other bird, fossil or living. It closely resembles that of small, bipedal carnivorous dinosaurs such as *Compsognathus*, except for the significantly greater length of the forelimb. These specimens demonstrate that birds could have evolved from small theropod dinosaurs some time in the Early Jurassic or Late Triassic. There is still much disagreement as to whether birds passed through an arboreal stage early in their evolution or evolved the capacity for flight as fast-running animals.

No birds other than *Archaeopteryx* are known from Jurassic rocks. The skeletons of birds from the Cretaceous and Cenozoic are much more highly specialized for flight than is the skeleton of *Archaeopteryx*. Except in birds that have secondarily lost the power of flight, the bones have become very thin, and a sternum (or breastbone) with a keel has evolved to provide attachment for the flight muscles. Both the muscles that lower and those that raise the wing are attached to the sternum. A tendon from the muscle that raises the wing extends dorsally and then passes through a pulleylike structure formed by the bones of the shoulder girdle. This reverses the direction of the muscle's force so that it can elevate the wing, although it originates on the ventral surface of the body. None of these features are evident in *Archaeopteryx*, which also has a long tail with a bony axis, lost in all other birds.

The best known Cretaceous birds belong to two extinct orders, the Ichthyorniformes and the Hesperornithiformes. Like *Archaeopteryx*, both retain teeth. The skeleton of *Ichthyornis* broadly resembles that of gulls, but it is not closely related to any modern birds. *Hesperornis* has a sternum, but the keel is missing and the forelimb is greatly reduced. This is the earliest example of a bird that has lost the power of flight. The rear limbs of *Hesperornis* are very large and resemble those of loons, which suggests that they were used for swimming (Fig. 11–30).

The modern bird orders appear in the fossil record only at the very end of the Cretaceous. Only three groups are currently recognized, the Charadriiformes (including the ancestors of gulls, auks, and flamingos), genera that are close to the ancestry of ducks, and the Procellariiformes (which include albatroses and petrels). It is uncertain whether the early appearance of these aquatic groups is the result of their origin before that of more terrestrial

Figure 11-30 *Hesperornis*, a diving bird from the Upper Cretaceous. (Modified from A. Feduccia, *The Age of Birds*.)

birds or if it reflects the bias of the fossil record toward the preservation of animals living in or near the water.

The avian fossil record in the early Cenozoic is still very incompletely known and provides little evidence of the time of evolutionary divergence of the major groups. The orders that include the kingfishers and the owls are both known from the Paleocene. The ancestors of the orders that include domestic fowl, pigeons, parrots, swifts, loons, pelicans, and penguins all appeared in the Eocene. Most of these early fossils can be placed in modern families and some in living genera. The basic way of life and general skeletal features of the modern bird orders and families were achieved early in the Cenozoic. Although cuckoos, birds of prey, and woodpeckers are not known as fossils before the Oligocene or Miocene, they probably evolved much earlier in the Cenozoic.

The most diverse and numerous of modern bird orders, the songbirds or passerines, appeared first in the fossil record of Europe in the latest Oligocene. By the early Miocene they became extremely numerous in Europe, but reached the New World only in the middle to late Miocene.

Only a few major groups of birds are known from the Cenozoic that are not represented in the modern fauna. These include two groups of gigantic terrestrial predators, the Paleogene Diatremiformes, known from Asia, North America, and Europe, and the Phorusrhacidae, which were common in South America (Fig. 11–31). Members of both groups were more than 2 meters tall, had gigantic skulls, and are thought to have been the dominant terrestrial predators prior to the radiation of the modern mammalian carnivores. Both these groups may be related to the cranes and rails.

Two other, very recently extinct orders of flightless birds are the elephant birds from Madagascar and the moas from New Zealand. Both were con-

Figure 11-31   Gigantic predatory birds of the early Cenozoic. A. *Phorusrhacus*, B. *Diatryma*. (A: Courtesy Department of Library Services American Museum of Natural History.)

temporaries of early man. Neither are known earlier than the Pleistocene, and their relationships with other bird orders have not been established. There is also continuing debate over the relationships of several orders of modern flightless birds, the rheas, cassowaries, kiwis, and ostriches. Only the ostriches have a significant fossil record, which demonstrates that they were present in Europe as early as the middle Eocene, and were widespread in Eurasia throughout the Cenozoic.

MAMMAL-LIKE REPTILES

Mammals are most common in the Cenozoic, but their fossils occur in rocks as old as Late Triassic in age, approximately 200 million years old. The ancestry of mammals can be traced even farther back to a diverse assemblage of more primitive amniotes, the synapsids or mammal-like reptiles, which were widespread in the late Paleozoic and early Mesozoic (Fig. 11–32).

Accompanying the earliest conventional reptiles

in the tree stump fauna of Joggins, Nova Scotia, are scattered remains of animals that are close to the ancestry of mammals, the pelycosaurs. The pelycosaurs are one of two orders of synapsids. They can be distinguished from other early reptiles by the presence of an opening in the skull behind the eye, the lateral temporal opening. This lies over the main jaw closing muscles and was further enlarged in later mammal-like reptiles and mammals. If we touch the area above our cheekbones, we can feel our jaw muscles contract through our temporal opening, which is directly homologous with that of the earliest pelycosaurs.

The primitive pelycosaurs are also distinguished from other early amniotes by their larger size. Their large body size and large teeth suggest that the early pelycosaurs were carnivores, feeding on large prey, in contrast to such genera as *Hylonomus*, which are thought to have fed primarily on insects. In the Late Carboniferous, the pelycosaurs diversified extensively to become the largest and most common reptiles. The early genera probably resembled such living lizards as varanids and iguanids in general

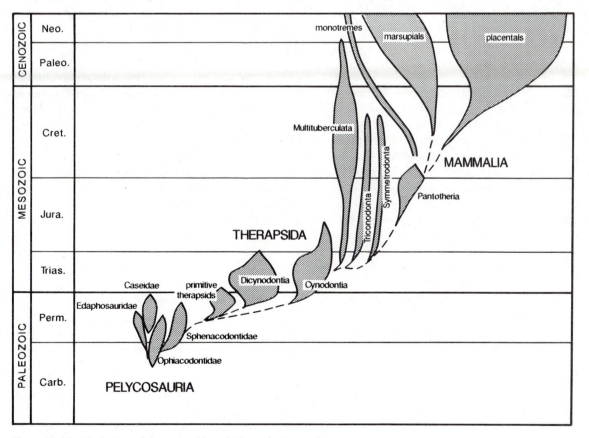

Figure 11-32   Phylogeny of the synapsids and Mesozoic mammals.

appearance. By the Early Permian, pelycosaurs included 70 percent of the known amniotes and some genera grew as long as 4 m.

Subclass—Synapsida
    Order—Pelycosauria
            Family—Ophiacodontidae
            Family—Sphenacodontidae
            Family—Edaphosauridae
            Family—Caseidae
    Order—Therapsida
        Suborder—Eotitanosuchia
            Family—Biarmosuchidae
        Suborder—Dicynodontia
        Suborder—Gorgonopsia
        Suborder—Cynodontia
            Family—Procynosuchidae
            Family—Galesauridae
            Family—Tritylodontidae
            Family—Chiniquodontidae
            Family—Tritheledontidae

Three major groups of pelycosaurs have been recognized in Late Carboniferous rocks: the primitive ophiacodonts, the large carnivorous sphenacodonts, and the edaphosaurs. The edaphosaurs were among the first herbivorous amniotes as indicated by their many small, blunt teeth. None of the modern amphibian groups are herbivorous as adults (the larval stages of most frogs are herbivorous but this is associated with a highly specialized anatomy of the mouth and digestive tract), and there is no strong evidence that any of the Paleozoic amphibians lived on plant material. Even among modern reptiles, most groups are carnivorous. Evidently, the digestion of plant material is metabolically complicated and developed relatively late in vertebrate evolution. *Edaphosaurus* was the only common edaphosaur. Later in the Early Permian, a second group of herbivorous pelycosaurs, the caseids, became common. They were taxonomically diverse and were the first group of terrestrial herbivores to dominate the fauna numerically. The caseid *Cotylorhynchus* may have weighed up to 600 kg and had a greatly expanded rib cage to hold food being digested.

The genus *Edaphosaurus* is also notable for the possession of a sail-like structure on its back, supported by long neural spines with short lateral processes (Fig. 11–33). Some sphenacodontids, including the very common genus *Dimetrodon*, also had long neural spines, although without the cross bars. The elongate neural spines were evidently connected by soft tissue since the pieces remain in place even after they have been fractured during life. Such a large area of tissue would certainly alter the physiological characteristics of the animal. In *Dimetrodon*, grooves at the base of the spine show the presence of large blood vessels. By controlling the flow of blood and the orientation of the body relative to the sun, *Dimetrodon* could use the sail to gain or lose body heat more rapidly than would an animal without a sail. Many modern reptiles use the heat of the sun to raise the body temperature above that of the surrounding air. The presence of long spines suggests that pelycosaurs also depended on an external source of heat and hence probably had as low a metabolic rate as living lizards (20).

Pelycosaurs were common in North America into the late Early Permian, but their remains became progressively rarer as the sediments show evidence of increasing aridity. The last evidence of this group comes from the lower beds of Late Permian age in Russia and South Africa. The next stage in the evolution of mammal-like reptiles is recorded in the Upper Permian of Russia and South Africa by the appearance of a much more diverse group, the therapsids. The presence of a peculiar process at the back of the lower jaw, the reflected lamina of the angular, in both therapsids and sphenacodontids, indicates that the major group of carnivorous pelycosaurs gave rise to the advanced synapsids (Fig. 11–34).

There is a gap in both time and morphology between the advanced sphenacodontids and the most primitive therapsids. When the therapsids first appeared, they were already greatly diversified with several distinct lineages of carnivores and herbivores. Therapsids were common fossils throughout the world from the Late Permian into the Late Triassic but are best known from flood plain deposits in southern Africa.

The most important of the herbivorous therapsids in the Late Permian and Early Triassic were the dicynodonts, which were the most common of all terrestrial vertebrates in southern Africa at this time and were present in most other parts of the world, including Antarctica. Dicynodonts ranged from the size of a squirrel to that of an ox. Most were terrestrial, but some had adaptations for burrowing, and the most common and widespread genus, *Lystrosaurus*, was aquatic (Fig. 11–35).

Figure 11-33 Pelycosaurs A. *Ophiacodon*, B. *Dimetrodon*, C. *Edaphosaurus*. All about 2 m long.

Dicynodonts must have had a wide range of diets, but all had a very similar, highly specialized feeding apparatus. Except for the earliest genera, dicynodonts had lost most of their teeth, but most had a large pair of canine teeth. In some species their presence or absence is an aspect of sexual dimorphism. The bone surface at the front of the jaws resembles that of turtles, indicating that the jaws were covered with a horny beak. The shape of the jaw articulation and the geometry of the areas for attachment of the jaw muscles indicate that the lower jaw extended forward as it opened, and moved back as it closed, so that the sharp edges of the horny coverings sheared against one another.

During the Late Permian and Early Triassic, several successive groups of carnivorous therapsids evolved. The first genera to appear were the biarmosuchids. They are known only from Russia. Their skulls greatly resemble those of the sphenacodonts, but the structure of the girdles and limbs indicates that they had a more erect posture than primitive reptiles and the limbs moved in a more nearly fore-and-aft direction. The biarmosuchids were succeeded in the Late Permian in Russia and South Africa by the common and diverse gorgonopsians.

The genus *Lycaenops* (Fig. 11–36) superficially resembled a modern mammalian carnivore, although the details of skeletal structure and especially the teeth were far more primitive.

The gorgonopsians were replaced in the Early Triassic by another group of carnivorous therapsids, the cynodonts. Most of the physiological and structural changes that led to the emergence of mammals occurred in this group. Early cynodonts were the size of cats or dogs, but they retained short limbs and a partially sprawling posture. The skull shows a number of features that presage the mammalian condition. One of the most conspicuous features of even the Late Permian cynodonts was the beginning of a secondary palate (Fig. 11–37). The secondary palate in mammals allows them to breathe while chewing their food. Most reptiles do not chew their food, but swallow it in large pieces. The food digests slowly this way, but since most reptiles require only about one tenth the food of a mammal of similar size, this is not a problem. Mammals chew their food thoroughly so that it can be digested more rapidly. Chewing is possible for mammals only if they can breathe at the same time, so a secondary palate is imperative.

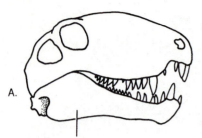

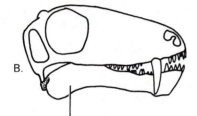

Figure 11-34 Skulls of the sphenacodont pelycosaur *Dimetrodon* and the early therapsid *Bairmosuchus*, showing the reflected lamina of the angular.

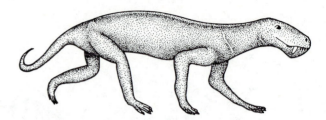

Figure 11-36 The Upper Permian gorgonopsid therapsid *Lycaenops*. (Modified from E. H. Colbert, *Bull. Am. Mus. Nat. Hist.*, v. 89.)

The secondary palate in early cynodonts indicates that their metabolic rate was similar to that of mammals. Cynodonts also show differentiation of the teeth into several different types, as do mammals. The anterior teeth are peg or chisel-shaped, like our incisors, and could be used to grasp and hold prey. Long, piercing canine teeth behind the

incisors are a heritage from the earliest amniotes. They were lost in other reptilian groups, but retained in the synapsids. The cheek teeth have multiple cusps, which broke up the food. In contrast with mammals, the cheek teeth of these therapsids were still continuously replaced so that occlusion (a specific pattern of contact between the upper and lower teeth) could not occur, limiting their effectiveness in breaking up food.

The shape of the skull in cynodonts shows that the pattern of the jaw muscles was approaching that of mammals. In pelycosaurs and primitive therapsids, the jaw muscles resembled those of primitive living reptiles. The major jaw closing muscle attached to the upper and inner surface of the lower jaw and pulled the jaws toward the midline as they were closed. Among the cynodonts, this muscle be-

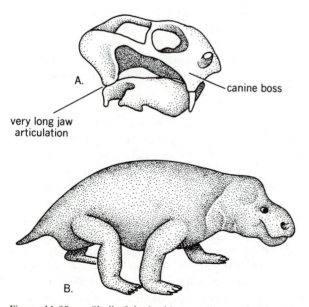

Figure 11-35 *a.* Skull of the herbivorous therapsid *Dicynodon*, *b.* The 3 meter long Lower Triassic dicynodon *Kannemayeria*.

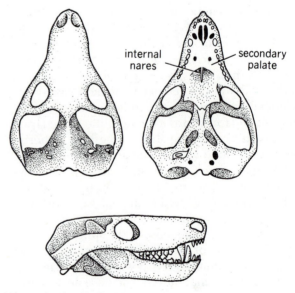

Figure 11-37 Skull of the Lower Triassic cynodont *Trinaxodon*, in dorsal, palatal, and lateral views. Note the extensive secondary palate and the very large openings for jaw muscles.

came subdivided into three components: the temporalis, which retained its primitive attachment to the inside of the back of the lower jaw, the deep masseter, which attached to the outside of the lower jaw and provided a balancing, laterally directed force, and the superficial masseter, which originated more anteriorly on the cheek bone, and pulled the jaw forward as well as upward. The differently oriented muscles acted like a sling to suspend the lower jaw, and released the pressure on the jaw articulation, which would have been severe if all the pressure had been directed inward and upward (21). This distribution of muscles was also necessary for the functioning of the complex molar teeth that are a hallmark of mammals.

The cynodonts also show a major modification in the structure of the braincase. The relative size of the brain probably did not change significantly within therapsids, but the bones that surround it were drastically altered, and the new configuration facilitated the expansion of the brain in their mammalian descendants. The changes in the braincase can be attributed to the expansion of the areas of muscle attachment and strengthening of the skull to resist the force of the increasingly massive jaw muscles. The bones forming the back and roof of the skull extended around the old, mostly cartilagenous, reptilian braincase and formed a nearly continuous external covering. The sidewall of the braincase was formed from a vertical bone, the epipterygoid, that was originally lateral to the base of the braincase, and new bone filled in the area medial to the eye. In advanced, mammal-like reptiles, the skull, including the braincase, had achieved a very mammalian appearance, although the brain itself remained small.

In the evolution from primitive therapsids through cynodonts, the lower jaw also changed significantly. In primitive reptiles, it consisted of a mosaic of large bones. The jaw articulation was formed by the articular, a bone that formed the core of the jaw in primitive fish. The teeth were borne by the dentary, the largest of the jawbones. The jaw muscles inserted on four other bones at the back of the jaw (Fig. 11–38). The fact that the forces exerted on the teeth, at the jaw articulation and by the jaw muscles, acted on different bones was clearly not mechanically efficient. In cynodonts, the dentary expanded dorsally and posteriorly to form a more effective area for muscle attachment. Among the most advanced cynodonts, the dentary approached the back of the skull, and in primitive mammals it formed a new jaw articulation (Fig. 11–39).

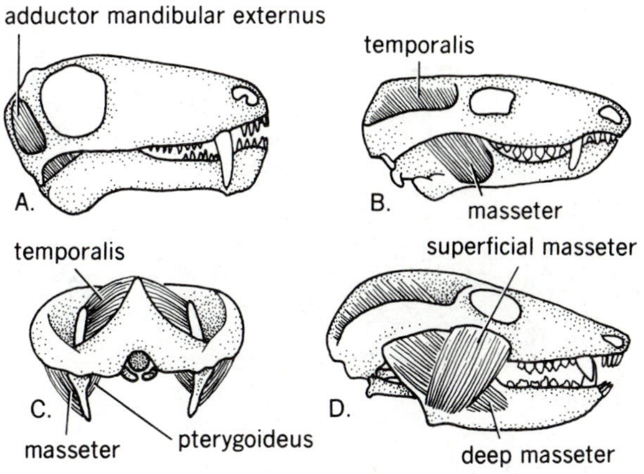

Figure 11-38   Evolution of the jaw muscles in therapsids. A. The primitive genus *Biarmosuchus*, B. *Thrinaxodon*, C. Posterior view of *Thrinaxodon* showing the balancing effects of the temporalis and masseter muscles, D. An advanced cynodont in which the masseter is divided into two components. (From *Vertebrate Paleontology and Evolution* by Robert L. Carroll. Copyright (©) 1987 W. H. Freeman and Company. Reprinted with permission.)

The dentary is the only jawbone in modern mammals, but not all the other bones of the lower jaw were completely lost. The articular bone of the lower jaw and the quadrate in the upper were very much reduced in late therapsids and incorporated into the middle ear as the malleus and incus in mammals (Fig. 11–40). The reflected lamina of the angular bone at the back of the jaw of pelycosaurs became gradually modified into a ring-shaped structure that supports the eardrum of cynodonts and mammals.

Changes in the rest of the skeleton also occurred among the therapsids. The vertebral column evolved so as to limit flexure in the horizontal plane, which was an important component of locomotion in primitive amniotes. The limbs came to move in a more or less fore and aft gait, but they were not brought directly under the body in even the most advanced mammal-like reptiles. A partially sprawling posture has been retained by many small living mammals, including rodents and insectivores.

The cynodonts diversified in the Early Triassic. Several families became specialized as herbivores. The most advanced were the tritylodonts, the skeleton of which had many mammalian features. Their dentition was specialized in the loss of canines and formation of very complex cheek teeth, and they were clearly not immediately ancestral to mammals (Fig. 11–41). The well-known chiniquodontids and rare trithelodonts retained a dentition indicative of

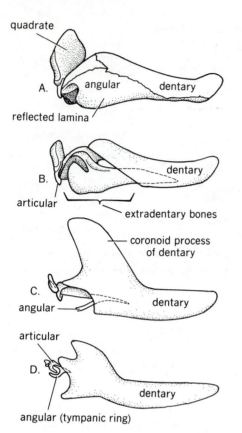

Figure 11-39   Evolution of the lower jaw in mammal-like reptiles. Note the gradual increase in size of the dentary relative to the other bones. (From E. F. Allin, *Journal of Morphology*, vol. 147.)

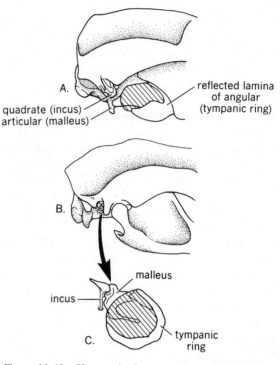

Figure 11-40   Changes in the proportions of the jawbones and ear ossicles between a primitive cynodont (A) and a primitive mammal (B). C. Expanded view of mammalian ear ossicles and tympanum (cross hatched). (Modified from A. W. Crompton & F. A. Jenkins, Jr., *Mesozoic Mammals: The First Two-Thirds of Mammalian History*, J. A. Lillegraven, Z. Kielan-Jaworowska, & W. A. Clemens (eds.).)

carnivorous habits, which is much closer to the pattern of early mammals. A lineage of small carnivorous cynodonts that was derived from either the chiniquodontids or the tritheledonts probably gave rise to mammals sometime in the Middle Triassic.

MESOZOIC MAMMALS

Therapsids persisted into the Middle Jurassic, but by the Late Triassic true mammals had appeared in the fossil record. They are recognized by the nature of the teeth and jaws and by their increased brain size (22).

The posterior cheek teeth of mammals, in contrast to whose of carnivorous therapsids, are not replaced, and the anterior teeth were replaced only once. The teeth in the upper and lower jaws maintain a fixed relationship with one another (a specific pattern of occlusion) so that the teeth shear past one another. In the earliest mammals, the wear

resulting from their contact established sharp cutting edges in the shape of reversed triangles (Fig. 11–42). The molar teeth of the most primitive mammals bear three large cusps oriented anteroposteriorly. The three-cusped nature of these molars is the basis for the name of the group to which they belong, the triconodonts.

Class—Mammalia
    Subclass—Prototheria
        Order—Monotremata
        Order—Triconodonta
        Order—Docodonta
    Subclass—Allotheria
        Order—Multituberculata
    Subclass—Theria
        Infraclass—Trituberculata
        Order—Symmetrodonta
        Order—Pantotheria
        Infraclass—Metatheria (marsupials)
        Infraclass—Eutheria (placentals)

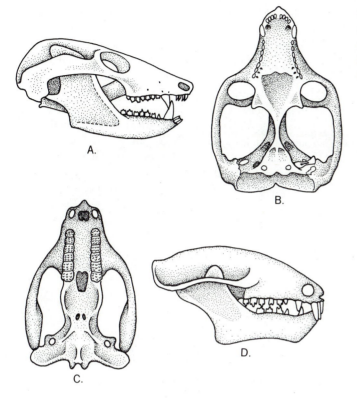

Figure 11-41   Skulls of advanced cynodonts. A, B. Lateral and palatal views of a Middle Triassic carnivorous chiniquodontid, C. Palatal view of a herbivorous tritylodont from the Lower Jurassic, D. A trithelodont from the Upper Triassic. (C: Courtesy of Dr. H.-D. Sues.)

The lower jaw is composed almost entirely of the dentary bone, which has extended posteriorly to form the principal jaw joint. The original reptilian jaw articulation between the quadrate and articular was retained, although these bones, now termed the incus and malleus, also served as ear ossicles. Nearly the entire skeleton of the early triconodonts is known from specimens in Europe, Africa, North America, and China (Fig. 11–43). They were small animals that probably weighed no more than 20 to 30 grams, close to the size of the smallest modern mammals, the shrews. The postcranial skeleton retained primitive features, similar to those of the most primitive living marsupials and placentals. The posture was still partially sprawling, as in shrews and rodents. The shape of the vertebrae indicates that the spinal column was flexed in the vertical plane, as in mammals, rather than horizontally as in more primitive amniotes.

The secondary palate and cheek teeth capable of chewing food into small particles suggest that these early mammals had as high a metabolic rate as primitive living mammals. Because they were tiny, they would have had a very high surface-to-volume ratio and would have lost body heat rapidly unless they were insulated. We assume, therefore, that they were covered with fur. In modern mammals, hair is associated with sweat and oil glands. The similarity of these to the structure and development of mammary glands suggests that the earliest mammals already suckled their young.

We do not have direct evidence of the pattern of reproduction of these early mammals, but the most primitive of living mammals, the monotremes, including the Australian echidna and platypus, may provide a useful analogue. Unlike all other modern mammals, they lay eggs. Unlike reptiles, the young hatch out at a very immature, essentially embryonic stage. The platypus builds a nest, but the echidna, like a marsupial, has a pouch. The young are tiny (less than 1 gram in weight), helpless, and completely dependent on their mothers for food, protection, and warmth. This seems a step backwards from the reptilian pattern in which the young hatch as minatures of the adults and can feed themselves immediately. How can the pattern of reproduction in primitive mammals be explained?

Mesozoic mammals were very small. Among living mammals, the metabolic rate per gram of body weight increases at a progressively higher rate in

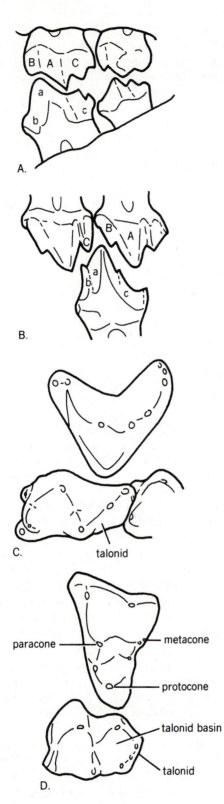

A.

B.

C.   talonid

paracone ———————— metacone

protocone

talonid basin

talonid

D.

Figure 11-42   Mammal teeth. A. Lateral view of the Upper Triassic triconodont *Morganucodon*. The upper and lower teeth occlude almost directly with one another. B. Molar teeth of the symmetrodont *Kuehneotherium*. The upper and lower teeth alternate with one another so that each tooth occludes with two successive upper teeth. C. Occlusal view of the upper and lower teeth of a pantothere. The upper tooth retains the triangular pattern of symmetrodonts, but the lower tooth has additional cusps forming a heel or talonid. D. Occlusal view of the upper and lower teeth of a therian mammal showing the tribosphenic pattern. The talonid has a central basin into which fits the apical cusp of the upper molar.

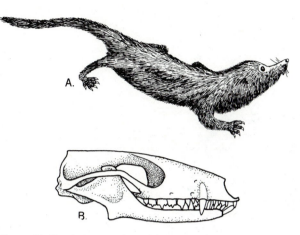

Figure 11-43   A. Restoration of the tiny Upper Triassic and Lower Jurassic mammal *Morganucodon*. It weighed 20 to 30 grams. B. Skull of *Morganucodon*. (A: Courtesy of A. W. Crompton.)

small species (Fig. 11–44). At this rate of increase, a mammal smaller than 2 to 3 grams would have such a high metabolic rate that it would require an infinite amount of food to maintain a constant high body temperature. Hence, no warm-blooded mammals can be smaller than 2 to 3 grams. Some shrews are only slightly larger than this, and their young are considerably smaller. How can they survive? They are born totally dependent on external heat sources, as are reptiles. All their heat and food come from their parents. All the food they consume goes

to growth, and none is used for heat production, which may account for 90 percent of the metabolic effort in small adult mammals. Even though modern monotremes are much larger than the earliest mammals, they retain this primitive reproductive pattern, which was almost certainly common to the earliest mammals. Marsupials, whose young are carried in a pouch, have not departed significantly from this primitive pattern either. Only the placental mammals have solved the problem of retaining the developing young within the body of the mother until they have reached an advanced stage of development. This is discussed in the next chapter.

The relative brain size of early mammals is about four times that of therapsids of comparable size. This can be attributed to the need to integrate greatly increased sensory input associated with increased acuteness of hearing, increased sense of smell, and increased tactile information provided by the vibrissae (such as the whiskers of a cat) and other hairs covering the body. The demands for more effective feeding required by the higher mammalian metabolic level must also have necessitated more sophisticated integration of the muscles used in feeding and running.

From such primitive triconodonts as *Morganucodon* (Fig. 11–43) arose a number of distinct lineages of mammals common in the Mesozoic. Among the most distinctive forms were the multitubercu-

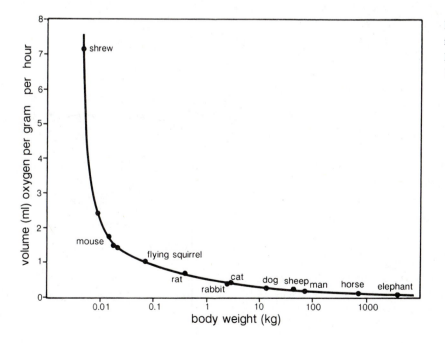

Figure 11-44   Progressive increase in the metabolic rate of small mammals. (From K. Schmidt-Nielsen, *Animal Physiology*. 1975 Cambridge University Press.)

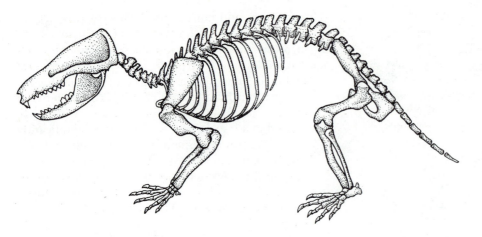

Figure 11-45   Skeleton of the triconodont *Gobiconodon* from the Upper Cretaceous. (By permission of F. A. Jenkins and C. R. Schaff and the *Journal of Vertebrate Paleontology*.)

lates (Fig. 11–46), which ranged from mouse to opossum-sized omnivores. Their teeth superficially resembled those of rodents with a wide gap, or diastema, between the incisors and cheek teeth. As in some marsupials, one of the lower cheek teeth was in the form of a large slicing blade. David Krause and Farish Jenkins (23) described features of the skeleton that indicate that many multituberculates were arboreal.

Typical multituberculates are recognized no earlier than the Late Jurassic, but they may be related to the haramiyids, known from isolated teeth as early as the Late Triassic. The youngest multituberculate fossils come from the Miocene.

The monotremes have been thought to be related to the multituberculates and to have had an ancestry very distinct from that of modern marsupials and placentals. Zofia Kielan-Jaworowska, Fuzz Crompton, and Farish Jenkins (24) suggest that a newly discoved fossil from the Lower Cretaceous of Australia may link monotremes to the base of the marsupial–placental lineage.

Other descendants of the morganucodontids, which were less diverse or long lived, include the docodonts and the gobiconodontids. The docodonts are known only from the Middle and Upper Jurassic. All were tiny, but the molars were specialized in having enlarged crushing and grinding surfaces. The gobiconodontids include Late Cretaceous genera from both Mongolia and North America that reached the size of the living opossum. Farish Jenkins and Chuck Schaff argue that they may have been scavengers (25)(Fig. 11–45).

Another group of mammals was already present

in the Late Triassic. The kuehneotherids are known only from teeth and jaws, but the configuration of the molars shows that they were related to the major groups of living mammals, the placentals and the marsupials (together termed therian mammals). The principal cusps are arranged in the pattern of an obtuse triangle, in contrast with their linear arrangement in triconodonts. The upper and lower teeth occlude so that the edges of the triangles shear against one another when they first erupt, rather than developing shear surfaces through wear, as in *Morganucodon*. The early mammals with a triangular pattern of the molar cusps are called sym-

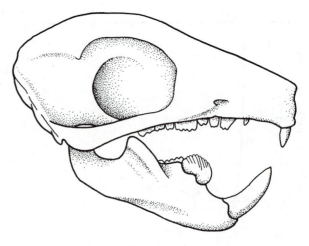

Figure 11-46   Skull of multituberculate, the most diverse group of Mesozoic mammals. The dentition superficially resembles that of rodents and some omnivorous marsupials. (Modified from Z. Kielan-Jaworowska, *Palaeontologia Polonica*, v. 25.)

metrodonts, because the upper and lower molar teeth are similar in shape. Symmetrodonts survived into the Cretaceous, but in the Jurassic they gave rise to a distinct lineage, the pantotheres, that were more advanced in the direction of marsupials and placentals. They were small, shrewlike forms that were diverse in the Jurassic. They are known primarily from their teeth. The upper molars were sharply acute triangles with the central cusp extending toward the tongue. The lower molars had a posterior heel, or talonid (Fig. 11–42D).

By the Early Cretaceous, the pantotheres had given rise to a lineage that included the ancestors of both marsupials and placentals. The teeth in these mammals are distinguished by the elaboration of a new major cusp in the upper molars, the protocone, that fits into an enlarged talonid in the lower molar. Teeth with this pattern are termed tribosphenic. This pattern provides very effective shearing surfaces that are implemented by a transverse component of jaw movement in primitive mammals.

The most primitive mammals with a tribosphenic dentition are known from little more than teeth and jaws. They cannot be allied specifically with either marsupials or placentals. Some may be close to the ancestry of modern therians, but others belong to one or more divergent lineages. Marsupials and placentals probably evolved in the Early Cretaceous, but only in the Late Cretaceous are they represented by fairly complete remains.

Mammals lived as contemporaries of the dinosaurs for most of the Mesozoic. Primitive living mammals are typically nocturnal and inconspicuous, and their ancestors probably did not come in direct contact with dinosaurs, which, like most living reptiles, presumably fed during the day. Brian Patterson referred to the Mesozoic as having a "lawn mower ecology," in which any mammals that were large or active during the daytime would be subject to predation by the large carnivorous dinosaurs.

REFERENCES

1. Gaffney, E. S., 1987, Triassic and Early Jurassic turtles. In: Padian, K. (ed.), The Beginning of the Age of Dinosaurs: Faunal Changes Across the Triassic–Jurassic Boundry: Cambridge, England, Cambridge Univ. Press, p. 183–187.

2. Whiteside, D. I., 1986, The head skeleton of the Rhaetian sphenodontid *Diphidontosaurus avonis* gen. et sp. nov. and the modernization of a living fossil: Philosophical Transactions of the Royal Society B, v. 312, p. 379–430.

3. Estes, R., 1983, Sauria terrestria, Amphisbaenia: Handbuch der Palaeoherpetologie, part 10A: Stuttgart, Gustav Fisher Verlag.

4. Rage, J., C., 1984, Serpentes. Handbuch der Palaeoherpetologie 11: Stuttgart, Gustav Fischer Verlag, p. 1–80.

5. Haas, G., 1980, Remarks on a new ophiomorph reptile from the Lower Cenomanian of Ein Jubrud, Isreal. In: Jacobs, L. L., (ed.), Aspects of Vertebrate History: Flagstaff, Museum of Northern Arizona Press, pp. 177–192.

6. Brinkman, D. B., and Sues, H. D., 1987, A staurikosaurid dinosaur from the Upper Triassic Ischigualasto Formation of Argentina and the relationships of the Staurikosauridae: Palaeontology, v. 30, p. 493–503.

7. Ostrom, J. H., 1969, Osteology of *Deinonychus antirrhopus*, an unusual theropod from the Lower Cretaceous of Montana: Bulletin of the Peabody Museum of Natural History, v. 30, p. 1–165.

8. Colbert, E. H., 1962, The weights of dinosaurs: American Museum Novitates, no. 2076, p. 1–16.

9. Dodson, P., 1975, Taxonomic implications of relative growth in lambeosaurine hadrosaurs: Systematic Zoology, v. 24, p. 37–54.

10. Horner, J. R., 1984, The nesting behavior of dinosaurs: Scientific American, v. 250 p. 130–137.

11. Galton, P. M., 1970, Pachycephalosaurids—dinosaurian battering rams: Discovery, v. 6, p. 23–32.

12. Sues, H. D., and Galton, P. M., 1987, Anatomy and classification of the North American Pachycephalosauria (Dinosauria:Ornithischia): Palaeontographica, v. l98, p. 1–40.

13. Wellnhofer, P., 1978, Pterosauria. In Wellnhofer, P. (ed.), Handbuch der Palaoherpetologie, Part 19: Stuttgart, Gustav Fischer Verlag, pp. 1–82.

14. Padian, K., 1985, The origins and aerodynamics of flight in extinct vertebrates: Palaeontology, v. 28, p. 413–433.

15. Langston, W., 1981, Pterosaurs: Scientific American, v. 244, p. 122–136.

16. Sues, H. D., 1987, Postcranial skeleton of *Pistosaurus* and interrelationships of the Sauropterygia (Diapsida): Zoological Journal of the Linnean Society, v. 90, p. 109–131.

17. Sues, H. D., 1987, On the skull of *Placodus gigas* and the relationships of the Placodontia: Journal of Vertebrate Paleontology, v. 7, p. 138–144.

18. Olson, S. L., 1985, The fossil record of birds. In Farner, D., King, J., and Parkes, K. (eds.), Avian Biology, v. 8, p. 79–238.

19. Hecht, M. K., Ostrom, J. H., Viohl, G., and Wellnhofer, P. (eds.), 1985, The Beginning of Birds: Willibaldsburg, Eichstatt, Freunde des Jura-Museums Eichstatt, pp. 382.

20. Hotton, N. III, MacLean, P. D., Roth, J. J., and Roth, C. (eds.), 1986, The ecology and biology of mammal-like reptiles: Washington/London, Smithsonian Institution Press, p. 326.

21. Bramble, D. M., 1978. Origin of the mammalian feeding complex; models and mechanisms: Paleobiology, v. 4, p. 271–301.

22. Lillegraven, J. A., Kielan-Jaworowska, Z., and Clemens, W. A., 1979, Mesozoic mammals: The first two-thirds of mammalian history: Berkeley, Univ. of California Press, 311 p.

23. Krause, D. W., and Jenkins, F. A. Jr., 1983, The postcranial skeleton of North American multituberculates: Bulletin of the Museum of Comparative Zoology, v. 150, p. 199–246.

24. Kielan-Jaworowska, Z., Crompton, A. W., and Jenkins, F. A. Jr., 1987, The origin of egg-laying mammals: Nature, v. 326, p. 871–873.

25. Jenkins, F., and Schaff, C. 1988, The Early Cretaceous mammal *Gobiconodon* (Mammalia, Triconodonta) from the Clovery Formation in Montana: Journal of Vertebrate Paleontology, v. 8, p. 1–24.

READINGS

Bakker, R., 1986, The dinosaur heresies: new theories unlocking the mystery of the dinosaurs and their extinction: New York, Morrow, 481 p.

Benton, M. J. (ed.), 1988. The Phylogeny and Classification of the Tetrapods: Clarendon Press, Oxford, 373 p.

Gaffney, E., Hutchison, J., Jenkins, F., and Meeker, L., 1987, Modern turtle origins: The oldest known cryptodire: Science, v. 237, p. 289–291.

Padian, K., (ed.), 1987, The beginning of the age of dinosaurs: fauna changes across the Triassic–Jurassic boundary: Cambridge, Cambridge Univ. Press.

Thomas, R. D. K., and Olson, E. C. (eds.), 1980, A cold look at the warm-blooded dinosaurs: AAAS Selected Symposium, 28: Colorado, Westview Press, 514 p.

Chapter 12

Cenozoic Vertebrates

The fossil record of all vertebrates groups is much more complete in the Cenozoic than in the Mesozoic or Paleozoic. The mode of life of fossil genera is also much easier to establish since most belong to living families.

The major groups of cartilaginous and bony fishes persisted from the late Mesozoic and the teleosts continued their extraordinary radiation. Even in the Cenozoic, the fossil record of amphibians is very incomplete, with only a small percentage of living species being recognized. Most of the families that are living today had probably originated by the late Mesozoic, and only a small number of families known from the early Cenozoic have become extinct.

Except for the dinosaurs, pterosaurs, and large marine reptiles, the major reptilian groups that were present in the Late Mesozoic have persisted into the Cenozoic, although the ranges of most families have been restricted as a result of the progressive cooling of the climate. Snakes underwent a major radiation in the later Cenozoic. The modern bird orders originated in the Late Cretaceous and Paleogene. Modern families and genera are known from the early Neogene.

We think of the Cenozoic as the age of mammals, for they are certainly the most conspicuous terres-

trial vertebrates. Like the reptiles of the Mesozoic, mammals also radiated extensively in aquatic environments, and the bats achieved flight.

CENOZOIC MAMMALS

Although placental and marsupial mammals are known primarily from the Cenozoic, the first stages in the differentiation of the modern groups began in the late Mesozoic. Marsupials and placentals had diverged from primitive therian mammals by the Early Cretaceous, and both groups had begun to radiate by the end of the Cretaceous.

Upper Cretaceous marsupials and placentals can be readily differentiated by their teeth, which provide the most consistent fossil record. Both had five upper and four lower incisors on each side. The numbers of the remaining teeth are the same in the upper and lower jaws. Both groups had a single canine on each side, but the marsupials had three premolars and four molars, while the placentals had four (primitively five) premolars and three molars. (The last premolar of placentals may be homologous with the first molar of marsupials.) All but the molar teeth are replaced in placentals, but

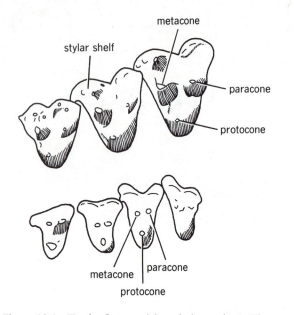

Figure 12-1   Teeth of marsupials and placentals. A. The three upper molar teeth of the Upper Cretaceous marsupial *Alphadon*. The area lateral to the paracone and metacone is wider than in placental mammals and may bear several small cusps. B. The four molar teeth of the Upper Cretaceous placental *Procerberus*.

only the third premolar is replaced in marsupials. The number of each type of tooth is typically written as a fraction (the tooth formula):

$$\frac{5}{4}\ \frac{1}{1}\ \frac{4(5)}{4(5)}\ \frac{3}{3} \text{ for placentals}$$

$$\text{and } \frac{5}{4}\ \frac{1}{1}\ \frac{3}{3}\ \frac{4}{4} \text{ for marsupials.}$$

The morphology of the individual molars can also be used to differentiate the two groups (Fig. 12–1).

By the early part of the Late Cretaceous, many placentals lived in Asia, but marsupials are absent from that fauna. By contrast, both groups inhabited North and South America in the latest Cretaceous. Marsupials radiated extensively in the early Cenozoic in South America, where placentals remained rare and little diversified.

Cenozoic Mammals, Plate Tectonics, and Zoogeography:

Late Mesozoic mammals encountered a world far different from that of their Triassic ancestors. In the early Mesozoic, the continents were still close together and rapid migration throughout the world was possible. The early mammal-like reptiles and even the Late Triassic and Early Jurassic mammals were part of a world fauna that was not separated into major endemic groups. In the late Mesozoic, the breakup of Pangaea was well advanced, and the current outlines of the continental blocks were beginning to be defined (Fig. 12–2). (See Chapter 18 for a more general discussion of biogeography.)

Continental breakup resulted in a regional differentiation of mammalian faunas that was considerably more pronounced than that of the Mesozoic biota. Separation of Pangaea into northern and southern continents had taken place by the Late Jurassic with the spread of the Tethys sea from the Caribbean through the Mediterranean region into the far east (Fig. 18–3). The North Atlantic Ocean began to form from the south about 130 million years ago, and by Late Cretaceous time (70 million years ago), Africa and South America were separating by the formation of the South Atlantic Ocean. Separation of Europe from North America was not complete in the north until late in the Eocene. South America and Antarctica were separated from one another by 35 million years ago, and Antarctica and Australia separated approximately 38 million years ago (1). North and South America were separated from sometime in the Late Cretaceous until the end of the Pliocene, approximately 3 million years ago. Africa retained some connection with Europe and Asia, but was apparently isolated to a degree from the beginning of the Cenozoic until the end of the early Miocene when a broad contact was made via the Arabian peninsula.

According to Malcolm McKenna (2), the Bering land bridge provided a broad connection between western North America and Eastern Asia from the Late Cretaceous throughout much of the Cenozoic, but it was at such a high latitude, even closer to the North Pole than today, that it was of limited use as a means of migration between the continents.

Eastern Asia was further isolated from the remainder of the world in the early Cenozoic because the Turgai Strait connected the Arctic Ocean in the north with the Tethys Sea to the south. As a result, an important endemic fauna developed in southeastern Asia, which remained in isolation until the end of the Eocene, at which time the marine barrier closed and mammals spread into North America and Europe. Europe and North America were close

Figure 12-2 Continental positions in the Late Cretaceous and early Cenozoic.

to one another in the early Cenozoic, and migration around the northern Atlantic was possible but intermittent (3).

The evolution of therian mammals was strongly influenced by the varying degrees of isolation of South America, Africa, Australia, and eastern Asia during the early Cenozoic.

MARSUPIALS

Mammalian evolution progressed in isolation in South America and Australia for most of the Cenozoic (Fig. 12–3). Both continents had a diverse marsupial fauna throughout this time—entirely without competition from placentals in Australia and with only limited interaction with placentals in South America.

There was marked evolutionary convergence between the marsupials in South America and Australia and the placentals in the other continents. This is most notable among the carnivorous forms including the wolverinelike Tasmanian devil and Tasmanian wolf in Australia and the cat, dog, and bearlike borhyaenids in South America.

There is continuing debate as to the specific nature of the relationships of the South American and Australian fauna because of the absence of any marsupials older than late Oligocene (about 28 million years ago) in the fossil record of Australia. By contrast, marsupials are known in Upper Cretaceous rocks of South America, and many genera have been described from upper Paleocene and lower Eocene beds.

Ten families of Cenozoic marsupials have been recognized from South America. They include 17 living genera and another 76 fossil genera. The Didelphidae, represented by the living opossum (a Pleistocene immigrant to North America), retain many primitive features of their Cretaceous ancestors. They are generally small, but some reach the size of a domestic cat. Most are arboreal and many are insectivorous. Twelve genera are known from Paleocene beds. These early forms may include the ancestors of all the other South American marsupials.

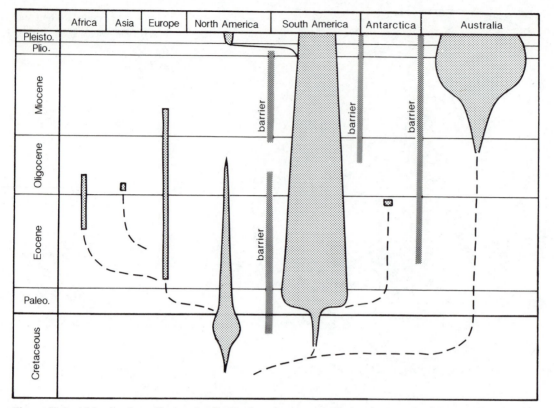

Figure 12-3 Major barriers affecting the distribution of marsupials during the Late Cretaceous and Cenozoic. (Data from A. O. Woodburne & W. J. Zinsmeister, *Journal of Paleontology*, v. 58(4).)

```
Infraclass—Metatheria
     Order—Marsupialia
     (New World marsupials)
          Suborder—Didelphoidea
          Suborder—Caenolestoidea
     (Australasian marsupials)
          Suborder—Dasyuroidea
          Suborder—Perameloidea (bandicoots)
          Suborder—Diprotodonta
             Superfamily—Phalangeroidea
             Superfamily—Phascolarctoidea (koalas)
             Superfamily—Vombatoidea
```

The Microbiotheriidae are known from only two, mouselike genera, one from the upper Oligocene and lower Miocene, and the other from the recent, that are characterized by a very large auditory bulla (a bony capsule surrounding the middle ear ossicles), and a foot structure resembling that of Australian marsupials. The Caenolestoidea are a large group of mouse and shrewlike forms, most common in the lower Miocene, but represented by three genera in the modern fauna. The first lower molar tooth is enlarged and bladelike, resembling the anterior premolar of multituberculates. The polydolopoids have a similar dentition, but with the last premolar modified as a shearing blade. Polydolopoids extend from the Paleocene into the Early Oligocene. The Eocene genus *Antarctodolops* is the only marsupial known from Antarctica.

The Oligocene to Pliocene family Argyrolagidae resemble the jumping mice in their body proportions and limb structure (Fig. 12–4). The Borhyaenidae and Thylacosmilidae evolved in convergence with placental dogs, bears, and saber-toothed "tigers." These families became extinct in the late Pliocene at the time placental carnivores reached South America from the north.

Recent analysis does not support earlier claims that one or more of the families of advanced South American marsupials had special affinities with others in Australia (4). The only phylogenetic connection that can be substantiated is between the base of the didelphid assemblage and the base of the Australian radiation.

The fossil record of marsupials in Australia began in late Oligocene, much later than that in South America, and long after their initial diversification. About 16 families are recognized in Australia. Most appear in the early Miocene and many of the early fossils resemble modern genera.

The Australian dasyuroids occupy an adaptive and phylogenetic position similar to that of the didelphoids in South America. They are generally primitive in their anatomy and include a range of mouse to cat-sized insectivores and carnivores. The Dasyuridae include the Tasmanian devil and many smaller genera. The Tasmanian wolf, which became extinct in the 1930s, is placed in the related family Thylacinidae. Its relatives first appeared in Miocene beds.

All other Australian marsupials are united by a peculiar specialization of the rear foot, termed syndactyly, in which the second and third digits are greatly reduced and united in a single sheath of tissue. The specialized digits are used for grooming. Two major groups of syndactylous marsupials are recognized: the Perameloidea, including only the rabbitlike bandicoots, and the much more diverse Diprotodontoidea. Among the living diprotodont families, the koalas, wombats, phalangerids, kangaroos, and gliding and pygmy possums can all be traced back to the Miocene. The Wynyardidae, from the early Miocene, combines primitive characters of phalangerids, kangaroos and wombats. Primitive kangaroos appear in the mid-Miocene, but the modern kangaroos and wallabies underwent a major radiation in the Pliocene, possibly associated with the spread of grasslands.

Some of the most interesting Australian marsupials are known only as fossils and have no similar modern representatives. Among the most enigmatic are the thylacoleonids (Miocene to Pleistocene). They are large relatives of the phalangerids in which the incisors are in the form of large canine teeth, and the last premolars form huge, slashing blades (Fig. 12–5). The skull appears superficially

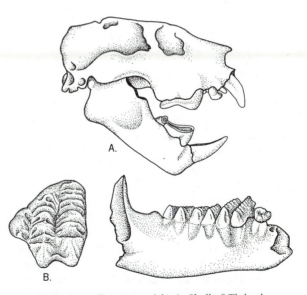

Figure 12-4   Restoration of the rat-sized jumping marsupial *Argyrolagus* from the Pliocene of Argentina.

Figure 12-5   Australian marsupials. A. Skull of *Thylacoleo*, a large phalangeroid with a shearing dentition superficially resembling that of carnivorous placental mammals, B. Upper molar and lower jaw of *Ektopodon*. Little else is known of this genus which is classified among the phalangeroids. (A: from M. E. Finch & L. Freedman, *Carnivorous Marsupials*, M. Archer (ed.), pp. 553–561. B: From N. S. Pledge, *Univ. Calif. Publ. Geol. Sci.*, 131. Courtesy of University of California Press © 1986 The Regents of the University of California.)

like that of a carnivore, but most phalangeroids are herbivores. The skeleton is the size of a leopard and shows arboreal adaptations, suggesting a similar predatory habit.

The Ektopodontidae (middle Miocene to Pliocene) includes three genera known primarily from their dentition. The molar teeth are extremely bizarre. They are greatly expanded transversely and marked by numerous closely placed ridges (Fig. 12–5B). They were initially thought to be related to the living monotremes, but additional material suggests affinities with the phalangerids, although no other marsupial has similar teeth.

The largest marsupials were the diprotodonts (late Oligocene to Pleistocene), some of which reached the size of a hippopotamus. The dentition and skull shape vaguely resemble those of large grazing placental mammals. Their closest affinities lie with the wombats.

Primitive, didelphoid marsupials spread from North America into Europe, Asia, and North Africa in the early Cenozoic but all of these lineages became extinct before the end of the Miocene without further diversification. These continents were the major areas for the diversification of the placentals.

PLACENTAL MAMMALS

Placental Reproduction:

In most features of anatomy and physiology, marsupials and placentals resemble one another closely, and they have adapted to similar ways of life. In their reproduction, however, they are clearly distinct. Although no marsupial lays eggs, the young are born at an immature stage of development equivalent to that of monotremes. The young of the giant kangaroo weighs less than a gram at birth and must complete most of its development in the pouch. Small placental mammals such as shrews and small rodents also give birth to incompletely developed, blind, and hairless young, but most newly born placentals are much more advanced; the young of large placental mammals are typically able to feed and move about by themselves soon after birth.

The short gestation period of monotremes and marsupials may be related to the incompatability of tissue between the fetus and the mother. Since the fetus combines the genetic information of the father and the mother, the mother would be expected to reject fetal proteins, much as we reject transplanted

organs. Rejection is not a problem in the many lizards and snakes that give birth to live young because of the lower degree of sophistication of their immune systems. Placental mammals have evolved a special embryonic tissue, the trophoblast, that forms a protective barrier between the fetus and the mother so that a rejection reaction does not occur.

Jason Lillegraven (5) points out that both the time necessary for complete development and the metabolic cost of maturation are greater for marsupials than for placentals of comparable size. Rapid maturation would give placentals a selective advantage in most environments and may explain the much greater success of placentals in the modern fauna.

Cretaceous Placentals:

The oldest placentals have been found in the Upper Cretaceous of Mongolia (Fig. 12–6). Complete skulls and much of the postcranial skeleton are known of the genera *Zalambdalestes, Barunlestes, Asioryctes,* and *Kennalestes.* They share primitive features with the most primitive living placentals, the Insectivora, and may have resembled shrews in their general appearance and way of life. The long rear limbs and feet of *Zalambdolestes* suggest that this genus hopped like modern jumping mice, but *Asioryctes* has more normal limb proportions. These early placentals lacked the specializations that characterize all the modern families belonging to the order Insectivora, and they are also excluded from all the other living orders.

Many additional species of early placentals have been found in the latest Cretaceous rocks of western North America, but most are represented by little more than jaws and teeth. These scant remains demonstrate the first stages in the radiation of the Cenozoic placental orders (Fig. 12–7). Among the fossils known from the latest Cretaceous are primates, insectivores, and ungulates (hoofed animals including the common ancestors of horses, elephants, and cattle). The ultimate ancestors of the carnivores and the common ancestors of rodents and rabbits were probably distinct by this time as well.

The precise pattern of the evolution of placental mammals in the Late Cretaceous and early Cenozoic has not been established. By the end of the Paleocene, all of the placental orders had differentiated. The exact sequence of their appearance and the groupings among the orders have not been

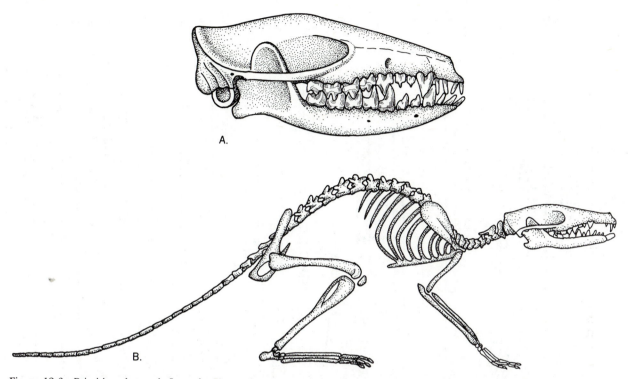

Figure 12-6 Primitive placentals from the Upper Cretaceous of Mongolia. A. Skull of *Kennalestes*, approximately 3 cm long, B. Skeleton of *Zalambdalestes*, approximately 15 cm long. (A: From Z. Kielan-Jaworowska. Courtesy of *La Recherche*, fevrier 1980, p. 152. B: From Z. Kielan-Jaworowska, *Palaeontologia Polonica*, v. 38.)

fully documented. Michael Novacek (6) has attempted a phylogenetic analysis based on a wide range of skeletal characters, but not all groups are securely assigned.

Primitive Cenozoic Mammals:

The most difficult placental mammals to classify are the primitive forms from the Upper Cretaceous of Mongolia and North America, since they cannot be assigned to any of the modern orders but might include the ancestors of all later placentals. They have been placed in a separate order, the Proteutheria. Several other early Cenozoic families are known that are sufficiently specialized to be recognized as distinct orders but show little diversity. They do not share significant derived characters with any of the better-known orders, but appear to share a common ancestry with the most primitive placentals. Some have been included in the order Proteutheria by some authors and treated as separate orders by others. These primitive placentals include the Apatemyidae, the Pantolestidae, the Taeniodonta, the Tillodontia, the Dinocerata, and the Pantodonta.

The dinoceratids (also called uintatheres) and

pantodonts are large, archaic herbivores with small brains and short, heavy limbs. The uintatheres were characterized by long canine teeth and bony processes extending from the skull. They were common in Asia and North America from the late Paleocene to the end of the Eocene. Eleven families of pantodonts are known, ranging in age from the Paleocene through the early Oligocene of North America and Asia. *Coryphodon* reached the size of a rhinoceros.

Tillodonts are poorly known forms of large size with feet well adapted for digging and greatly enlarged incisors. They first appeared in the early Paleocene of Asia and ranged into the Eocene of North America and Europe.

The taeniodonts are limited to the lower Paleocene through lower Eocene beds of western North America. They increased from opossum to bear size, and evolved gigantic canine teeth for crushing. All of the teeth grew throughout the life of the animal.

The pantolestids (Paleocene to early Oligocene in North America, extending into the Oligocene in Africa) evolved a dentition and body form resembling those of the sea otter. It is probable that they fed on fish and molluscs. The apatemyids (early

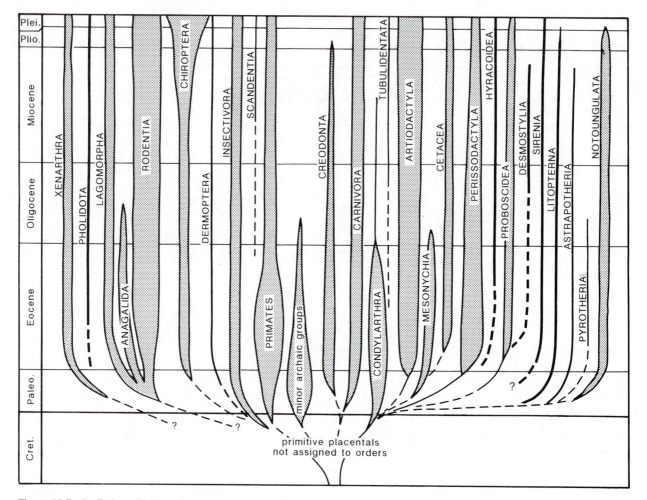

Figure 12-7   Radiation of placental mammals during the Cenozoic.

Paleocene to Oligocene in North America and Europe) had greatly enlarged incisors, suggesting a rodentlike way of life. Nothing has been known of this group except the skull and dentition, but a complete skeleton has recently been found in the mid-Eocene of Europe.

In addition to these archaic and short-lived groups, the ancestors of all the modern placental orders can be recognized in Upper Cretaceous or lower Cenozoic beds.

EDENTATES

Among the most phylogenetically isolated of Cenozoic mammals are an assemblage termed the ed-

entates, which have much reduced or entirely lost their teeth. The edentates include three primarily South American groups: the armadillos, sloths, and anteaters; the Asian and African pangolins; and numerous genera from the early Cenozoic of North America, the Palaeanodontia.

Michael Novacek and Malcolm McKenna (6,7) argued that the edentates were the first group to diverge from the ancestral placental stock because they retain primitive features that are absent in all other placental groups, including an extra bone in the nasal region, the septomaxilla, and primitive aspects of the reproductive system. No fossils of edentates are known from the Cretaceous, and no characters have been recognized that are uniquely shared with any other placental genera that are known to have diverged in the Cretaceous.

Fossils of edentates are known as early as the middle Paleocene in South America, the upper Paleocene in North America and the middle Eocene in Europe. We do not know whether these geographically separated groups shared a common ancestry with one another, or evolved from three separate lineages of primitive placentals. The North American palaeanodonts appear to be the most primitive edentates and, according to Ken Rose (8), share some derived skeletal features with the pantolestids. The aquatic way of life of the pantolestids was, however, far different from the adaptation of the early edentates.

South American Edentates:

The South American edentates, including the living tree sloths, armadillos, and myrmecophagid anteaters, and the extinct ground sloths and glyptodonts, clearly form a natural group united by similar specializations of the pectoral and pelvic girdles and extra articulating surfaces between the vertebrae. The extra or strange (Greek *xenos* = strange) vertebral joints are the basis for the ordinal name of the group, the Xenarthra.

Edentates
  Order—Xenarthra
    Infraorder—Loricata
      Superfamily—Dasypodoidea
        (armadillos)
      Superfamily—Glyptodontoidea
    Infraorder—Pilosa (sloths)
      Superfamily—Megalonychoidea
      Superfamily—Mylodontoidea
      Superfamily—Megatherioidea
    Infraorder—Vermilingua (South
      American anteaters)
  Order—Palaeanodonta
  Order—Pholidota

The peculiar nature of the vertebrae and girdles of the xenarthrans may have evolved in relationship to digging. Many members of this group, possibly including the oldest known genera, fed primarily on such insects as ants and termites that are dug from the ground. Unlike primitive insectivorous mammals, the prey are not chewed but swallowed whole, and the teeth are reduced or lost entirely.

The oldest known armadillos are from middle Paleocene deposits. By Eocene time they had already evolved most of the characters by which we recognize the modern members of this group, although the armor was not as well consolidated. From the armadillos evolved the even larger and more heavily armored glyptodonts, which survived into the Pleistocene. Their distinctive teeth are recognized in the middle Eocene. They are thought to have fed on soft vegetation, although they had eight, very long-rooted cheek teeth that grew throughout the life of the animals. Both armadillos and glyptodonts migrated into North America in the Pliocene and Pleistocene (Fig. 12–8).

The sloths apparently evolved from ancestors resembling the early armadillos and some genera have nodules of dermal armor in their skin. Three major groups are recognized. The Mylodontoidea appeared in the early Oligocene. They were bear-sized terrestrial animals. The megalonychoids and megatherioids were common from the early Miocene. The early members were small and show arboreal adaptation. Each of these superfamilies may be ancestral to one of the two modern tree sloth genera. Other members of these two superfamilies became specialized for terrestrial locomotion and achieved large size. The terrestrial megalonychoids and megatherioids reached North America in the late Cenozoic and some genera extended as far north as Alaska. Only the small arboreal genera survive in the modern fauna.

The highly specialized myrmecophagid anteaters appeared in South America as early as the middle Miocene, but an essentially modern genus has been described from the middle Eocene of Europe (9). It is difficult to understand how this group could have migrated from Europe to South America, or vice versa, in the early Cenozoic, and there is some question as to the affinity of the European genus. This genus, *Eurotamandua*, is part of an extraordinary fauna from Messel, Germany, that is preserved in an oil seep. It includes a diverse assemblage of mammals and birds, some including such soft tissues as hair, feathers, and gut contents. Government authorities plan to use the Messel locality for a garbage dump and prevent further collecting.

The scaly pangolins (order Pholidota) are Asian and African anteaters that resemble the South American myrmecophagids in the long tubular shape of the skull and the complete absence of teeth, but lack the specializations of the girdles and vertebrae that typify the South American xenarthrans. Pan-

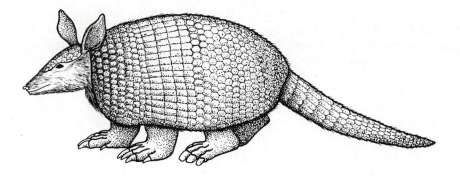

Figure 12-8 The giant armadillo *Holmesina*, approximately 3 meters long, from the late Pleistocene. (From G. Edmund, *The Evolution and Ecology of Armadillos, Sloths, and Vermilinguas*, G. G. Montgomery (ed.), pp. 83–93. Smithsonian Institution Press. © 1985 Smithsonian Institution, Washington, D.C.)

golins are also known from the Messel fauna (Fig. 12–9).

Two families from the early Cenozoic of North America (order Palaeanodonta) have been associated with both the xenarthrans and the Pholidota. The palaeanodonts are archaic forms, showing various degrees of digging specialization (Fig. 12–10). They have reduced but not completely lost their teeth, but they do not show the other skeletal specializations of the South American edentates.

INSECTIVORA

Modern members of the Insectivora have long been thought to be the most primitive living placental mammals on the basis of their small brain size and primitive skeleton. For a time, the concept of this order expanded to include a variety of primitive Late Cretaceous and early Cenozoic genera

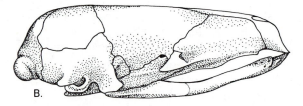

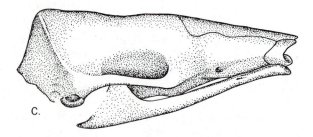

Figure 12-9 Skulls of anteaters. A. *Palaeomyrmedon*, a myrmecophagid from the Pliocene of South America. B. *Manis*, the modern Asian and African pangolin. C. *Eurotamandua*, from the Eocene of Europe. (B & C: From G. Storch, *Senckenbergiana lethaea*, v. 61.)

Figure 12-10 Restoration of *Metacheiromys*, a palaeanodont from the Eocene of North America. This genus may be related to the South American, Asian, and African anteaters, or represent a distinct lineage that independently specialized for feeding on subterranean insects. Approximately 1 meter long. (From G. Edmund, *The Evolution and Ecology of Armadillos, Sloths, and Vermilinguas*, G. G. Montgomery (ed.), pp. 83–93. Smithsonian Institution Press. © 1985 Smithsonian Institution, Washington, D.C.)

otherwise placed in the Proteutheria. Shrews, moles, and hedgehogs are now believed to share a common ancestry that is distinct from that of the other orders of Cenozoic placental mammals. Insectivores can be traced back to the latest Cretaceous. The ancestors of the modern families can be recognized by specialized features including the relatively small size of the eye socket, the reduction or loss of the jugal bone of the cheek, pecularities of the base of the skull, and loose attachment of the halves of the pelvic girdle. The oldest members of the group including shrews and moles, the Soricomorpha, are found in Upper Cretaceous rocks. The modern families Soricidae (shrews) and Talpidae (moles) both appear in the late Eocene. Several other extinct families are known from the early Cenozoic. These include the Geolabididae (Eocene to Miocene of North America), Apternodontidae (Eocene to Oligocene of North America), and Nyctitheriidae (Paleocene to Oligocene of North America and Europe).

```
Order—Insectivora
    Suborder—Erinaceomorpha (hedgehogs)
    Suborder—Soricomorpha
        Superfamily—Soricoidea
            (shrews and moles)
    Suborder—Zalambdodonta
        Superfamily—Tenrecoidea (tenrec)
        Superfamily—Chrysochloroidea
            (golden mole)
```

The hedgehogs (Erinaceomorpha) may be represented as early as the mid-Late Cretaceous by the genus *Paranyctoides*. Its dentition is similar to that of primitive insectivorous placentals except for a blunting of the cusps and a more squared outline of the molars, suggestive of crushing and grinding rather than piercing. The Paleocene to Oligocene Adapisoricidae link *Paranyctoides* with the living erinaceids. The specific origins of other families of living insectivores remain controversial. The Solenodontidae from the Pleistocene and Recent of the West Indies may have evolved from the Apternodontidae or the Geolabidae. The tenrecs and golden moles are known as early as the lower Miocene in Africa, but even the earliest members are distinct from the other major insectivore groups.

BATS

Bats are among the most specialized of mammals in their highly developed flight capabilities and the capacity of most species to use echolocation for capture of prey and navigation in the dark. The fossil record shows that bats had already achieved a nearly modern skeleton by the early Eocene (Fig. 12–11) and the structure of the inner ear of these fossils is similar to modern bats that echolocate (10). No fossils illustrate the early stages in the evolution of their flight. Specializations of the molar teeth in early bats are similar to those of nyctitheriid insectivores which are unfortunately known from no remains other than jaws and teeth. The skeleton of bats probably evolved from those of primitive insectivores, but we have no idea how long this transition may have taken. Fossils of animals that could be the ancestors of bats are known from 8 to 25 million years before the first bats.

Modern bats are divided into two major groups, the megachiropterans or flying foxes, which feed primarily on fruit and do not echolocate, and the primarily insectivorous microchiropterans, which do. These groups probably diverged in the Paleocene. Only two fossils of megachiropterans have been found, one from the Oligocene of Europe and the other from the Miocene of Africa. There are 18 living families of microchiropterans, six of which are represented as fossils as old as the Eocene or early Oligocene. Several living genera appear in the Oligocene.

TREE SHREWS, ELEPHANT SHREWS, AND FLYING LEMURS

Several other groups of generally primitive and little diversified placentals were long thought to be descended from the base of the insectivore lineage, but no unique derived features have been recognized that unite them with the common ancestors of the modern insectivore families. All are now placed in separate orders, whose specific relationships among primitive placentals remain uncertain (11).

The tree shrews are represented by five living genera from southeast Asia that resemble tree squirrels in their general appearance, but most features of the skeleton are close to the pattern of Late Cretaceous or early Paleocene placentals (Fig.

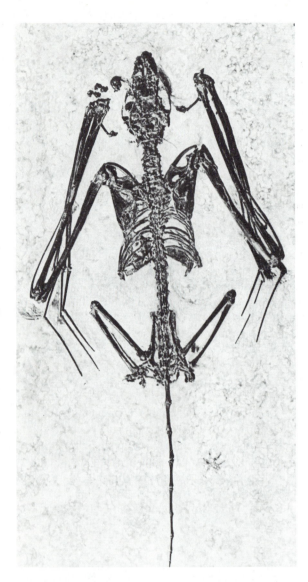

Figure 12-11   Skeleton of the Early Eocene bat *Icdaronycteris*. It is nearly identical to some modern bats. (Photograph courtesy of Dr. John Ostrom and Peabody Museum of Natural History, Yale University.)

Figure 12-12   Primitive living placental mammals. A. Tree shrew, B. Elephant shrew, C. Flying lemur. All have been hypothesized as being closely related to insectivores, bats and/or primates. All retain primitive characters and cannot be allied with any of the other placental orders. Miocene fossils of tree shrews and elephant shrews resemble modern members of these groups and do not contribute to establishing their relationships. Possible ancestors of the flying lemur are known from the Eocene.

12–12). Features that were long used to suggest affinities with primates are also shared with other early placentals. Fossils of trees shrews have been found in rocks as old as Miocene, but they do not differ significantly from living genera.

The order Dermoptera consists of a single living genus from tropical Asia and Indonesia. The front and hind limbs are very long and support a gliding membrane like that of "flying" squirrels. They feed entirely on plant material and symbiotic bacteria in the caecum break down cellulose. The modern family has no fossil record, but remains from the Paleocene and Eocene assigned to the Plagiomenidae have a very similar dentition. Plagiomenids are especially common in the early Eocene of Ellesmere Island in the Canadian Arctic. Recently discovered cranial remains may make it possible to establish the origins of this groups and test its relationship to the modern genus (12).

Elephant shrews (order Macroscelidae) have been restricted to Africa throughout their known history. They are small, mouse- to rabbit-sized insectivorous animals that are characterized by a long, flexible snout. The long hind limbs are specialized for hopping. Their fossil record begins in Oligocene beds but does not demonstrate specific affinities with other orders of placental mammals. Similarities of dentition and jaw structure suggest affinities with the lower Cenozoic Asian anagalids (discussed with the rodents), but no intermediate forms are known and their geographical separation suggests that these groups probably evolved independently since the Late Cretaceous.

PRIMATES

Because they include our own genus, *Homo*, primates are the most intensively studied mammalian group. Abundant primitive primates are known in North America and Europe from rocks ranging in age from early Paleocene into early Oligocene. The major modern groups probably originated in the late Eocene, but few of their fossils are known from rocks older than Miocene in age. The lemurs are first represented in Pleistocene rocks. No fossils are known of the living great apes from Africa, the gorilla and chimpanzee.

The irregular distribution of primate fossils in time is due to their habitat and the changing climate of the Cenozoic. Nearly all modern primates live in tropical jungles and forests. Animals living in these environments generally have a poor fossil record because sediments accumulate at a low rate, and organic decay is rapid. Collection of fossils in the tropical areas today is hampered by heavy vegetation and rapid chemical and physical weathering of the exposed sediments. In the early Cenozoic, the tropics were much more extensive than they are today, reaching far into the latitudes that are now temperate in the North. Areas that were tropical in the early Cenozoic but are now within the temperate zone have yielded most of the fossil primates. The end of the Eocene and beginning of the Oligocene was a time of increasing seasonality and decreasing temperatures, greatly restricting the tropical regions. Few primates lived in North Amer-

ica or Europe north of the Mediterranean region after this time.

Fossil primates have not been collected in the early Cenozoic rocks of either South America or Africa. Primates may have entered Africa from Eurasia at an earlier time to account for the very primitive genera that now live in Madagascar, but their fossil record does not begin in Africa until the Lower Oligocene. Primates appear in South America in the Middle Oligocene. Whether they reached there from North America or Africa continues to be disputed (13). South American primates evolved in isolation during the remainder of the Cenozoic.

```
Order—Primates
    Suborder—Plesiadapiformes
    Suborder—Prosimii
        Infraorder—Adapiformes
        Infraorder—Lemuriformes
        Infraorder—Tarsiiformes
    Suborder—Anthropoidea
    Infraorder—Platyrrhini
    Infraorder—Catarrhini
        Superfamily—Cercopithecoidea
        Superfamily—Hominoidea
        Family—Hylobatidae
        Family—Pongidae
        Family—Hominidae
```

Primates are common in the Paleocene and Eocene beds of the northern continents, and a single tooth has been described from the Upper Cretaceous. It is a lower molar, assigned to the early Paleocene genus *Purgatorius* on the basis of the distinctive pattern of the blunt molar cusps. The dominant Paleocene primates, the plesiadapiforms, are much more primitive than later primates in the structure of the ear region and the pattern of circulation at the base of the skull. Except for primitive features shared with insectivores and tree shrews, the plesiadapiforms show no features that support specific affinities with any group except the most primitive placentals. Most plesiadapiforms are specialized in the large size of the incisors and the presence of a long gap, or diastema, between them and the cheek teeth which gives them a very rodentlike appearance (Fig. 12–13). In these features they differ markedly from later primates. It has been sug-

B.

A.

C.

Figure 12-13 Primitive primates. A. Single lower molar tooth of a primate from the Upper Cretaceous, B. Restoration of the Paleocene primate *Plesiadapis*, C. Skull of *Plesiadapis*; the dentition resembles that of rodents in having prominent incisors and a large gap anterior to the molar teeth. (A: From F. S. Szalay & E. Delson, *Evolutionary History of the Primates.* © 1979 Academic Press. C: Courtesy of P. D. Gingerich. Reprinted by permission from *Nature* Vol. 232, p. 566. Copyright © 1971. Macmillan Magazines Ltd.)

gested that rodents evolved from these early primates, but the similarity of the dentition is almost certainly an example of convergence. The radiation of early rodents corresponds in time with the decline of the early, rodentlike primates.

In the early Eocene, two other groups of primates, the Adapidae and the Omomyidae, appeared in the northern continents. Both resembled more advanced primates in the nature of the ear region and the larger size of the brain, and showed features in the postcranial skeleton indicative of arboreal life (Fig. 12–14). The Adapidae were larger forms that broadly resembled the living lemurs and lorisids. Adapids lingered into the Miocene, but no fossils demonstrate the transition to the modern groups. The fossil record of lemurs is limited to a diverse assemblage of Pleistocene and sub-Recent remains from the island of Madagascar. Fossils of lorisids appear first in the Miocene in East Africa.

The omomyids are typically smaller than the adapids and consequently have a relatively larger braincase and a skull that more closely resembles that of advanced modern primates. They most closely resemble the tarsioids whose fossil record is limited to a single jaw from the Oligocene of North Africa.

The tarsioids, lemuroids, and their fossil relatives have been grouped as the Prosimii, and the more advanced primates (monkeys, apes, and man) termed the Anthropoidea. Teeth and jaws from the upper Eocene of Burma may have belonged to ancestors of the anthropoids, but this group is positively identified only from the Oligocene onwards. There is continued debate as to whether their ancestors were closer to the Adapidae or to the Omomyidae.

The anthropoids diversified early in their evolution; the more primitive Platyrrhini (New World monkeys) reached South America in the Oligocene. The remaining members of the Anthropoidea, the Catarrhini, remained restricted to the Old World until our own species evolved in the Pleistocene. The New World monkeys are limited to 16 living genera; eight more are known in the Cenozoic. They are more primitive than Old World Monkeys in the retention of three premolar teeth, and have greatly expanded auditory bullae.

Primitive members of the Catarrhini have been

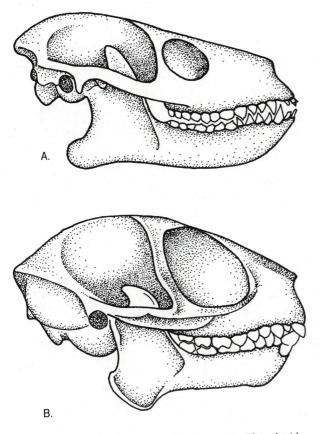

Figure 12-14 Skulls of Paleogene primates. A. The adapid *Notharctus*, B. The omomyid *Necrolemur*. Both the eye and the braincase are relatively larger in *Necrolemur*, but this may be a result of the absolutely smaller size of the skull.

consul (Fig. 12–15) probably represents the group that gave rise to both the great apes and man. It was a baboon-sized, arboreal quadruped.

In the later Miocene, primitive hominoids radiated within Africa, and in the late Miocene their descendants spread into Europe and Asia. Their Asian descendants, now placed in the genus *Sivapithecus*, are close to the ancestry of the living orangutan, *Pongo*. The contemporary genus *Ramapithecus*, which was once thought to be close to the ancestry of the hominids on the basis of the heavy layer of enamel on the teeth, is now considered to be a member of the genus *Sivapithecus*. The European branch of this radiation is represented by *Dryopithecus*.

The ancestry of our own family, the Hominidae, may stem from a poorly known East African genus *Kenyapithecus* from the mid-Miocene. Unfortunately, the fossil record of hominoids in Africa is essentially nonexistent from 4 to 14 million years ago. There is no fossil record at all for the gorilla or chimpanzee, probably because they were restricted to heavily forested habitats. Hominids appeared 4 million years ago in East Africa. The earliest known species, *Australopithecus afarensis*, is represented by several specimens, including one called "Lucy," which includes most of the skeleton (15). Members of this species were bipedal, as shown

found in lower Oligocene beds of North Africa. These fossils may include the ancestors of both the Old World Monkeys (Cercopithecoidea) and the great apes and man (Hominoidea). The Old World monkeys, which are distinguished by squared molars dominated by four prominant cusps, are not clearly distinct until the Miocene. The Oligocene genera from North Africa may include the specific ancestors of the gibbons (Hylobatidae) which are now restricted to southeast Asia, but this group otherwise has no fossil record before the Pleistocene.

Elwin Simons (14) has argued that the early Oligocene genus *Aegyptopithecus* may be included in the Hominoidea. It lacks the dental specializations of the Old World monkeys, and the postcranial skeleton indicates a quadrupedal, arboreal mode of locomotion. Other members of the Hominoidea are known from the early Miocene of East Africa. *Pro-*

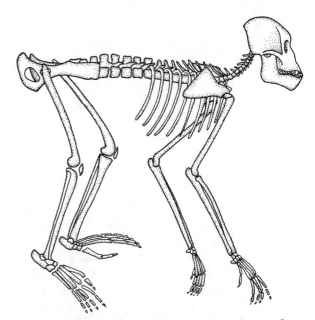

Figure 12-15 Skeleton of *Proconsul* from the Miocene of Africa. This baboon-sized primate may include the ancestors of the chimpanzee, gorilla, and humans.

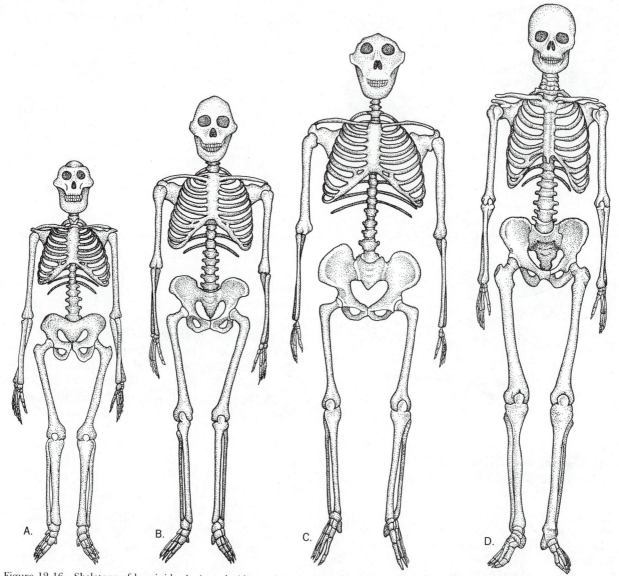

Figure 12-16 Skeletons of hominids. A. *Australopithecus afarensis*, B. *Australopithecus africanus*, C. *Australopithecus robustus*, D. *Homo sapiens*.

by the configuration of the skeleton and by contemporary footprints, although the arms were relatively longer and the legs relatively shorter than in modern humans. The bones of the hands and feet have large processes for the attachment of muscles, which suggests that they retained the capacity for effective arboreal locomotion. The brain size of approximately 500 cc is 20 percent to 30 percent larger than that of a chimpanzee with a similar body weight (25 to 50 kg). Both the front teeth and the molars of *Australopithecus afarensis* are large and apelike, suggesting that this early member of our family fed primarily on plant material (Fig. 12–16).

Several other species of *Australopithecus* lived during the next 2 million years in Africa, including *A. africanus*, *A. boisei*, and *A. robustus*. The latter two species greatly elaborated the size of the molar teeth and the area of the jaw muscles. Their massive teeth may have been used to break up hard seeds and nuts of plants that were common in the increasingly arid areas in eastern Africa where most of the fossils of early hominids have been found.

The first member of the genus *Homo, H. habilis,* appeared in the fossil record approximately 2 million years ago. The braincase had increased to approximately 700 cc and the tooth row was reduced to give a more human face. The proportions of the limbs, however, remained similar to those of the early australopithicines, with long arms and short legs. The relationship of *Homo* to the australopithicine species remains controversial. What is increasingly clear is that several distinct hominid lineages were present simultaneously in eastern Africa (Fig. 12–17).

Homo habilis was succeeded by *Homo erectus* 1.75 million years ago. *Homo erectus* is conspicuously advanced in the achievement of modern limb proportions, and the cranial capacity has increased to between 850 and 1000 cc. Approximately one million years ago, *Homo erectus* spread out of Africa into Asia. *Homo erectus* apparently gave rise directly to our own species, *Homo sapiens,* approximately 300,000 years ago (Fig. 12–18).

The term "archaic" *Homo sapiens* is applied to members of our species living between 300,000 and 30,000 years ago. The neanderthals are included in this category. They were common in Europe and southwest Asia between 70,000 and 30,000 years ago. They had the cranial capacity of modern humans (1300 cc), but heavier facial features, including brow ridges, and a shorter stature. As in the case of all the hominid species, there is not a clear demarcation between archaic and modern *Homo sapiens,* and populations from different geographical areas evolved at different rates, so that there were significant periods of overlap between forms with ancient and modern morphology. Modern cranial features were evident in some African populations as long as 100,000 years ago, while the more archaic neanderthals remained dominant in central Europe for another 70,000 years.

CARNIVORA

The carnivores—lions, tigers, dogs, hyaenas, and bears—are among the most conspicuous members of the modern mammalian fauna. Surprisingly, this group was one of the last mammalian orders to become diverse in the early Cenozoic.

The teeth of primitive placental mammals were specialized for piercing and breaking up insect cuticle, with sharp cusps and many small shearing surfaces. All primitive placental mammals were small, as are modern shrews and rodents.

With the extinction of carnivorous dinosaurs at the end of the Mesozoic and the emergence of diverse medium and large-sized mammals in the early Paleocene, an adaptive zone for large mammalian carnivores became available. The vast majority of

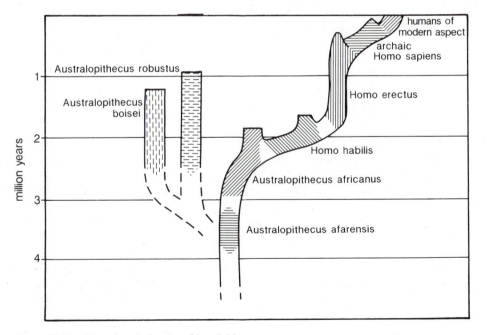

Figure 12-17   Tentative phylogeny of hominids.

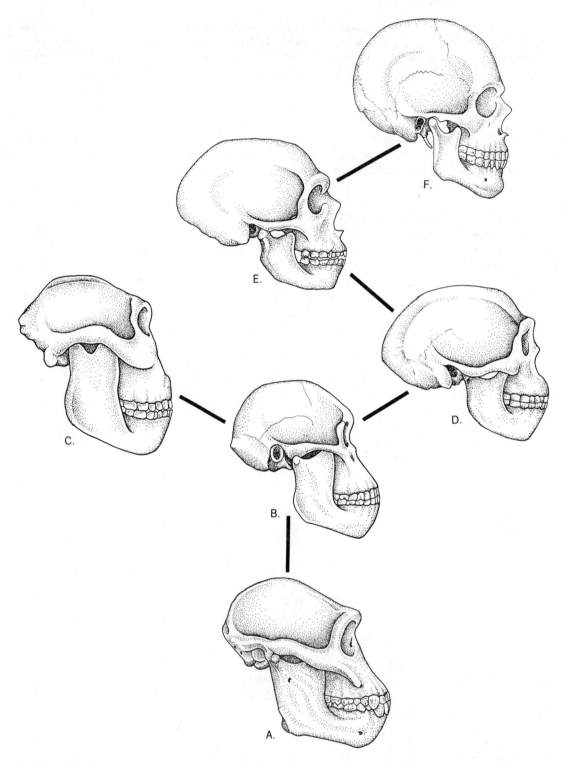

Figure 12-18   Hominid skulls. A. *Australopithecus afarensis*, B. *Australopithecus africanus*. C. *Australopithecus robustus*, D. *Homo erectus*, E. *Homo sapiens* (neanderthal), F. *Homo sapiens* (modern human).

early Cenozoic placentals show specializations of their dentition for a herbivorous diet, and large carnivores remained relatively rare. Gigantic terrestrial birds appear to have occupied a significant portion of the predator adaptive zone during the early Cenozoic of both the northern continents and South America.

Among mammals, feeding on vertebrate prey requires the evolution of teeth with large shearing surfaces to slice through the muscle fibers. This is most effectively achieved through the evolution of bladelike carnassial teeth near the back of the jaw where the muscles can be applied most forcefully (Fig. 12–19). In contrast with the transverse jaw movements of primitive mammals, the jaws of carnivores must move primarily in the vertical plane to give a scissorlike contact between the upper and lower carnassial teeth. Initiation of carnassial teeth required only a slight change from the tooth shape in the primitive insectivorous placentals that are known from the Late Cretaceous.

Two major groups of carnivores diverged from primitive placentals—the order Carnivora, which includes all the modern carnivores, and a second, now extinct assemblage, the Creodonta. The order Carnivora is characterized by the elaboration of the last upper premolar and the first lower molar as carnassial teeth. The creodonts specialized upper molars 1 and/or 2, and lower molars 2 and 3. The posterior position of the carnassial teeth in creodonts presumably provided greater leverage for the jaw muscles than that in the order Carnivora, but the specialization of the posterior molar teeth as shearing blades reduced the area available for grinding and crushing, which may have limited the type of food that they could eat.

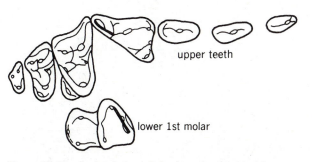

Figure 12-19   Occlusal view of cheek teeth of a primitive member of the order Carnivora. Thick lines emphasize the shearing blades of the carnassial teeth. (Reproduced with permission from R. J. G. Savage and *Palaeontology*, v. 20(2).)

of North America and Eurasia, include primarily medium-sized weasel and catlike forms, but also the bear-sized *Sarkastodon*. Hyaenodontids, which appeared in the late Paleocene and lingered into the late Miocene of Asia and Africa, may be divided into three major groups, the civit and doglike proviverines, the canid and hyaenidlike hyaenodontines, and the Machaeroidinae, which had long canine teeth resembling the saber-toothed cats. The genus *Hyaenodon* is known from North America, Eurasia, and Africa (Fig. 12–20).

The Oxyaenidae retained a primitive, plantigrade posture (walking flat on the soles of their feet), but the Hyaenodontidae were long-limbed with digitigrade posture (walking on their toes). All creodonts remained primitive in the flexible nature of the wrist joint, composed of three separate bones, whereas these bones fused in advanced members of the Carnivora. Creodonts were also primitive in the absence of ossification of the auditory bulla (the tissue that surrounds the ear ossicles).

Although the creodonts were the dominant carnivores in the Paleocene and Eocene, they were largely replaced by the early Oligocene by members of the order Carnivora. The ancestors of the modern carnivores appeared as early as the early Paleocene. The early members were small, weasel-like forms, referred to as miacids.

Two groups are included within the miacids, the Viverravidae, which appeared in the Early Paleocene but were already specialized by the absence of the last molar tooth, and the Miacidae, which appeared in the early Eocene on all the northern continents. The Miacidae may have given rise to all later members of the Carnivora. The miacids retained a primitive posture, as did the early creodonts, and lacked fusion of the carpals and ossification of the auditory bulla.

Order—Creodonta
    Family—Hyaenodontidae
    Family—Oxyaenidae
Order—Carnivora
    Superfamily—Miacoidea
    Superfamily—Aeluroidea
    Superfamily—Arctoidea
    Superfamily—Otarioidea

The creodonts were never as diverse as the Carnivora. They included only two families, the Oxyaenidae and the Hyaenodontidae, and about 150 genera, compared with 400 genera of Carnivora. The Oxyaenidae, from the Paleocene and Eocene

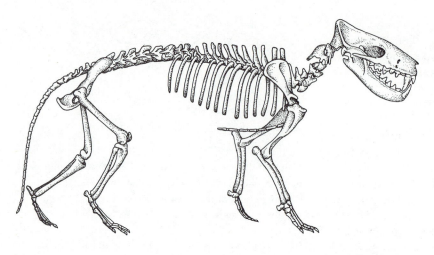

Figure 12-20   The hyaenalike creodont *Hyaenodon*. (From: W. K. Gregory, *Evolution Emerging*. Courtesy Department of Library Services American Museum of Natural History.)

All the modern carnivore families had appeared by the early Oligocene, and their ancestors may have diverged by the late Eocene. There is continuing controversy over their specific origin and interrelationships, but they are typically grouped into two superfamilies, the Arctoidea, including dogs, bears, mustelids (such as weasels and skunks), and procyonids (such as raccoons), and the Aeluroidea—viverrids (civets and mongooses), felids, and hyaenids. Unlike the miacids, the auditory bulla is ossified, but according to a different pattern in each group.

Two major groups of advanced carnivores are extinct. The Amphicyonidae or bear-dogs, which are related to the canids, were common in the late Eocene and continued into the Pliocene. They were generally short-limbed, flat-footed, and relatively large. A second extinct group are the nimravids, which were primitive, catlike forms. They have long been associated with the Felidae, but are primitive in the incomplete ossification of the bulla, and specialized in the ankle structure (16). Nimravids may have evolved separately from the miacid level, independent of both arctoids and aeluroids. They became extinct about 7 million years ago. Many nimravids (also called paleofelids) developed long, laterally compressed canine teeth. This same "saber-toothed" specialization evolved separately among the true felids, typified in the Pleistocene by *Smilodon*, which is found in large numbers in the La Brea tar pits (Fig. 12–21).

The seals, sea lions, and walruses have long been

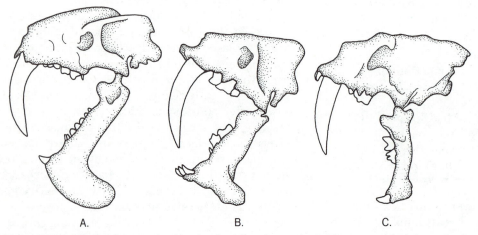

Figure 12-21   Skulls of mammals with sabershaped canine teeth. A. The marsupial *Thylacosmilus* from the Miocene of South America, B. The nimravid "cat" *Barbourofelis*, from the Miocene of North America, C. *Smilodon*, a true felid from the Pleistocene of North America.

recognized as aquatic derivatives of the terrestrial carnivores. They have typically been placed in a separate suborder, the Pinnipedia, but their specific relationships remain subject to dispute. All appear to have diverged from the arctoid rather than the aeluroid assemblage. The sea lions and walruses are generally thought to share close affinities with the bears, but the true seals are typically allied with the mustelids. The Miocene genus *Potamotherium*, which resembles an otter, appears as a structural intermediate between mustelids and true seals. The Late Oligocene *Enaliarctos* unites the base of the sea lion and walrus lineages with primitive bears. True seals appear to have originated along the margins of the Atlantic Ocean and Mediterranean Sea. The origin of the seal lions and walruses seems to have occurred in the North Pacific.

RODENTS AND RABBITS

The small, gnawing rodents and rabbits constitute the largest assemblage of modern placental mammals, together encompassing more then 50 families and 1700 species. Rodents are by far the larger group, with rabbits being restricted to 2 modern familes and 46 species. There has been continuing controversy regarding the origin and relationships of the orders Rodentia and Lagomorpha, but there is growing evidence that both evolved from an early group from China, the Anagalida (17). The anagalids were members of the large endemic east Asian fauna of the early Cenozoic. The anagalids, in turn, may be traced to Upper Cretaceous placentals related to *Zalambdelestes*. Anagalids include genera that had several very distinct tooth patterns. The molars were expanded medially, which indicates a particular pattern of side-to-side chewing. Advanced anagalids, like rodents and rabbits, evolved a long diastema between the incisors and the cheek teeth, and the incisors grew throughout their life. Some anagalids had only a single pair of upper incisors, like rodents, and others, like the lagomorphs, retained two pairs. Some Paleocene genera have been classified among both the anagalids and the lagomorphs (Fig. 12–22).

Rabbits appeared in China in the Paleocene, spread to North America in the late Eocene, and entered Europe by the beginning of the Oligocene. They reached Africa and South America in the late Cenozoic.

The basic body form, with long rear limbs suited for hopping and a short tail, has remained conservative throughout the history of lagomorphs. Rabbits are more primitive than rodents in retaining two upper incisors, but the second pair is immediately behind rather than lateral to the first. The cheek teeth in advanced rabbits grow throughout their life, but they retain a similar morphology throughout the group. Transverse movements of the jaws are important in chewing. The pattern of the teeth and jaw movements make it possible for rabbits to feed on a wide range of plant material, which may explain why they have not speciated to the extent of their close relatives, the rodents.

Order—Lagomorpha (rabbits)
Order—Rodentia
   Suborder—Sciurognathi
      Infraorder—Protrogomorpha
         Superfamily—Ischyromyoidea
            (primitive rodents)
         Superfamily—Aplodontoidea (mountain
            beavers)
      Infraorder—Sciuromorpha
         Superfamily—Sciuroidea (squirrels)
      Infraorder—Castorimorpha (beavers)
      Infraorder—Gliroidea (dormice)
      Infraorder—Myomorpha
         Superfamily—Muroidea (mice and rats)
   Suborder—Hystricognathi
      Infraorder—Caviomorpha (South
         American rodents)

Rodents are far more numerous and varied than rabbits. They are united by the retention of only a single pair of upper incisors that have two layers, rather than a single layer, of enamel on the anterior surface. Fore-and-aft movement of the jaws is very important when feeding. The lower jaw is moved anteriorly for gnawing with the incisors and posteriorly to chew with the cheek teeth.

Rodents may have evolved from the Asian anagalids, but they first appear in the fossil record in North America in the upper Paleocene. They radiated extensively in the early Eocene, eventually giving rise to approximately 50 families. Their primary radiation was in the northern continents, but both Africa and South America became secondary centers of diversification. Rodents may be divided into two major groups on the basis of jaw structure (Fig. 12–23). In the primitive, sciurognathous pattern, the angle of the jaw arises below the base of

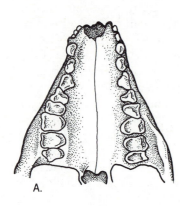

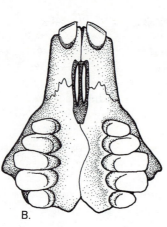

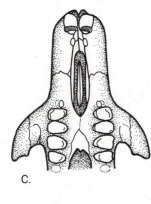

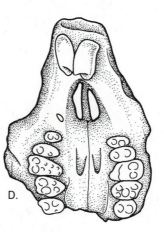

Figure 12-22   Upper dentition of anagalids: rodentlike and rabbitlike mammals from the Paleogene of China. A. *Anagale*, a primitive genus with a complete dentition. The cheek teeth are expanded from side to side indicating transverse movement of the lower jaws. B. *Eurymylus*. A gap, or diastema has developed between the rodentlike incisors and the rabbitlike cheek teeth. C. *Mimolagus*. Like rabbits, two pairs of incisors are retained. D. *Heomys*. Both the incisors and the cheek teeth show a pattern like that of rodents.

the root of the lower incisors. In the hystricognathous pattern, which is expressed in all the native South American rodents and a few African families, the angle of the jaw originates lateral to the plane of the incisor. A further division is based on the pattern of the jaw muscles (Fig. 12–24).

The most primitive group of living rodents is the mountain "beaver" or sewellel from the western United States, with a fossil record going back to the late Eocene. The Mylagaulidae is a related group from the middle and late Cenozoic that is distinguished by the presence of hornlike extensions from the nasal bones (Fig. 12–25). Squirrels appeared in the early Oligocene, and by the Miocene an essentially modern morphology was achieved. Beavers also appeared in the Oligocene. The Pleistocene genus *Castoroides* was as large as a bear.

The extremely diverse group that includes the

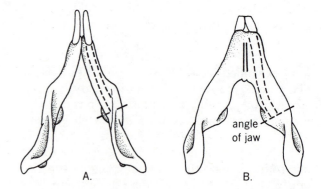

Figure 12-23   Distinguishing features of the lower jaws of the two major divisions of rodents. A. Sciurognath pattern of a marmot in which the angle of the jaw arises below the incisor. B. Hystricognath pattern of a porcupine, in which the angle originates lateral to the plane of the incisors. (*Source*: L. L. Jacobs, *Rodentia*. From L. L. Jacobs, *Univ. Tennessee Studies in Geol.*, v. 8.)

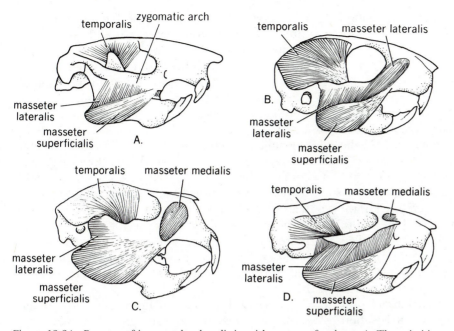

Figure 12-24   Patterns of jaw muscles that distinguish groups of rodents. A. The primitive, protrogomorph pattern of the oldest known rodents in which the masseter muscle originates entirely on the zygomatic arch. B. The sciuromorph pattern of modern squirrels and beavers in which the anterior part of the masseter lateralis originates on the rostrum. C. The hystricomorph pattern of the porcupine, in which the anterior part of the masseter medialis originates largely on the rostrum and passes through the enlarged infraorbital foramen. D. The myomorph pattern of rats and mice in which the masseter superficialis originates on the rostrum and the anterior part of the masseter lateralis originates on an anterior extension of the zygomatic arch. The infraorbital foramen is narrow. It was once thought that each of the advanced patterns originated only once and were diagnostic of major groups. It is now recognized that each pattern evolved several times and one pattern may give rise to another.

rats and mice, the Muroidea, diverged very early in the Eocene. More than a thousand living species are included in this superfamily. The distinct patterns of cusps and lophs on the molar teeth provide an excellent basis for stratigraphy and for establishing evolutionary patterns and processes (18).

Hystricomorph rodents entered South America at the same time as the New World monkeys, in the middle Oligocene. There is much debate as to whether they originated in North America or Africa. Fourteen families of native rodents, the Caviomorpha, are now recognized in South America.

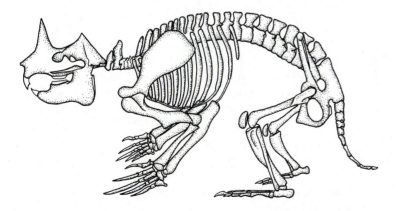

Figure 12-25   Skeleton of a mylogaulid rodent, mid-Cenozoic of North America, 20 cm long. (Modified from J. W. Gidley, *Proc. U.S. Natl. Mus.*, v. 32.)

In addition, there has been an extensive radiation of muroid species of North American ancestry since the late Pliocene.

HOOFED MAMMALS

Ungulates, or hoofed mammals, including the ancestors of horses, cows, and elephants, were the most diverse placental mammals in North America at the end of the Cretaceous and the beginning of the Paleocene (19). They may have originated in North America, but very little is currently known of mammals of this age in other parts of the world. Primitive ungulates, belonging to a group called the Condylarthra, spread into Asia, South America and (perhaps somewhat later) Africa. Either in the latest Cretaceous or earliest Paleocene, South America became so completely separated from North America that its ungulate fauna evolved in isolation until the end of the Pliocene, when land contact was established across the isthmus of Panama. Africa was also partially isolated until the end of the early Miocene.

Ungulates were able to travel between eastern North America and western Europe intermittently in the early Cenozoic but the northern land connection was broken by the end of the Eocene. The Bering land bridge may have been present for most of the Cenozoic, but it was too far north to allow more than occasional passage. Although some groups, such as the early artiodactyls (including cows and other cattle) and perissodactyls (including the horses, tapirs, and rhinos) appear to have spread freely across the northern continents, many groups were largely restricted to either North America or Eurasia.

Primitive ungulates, the Condylarthra, include ten families, several of which are close to the ancestry of the more advanced orders. Unfortunately, the earliest condylarths are known primarily from scattered remains of jaws and teeth, which are not sufficient to establish their specific relationships to more advanced ungulates (20). The order Condylarthra will probably be gradually dismantled as the relationships of the constituent families become better known.

As the Order Condylarthra is currently understood, its members are recognized primarily by primitive features that distinguish them from the advanced Cenozoic orders. Their brains were small, and the area of the jaw muscles was not separated from the orbits by a postorbital bar. In most species, most of the teeth are retained, without the development of a diastema or gap between the incisors and cheek teeth that is common in advanced groups. The molar teeth bear distinct, low cusps. The body is small, and the limbs are short. There is no fusion

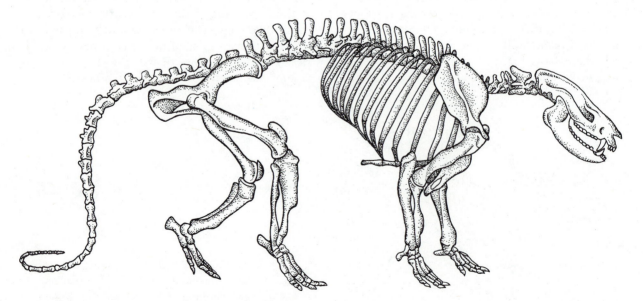

Figure 12-26   Skeleton of *Ectoconus*, a primitive Paleocene ungulate, approximately 2 meters long. (*Source*: W. K. Gregory, *Evolution Emerging*. Courtesy Department of Library Services American Museum of Natural History.)

or loss of bones from the primitive placental pattern. Five toes are present in the front and rear feet (Fig. 12–26).

The following are well-known condylarth families:

Arctocyonidae. The most primitive of all and probably including the ancestors of all later ungulate orders. *Protungulatum* is the best-known member of this group from the Upper Cretaceous (Fig. 12–27).

Phenacodontidae. Long-limbed genera, almost certainly including the ancestors of the perissodactyls.

Hyopsodontidae. Short-limbed animals that were possibly the ancestors of the South American ungulates.

Meniscotheriidae. Distinguished by their complex cheek teeth but not close to the ancestry of any later ungulates.

Periptychidae. Characterized by the crenulated pattern of the molar enamel. Some reached large size, but were not close to the ancestry of the later groups.

Didolodontidae. The only South American condylarths and probably including the immediate ancestors of most if not all ungulate orders that evolved on that continent.

Phenacolophidae. Poorly known Asian group, probably including the ancestors of the elephants, a related aquatic order the Desmostylia, and possibly an extinct African group, the Embrithopoda.

A further family, the Mesonychidae, long included among the condylarths, is ancestral to the whales.

The initial diversification of all these groups occurred within 1 to 2 million years after the appearance of the first fossil ungulates in the Late Cretaceous. Although differences in the dentition allow us to recognize the lineages leading to the advanced ungulate orders as early as the Late Cretaceous, typical members of these groups are generally not known until the early Eocene. Most of the major South American orders appear first in middle to upper Paleocene beds, but the fossil record of artiodactyls, perissodactyls, whales, and elephants all begins near the base of the Eocene.

South American Ungulates:

A few enigmatic remains of placental mammals are known from the Upper Cretaceous of Peru, but the fossil record in South America begins in earnest only in the middle to late Paleocene. Four major orders are unique to South America. All diverged by the end of the Paleocene, but none survived the end of the Pleistocene. The didolodonts, known from the late Paleocene into the middle Miocene, link North and South American ungulates. They are classified among the condylarths, but show especially close similarities with the South American order Litopterna. The Didolodontidae are known primiarily from dental remains, but the Litopterna are represented by more complete material, representing two major lineages, the horselike proterotheres and the camel-like macraucheniids (Fig. 12–28). In the Miocene the proterotheres had achieved a greater degree of reduction of the lateral digits than that reached in the Pliocene by true horses.

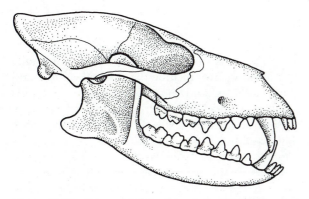

Figure 12-27 Reconstruction of the skull of the oldest known ungulate, *Protungulatum* from the Upper Cretaceous. The skull proportions are similar to those of later carnivores, but the teeth are specialized for grinding plant food. (*Source:* F. S. Szalay, *Evolution* v. 23(4).)

South American—Ungulates
    Order—Condylarthra
        Family—Didolodontidae
    Order—Litopterna
    Order—Pyrotheria
    Order—Astrapotheria
    Order—Notungulata
        Suborder—Notoprongonia
        Suborder—Toxodontia
        Suborder—Typotheroidea
            Superfamily—Typotheroidea
            Superfamily—Hegetotheroidea

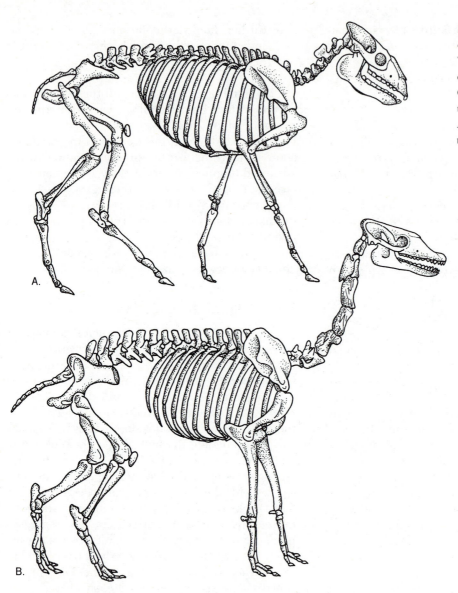

Figure 12-28 Litopterns, a groups of South American ungulates. A. The horselike *Diadiaphorus*, B. The camel-like *Theosodon*. (*Source*: W. K. Gregory, *Evolution Emerging*. Courtesy Department of Library Services American Museum of Natural History.)

A.

B.

Two other groups also show a pattern of convergent evolution with the ungulates of the other continents. The pyrotheres were robust animals with a somewhat elephantine dentition; the shape of their skull suggests the presence of a trunk. The astrapotheres somewhat resembled rhinos.

Most South American ungulates are placed in a single diverse order, the Notungulata. Primitive notungulates resembled the large archaic herbivores of the northern continents, such as the uintatheres and pantodonts. They retained five toes on the front and rear limbs and a planigrade posture. They are distinguished from other South American ungulates by the pattern of the molar cusps and the presence of extra chambers in the auditory bulla. The notungulates include the sheeplike and rhinolike toxodonts, and the rabbitlike and rodentlike typotheres and hegetotheres. Advanced members of several lineages evolved evergrowing incisors (Fig.12–29).

Several genera from North America and Asia, grouped in the Arctostylopidae, have a dentition like that of some notungulates and it has been suggested that they are related. If such a relationship could be demonstrated, it would imply that the notungulates either originated in the northern continents or migrated there from South America (21). A further, very distinct South American genus, *Car-*

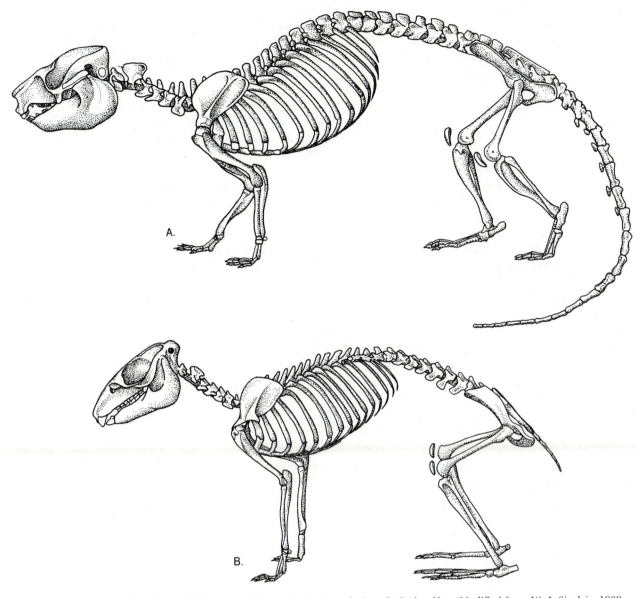

Figure 12-29   Notungulates from the Miocene of Patagonia. A. *Interatherium*, B. *Pachyrukhos*. (Modified from W. J. Sinclair, 1909. *Reports of the Princeton University Expeditions to Patagonia.*)

odnia has been proposed as a close relative of the North American and Asian Dinocerata, providing some evidence of the movement of ungulates into or out of South America during the Cenozoic.

The greatest diversity of South American ungulates occurred in the late Paleocene and early Eocene time, in which interval 14 or 15 families have been recognized. There is a gradual reduction in diversity at the family level throughout the Cenozoic. Six families extended into the Pliocene, and four lingered into the Pleistocene. This assemblage was already very much reduced prior to the emer-

gence of the isthmus of Panama and the influx of advanced ungulates and placental carnivores from the north (22).

South America in its isolation illustrates a pattern of evolution that parallels that of the northern continents and Africa. In the Late Cretaceous and early Paleocene, only a few of the many lineages of placental mammals that were present in the north managed to enter South America. The marsupials occupied many of the adaptive zones filled by placental carnivores, insectivores and rodents on other continents. The notungulates, litopterns, astrapotheres, and pyrotheres filled the roles of artio-dactyls, perissodactyls, and proboscidians as well as evolving rodent and rabbitlike forms. Therian mammals seem to have the potential for a spectrum of adaptive types that may be expressed by different taxonomic groups, depending on the availability of habitats and the presence or absence of competing forms.

African Mammals:

Elephants are today limited to two genera, one in central Africa, and the second in southern Asia. As recently as the Pleistocene, members of the elephant order, Proboscidea, were widespread in Europe, northern Asia, and both North and South America. Their bones are so common in some areas of Siberia that they were used to form the frameworks of the dwellings of primitive tribes.

Incomplete remains of primitive ungulates belonging to the family Phenacolophidae from the Upper Paleocene of China may demonstrate an Asian origin of the Proboscidea. These animals are already of large size, and the molar teeth have prominent cross ridges or lophs.

Fossils from the early Eocene of Algeria clearly show the specialization of the anterior teeth as tusks. The configuration of the skull indicates the presence of a trunk, and the limbs are massive to support a heavy body. Several elephant lineages radiated within Africa in the Oligocene and Early Miocene. At the end of the early Miocene, Africa became joined with Asia across the Arabian peninsula, and elephants spread into Asia and Europe. By the late Miocene, several families had entered North America. All the advanced elephant lineages evolved huge molar teeth, with many cross lophs. In advanced members of the family Elephantidae, only one and a half pairs of teeth are functional at any one time. Vincent Maglio (23) has traced detailed evolutionary changes among the various lineages of African elephants (Fig. 12–30).

Moeritherium is an elephantlike form common in northern Africa in the upper Eocene and Oligo-

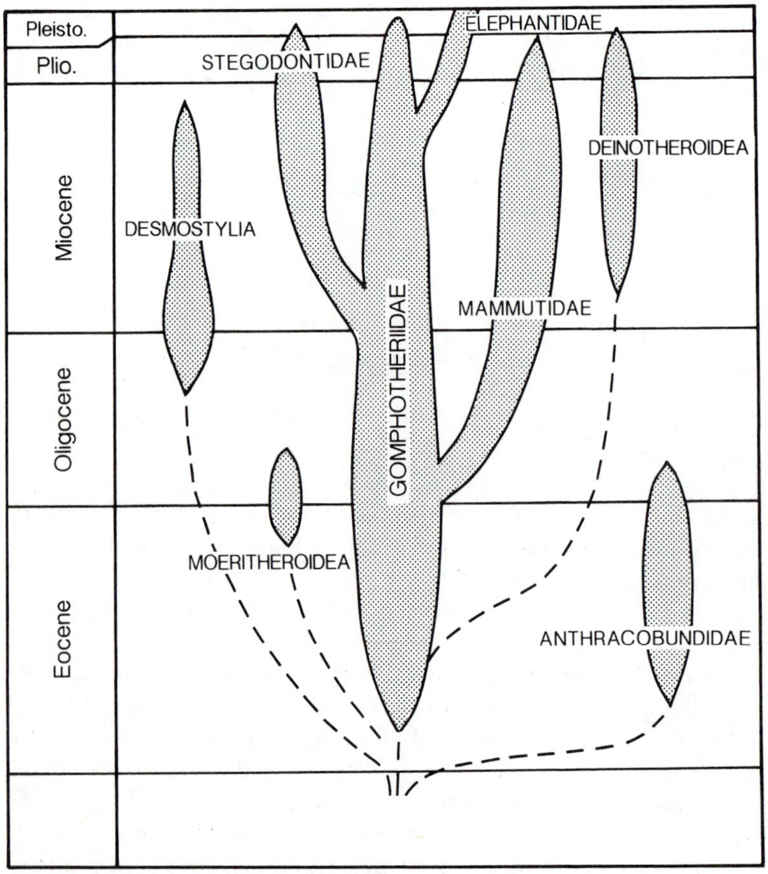

Figure 12-30   The phylogeny of elephants.

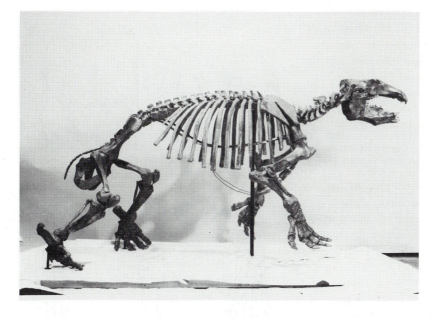

Figure 12-31   Skeleton of *Palaeoparodoxia*, a Miocene desmostylian common on both sides of the Pacific Basin. (Photograph courtesy of Dr. Yoshikazu Hasegawa.)

cene. The marine deltaic deposit in which it is found indicates that it was amphibious. This way of life suggests affinities with a group of more aquatic mammals of the upper Oligocene and Miocene, the desmostylians (24). Their dentition vaguely resembles that of elephants in having tusklike anterior teeth, and complex molars. Desmostylians are confined primarily to the margins of the North Pacific Ocean. They are thought to have fed on algae growing in the intertidal zone (Fig. 12–31) (24).

The sirenians are a second group of aquatic herbiores that have often been linked with the elephants. Many early sirenians are known in northern Africa, which was also the locale of early elephant fossils. However, early sirenians are also known in the West Indies and eastern Europe and show no obvious similarities with early elephants. An important feature by which early sirenians differ not only from proboscideans but all other Cenozoic mammals is the presence of a fifth molar tooth, which is probably a relict of Cretaceous mammals.

Sirenians resemble whales in having a fusiform body with the forelimbs reduced to paddles and the tail flattened as a horizontal fluke (Fig. 12–32). Unlike whales, sirenians have not developed sonar and do not actively dive. Their body weight is increased by the presence of pachyostotic (greatly thickened and solidly ossified) ribs, which are also present in several groups of secondarily aquatic reptiles. The fossil record of this group is fairly well known throughout the Cenozoic, but there are only two surviving genera. The dugong lives in the In-

dopacific basin, and the manatee extends from the Amazon basin through the Caribbean and along the west coast of Africa. A third genus, Stellar's sea cow, was present along the northern rim of the Pacific Ocean basin until the middle of the eighteenth century when it was exterminated by hunters. Dungongs have reduced their cheek teeth, and these teeth are lost completely in Stellar's sea cow, but the manatees, which feed on hard, siliceous sea grasses, have up to eight pairs of functional cheek teeth at a time, and they are continually replaced, a pattern found in no other mammal.

Associated with the proboscideans and sirenians because of their radiation in Africa are the hyraxes, or conies, and the aardvark. The conies are small to medium-sized, rabbitlike animals that appeared in the Eocene. The dentition of some early genera shares features with that of horses. Their taxo-

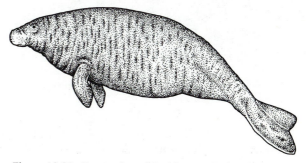

Figure 12-32   Restoration of *Dusisiren*, an Early Miocene relative of Stellar's sea cow.

nomic position remains a source of contention. Aardvarks have a sporadic fossil record in Africa and adjacent margins of the Mediterranean sea. Most are specialized for digging and, like the edentates, have reduced their dentition in relation to an ant-eating habit. They are usually thought to have affinites with the condylarth assemblage.

Artiodactyls:

While South America, and to a lesser degree Africa, were isolated during the early Cenozoic, the northern continents were dominated by artiodactyls and perissodactyls. Both were widely distributed in the early Eocene of North America and Eurasia. In the later Cenozoic, migrations between the northern continents were variably restricted. Some families were limited to a single continent, but others were more widely distributed. After the early Miocene, many families spread into Africa, and some reached South America in the Plio–Pleistocene.

Artiodactyls are the most numerous and diverse modern ungulates, including antelopes, cattle, deer, sheep, pigs, and hippos. The ancestry of this assemble can be traced to the earliest Eocene. The lower Eocene genus *Diacodexis* (Fig. 12–33) is distinguished from all more primitive ungulates by the great length of the limbs and their high degree of specialization for running. All artiodactyls can be recognized by the double pulley shape of the astragalus bone in the ankle. In contrast with the perissodactyls, most of the weight of the body is supported by two digits, 3 and 4 (Fig. 12–34).

Most of the early Eocene artiodactyls can be in-

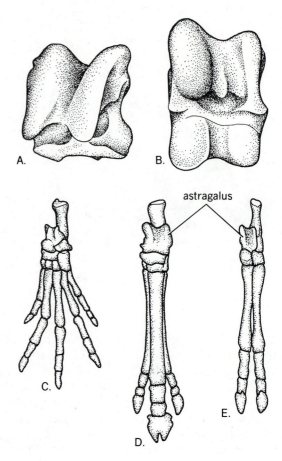

astragalus

Figure 12-34   Feet of artiodactyls and perissodactyls. A. Astragalus of a perissodactyl (the horse). The upper surface is in the shape of a pulley. B. Astragalus of an artiodactyl (a deer). Both the upper and lower surfaces are pulleyshaped to facilitate articulation between the lower limb, ankle, and foot. C. Foot of the primitive ungulate *Tetraclaenodon*, with five digits, D. Foot of *Plagiolophus*, a primitive perissodactyl in which most of the weight is supported by the middle digit, E. Foot of *Protoceras*, a camel-like artiodactyl in which the weight is borne by digits 3 and 4. (B) Courtesy of J. Piveteau, *Traité de Paléontologie*, VI, © MASSON 1961. (C) *Source*: W. K. Gregory, *Evolution Emerging*. Courtesy Department of Library Services American Museum of Natural History. (D) Courtesy of J. Piveteau, *Traité de Paléontologie*, VI, © MASSON 1958. (E) *Source*: W. K. Gregory, *Evolution Emerging*. Courtesy Department of Library Services American Museum of Natural History.

Figure 12-33   Restoration of *Diacodexis*, a rabbit-sized artiodactyl known in North America, Europe, and Asia in the Early Eocene. (*Source:* Modified from K. D. Rose, *Science*, v. 216.)

cluded in a single family. By the end of the Eocene, this group had diverged into 16 distinct lineages. Artiodactyls have continued to diversify throughout the Cenozoic.

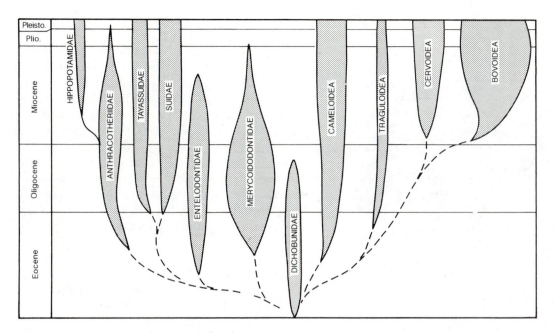

Figure 12-35   Phylogeny of major artiodactyl groups.

Order—Artiodactyla
    Suborder—Palaeodonta (including *Diacodexis*)
    Suborder—Suina
        Superfamily—Entelodontoidea
        Superfamily—Suoidea (pigs and peccaries)
        Superfamily—Hippopotamoidea
    Suborder—Tylopoda
        Superfamily—Merycoidodontoidea
        Superfamily—Cameloidea (camels)
    Suborder—Ruminantia
        Infraorder—Traguloidea
        Infraorder—Pecora
            Superfamily—Cervoidea (deer)
            Superfamily—Bovoidea (cattle)

The artiodactyls are divided into several major groups: the suborder Suina, including pigs and hippos in the modern fauna; the Tylopoda, including camels and similar extinct forms; and the ruminants, including the primitive tragulids (deerlike forms but without antlers), and the advanced pecorans—deer, sheep, cattle, and antelopes (Fig. 12–35).

All the major groups had diverged by the end of the Eocene. The suina generally retain the primitive tooth pattern of the ancestral artiodactyls, with distinct rounded cusps. The limbs are only rarely specialized, and most genera are less highly adap-

tated for running than the most primitive artiodactyl, *Diacodexis*.

The Hippopotamoidea are represented in the Eocene by the anthracotheres, heavy terrestrial animals with primitive, short limbs. They are common throughout the Old World and North America in the early Cenozoic, but by the Miocene they were reduced in numbers and their manner of preservation suggests that some were adapting to a more amphibious mode of life. The Hippopotamidae appears as a divergent group in the early Miocene, with the anthracotheres becoming extinct in the Pleistocene.

A number of piglike forms lived in the early Cenozoic. The entelodonts, common from the mid-Eocene to the Late Miocene, were distinguished by their large size and the presence of bony protruberances from the lower jaw and cheek region (Fig. 12–36). Two familes of modern pigs are recognized. The Suidae, including the domestic hog, are restricted to the Old World throughout their history. They are distinguished by the outward and upward curvature of the upper canine. The peccaries (Tayassuidae), appeared first in North America in the Oligocene and spread progressively into Europe, Asia, Africa, and South America. Unlike the Suidae, their upper canines are straight.

The remaining artiodactyl groups evolved cheek teeth in which the cusps were joined to form crescentic shearing surfaces, a condition termed selen-

Figure 12-36   Restoration of an entelodont, a piglike artiodactyl group, common from the Eocene into the Miocene.

odont (Fig. 12–37). The teeth also became higher crowned to provide a longer-lasting wear surface. Among the more primitive selenodont artiodactyls, the Merycoidodontoidea (commonly termed oreodonts) were extremely common in North America from the middle Eocene into the Pliocene. Two families and 11 subfamilies are recognized, ranging from piglike and sheeplike to hippolike forms. Their

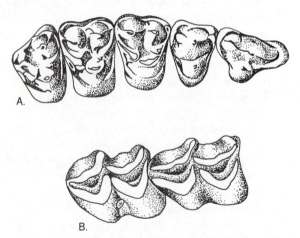

Figure 12-37   Artiodactyl teeth. A. The blunt cusp or bunodont condition of primitive and piglike artiodactyls, B. The selenodont condition of more advanced artiodactyls in which the primitive cusps have joined to form crescentic lophs that shear past one another during chewing.

remains are found by the thousands in the Oligocene badlands of South Dakota.

Four families of camel-like artiodactyls are known in the late Eocene, two of which were short-lived, but the Protoceratidae continued into the Pliocene. They resembled modern camels in the postcranial skeleton, but the skull was characterized by the evolution of a variety of bony protuberances (Fig. 12–38).

The family Camelidae was already distinct in the late Eocene. This group evolved very long limbs and lost the lateral digits early in the Cenozoic. Footprints from the Miocene show the early appearance of the spreading toes that distinguish the group and the peculiar gait, referred to as pacing, in which the legs on one side of the body move in unison, rather than alternating. Camels originated in North America but became extinct there in the Pliocene, surviving only in South America, western Asia and northern Africa where they migrated in the Pliocene.

Most living artiodactyls have more complex stomachs than other ungulates, but their elaboration is most pronounced in the ruminants, including deer, cattle, and antelopes, in which there is a four chambered stomach where bacterial fermentation assists in the breakdown of cellulose. Ruminants also show advances in the degree of fusion of the bones of the ankle that distinguish their fossil remains as early as the late Eocene. A sequence of more advanced animals are known in the Oligocene and Miocene. The most primitive of living ruminants are the tragulids, small deerlike forms from Africa and southern Asia that lack horns but have elongate canine teeth. Other ruminants are grouped as the Pecora. They may have diverged as early as the late Eocene but did not appear as fossils until the late Oligocene, and the modern families did not emerge until the early Miocene.

The late Oligocene to Pleistocene families Dromomerycidae (North America) and Palaeomerycidae (Old World), include a variety of deerlike and giraffelike animals with complex antlerlike structures. They are generally thought to include the ancestors of both deer and giraffes (Fig. 12–39).

All the living pecoran families appeared in the early Miocene, but their specific interrelationships have not yet been established (25). Each family is characterized by the particular nature of the bony processes arising from the forehead; the antlers of cervids are shed and the varied "horns" of the other groups are not.

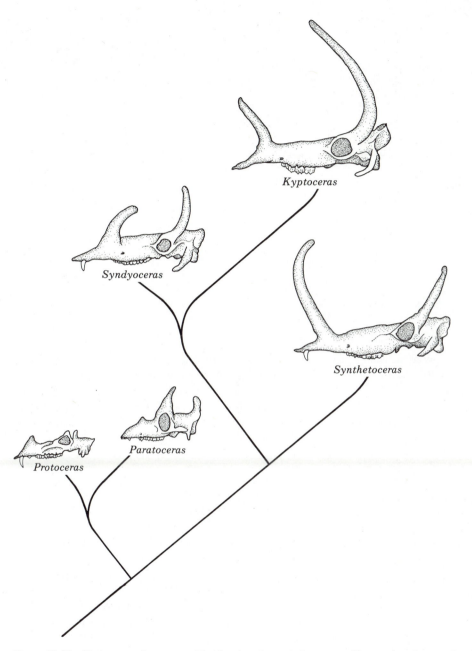

Figure 12-38   Phylogeny of protoceratids, showing the varied pattern of horns that characterize this group. The remainder of the skeleton resembles that of camels. (By permission of S. D. Webb and the *Journal of Vertebrate Paleontology*. v. 1(3–4). 1981.)

Deer (Cervidae) are a primarily North Temperate group that spread from Eurasia into North and South America late in the Cenozoic. The pronghorn antelope (Antilocapridae) is a relict of a much larger mid-Pliocene radiation in North America. The Giraffidae are all Old World. The bovids are primarily Old World, with the exception of mountain sheep and goats, bison, and musk ox, which were sufficiently tolerant of the cold to cross into North America from the north. The center of the bovid radiation that included cattle, sheep, goats, and antelopes is in Africa, where there is an extremely rich late Cenozoic fossil record (26).

Perissodactyls:

Horses, tapirs, and rhinosceroses are included in the order Perissodactyla, which also embraces

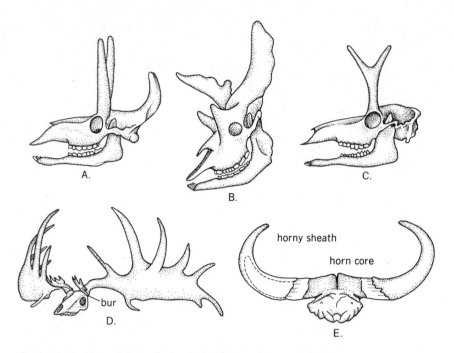

Figure 12-39 Frontal appendages of advanced ruminant artiodactyls. A. *Cranioceras*, a dromomerycid, an extinct group with bony processes extending from the skull bones, B. *Sivatherium*, a Pliocene relative of the giraffe. The bony appendages, termed ossicones, are not shed. C. *Meryceros*, a pronghorn antelope. The forked bony structure is not shed, but it was surrounded by a horny sheath that was. D. The irish elk *Megaceros*. Like other deer, the bony antler is shed annually. It breaks off at the structure labelled bur. E. *Bison*. The horns of African antelope and cattle are formed of an inner bony horn core and an external horny sheath. Neither is shed. These appendages may have evolved separately in each group.

a number of extinct groups (Fig. 12–40). In contrast with the artiodactyls, horses and their kin emphasize the middle digit, a pattern termed mesaxonic, rather than digits three and four, a pattern termed paraxonic. Horses show progressive reduction of the lateral digits, with the modern species supported solely by digit 3 in the front and hind feet.

Like artiodactyls, the perissodactyls appeared first at the very base of the Eocene, but their affinity with earlier condylarths, specifically the family Phenacodontidae, has long been recognized. Early perissodactyls are distinguished from condylarths by the specialization of the ankle joint, which facilitates anterior–posterior flexion but reduces lateral movement. They were most diverse and widespread throughout the northern continents in the early Cenozoic. Twelve families are recognized in the Eocene. The greatest diversity was among the tapiroids, which survive today as the most conservative of the horse's relatives.

Order—Perissodactyla
    Suborder—Hippomorpha
        Superfamily—Equoidea (horses)
        Superfamily—Brontotherioidea
    Suborder—Ancylopoda (chalicotheres)
    Suborder—Ceratomorpha
        Superfamily—Tapiroidea
        Superfamily—Rhinocerotoidea

Relatives of the modern rhinoceroses radiated extensively in the mid-Cenozoic, giving rise to fast-running forms, and others that resembled hippopotamuses. The late Oligocene and early Miocene genus *Baluchitherium* was the largest land mammal that ever existed, standing 5 meters tall at the shoulders.

In addition to the ancestors of the modern horses, tapirs, and rhinos, several important extinct groups diverged in the Eocene. Closely related to the horses are the paleotheres—large, European genera with

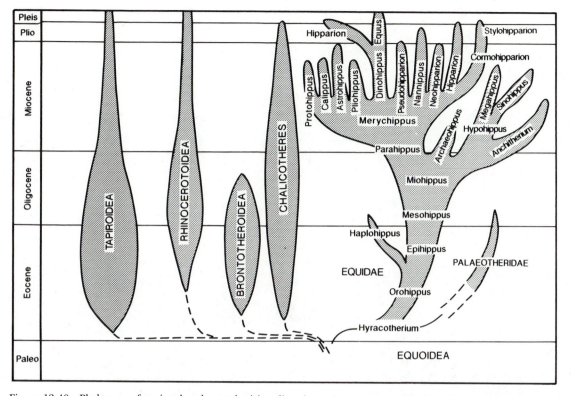

Figure 12-40   Phylogeny of perissodactyls, emphasizing diversity at the generic level in the horse family Equidae.

a dentition somewhat resembling that of artiodactyls—that became extinct in the early Oligocene.

The brontotheres were large rhinolike forms, common in North America and Europe in the Eocene and Oligocene. Among the strangest relatives of the horse are the chalicotheres, which arose in the early Eocene and survived into the Pleistocene. They are characterized by the possession of claws. *Chalicotherium* from the Miocene of Europe had forelimbs considerably longer than the rearlimbs, giving it a gorillalike stance (Fig. 12–41). Chalicotheres are thought to have fed on the leaves of tall trees which were pulled down to the mouth.

The best known of the early perissodactyls is *Hyracotherium*, whose remains are found in Europe, North America, and Asia. Although primitive in most features—the presence of four front toes, and three on the rear foot, and a complete dentition— *Hyracotherium* has specializations of the braincase that demonstrate relationships with the Equidae, which includes the modern horses.

Horses have had a complex pattern of evolution (Fig. 12–42). In the Eocene and Oligocene, a single lineage evolved in North America that showed increase in size and loss of the lateral toe in the front

foot. The Miocene was a time of major radiation. The evolution of more complicated patterns of the cheek teeth during the Miocene may be associated with feeding on hard grasses that were becoming

Figure 12-41   Restoration of a chalicothere.

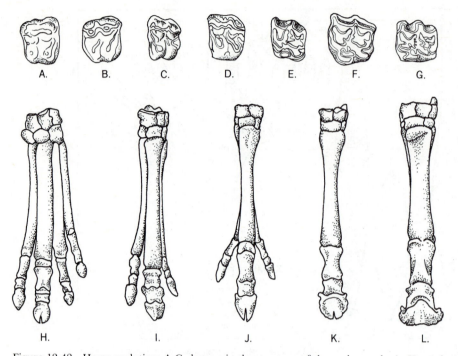

Figure 12-42 Horse evolution. A-G changes in the structure of the molar teeth. A. *Hyracotherium* (Eocene), B. *Orohippus* (Eocene), C. *Mesohippus* (Oligocene), D. *Miohippus* (Oligocene), E. *Merychippus* (Miocene), F. *Pliohippus* (Pliocene), G. *Equus* (Pleistocene and recent). In addition to gradual size increase and increased complexity of the cusp pattern of the molar teeth, the premolar teeth also become molariform and the teeth become higher crowned (from root to occlusal surface) so that they take longer to wear down. H–L. Changes in the feet. H. *Hyracotherium*, I. *Miohippus*, J. *Parahippus*, K. *Pliohippus*, L. *Equus*.

widespread in the North Temperate zone at this time. Further evolution in the late Miocene and Pliocene led to the progressive reduction of the lateral toes, increased size, further tooth complexity, and the origin of many distinct lineages. Most of the lineages died out by the end of the Pliocene, with a single major group continuing into the Pliocene and Pleistocene. Only the genus *Equus* survived into the Pleistocene, but it spread into South America, Asia, and Africa from its origin in North America. Surprisingly, *Equus* became extinct in the New World, although it thrived when it was reintroduced into North America by Europeans. Evolution of the horse was centered in North America, but they have survived to the present only because they were able to migrate into the Old World.

CETACEANS

The largest of all mammals and in fact the largest of all vertebrates are the whales. Whales also have

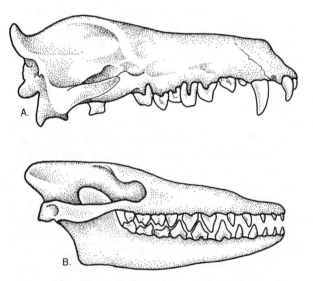

Figure 12-43 A. Skull of the Eocene terrestrial carnivore *Andrewsarchus*, nearly 1 meter long. It is close to the ancestry of whales. B. Restoration of the skull of *Pakicetus*, an early Eocene whale. (Courtesy of P. D. Gingerich and *Science*, vol. 220, pp. 403–406. Copyright 1983 by the AAAS.)

Figure 12-44   The Upper Eocene whale *Basilosaurus*, 25 meters long.

the largest brains of all vertebrates, and the toothed whales have a system of sonar that they use for locating prey and avoiding obstacles. Whales were already highly adapted to an aquatic way of life in the late Eocene. The body was greatly elongated, the forelimbs were specialized as paddles, and the rear limb and girdle were greatly reduced (Fig. 12–44). Incomplete skull remains from the early and middle Eocene show a transition between primitive terrestrial carnivores and early marine whales. These remains are found with terrestrial mammals, suggesting that the immediate ancestors of whales may have been amphibious. Early Eocene whales probably still came to land to reproduce (27).

The closest terrestrial relatives of whales were the mesonychids, which are closely related to the primitive condylarths. The mesonychids were the largest of the early Cenozoic carnivores, with skulls nearly a meter long but with molar teeth with elongated crowns that did not form the effective carnassial blades that charcterize the order Carnivora (Fig. 12–43).

The Eocene whales are united in a single suborder, the Archeoceti. Fossils of whales are rare in Oligocene rocks, but the two major groups of modern whales, the toothed whales, Odontocete, and the whale bone or baleen, whales, the mystocetes, had begun to diverge. All the major families of modern whales had become distinct by the early Miocene (Fig. 12–45).

Evidence of the evolution of sonar in the toothed whales can be determined by the configuration of the skull. The external narial opening became single through an asymetrical twisting of the bones of the skull. This twisting occurred separately in several different lineages, indicating the elaboration of sonar occurred six to ten times independently.

REFERENCES

1. Woodburne, M. O., and Zinsmeister, W. J., 1984, The first land mammal from Antarctica and its biogeographic implications: Journal of Paleontology, v. 58, p. 913–948.

2. McKenna, M. C., 1983, Cenozoic paleogeography of North Atlantic land bridges. In Bott, M. H. P., Saxon, S., Talwani, M., and Thiede, J. (eds.) Structure and Development of the Greenland–Scotland Ridge: New York, Plenum, pp. 351–399.

3. Savage, D. E., and Russell, D. E., 1983, Mammalian Paleofaunas of the world: London, Addison–Wesley, pp. 432.

4. Archer, M. (ed.), 1982, Carnivorous Marsupials: New South Wales, Royal Zoological Society.

5. Lillegraven, J. A., Thompson, S. D., McNab, B. K. and Patton, J. L., 1987, The origin of eutherian mammals: Biological Journal of the Linnean Society, v. 32, p. 281–336.

6. Novacek, M. J., 1986, The skull of leptictid insectivorans and the higher-level classification of eutherian mammals: Bulletin of the American Museum of Natural History, v. 183, p. 1–112.

7. McKenna, M. C., 1975, Toward a phylogenetic classification of the Mammalia. In Luckett, W. P., and Szalay, F. S. (eds.), Phylogeny of the Primates, New York, Plenum, p. 21–46.

8. Rose, K. D., 1981, A new Paleocene palaeanodont and the origin of the Metacheiromyidae (Mammalia): Breviora, no. 455, p. 1–14.

9. Storch, G. 1981, *Eurotamandua jorensi*, ein Myrmecophagide aus dem Eozän der "Grube Messel" bei Darmstadt (Mam-

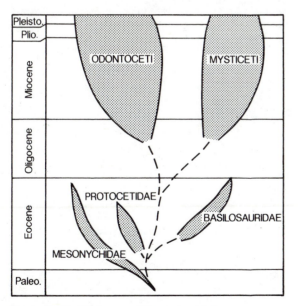

Figure 12-45   Phylogeny of the major whale groups.

malia, Xenarthra): Senckenbergiana Lethaea, v. 61, p. 247–289.

10. Novacek, M. J., 1985, Evidence for echolocation in the oldest known bats: Nature, v. 315, p. 140–141.

11. Luckett, W. P. (ed.), 1980, Comparative Biology and Evolutionary Relationships of Tree Shrews: New York, Plenum, 314 p.

12. Thurston, H., 1986, Icebound Eden: Equinox, number 27, p. 72–85.

13. Ciochon, R. L. and Chiarelli, A. B., 1980, Evolutionary Biology of the New World Monkeys and Continental Drift: New York, Plenum, 528 p.

14. Simons, E. L., 1985, Origins and charcteristics of the first hominoids. In Delson, E. (ed.) Ancestors: The Hard Evidence: New York, Alan R. Liss, Inc., pp. 37–41.

15. Johanson, D. C., and Edey, M. A., 1981, Lucy. The Beginnings of Humankind. New York, Simon and Schuster, 409 p.

16. Hunt, T. M. Jr., 1987, Evolution of the aeluroid Carnivora: Significance of auditory structure in the nimravid cat *Dinictis*: Novitates, no. 2886, p. 1–74.

17. Luckett, W. P., and Hartenberger, J. L., 1985, Evolutionary relationships among rodents: New York, Plenum, 721 p.

18. Chaline, J., Mein, P., and Petter, F., 1977, Les grandes lignes d'une classification evolutive des Muroidea: Mammalia, v. 41, p. 245–252.

19. Rose, K. D., 1981, Composition and species diversity in Paleocene and Eocene mammal assemblages: An empirical study: Journal of Vertebrate Paleontology, v. 1, p. 367–388.

20. Van Valen, L., 1978, The beginning of the age of mammals: Evolutionary Theory, v. 4, p. 45–80.

21. Gingerich, P. D., 1985, South American Mammals in the Paleocene of North America. In Stehli, F. G. and Webb, S. D., (eds.) The Great American Biotic Interchange: New York, Plenum, p. 123–137.

22. Webb, S. D., 1985, Late Cenozoic mammal dispersals between the Americas. In Stehli, F. G., and Webb, S. D., (eds.) The Great American Biotic Interchange: New York, Plenum, p. 357–386.

23. Maglio, V. J., 1973, Origin and evolution of the Elephantidae: Transactions of the American Philosophical Society. New Series, v. 63, p. 1–149.

24. Domning, D. P., Ray, C. E., and McKenna, M. C., 1986, Two new Oligocene Desmostylians and a discussion of tethytherian systematics: Smithsonian Contributions in Paleobiology, v. 59, p. 1–56.

25. Janis, C. and Scott, K., 1987, The interrrelationships of higher ruminant families with special emphasis on the members of the Cervoidea: Novitates, no. 2893, p. 1–85.

26. Vrba, E. S., 1984, Evolutionary pattern and process in the sister-group Alcelaphini–Aepycerotini (Mammalia: Bovidae). In Eldredge, N., and Stanley, S. M. (eds.), Living fossils. New York/Berlin, Springer-Verlag, p. 62–79.

27. Gingerich, P. D., Wells, N. A., Russell, D. E., and Shah, W. M. I., 1983, Origin of whales in epicontinental remnant seas: new evidence from the Early Eocene of Pakistan: Science, v. 220, p. 403–406.

SUGGESTED READINGS

Benton, M. J. (ed.), 1988, The Phylogeny and Classification of the Tetrapods. V. 2: Mammals. The Systematics Association special volume No. 35B. Clarendon Press, Oxford, 326 pp.

Delson, E. (ed.), 1985, Ancestors: the Hard Evidence: New York, Alan R. Liss, Inc., 366 p.

Savage, D. E., and Long, M. R., 1986, Mammal evolution: an illustrated guide: New York/Oxford, Facts on File Publication. 259 p.

Shipman, P., 1987, Dumping on science: Discover, v. 8, p. 60–66.

Stehli, R. G., and Webb, S. D., 1985, The Great American Biotic Interchange: New York, Plenum, 532 p.

Woodburne, M. O. (ed.), 1988, Cenozoic mammals of North America: geochronology and biostratigraphy: Berkeley, Univ. of California Press, 336 p.

PART THREE

LESSONS FROM THE RECORD

Chapter 13

Biostratigraphy

DIVISION OF SEDIMENTARY SUCCESSIONS BY FOSSIL RANGES

Stages and Zones:

Through the work of William Smith in Britain and Georges Cuvier and Alexandre Brongniart in France the association of specific fossils with specific layers or strata of the sedimentary record was becoming widely recognized in the early decades of the nineteenth century. Smith's work was largely practical and he does not appear to have been interested in causes, but Cuvier and Brongniart postulated that each of their faunal units represented a new creation and was terminated by an event of catastrophic proportions which they called a revolution, perhaps recalling the events that had overturned the French social system in their lifetimes. Where they worked in the environs of Paris the Cenozoic succession is an alternation of marine and nonmarine deposits and the faunas they studied replaced each other suddenly at the abrupt changes in environment. Cuvier's work was continued by Alcide d'Orbigny who called the strata bearing a distinct fauna a stage (étage) and established ten stages in the Jurassic System. D'Orbigny's stages were named after places where the fossils characteristic of that age could be collected and were defined on the basis of unique associations of fossils.

The use of fossils in dating rocks began with the succession of species found in superposed beds in a single section, such as a canal trench. The conclusion that the species found in the trench represent a succession of faunas in time, which could be inferred from their vertical superposition, was reinforced as similar sections elsewhere showed the same sequence of species without reversals of order. If their order of appearance were fortuitous and not controlled by the occurrence of the species or fauna in past time, we would expect it to be reversed or scrambled in other sections. The succession of faunas was extended in exposures that duplicate the interval of the initial section but also include beds that are younger or older, and that contain different fossils. Eventually a history of life throughout the Phanerozoic eras was built up, and its reliability was reinforced repeatedly as new sections containing segments of the history of life were described. We now recognize the sequence of life forms as being the result of the process of evolution but the outlines of the succession were established and used to organize stratigraphic knowledge long before Darwin. During the early part of the nineteenth century, naturalists who believed that the history of life was a continuous progression from simple ancient organisms to modern complex ones were in the minority and the fossil record, as then known, did not support this view. Charles Darwin advanced this position in the middle of the century by proposing a mechanism (natural selection) that explained how the evolution of life occurred.

In the 1840s and 1850s after most of the systems of the relative time scale had been set up, Albert Oppel carefully determined the ranges of fossils in Jurassic rocks of Europe and found that they were not separated into discrete units as d'Orbigny had proposed but overlapped in time. The ranges of some fossils were short, some long, but they were not all terminated at the end of a stage. He proposed to divide the stages into zones characterized by sets of fossils whose ranges overlapped in that interval and whose boundaries were drawn where new species appeared. Oppel named zones after the species of fossil that occurred within the zone, but the named species were not necessarily confined to their zone.

This work of paleontologists in northern Europe developed a scheme for the division of stratigraphic successions into stages and zones on the basis of the vertical or time ranges of fossils. It is a biostratigraphic classification because it divides rock strata into units on the basis of the life-forms contained

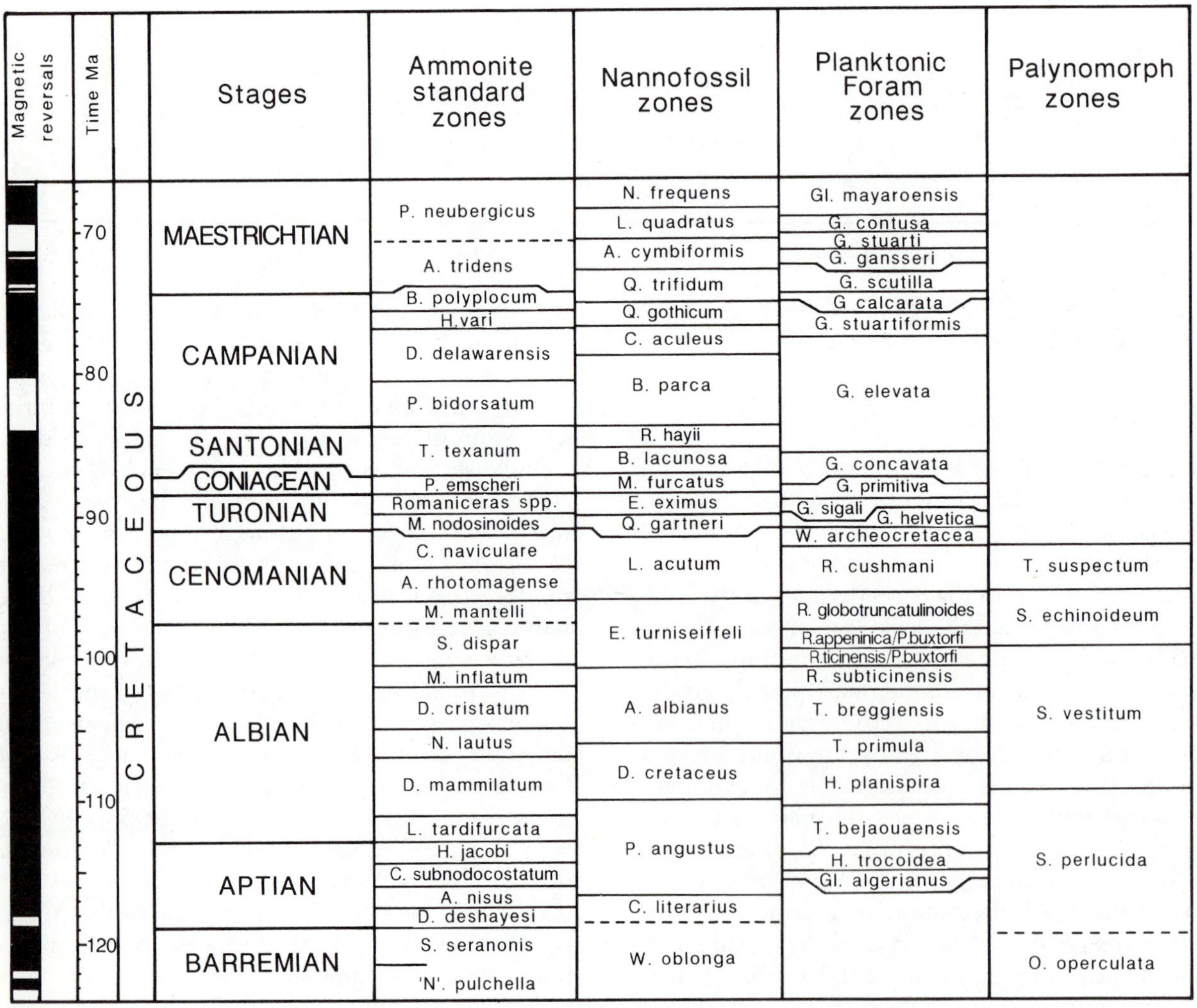
Magnetic reversals
Time Ma
Stages
Ammonite standard zones
Nannofossil zones
Planktonic Foram zones
Palynomorph zones
70
80
90
100
110
120
CRETACEOUS
MAESTRICHTIAN
CAMPANIAN
SANTONIAN
CONIACEAN
TURONIAN
CENOMANIAN
ALBIAN
APTIAN
BARREMIAN
P. neubergicus
A. tridens
B. polyplocum
H. vari
D. delawarensis
P. bidorsatum
T. texanum
P. emscheri
Romaniceras spp.
M. nodosinoides
C. naviculare
A. rhotomagense
M. mantelli
S. dispar
M. inflatum
D. cristatum
N. lautus
D. mammilatum
L. tardifurcata
H. jacobi
C. subnodocostatum
A. nisus
D. deshayesi
S. seranonis
'N'. pulchella
N. frequens
L. quadratus
A. cymbiformis
Q. trifidum
Q. gothicum
C. aculeus
B. parca
R. hayii
B. lacunosa
M. furcatus
E. eximus
Q. gartneri
L. acutum
E. turniseiffeli
A. albianus
D. cretaceus
P. angustus
C. literarius
W. oblonga
Gl. mayaroensis
G. contusa
G. stuarti
G. gansseri
G. scutilla
G calcarata
G. stuartiformis
G. elevata
G. concavata
G. primitiva
G. sigali
G. helvetica
W. archeocretacea
R. cushmani
R. globotruncatulinoides
R.appeninica/P.buxtorfi
R.ticinensis/P.buxtorfi
R. subticinensis
T. breggiensis
T. primula
H. planispira
T. bejaouaensis
H. trocoidea
Gl. algerianus
T. suspectum
S. echinoideum
S. vestitum
S. perlucida
O. operculata

Figure 13-1 Biostratigraphic zonation of the Upper and part of the Lower Cretaceous on the basis of ammonites, nannofossils, foraminiferans and spores. (*Source:* Modified from D. V. Kent and F. M. Gradstein, *Bulletin of the Geological Society of America*, v. 96, pp. 1424–1425.)

in them. Because the ranges of fossils can be used not only for subdivision but also for estimating whether beds in different regions were deposited at the same time (that is, for correlation), these divisions could be extended from northern Europe to other continents as knowledge of the stratigraphic record of the world expanded. The systems that are the basis of major divisions of the standard time scale were originally proposed as successions of rocks of distinctive character, but away from the areas where they were first described, they could only be identified by their fossil content. Systems and stages were based on successions of rocks at localities where they were first distinguished and where their fauna was thought to be typical of that interval. Zones were defined on the basis of fossil ranges only.

The biostratigraphic division of the Upper and part of the Lower Cretaceous System as it is now accepted is shown in Figure 13–1. In this system ammonites are the most useful fossils for defining zones, and the standard divisions of the system into stages depends on the ammonites' ranges. Zonal schemes using other groups, as shown in the chart, are used for strata that do not contain ammonites.

At present the term zone is used in several ways and modifiers are necessary to distinguish them. Interval zones have boundaries based on the first and last appearance of fossils. If the boundaries are based on the first and last appearance of a single fossil taxon, then the interval zone is a range zone. If the interval is defined on the overlap of the last appearance of one taxon and the first appearance of another, it is a concurrent range zone. Usually zones are based on the overlapping occurrence of three or more species and their boundaries are not fixed by first or last appearances. Such zones are called assemblage zones or, more formally, coenozones.

Controls on Fossil Ranges:

The total interval in rock strata in which a fossil species occurs from the time it evolved from its ancestor to its extinction, or its passage into one or more descendant species, is its biozone.

An example will illustrate the problems faced by paleontologists in defining biozones of fossil species (Fig. 13–2A). In a marine succession of limestone, sandstone and shale most species are confined to a certain lithology because organisms are adapted to specific environments and each environment on the

seafloor is represented by a type of sediment. We deduce that species b is probably a swimming or floating animal because it ranges through the succession, and its occurrence is independent of changes in lithology that signal changes in bottom conditions. Species c, d, and e are confined to the thick limestone unit I, and the ranges of d and e end abruptly at its top. Did they become extinct at this level owing to the environmental change or did they migrate to follow the limestone environment elsewhere when it was replaced by the muddy bottom on which the shale was deposited? The question can only be answered by looking for other sections of the same age in which limestone deposition continued into the time represented by the shale unit II in this section. We cannot define the top of the biozones of a species from this section alone but as its ranges in other sections are examined, we can come closer to defining it. We can never be sure that another section will not be found where its range extends higher than current findings indicate, some place where it continued to live after most of the individuals of the species became extinct.

The term facies means aspect and refers to the sum of the characteristics of a rock that relates to its origin. The term is also used with the prefixes litho- and bio-. Different lithofacies are distinguished on the basis of the physical aspects of rocks, and biofacies, on the fossil content. Stratigraphers would recognize unbedded porous limestones and dolomites as the reefal lithofacies. Paleontologists would recognize an assemblage of corals, stromatoporoids, and algae as the reefal biofacies.

In Figure 13–2A species such as g, a mud lover occurring only in shales, and h, a sand dweller found only in sandstones, are called facies fossils because they are confined to a certain facies. The occurrence of species g in this local section tells us little about its biozone. For species like g and h we can be certain that their comings and goings are migration events dependent on changes in local conditions. Their appearance may still be useful for correlating rocks or placing them in the relative time scale but the boundaries of their local ranges are unlikely to be origination or extinction events.

The ranges of such species as a and b that are not bounded by lithologic changes may be controlled by a variety of factors. Not all changes in environment will leave their mark; a change of temperature might have little effect on the accumulating sediment yet control the appearance or disap-

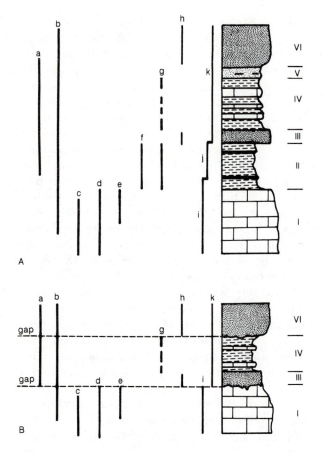

Figure 13-2 Hypothetical example to show the influence of lithology and gaps in sedimentation on the ranges of fossils. A. A section of nearly continuous deposition. B. The same section interrupted by two disconformities.

pearance of a species. A barrier that prevented a species from reaching this area could have been cleared away to allow immigration. For land animals such a barrier might be a strait of water that was blocked by rising land; for marine animals, an isthmus of land that became submerged.

If, in a section of continuous deposition, such as Fig. 13–2A, the paleontologist can identify a series of ancestors and descendants, then biozones of species, such as species j, may be precisely defined. Within the shale unit II, j evolved from its ancestor species i, and at the beginning of unit III it evolved into its descendant species k. Because such evolutionary events are thought to have occurred in small populations occupying limited areas, the chances of such an event being preserved in the record is very small. Ideally the division of strata into zones and stages should be on the basis of ranges estab-

lished from such evolutionary successions, but unfortunately such successions can be recognized only rarely. Zones established in such successions have been called phylozones.

In Figure 13–2A continuous deposition was assumed, but few sections are uninterrupted by pauses in deposition or times of erosion called disconformities. In Figure 13–2B two gaps in deposition that eliminate units II and V are identified. The effect of the gap between I and III is to concentrate the tops and bottoms of the ranges of species a, h, i, and k at this level giving it an apparent significance in the evolution of the fauna that is unjustified. Species f and j are entirely eliminated and the transition from i to k is no longer evident. On the other hand, the gap between IV and VI makes little difference to the ranges of species.

Most of the beds from which the marine invertebrate record has been collected were deposited in seas that flooded the interiors of continents. Because deposition in these epicontinental seas was discontinuous, more time is represented by the erosional and nondepositional gaps between beds than by the sedimentary layers themselves. In the alluvial deposits that contain much of the vertebrate record, deposition is also episodic and interrupted. Such gaps obviously affect our knowledge of the ranges of species determined from a succession of sedimentary rocks. Because most of these gaps in the record are of local extent and are filled by sediments at some other location, the discontinuities in the ranges of fossils can be filled by patient regional work. Over the past 150 years this type of work has produced such detailed zonal schemes as illustrated in Figure 13–1.

Paleontologists who work with samples from the drilling of oil wells have several problems in the recognition of zones. The fossils are contained in fragments of rock commonly less than 1 cm across that are chipped from the rock at the bottom of the hole and brought to the surface by mud that is circulated in the hole. In the cuttings brought up in the mud the paleontologist encounters new fossils from the top of their ranges, not as they appeared successively in time. Most stages are defined by beginnings of ranges but subsurface samples can rarely define these beginnings, which would be signalled by the absence of a fossil found higher in the hole. This is because fragments of fossils from higher in the hole mix with newly cut chips as they are flushed to the surface. The paleontologist learns to look for new kinds of fossils appearing in successive samples as the top of the fossil zones are penetrated and ignores fossils that have already been recorded from higher in the hole whose presence or absence may be fortuitous.

CORRELATION

General Principles:

Nearly all stratigraphic information is available only in geographically separated areas that are isolated successions of sedimentary beds exposed at the surface or encountered in drilling wells. Within an ancient basin where the sedimentary beds were deposited, the strata were originally laterally continuous, but subsequent erosion removes parts once deposited and later deposits limit our access to lower layers to wells drilled considerable distances apart. To determine ancient environments of deposition, stratigraphers must take the digitized data provided by wells and outcrop and compare lithology, fossil ranges, and other features that indicate that sediments were deposited contemporaneously to reconstruct the original continuity in the depositional setting. This comparison is known as correlation.

The word correlation is used by paleontologists and stratigraphers in several different senses and its precise meaning is not always obvious. We can distinguish three types of correlation. Lithostratigraphic correlation is the comparison of lithofacies between outcrop sections and wells to establish the depositional continuity of these units. The units of lithology, or formations, are traced from well to well or outcrop to outcrop. Time correlation is the comparison of successions in order to establish equivalence in time Sets of strata laid down in the same interval are said to be time correlative. Biostratigraphic correlation is the comparison of the ranges of fossils in successions to determine time correlation. In most stratigraphic studies, biostratigraphic correlation is as close as a stratigrapher can get to time correlation because few criteria of contemporaneity are preserved in sedimentary rocks. Sedimentary beds that were deposited in a short time over a large area, such as the ash layers that result from a volcanic eruption, are invaluable for establishing time correlation in both marine and nonmarine successions but are relatively rare in many depositional environments. They have the additional advantage that they can be dated isotopically in years. The value of fossils as time indicators is limited by factors discussed below but in most sections they are the best criteria available. In most stratigraphic studies time correlation is an unattainable ideal.

Since the end of the eighteenth century when faunal succession was discovered, similarity of fossils from two locations has been taken to indicate that beds containing the fossils were deposited at the same time. The precision of the correlation will depend on the accuracy with which the ranges of the fossil species are known and the length in time of those ranges. Species that lived for a short time and are abundant enough that their ranges have been confirmed in many areas are more valuable for establishing precise correlation than are species that lived for a long time or those that are so rare that the tops and bottom of their ranges have not been confirmed.

Lack of similarity of fossils does not indicate that the rocks were laid down at different times. The kinds of fossils in a bed depend on the environment in which they were deposited as much as on the time of deposition. Contemporaneous beds of shallow-water limestone and deep-water shale may contain no fossil species in common. Attached, crawling, and burrowing organisms are dependent on bottom conditions and sediment, but swimmers and floaters can live above many different bottom environments and fall onto them when they die. The fossils of such organisms occur in many different facies and can be used to correlate rocks of different lithology. With few exceptions, bottom-dwelling organisms are facies fossils, associated with a specific rock type but nonetheless useful for correlating beds of that facies.

Relative Value of Fossils:

The most useful fossils for correlation are those that are abundant, widespread, rapidly evolving (and therefore of short range), and distributed in a variety of facies. For different times in the past different groups are useful. In the Cambrian, trilobites were the only organisms that fulfilled these requirements and within that system, division into zones and correlation of these zones is based on trilobite ranges. Graptolites and conodonts are the most useful fossils for correlating rocks of Ordovician, Silurian, and Devonian ages. The ammonoids appear in Upper Devonian rocks and between that series and the top of the Cretaceous System they have proved to be invaluable for correlation and zonation. Foraminiferans are useful zone fossils for the late Paleozoic when the fusulines were abundant and for post-Jurassic time when planktonic foraminiferans spread widely through the oceans.

This short list should not give the impression that other fossils are useless. Practically every group has been used for the division and correlation of some group of rocks. Almost all fossils are of some use in dating but relatively few, because their ranges are short and our knowledge of their succession in time is extensive, are capable of giving precise ages. The precision in years attained by fossil dating varies widely in time and place but for the Jurassic and Cretaceous rocks whose fossils have come under the closest scrutiny for the longest time, a precision of a few hundred thousand years is possible.

Dispersal and Migration:

If dispersal of organisms from their site of first appearance is slow, new species will appear in the sedimentary record at successively later times away from the population from which they evolved. Paleontological correlation is based on the assumption that dispersal rates are rapid, in fact essentially instantaneous, when compared to the units of time that can be distinguished by fossil ranges. How valid is this assumption?

If dispersal rates were measurable in millions of years, the order in which species appear in widely separated sections of a given time interval would be different. Starting from various dispersal centers they would arrive at distant sites in different sequences (Fig. 13–3). In fact this is not observed. Reversals in order of appearance of fossil species are uncommon enough to require special investigation, such as the search for environmental barriers that may have retarded the arrival of a species in one region to a time after its appearance in neighboring regions.

Marine animals that live in the water rather than on the bottom can disperse rapidly on currents or through their own swimming power. Most bottom-dwelling invertebrates pass through a larval stage in which they are free-swimming or floating. During this stage they may be dispersed widely depending on its length and the strength of currents. Although most invertebrate larvae must find a site to settle on the bottom within a few days or weeks, some can postpone settlement if no suitable substrate is available. Larvae that feed on plankton are

capable of more extensive dispersal than those that do not feed during the larval stage. Such larvae can travel for several months without settling to the bottom. These larvae are capable of crossing large stretches of deep water such as the North Atlantic Ocean on currents. The rate of spreading of molluscs along coastlines to which they have been introduced by shipping or currents in historical times has been traced by shell collectors. At these rates they would disperse around the world in a few hundred years. Marine organisms are not universally dispersed because their larvae face barriers in their search for a place to grow to adulthood. They have the potential for geologically instantaneous dispersal, but dispersal may be blocked by ecological or geographic barriers that delay or prevent their appearance in all areas at the same time. For correlation those animals that are likely to be least subject to such barriers are therefore preferable.

Quantitative Correlation:

Two collections of fossils from different areas are unlikely to contain exactly the same species; some will be common to the two collections, others will be found in only one. Some paleontologists have argued that the reliability of the correlation or the closeness in age of the two collections can be estimated by the proportion of species that are common to those that are different. Expressions the degree of difference between collections are used in many statistical procedures. Several similarity coefficients for comparing collections are listed below. They are improvements on simple counts of the number of species common to the collections in that they take account of the difference in size of the collections. In all of these C is the number of items in common, N_1 is the number of species in the smaller sample, and N_2 is the number in the larger sample.

SIMILARITY COEFFICIENTS

$$\text{Simpson coefficient} = C/(N_1 + N_2)$$
$$\text{Jaccard coefficient} = C/(N_1 + N_2 - C)$$
$$\text{Dice coefficient} = 2C/(N_1 + N_2)$$
$$\text{Otsuka coefficient} = C/\sqrt{N_1 N_2}$$

The larger the values of the coefficients calculated from two faunas being compared, the closer

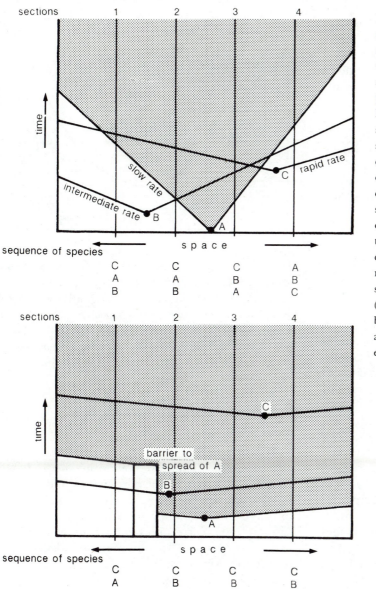

sequence of species

	1	2	3	4
	C	C	C	A
	A	A	B	B
	B	B	A	C

sequence of species

	1	2	3	4
	C	C	C	C
	A	B	B	B
	B	A	A	A

Figure 13-3   Hypothetical example to show the effect of the rate of dispersal on the order of appearance of fossil species in four sections (1, 2, 3, 4). The species originate at the dots marked A, B, and C, and their dispersal outward from these points is represented by the upwardly diverging lines. The increasing area occupied by species A is shaded for emphasis. When the rate of dispersal is slow, these lines form an acute angle with the apex of origin; when it is fast, the angle is obtuse. The order in which the species appear is listed below each section. If some species disperse at relatively slow rates, the species will appear in different orders in the different sections (top diagram). As this is not commonly observed in the paleontological record, we deduce that rates of dispersal were rapid, as in the lower diagram. This lower diagram shows that a barrier to the spread of one species (species A), which is present for a limited time, may block its spread into section 1 and delay its appearance there until after the appearance of a later evolved species B.

in age they are considered to be. In merely summing matches and mismatches the paleontologist is assuming that all species affect the accuracy of the correlation equally. From what has been indicated in the discussion above this is obviously not true; species with short ranges are more valuable than others and species with very long ranges are practically valueless. Attempts have been made to apply a weighting factor that would modify the impact on the similarity coefficient of each species on the basis of its geographic and stratigraphic range but these factors inevitably have subjective aspects.

Paleontological correlation has so far resisted reduction to a mathematical expression, and fossil biotas must be compared by an expert who knows the value of each fossil as a time indicator before correlations can be made.

Depth-Controlled Communities and Correlation:

Sedimentary successions all over the world record episodes of sea level rise and fall. The magnitudes of these oscillations range from those that caused oceans to become a few meters deeper to those that inundated large parts of continents. The larger changes in sea level, such as the one that caused extensive continental seas to flood the continents in Late Cretaceous time, have been known for many years to have affected much of the earth. Only recently have many stratigraphers become convinced that the smaller variations that occurred 10 to 20 times during a geological period are also worldwide. Once the times of these sea level changes have been established on the basis of normal techniques of paleontological dating, the changes become useful for correlation. Large-scale sea withdrawal will be marked by erosion surfaces in the sedimentary succession, and large-scale transgression by the appearance of deep-water sediments in areas of shallow-water deposition, but smaller changes may be reflected only in fossil faunas that are sensitive to water depth.

The composition of modern and fossil invertebrate communities changes with depth. For an interval in the past, the sequence of fossil communities can be established from the reconstruction of an offshore, downslope gradient at the margin of ancient continental platforms. The vertical (temporal) changes in communities in stratigraphic sections can be interpreted in terms of changes of sea level and transgressive and regressive episodes can be identified. An example will illustrate the method. In England and Wales a transition can be traced in Lower Silurian rocks from shallow-water environments to the deep-water graptolitic shales of a marginal trough. The transition is marked by changes in fossil faunas that have been divided into five depth-controlled communities. Alfred Zeigler and his co-workers distinguished them on the basis of distinctive brachiopods (1) (Fig. 13–4). Markes Johnson and his co-workers have used a wider spectrum of organisms to identify similar depth-controlled communities (2). Deep communities are recognized by the occurrence of pentamerid brachiopods, intermediate ones by corals and algae and shallow-water ones by stromatolites and ostracodes. When the sequence of these communities is recognized in stratigraphic sections, a curve showing the relative change of sea level during the deposition of the section can be constructed (Fig. 13–5). The curves of many sections show four periods of deepening water in Early Silurian time. If these deepening events can be identified in new areas, correlation between two sections can be established without the use of the ranges of fossils. The fossils themselves are not matched from sec-

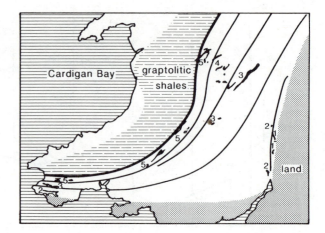

Figure 13-4 Depth-controlled assemblages of Early Silurian brachiopods in southern Wales. The shelf between the deeper water graptolitic shales and the land has been divided into 5 assemblages indicated by numbers. 1: *Lingula* assemblage; 2: *Eocoelia* assemblage; 3: *Pentamerus* assemblage; 4: *Stricklandinia* assemblage; 5: *Clorinda* assemblage. The areas in which the shelf assemblages are exposed and identified are indicated by the irregular black patches. (*Source:* Modified from A. M. Zeigler, *Nature*, v. 270, p. 271.)

tion to section and sections can be correlated that have no communities in common.

Transgression and regression of seas not only change depth communities but influence, and may control, the evolution of new species and the extinction of old. The bases of zones marked by the introduction of new species of ammonoids in the Carboniferous correspond to times of sea level rise and transgression (3). Events that expanded or contracted the living space of marine organisms can be expected to have had a profound effect on their rate of evolution and extinction. Within the last decade stratigraphers have established that worldwide changes of sea level occurring in superposed cycles of as many as five or six orders have affected the transgression and regression of the sea and presumably also the life of the sea. The subject is discussed further in Chapter 16.

STRATOTYPES: DRIVING THE GOLDEN SPIKE

Type Sections:

The stratigraphic sections from which the systems were originally described became standards for the recognition of these rocks and fossils because they

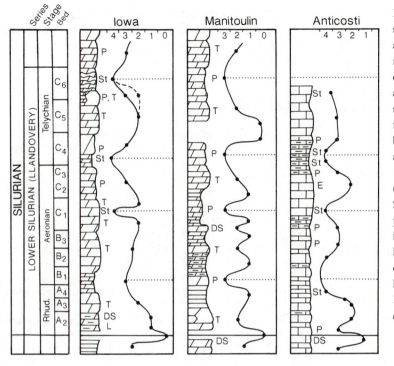

Figure 13-5   Correlation of three stratigraphic sections on the basis of change of water depth, as deduced from depth-associated fossil communities in the Lower Silurian of central and eastern North America. The four benthonic assemblages are scaled at the top of the graphic sections in order of increasing depth (1-4). The fossil assemblages indicating depth are indicated by the letters to the right of the graphic sections as follows: L = *Lingula* assemblage (assemblage 1); DS = domal stromatolites or stromatoporoids (assemblage 2); T = tabulate corals (assemblage 2); Eo = *Eocoelia* fauna (assemblage 2); P = pentamerid fauna (assemblage 3); St = stricklandiid fauna (assemblage 4). The continuous sinuous line represents the inferred relative rise and fall of sea level and the dotted lines join correlative levels of high water. (*Source:* Modified from Markes Johnson, *Geological Society of America Bulletin*, v. 96, pp. 1484–1497.)

were at first the only ones available. For example, in North America rocks were recognized as deposited during the Silurian Period by comparison with fossils that had been collected by Roderick Murchison when he set up the Silurian System in Wales. When d'Orbigny set up the stages as subdivisions of the Jurassic System, he gave them geographic names from localities where the units could be examined and the fossils that defined their place in the geological time scale could be collected. The Bathonian Stage was typically exposed at Bath in England and the Hettangian Stage at Hettange in France. Unfortunately d'Orbigny did not always locate his type sections at the locality that gives its name to the stage; for example, the Oxfordian Stage is defined by cliff sections in Yorkshire, not at Oxford. Biostratigraphic units must have a type section where their boundaries are designated and displayed not only by fossil ranges, but also at specific levels in the section. These sections are called type sections, or stratotypes.

Because the systems were described originally from various areas across northern Europe at a time when paleontological correlation was in its infancy, there was no way of knowing whether time gaps existed between the systems or whether they overlapped. For example, when the Permian System was first described in Russia it inappropriately included in its base rocks of the same age as those that had already been assigned to the Carboniferous System in England. Detailed correlation on the basis of fossils brought gaps and overlaps (such as the Carboniferous–Permian one) to light, and international commissions of paleontologists were set up to make decisions on where the boundaries should be placed.

Some of the regions where the systems were first described were far from ideal as standards for worldwide correlation. The Devonian rocks of Devon are complexly folded and faulted, and the stratigraphic succession there is still uncertain. As a result the rocks of the Ardennes region of Belgium have been used since the last century as a standard for Devonian correlation.

Boundary Stratotypes:

Recently commissions of the International Union of Geological Sciences have concentrated on designating boundary stratotypes, that is, sections to act as standards in correlation of boundaries between biostratigraphic units. If the boundaries of the unit are well defined in terms of fossil ranges, then the whole unit is precisely defined as the rocks lying between these boundaries. Recently a commission has designated a section at Klonk in Czechoslovakia as the stratotype of the Silurian–Devonian boundary and the horizon at the top of the zone of the graptolite *Monograptus uniformis* as the boundary itself. The boundary between the Ordovician and Silurian systems has recently been set by the vote of a committee at Dobb's Linn in Scotland at the base of the *Parakidograptus acuminata* graptolite zone.

Informally the process of choosing boundary stratotypes has been called "driving the golden spike." By setting for all time the plane of the boundary, the commission figuratively drives a marker into the section. Such boundaries should be in sections that are accessible, rich in fossils, and of essentially uniform lithology so that changes in fauna cannot be ascribed to local changes in environment. Although the boundary position is usually chosen at the base of some fossil zone, the boundary is defined as a position in the rock sequence, not the base of the zone.

The Stratigraphic Code:

Efforts to codify stratigraphic procedure culminated in the publication of the International Stratigraphic Guide in 1976, and the North American Stratigraphic Code in 1983 (4,5). Although not all the recommendations on biostratigraphic units are accepted by all stratigraphers and paleontologists, most follow their major recommendations. The American code makes the distinction between stratigraphic units defined by time criteria, such as isotopic ages (chronostratigraphic units), magnetic reversals (magnetopolarity units), and those defined by fossils (biostratigraphic units). Systems and series are considered by those that drafted the Codes to be chronostratigraphic units and only biozones are biostratigraphic. For most sedimentary rocks, however, fossils are the only means available to tell time,

and the original divisions of sedimentary rocks into systems and series were originally, and continue to be, on this basis. In stratigraphic studies, paleontologists "hold the watch" (6).

REFERENCES

1. Zeigler, A. M., Cocks, L. R. M., and Bambach, R. K., 1968, The composition and structure of Lower Silurian marine communities: Lethaia, v. 1, p. 1–27.
2. Johnson, M. E., and Campbell, G. T., 1980, Recurrent carbonate environments in the Lower Silurian of northern Michigan and their inter-regional correlation: Journal of Paleontology, v. 54, p. 1041–1057.
3. Ramsbottom, W. H. C., and Saunders, W. B., 1985, Evolution and evolutionary biostratigraphy of Carboniferous ammonoids: Journal of Paleontology, v. 59, p. 123–139.
4. Hedberg, H. (ed.), 1976, International Stratigraphic Guide. International subcommission on stratigraphic classification: New York, Wiley, 200 p.
5. North American Commission on Stratigraphic Nomenclature, 1983, North American Stratigraphic Code: American Association of Petroleum Geologists Bulletin, v. 67, p. 841–875.
6. Ross, R., 1986, Who's holding the watch?: Palaios, v. 1, p. 1.

SUGGESTED READINGS

Berner, R. L., and McHargue, T. R., 1988, Intregrative Stratigraphy, Chapter 9: Englewood Cliffs, N.J., Prentice–Hall, 419 p.

Berry, W. B. N., 1987, Growth of a Prehistoric time scale, Revised edition: Boston, Blackwell Scientific Publications, 202 p.

Harper, C. W., 1980, Relative age inference in paleontology: Lethaia, v. 13, p. 109–118.

Hay, W. W., and Southam, J. R., 1978, Quantifying biostratigraphic correlation: Annual Review of Earth and Planetary Sciences, v. 6, p. 353–375.

Kauffman, E. G., and Hazel, J. E. (eds.), 1977, Concepts and methods of biostratigraphy: Stroudsburg, Pa., Dowden, Hutchison and Ross, 658 p.

Chapter 14

Adaptation and
Functional Morphology

ADAPTATION

Stephen Gould and Richard Lewontin defined adaptation simply as the "good fit of organisms to their environment" (1). In another place Lewontin described adaptation as the process by which organisms provide better and better solutions to the problems posed by the environment (1). The term *adapt* implies movement towards an appropriate or apt relationship to environmental demands and is commonly linked to the term *fitness* in discussions on the relationships of organisms to their environments. Evolution has been explained in terms of the "survival of the fittest," meaning the best adapted organisms.

The concept of the good fit of organisms to their environments can be traced back to antiquity. It was most clearly stated by the natural theologians of the eighteenth century. The world and its inhabitants were regarded by natural theologians as expressions of the perfection of their Creator and therefore in themselves perfect. The Creator perfected the form and behavior of organisms to fulfill their functions in the world and His creation could be no less perfect than Himself. Charles Darwin asked "How have all those exquisite adaptations of

one part of the organization to another part, and to conditions of life, and of one organic being to another being been perfected?" and answered that it was not by God but by competition, natural selection, and the survival of the fittest (2).

If only the fittest survive in the competition, levels of fitness must exist among the competitors. Several possible measures of fitness have been proposed among which are survival, reproductive success, and breadth of environments successfully occupied. The organism that survives when its competitors die for other than chance reasons must be considered better adapted to its environment. The organism that produces more offspring than its competitors will in time displace the competitors in their home range. Their greater fecundity will then give them the opportunity to spread to adjacent areas to increase the range of the species. Those organisms that are more fit can be described as those that are more successful under natural selection. Fitness then becomes a description of the biological properties of an organism that result in its survival and perpetuation.

To some evolutionists only whole organisms can be considered as adapted to their environments; to others, the features or parts of the organisms are adaptations. One method of studying adaptation is to divide the organism into traits, either morphologic features or patterns of behavior, and to consider separately how each of these is related to environmental factors. Those who take a more holistic approach say that the traits of an organism so interact that the organism cannot be reduced to its component parts but must be considered as a whole (1). Paleontologists are concerned with adaptation because they need to know how ancient organisms lived. Most of the evidence they use is confined to the preserved parts of the body. Body parts adapted to fit the environment should in their design give evidence of their use and help us to interpret the lifestyle and habitat of the organism.

Becoming Adapted:

Darwin's concept of competition among individuals leading to progressively better adaptation implies that an organism is never completely or perfectly adapted and that its gains in solving the problems of the environment are at the expense of other organisms. Darwin used the metaphor of a group of wedges driven into a limited space. Each species is a wedge hit occasionally by a hammer but it can only make progress into the hole by squeezing out other wedges. Leigh van Valen (3, 4) formulated a similar idea as the Red Queen hypothesis. In "Through the Looking Glass" the Red Queen told Alice that in her country you had to keep running just to stay where you are. Van Valen postulated that, because a species competes for resources with other species, even in a stable environment its relationship with the environment steadily deteriorates unless it is continuously improving its adaptation. Those that fall behind in this race become extinct.

The Red Queen hypothesis assumes a stable physical environment but geologists know that in the long run no environment is entirely stable. Paleontologists conceive of evolutionary changes as repeated responses by organisms adapting to changes in their physical and biological environments. Major steps in evolution discussed in the next chapter may involve other mechanisms, and mass extinction events that cause major shifts in the history of life also disturb the process of adaptation. If adaptation is viewed from this longer geological perspective, better adaptation can be measured in terms of survival of the species through the greatest number of environmental changes or the greatest length of geological time. To achieve maximum capacity to adapt according to this longer time scale, a species would have to maintain a large stock of genetic variability within the population in order to be able to respond repeatedly to changes in the environment, with morphologies that were successful in terms of short-term survival of the individual. Alternatively, species may be able to survive for millions of years as a result of extensive capacity for physiological adaptation that is not based on genetic variability.

Imperfection of Adaptation:

Animals of different ancestry adapted to the same environment may have similar forms. This is the phenomenon of evolutionary convergence, and it is good evidence of the reality of adaptation. A classical example of convergence from the paleontological record is that of the ichthyosaur of the Mesozoic and the dolphin of today. One was, and the other is, a pelagic, fast-swimming predator. They have in common streamlined bodies, the arrangement of fins, and a pointed snout with teeth (Fig. 14–1). Yet the first was a reptile with reptilian skeletal structures and physiology and the second, a mammal with mammalian skeleton and reproductive system. Although molded by the requirements of the environment to superficial resemblance, they maintain many internal features of their ancestors. That is, their adaptation to the same environment at different periods of life history through natural selection has not produced identical organisms.

Even casual observation shows that parts of complex organisms are used for a variety of functions and usually are not adapted for a single particular function. In birds the beak may be used to obtain a certain food but, since it must serve also for feeding the young, grooming, fending off rivals, etc., its shape must be a compromise between the forms that are ideal adaptations for each of these uses. The study of adaptation is simpler for invertebrates that live attached in one place utilizing one special food resource, but the organs of even these simple organisms cannot be optimized through natural selection for a single feeding condition because they are likely to encounter a range of conditions that follow each other in time.

The constraints of ancestry and the multiple use of organs are two reasons why traits or individuals

Figure 14-1   Evolutionary convergence illustrated by adaptation to fast swimming. A. Swordfish (a bony fish). B. Shark (a cartilaginous fish). C. Dolphin (a mammal). D. Ichthyosaur (a reptile). Each of the groups independently evolved a spindle-shaped body and a high lunate tail which enables it to swim rapidly (approximately 40 km/hour). Dolphins and ichthyosaurs evolved from terrestrial ancestors. Primitive bony and cartilaginous fish had elongate bodies and were only capable of swimming at much slower speeds. In contrast with their similar body shapes, details of their skeleton reveal their distinct ancestries. Typical of bony fish, the swordfish has a large operculum covering the gill region; sharks have several separate openings for the gills. Both dolphins and ichthyosaurs breathe through nostrils. The vertebral column of sharks supports the dorsal lobe of the tail; in ichthyosaurs, it supports the ventral lobe. The vertebral column does not extend into the tail in bony fish. In contrast with all other groups, the tail of the dolphin is horizontal, not vertical.

cannot be fully optimized by natural selection to take advantage of their environments (5). Three other reasons, among many, can be mentioned here briefly.

1. Features that are nonadaptive for everyday life may be essential for attracting mates. Flamboyant plumage in male birds may attract hawks but also attracts females and is therefore essential to the survival of the species.
2. Natural selection operates in a world where chance events—storms, earthquakes, and droughts—may carry away the fittest and leave the less fit, thereby interrupting the selective processes that tend to optimize adaptive features.
3. If the adaptive value of a trait is neutral or even slightly deleterious, the genes that determine its appearance in a population may still become widespread in a small population, not by selection, but merely by chance interbreeding. This phenomenon, called genetic drift, may perpetuate traits that are of relatively low selective value to the organism.

FUNCTIONAL MORPHOLOGY

Functional morphology is the study of the relationship of form to the functions that organisms or their component parts perform. To determine how a fossil organism lived the paleontologist analyzes the possible functions of the parts that are preserved. Adolf Seilacher (6) proposed that form can be understood in terms of the interaction of three factors: adaptation, growth, and phylogeny (Fig. 14–2).

Phylogeny:

Organisms do not have unlimited capacity to generate new adaptations. They modify for new functions the structures that they have inherited from their ancestors. The inherited structures were selected to meet environmental demands which were faced by ancestors but are possibly irrelevant to the way of life of the descendant as it has entered new environments. Each organism must use imperfect modifications of the adaptations of its forebears to meet its own environmental challenges. Stephen

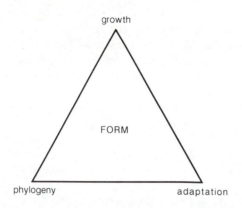

Figure 14-2 The three factors controlling morphology suggested by Adolf Seilacher.

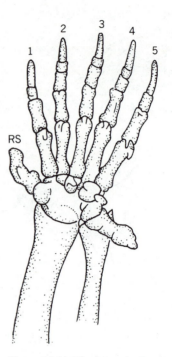

Figure 14-3 The bones in the panda's forelimb. In addition to the five digits (numbered), the panda has a "thumb" formed by modification of the radial sesamoid bone (RS) used for stripping leaves from bamboo. (*Source:* Modified from S. J. Gould, *The Panda's Thumb.* 1980 New York, Norton)

Gould (7) has pointed out that these types of imperfect adaptations are strong evidence for evolution. His favorite example is the panda's thumb, which is used to strip leaves from young bamboo shoots—the panda's only food. The panda is descended from bears whose front feet were specialized for running and were not able to grasp objects between their thumb and other fingers as primates do. The panda inherited five digits but the thumb was not opposable or useful for stripping bamboo. In adapting to the bamboo diet, the panda developed a thumb-substitute by a modification of one of the bones that is normally a small part of the wrist (Fig. 14–3). Although this sixth digit is not an ideal adaptation, it evolved to become an efficient leaf stripper because the panda was constrained by inherited fore-foot anatomy in its adaptive pathways.

The basic designs of the invertebrate phyla were established at the beginning of the Paleozoic Era as organisms diversified into different habitats in the oceans. Specialization within, and progression beyond, these original adaptations may have obscured the original adaptation, but the potential to develop new adaptations has since been constrained by these early events. When a group of bivalves, the scallops, diversified into the swimming niche they did not have the materials to grow fins as did fish but could only modify the shell and mantle inherited from distant ancestors adapted to burrowing in sediment. The scallop's adaptation to swimming is not comparable in perfection to that of fish or squid, which from their early history entered this niche, but it is apparently an effective method of escape from predators. The form of the scallop is based in its history but was modified by adaptation to a lifestyle much different from that of its ancestors.

Growth and Allometry:

The form of the fossil skeleton will be a product of how it grows. Some organisms form a skeleton on the exterior of their bodies (bivalves, brachiopods), others within the tissue (vertebrates, echinoderms). In some it grows continuously (sponges), and in others it is periodically molted (arthropods).

If, in the growth of an individual, the parts all grow at the same rate, their proportions will be in the same in the adult as in the juvenile form. Such growth is said to be isometric but this is not a common mode of growth. In most higher organisms, some parts grow faster or slower than others so that as the organism increases in size its proportions change. This is allometric growth. The relation between the size of a part and the whole organism can commonly be expressed in an equation of the following type:

$$X = a\,Y^b,$$

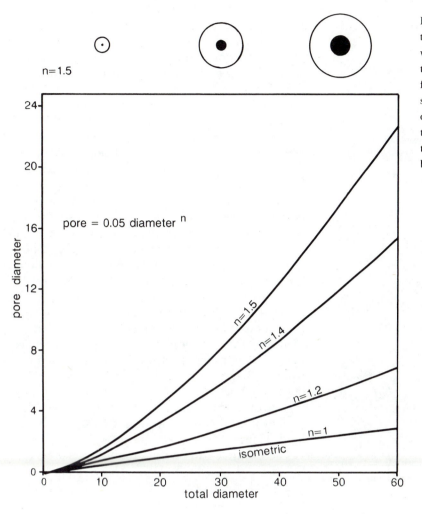

n=1.5

pore = 0.05 diameter^n

pore diameter (y-axis)

n=1.5
n=1.4
n=1.2
n=1
isometric

total diameter (x-axis)

where X is the dimension of the part growing allometrically, Y is a measure of the size of the organism, and a and b are constants (Fig. 14–4). If the exponential constant b is large, the increase in X for a small change in size (Y) may be very large. If b is 1, growth is isometric. Parts that increase in size faster than the animal grows are said to have positive allometry and b is greater than 1. An example is the head of the human foetus, which grows faster than the body. Those that decrease relatively in size (b is less than 1), have negative allometry. After birth the human head grows at a rate slower than that of the body and is therefore an example of both positive and negative allometry at different times of its development.

Allometry has been used to explain features that are not only nonadaptive but apparently harmful. The most famous example of this supposed rela-tionship is the extinct Irish Elk (more properly, a deer) whose antlers are found in abundance in Pleistocene peat bogs in northern Europe (Fig. 14–5). The antlers of this deer reach spreads of 3.5 meters and weights of 40 kg. They are so large that many students of evolution have thought that they must have impeded the mobility of the animal and drained its energy as they were formed each year only to be discarded after the mating season. In the deer family (Cervidae) antler length (A) is allometrically related to the overall size expressed as the height at the shoulder (H) by the equation:

$$A = 0.0754H^{1.63}$$

In the skulls of fossil Irish Elks the relationship between the antler's length (A) and the basal length

Figure 14-5   Reconstruction of the extinct Irish Elk *Megaloceras*. (Source: J. G. Millais, British Deer and Their Horns, 1897.)

of the skull (*L*, used as a measure of overall size) is shown by the equation:

$$A = 0.00025L^{2.115}.$$

Both in deer in general and in the Irish Elk small increases in overall size led to large increases in the size of the antlers. Possibly increase in overall size was so advantageous to the elk that the harmful effect of the large antlers was tolerated or possibly the antlers were even, at their largest, adaptive in the competition for mates (8).

THEORETICAL MORPHOLOGY

Paradigm Method:

The term *paradigm* means a model, a pattern or example in general usage. In science, it has come to mean a way of thinking about, or solving, a problem. The paradigm method can be applied to problems of functional morphology through proposing a model for the function of a structure in question. For example, the zigzag folding of some brachiopod shells, particularly rhynchonellids, could be postulated to have evolved to strengthen the shell (Fig. 6–53). A paradigm or model is then defined which is the structure that would be capable of fulfilling the supposed function with the maximum efficiency under limitations imposed by the nature of the materials. In this case the paradigm could be a surface repeatedly folded, or corrugated, along axes perpendicular to the line along which it might break or deform; a surface like a corrugated sheet of iron. In brachiopods, because the stress is along the roughly semicircular open margin where the valves are pressed together, the corrugations would ideally be in the form of a fan. The fossil shell is then compared with the model and the extent of their similarity determines the degree of confidence that the true function of the feature has been discovered. Because the radially folded brachiopod shell does closely resemble a sheet corrugated to increase its resistance to bending, the conclusion that the shell evolved this shape to resist breaking or bending is logical.

The example can be used to illustrate some difficulties with the method. The zigzag shape of the aperture has also been interpreted as adapted either for directing feeding and respiratory currents to the lophophore within, or to preventing sand grains from entering with the incoming current. It is effective in keeping out sand because the brachiopod does not need to open such an aperture as widely as a straight one. A zigzag aperture lets in the volume of water required by the animal at a smaller gape than a straight one because it is longer. The folding of the brachiopod shell could be an adaptation for one or all of these three requirements. Most parts have more than one use and their form is not specifically engineered for any one function.

The paradigm approach can also be applied to exploring the function of astrorhizal systems of Paleozoic stromatoporoids. Stromatoporoids commonly have on successive growth surfaces radially arranged canals that lead to a central opening (Fig. 6–19). Different paleontologists have proposed that these canal systems, called astrorhizae, housed the exhalent canals of an encrusting sponge, canals for diffusion of nutrients between polyps, filamentous juveniles of cyanobacteria, and parasitic animals. Their interpretations were based on the different phyla to which each assigned the stromatoporoids. The canals join and increase in size toward the center. Michael LaBarbera (9) has pointed out that if water was gathered from two smaller canals into a larger one and the flow was laminar in nature, the laws of fluid flow indicate that the sum of the cubes of the radii of the tributary canals should equal the cube of the radius of the canal to which they lead (that is, $R^3 = r_1^3 + r_2^3$). This ideal relationship or paradigm can be tested in stromatoporoids in which the canals are empty and can be filled with plastic to make internal molds, released from the skeleton by dissolving it with acid, and measured. The average difference between the expected and the observed values derived from the equation was 13 percent. The closeness of this value to the one predicted by the flow equation suggests that the tubes were capable of carrying a current of water. The geometry of the canal system would therefore allow its use as an exhalent system of an encrusting sponge and the similarity of its form to the canals of living sponges suggests that this is how it was used (Fig. 6–20).

Computer Simulation of Fossil Form:

The growth of some organisms and their shells is simple and regular enough to be expressed by a mathematical equation. For such organisms variations in morphology can be expressed as changes in the parameters of the growth equation. The growth formula can be programmed for a computer and the machine can be asked to draw the range of forms that result from changes in the parameters. Such studies help paleontologists to understand the factors that determine form and the relation between animals of different forms. The method works best for invertebrates of simple skeletons; the growth of complex skeletons and shells cannot be reduced to equations.

Many simple invertebrates, such as bivalves, gastropods, cephalopods, brachiopods, and foraminiferans, secrete a shell in the form of a logarithmic spiral. Although the correspondence of the gross form of their shells to the spiral was recognized early in the last century, David Raup was the first to use a computer to illustrate the range of forms

that can be generated by varying the parameters (10). The logarithmic spiral growth pattern provides for growth of the shell without change in its shape.

The basic form of all these shells is a coiled cone growing and expanding at its aperture. The various shapes of cones in cross section are not easily expressed mathematically. However, variations in gross shell form can be studied by making the cone circular in section. If this is done, the shell form can be described in terms of three parameters: (1) the whorl expansion rate (W), which expresses the amount that the tube expands during one turn (whorl) of the spiral (Fig. 14–6); (2) the distance from the coiling axis of the tube (D); 3. the translation rate down the axis of coiling in one whorl (T). Shells in which the translation rate is high will have high spires, like many gastropods, and those in which it is zero will be planispiral, like most ammonites. Shells in which the distance from the coiling axis D is large will be loosely coiled with central cavities, and those in which it is low will have overlapping whorls that cover older parts of the shell. Variations in the three parameters can be plotted on the three dimensions of a cube (Fig. 14–7). The plot shows that not all possible shell variations have

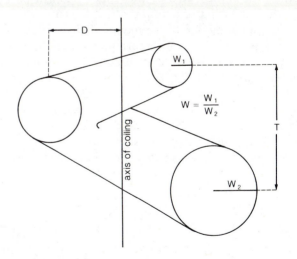

Figure 14-6 Parameters of the geometry of spiral shells. W = whorl expansion rate, the ratio between diameters of the expanding cone at the beginning (W_1) and end of one whorl (W_2). T = translation rate, distance along the axis of coiling travelled in one whorl. D = distance of the center of the cone from the axis of coiling. (*Source:* Modified from D. M. Raup, *Journal of Paleontology*, v. 40, pp. 1178–1190.)

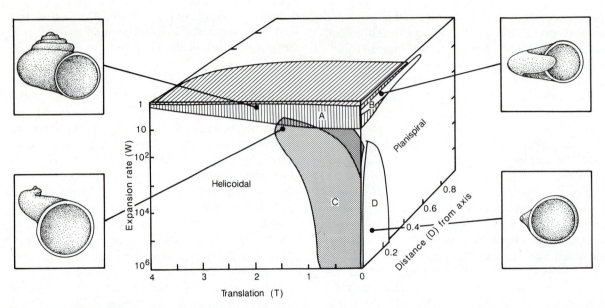

Figure 14-7   Block diagram defined by the three parameters of a spiral shell showing the areas occupied by gastropods (A), coiled cephalopods (B), bivalves (C), and brachiopods (D). The form of the shell represented by the four points marked by dots is shown in the four boxes around the block. Helicoidal shells plot on the front face of the block and planispiral shells on the side face. (*Source:* Modified from D. M. Raup, *Journal of Paleontology*, v. 40, pp. 1178–1190.)

evolved, for parts of the solid are not represented by known shells, and also that certain forms are favored by certain animal groups. Most gastropod shells are represented by the upper forward part of the cube where expansion rate is low and translation rate is moderate. Ammonite shells occupy an area on the right face where translation rate is zero, expansion rate is low, and distance from the axis (*D*) is variable. If *D* is low, the adult whorls completely overlap earlier ones as in *Nautilus* (Fig. 7–3); if it is high, all the inner whorls can be seen in adult shells (Fig. 10–2). Bivalves and brachiopods have shells with such high expansion rates that they are bowl-shaped. Shells of this form have such wide

apertures that they are of little use to the animal for protection unless the aperture is closed either by another similar shell, as in bivalves and brachiopods, or by clamping it against a rock surface as in limpet gastropods. In all bivalved shells the expansion rate for each valve must be more than 10 and additional growth must take place on the side of the aperture near the apex if the shell is to open without the apexes interfering with each other (Fig. 14–8).

Richard Fortey and Adrian Bell (11) have simulated graptolite colonies by a computer program. They have shown that the form and branching pattern of colonies depended on the even placing of

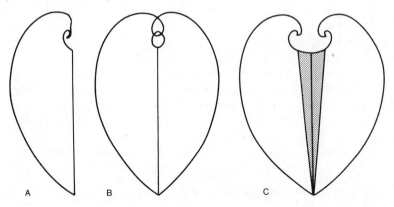

Figure 14-8   Cross sections of hypothetical bivalve showing the necessity of interumbonal growth if the valves are of equal size. A. A single valve with whorl expansion rate of 10. B. If two similar valves are placed together their umbos overlap, an impossible relationship. C. By growth of the shaded part of the valves, a functional bivalve shell can be formed. (*Source:* Modified from D. M. Raup, *Journal of Paleontology*, v. 40, pp. 1178–1190.)

individuals on a horizontal plane so that they could efficiently harvest planton.

COMPARISON WITH LIVING ORGANISMS

The adaptations of fossil organisms may be understood by comparisons with living relatives whose habitat and behavior are known. The most reliable inferences can be drawn from structures in related animals that are derived from the same organ. Such structures are said to be homologous. The septa of ammonites and living *Nautilus* are homologous structures. Structures that serve the same function but are derived from different body parts are said to be analogous. The panda's thumb is analogous to the thumbs of primates.

Functional Morphology of Ammonoid Cephalopods:

Extensive studies have been made of *Nautilus* to help paleontologists understand the adaptations of the extinct nautiloids and ammonoids. *Nautilus* secretes a coiled shell divided into gas-filled chambers by septa (Chapter 7). The animal lives in the last, or living, chamber and communicates with all the others through a thread of tissue bearing arteries and veins that passes through each one in a tube called a siphuncle. Before a septum is secreted the

body moves forward, water enters to fill the space between the body and the last septum, and the back of the body secretes a new septum, partitioning off a new chamber filled with water (Fig. 14–9). This water is then slowly pumped out of the chamber through the siphuncle using an intracellular mechanism based on osmosis. The pumping action may take weeks and during this time the shell buoyancy slowly increases. Displacement of the water from the chambers compensates for the growth of body and shell weight, maintaining the whole animal at near neutral buoyancy. Some of the water is out of contact with the siphuncle and may stay in the corners of the chamber for some time. The chambers are filled with gas at less than 1 atmosphere pressure by diffusion from the siphuncle. When the animal reaches the adult stage after several years of growth, new chambers are no longer added.

In addition to compensating in juvenile stages by long-term changes in buoyancy for the growth of the animal, the shell of *Nautilus* allows the animal to migrate vertically in the water in youth and adulthood. *Nautilus* lives at 200 and 300 meters during the day and rises to about 150 meters at night to feed on the coral reefs of the South Pacific. Its movements have been directly observed and individuals tagged with electronic devices have been traced by radio. In this depth range the ambient pressure changes by about 15 atmospheres. A flexible cavity with gas at 1 atmosphere pressure would be collapsed as the animal descended by an increase

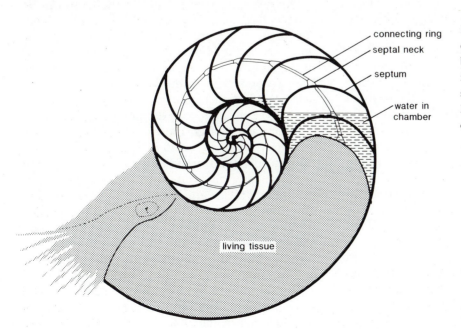

connecting ring
septal neck
septum
water in chamber
living tissue

Figure 14-9 Juvenile *Nautilus* in cross section, showing the last chamber filled with water and recently formed chambers still containing some water. (*Source:* Modified from E. J. Denton, *Proceedings*, Royal Society, London, v. B185, p. 288.)

in pressure of this magnitude, but the *Nautilus* shell is a rigid pressure vessel and, until the seal is broken, will maintain constant buoyancy. The depth at which the animal can live over a prolonged period is limited by the ability of the permeable siphuncle to pump out the chambers against the ambient hydrostatic pressure as they form and, after they are all formed, to prevent seawater from forcing its way into the chambers against the osmotic pump. This limit is at about 300 meters where the pressure is approximately 30 atmospheres.

The depth to which the animal can penetrate for short periods is theoretically limited by the strength of its shell to resist hydrostatic pressure. Animals forced to depths of over 750 m in cages implode and die. Although failure of the shell might be expected at the siphuncle, it is as resistant to short-term stress as the rest of the shell and failure in pressure tank experiments commonly occurs in the septa, in the walls, and the body chamber. Gerd Westermann estimated the strength of the septa of fossil cephalopods on the basis of simple measurements of thickness and curvature and deduced that Paleozoic cephalopods with wide apical angles and septa of low curvature would have imploded at depths of 250 meters. Long shells with lower apical angles and thicker, more widely spaced septa could have lived at greater depth (12, 13).

We can apply lessons learned from observing *Nautilus* and experimenting with its shell to its extinct relatives. Several features of ammonite shells appear to be adaptations to living in an environment of changing pressures as the animal rose and fell in the sea. The intricately folded sutures of ammonites must have strengthened the line of potential failure between septum and wall and also buttressed the wall itself against outside pressure by extending the line of contact with the septum. Shells strengthened in this way would have allowed the animal to live over a much larger depth range than animals with weaker shells. For a given strength and shape the shell could be lighter and therefore more easily moved and easier to secrete. The cross section of the whorl can also be interpreted in terms of pressure resistance. Cross sections that approach a segment of a circle would have been more effective in resisting pressure changes than those with flat sides because a sphere is the most efficient pressure vessel. Ammonites with flat or complexly folded sides inhabited shallow waters and those with smooth surfaces and rounded whorl cross sections were capable of diving to greater depths like *Nautilus*.

THE METABOLISM OF DINOSAURS

The function of the ammonite shell can be reconstructed with some degree of assurance on the basis of its living relative *Nautilus*. The way of life of extinct organisms without living counterparts is more difficult to reconstruct. For example, how can we discover the biological factors that explain how the dinosaurs achieved and maintained dominance in most major terrestrial environment for a period of more than 100 million years?

When dinosaurs were first recognized in the early nineteenth century, they were compared with modern reptiles such as lizards and crocodiles. They do share a common ancestry with crocodiles, but does the crocodile provide an accurate basis for establishing the nature of this extremely diverse assemblage of Mesozoic reptiles?

Anatomists have long recognized that all dinosaurs clearly differed from modern reptiles in having a relatively erect posture of the limbs. Whether they were bipedal or quadrupedal, the joint structures show that the body was habitually held off the ground. In contrast, nearly all modern reptiles have a sprawling posture and are active for only short periods of time. Crocodiles and some lizards can draw their limbs under the body and move them in a fore and aft direction, but most of the time the body rests on the ground. Such a posture would be impossible for dinosaurs. The joints are oriented so that the limbs must be held nearly vertically.

The nearly vertical posture of the limbs can be explained in many dinosaurs by their great bulk. Neither bone nor muscle is strong enough to support and lift the body if their limbs were sprawled to the side. Even the elephant, which is small by comparison with many dinosaurs, must keep its legs nearly stiff when running. If the great weight of the large dinosaurs provide the primary explanation for their upright posture, one might expect the small dinosaurs would have a sprawling stance like crocodiles and large living lizards. In fact, some of the smallest dinosaurs, and some of the earliest members of the group, have a skeletal anatomy that can only be interpreted as indicating an upright posture, and many early genera must have been obligatorily bipedal, resembling large terrestrial birds such as the ostrich much more closely than any living lizards.

The skeleton of most dinosaurs suggests a level of locomotor activity approaching more closely that of large modern mammals than that of any modern reptiles. One of the most fundamental differences between modern reptiles and mammals is in their capacity for sustained activity. No reptile can run actively for more than a few minutes, after which it must rest for a long period of time before becoming active again. Many mammals, in contrast, can remain active for many hours without fatigue. The differences in their capacity for continual activity lies in the metabolism of the muscles. In reptiles, most of the energy for contraction of the muscles results from the fermentative breakdown of glycogen, giving rise to lactic acid. Lactic acid is a toxic material that must be removed from the muscles. The slow breakdown of lactic acid requires long periods of rest between bouts of activity. Among mammals, most of the energy for muscle contraction is achieved by oxidative metabolism in which the only waste products are water and carbon dioxide which can be rapidly removed by the blood. Moderate activity may be maintained for hours without fatigue.

The capacity for continuous muscle activity explains the selective advantage for an upright posture in even the smallest dinosaurs. Most paleontologists now assume that dinosaurs must have relied primarily on oxidative metabolism for muscle contraction, and had achieved a roughly mammalian level of activity.

Other aspects of dinosaur metabolism remain more controversial. Modern mammals have approximately ten times the metabolic rate of reptiles of comparable body size (Fig. 14–10). Did dinosaurs have a nearly mammalian metabolic rate, or was it considerably lower than that of mammals and yet significantly above that of typical reptiles?

Robert Bakker and others (14) have suggested that dinosaurs may have had a truly mammalian metabolic level, but there is some evidence to the contrary. Maintenance of a high metabolic rate requires a great deal more food than is necessary for modern reptiles and a more effective feeding strategy, a well as more effective integration of sensory input and motor coordination. In birds and mammals this is associated with a significantly larger brain size than that of modern reptiles. James Hopson (15) has demonstrated that some dinosaurs had a brain size (as measured relative to body weight) approaching that of mammals, but most did not, and the average brain to body weight ratio of dinosaurs may not have been much different from that of their nearest living relatives, the crocodile. Though the metabolism of dinosaurs was almost

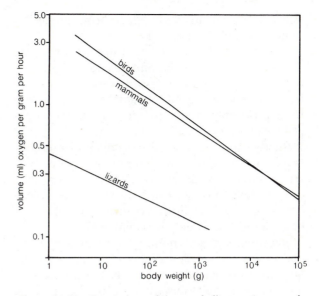

Figure 14-10   Comparison of the metabolic rates (expressed as oxygen consumption) of mammals, birds, and lizards. Note that the scales are logarithmic. For the same body weights, the reptiles have a rate of consumption of oxygen about one tenth of that of birds and mammals. (*Source*: A. F. Bennett & W. R. Dawson, *Metabolism*. Academic Press 1976. Courtesy of A. F. Bennett and W. R. Dawson, *Biology of Reptiles*, v. 5, p. 158, Academic Press.)

certainly different from that of living turtles, lizards, and crocodiles, it should not be considered exactly like that of modern mammals or birds.

A very important fact that should be considered in evaluating the metabolism of dinosaurs is the almost complete absence of small dinosaurs. In small mammals, with weights up to a few kilograms, much of the metabolic energy goes toward maintaining a high, constant body temperature. As was pointed out in Chapter 11, the smallest mammals have an exceedingly high metabolic rate because of the problem of heat loss. Dinosaurs are exceptional compared to all other major groups of vertebrates in the great body size of most species. Only two or three species are known to have had an adult size of less than about 10 kg. In contrast, the vast majority of both birds and mammals are of small size.

Adult dinosaurs would not have had to maintain the extremely high metabolic rate of most birds and small mammals in order to have a high, constant body temperature. In fact, J. R. Spotila (15) has demonstrated that many dinosaurs could have maintained a high, constant temperature without any metabolic expenditure, as long as the ambient temperature remained as high and constant as it is today over large areas of the tropics.

Modern reptiles are active primarily during the day, rather than at night, and this was presumably the case in the Mesozoic, so that dinosaurs could have taken advantage of the radiant heat of the sun to help maintain a high temperature for most periods when they needed to be active.

Reptiles the size of adult crocodiles have no difficulty maintaining a constant, high body temperature. They may, in fact, be tied to a semiaquatic way of life because the water provides an easy way to lose excess heat to the environment. By their bulk alone, most dinosaurs could have maintained a body temperature comparable to that of mammals. Only the smallest dinosaurs would have required a high metabolic rate for this purpose. Maintenance of a high constant body temperature by being large is, however, highly dependent on the relative constancy of the external environment.

There is much evidence from the distribution of other animals and plants that the climate during the Mesozoic was far different from that in the latter half of the Cenozoic. The temperature was apparently well above freezing for all of the year as far north as the Arctic Circle. The low level of the continents and the vast extent of the inland seas would have provided a maritime climate over much of the globe. The major land masses were close together, making migration easy, if seasonality were sufficiently severe in any one area to interfere with maintenance of a regular body temperature. Under such conditions, lasting scores of millions of years, there would be little selective advantage for evolving a high metabolic rate that would make it possible for a large reptile to warm up rapidly or maintain a high temperature in an occasionally cold environment. As long as the climate remained relatively stable, maintenance of a high, constant body temperature through large size alone would have been adequate for the vast majority of dinosaurs during the Mesozoic. They would, however, have been extremely vulnerable to any long-term and widespread decrease in average temperatures, and especially to rapid temperature increase or decrease. Many reptile groups that were prominent in the Mesozoic did not survive into the Cenozoic, but there appears to have been a relatively gradual extinction of the dinosaurs toward the end of the Mesozoic rather than a sudden catastrophic extinction (see Chapter 16). This accords with the current

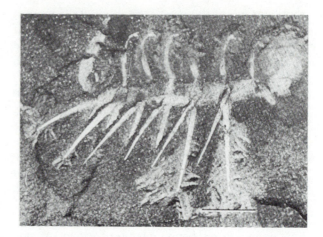

Figure 14-11   The enigmatic fossil *Halucigenia sparsa* from the Burgess Shale. The scale bar is 2 mm. The head is presumed to be the blob on the right and the pairs of spines are thought to have acted as legs. The function of the tentacles along the back is problematic. (Courtesy of S. Conway Morris and Blackwell Scientific Publications, *Geology Today*, p. 91, May–June 1987.)

understanding of dinosaur metabolism, which would have made them more susceptible to climatic change than either mammals or other reptile groups.

Enigmatic Extinct Organisms: *Halucigenia*:

Some fossil animals are so unlike any living organism that their interpretation by the principles of functional morphology is difficult. The small fossil *Halucigenia* from the Burgess Shale is such an animal (Fig. 14–11). At one end is a ball-like structure that has been thought to be the head; at the other the pipelike body is turned into a structure like a chimney. The body has seven pairs of spines that are reconstructed in the figure as legs. Along the opposite side of the body are seven tentacles which split into two parts at their tips. Behind them are a set of six smaller appendages. No mouth or anus has been identified, and both the front to back and top to bottom orientation may be incorrectly represented in Figure 14–11. The organism has been reconstructed as a bottom dwelling scavenger that walked on its spines but, in the absence of similar living organisms, or similar structures in living organisms, these reconstructions are as insubstantial as hallucinations.

REFERENCES

1. Gould, S. J., and Lewontin, R. C., 1979, The spandrels of San Marco and the Panglossian paradigm: a critique of adaptationist programme: Proceedings Royal Society of London, Series B, v. 205, p. 581–598.

   Lewontin, R. C., 1984, Adaptation. In Sober, E. (ed) Conceptual Issues in Evolutionary Biology: Cambridge, Mass, MIT Press, p. 235–251.

2. Darwin, C., 1859, The origin of species. Chap. 3, Natural Selection: London, John Murray.

3. Van Valen, L., 1973, A new evolutionary law: Evolutionary Theory, v. 1, p. 1–30.

4. Krimbas, C. B., 1984, On adaptation, neo-Darwinian tautology and population fitness: Evolutionary Biology, v. 17, p. 1–57.

5. Carroll, R. L., 1986, Physical and biological constraints on the pattern of vertebrate evolution: Geoscience Canada, v. 13, p. 85–90.

6. Thomas, R. D. K., 1979, Morphology, Constructional. In Fairbridge, R., and Jablonski, D. (eds.), Encyclopedia of Paleontology: Stroudsburg, Pa., Dowden, Hutchison and Ross, p. 482–487.

7. Gould, S. J., 1980, The Panda's Thumb: New York, Norton.

8. Gould, S. J., 1974, The origin and function of "bizarre" structures: antler size and skull size in the "Irish Elk": Evolution, v. 28, p. 191–220.

9. LaBarbera, M. (personal communication)

10. Raup, D. M., and Stanley, S. M., 1978, Principles of Paleontology, 2nd ed: San Francisco, Freeman.

11. Fortey, R. A., and Bell, A., 1987, Branching geometry and function of multiramous graptoloids: Paleobiology, v. 13, p. 1–19.

12. Westermann, G. E. G., 1982, The connecting rings of *Nautilus* and Mesozoic ammonoids: implications for ammonoid bathymetry: Lethaia, v. 15, p. 373–384.

13. Chamberlain, J. A., and Chamberlain, R. B., 1985, Septal fracture in *Nautilus*, implications for cephalopod bathymetry: Lethaia, v. 18, p. 261–270.

14. Bakker, R. T., 1986, The dinosaur heresies: New York, Morrow, 481 p.

15. Thomas, R. D. K., and Elson, E. E. (eds.), 1980, A cold look at the warm-blooded dinosaurs: American Association for the Advancement of Science Symposium 28.

SUGGESTED READINGS

Mayr, E., 1983, How to carry out the adaptationist program: American Naturalist, v. 121, p. 324–334.

Paul, C. R. C., 1975, An appraisal of the paradigm method: Lethaia, v. 8, p. 15–21.

Savazzi, E., 1987, Geometric and functional constraints on bivalve shell morphology: Lethaia, v. 20, p. 293–306.

Vermeij, G. J., 1987, Evolution and Escalations: an Ecological History of Life. Princeton, N. J., Princeton Univ. Press, 527 p.

Chapter 15

The Mechanisms of Evolution

INTRODUCTION

The term *evolution* is used to refer to the processes by which life has changed through the geological ages. In Chapter 1 it was described as the chain that unites all organisms through the history of life. Implicit in the term evolution is the idea that life has progressed—that, in some way that is difficult to define, life is better adapted and more efficient now than it was at its beginning. That life of the present is more complex than that of the Archean would be difficult to deny, but that modern organisms are better adapted, or more fitted, to their environment than were the bacteria of the Archean would be difficult to demonstrate. Stephen Gould has claimed that substantial progress in establishing new and more complex life forms ended early in the Paleozoic and since then evolution has consisted mostly of fine adjustments in the adaptation of these basic architectures. Viewed from a long perspective, life has become more complex, has colonized a greater variety of environments, has divided the earth's resources more finely, and in exploiting these smaller parcels has evolved more specializations. Whether these trends are considered progressive depends on our understanding of the meaning of that word.

Evolutionary theory today is in a state of rapid flux. Wholesale reexamination of the basis of Darwin's mechanism of evolutionary change is taking place. The conflicts inherent in this reformulation of evolutionary theory have been exploited by conservative religious groups in a new attack on the whole concept of evolution. Such attacks extend from the models that we use to explain the paleontological record to the reality of the record itself. In discussing the validity of evolution, a clear distinction must be made between the record of life preserved in rocks and the models formulated by scientists to explain that record. Scientists disagree on the details of the mechanisms, but the record of life described in the first 12 chapters of this book is verifiable by a diligent observer and documented by nearly 200 years of scientific investigation. That the large-scale continuity of this record is the result of descent of one life-form from another is a conclusion that cannot be avoided except by logical contortions. Yet pressure of religious groups on the educational system has meant that adequate discussion of the history of life and mechanisms of evolution has been excluded from many school biology and natural history textbooks for many years. Disagreement on evolutionary theory should not be interpreted as indicative of doubts within the biological and paleontological community as to the central truths of evolution: that all life is connected by ancestor–descendant relationships and that sorting of the variations inherent in the reproductive process has resulted in its present diversity. These disagreements are part of any science that is moving rapidly to new knowledge of nature.

DARWINISM AND THE MODERN SYNTHESIS

Although Darwin's book was called *The Origin of Species*, it is not about how species are produced but rather about the origin of adaptations or how species change. Darwin apparently believed that the two were so closely linked that explanations of adaptation were sufficient to explain the origin of species. From the evidence that organisms produce more offspring than survive, and individuals carry different heritable variations, he deduced that in the inevitable conflict for resources, living space, and mates, those with the variations that fit best with the environment would preferentially survive. The individual that survived would reproduce in its descendants the variation that made it better adapted. The process of selection is measured by differential reproductive success. Darwin distinguished two types of selection: natural and sexual. Natural selection depended on the fitness of the organism to its environment; sexual selection on its attractiveness to mates. He recognized that the latter, as in the case of brightly colored birds mentioned in the last chapter, could work against the former.

Darwin postulated that the variations favorable to an organism would accumulate by small increments, generation by generation, until an organism had evolved that was sufficiently different from its ancestors to justify its recognition as a new species. He hoped that the paleontological record, then just being brought to light, would substantiate the continuity of species in time through transitional series of forms. What was then known about the record did not support this hypothesis, and he attributed the failure to the imperfection of the fossil record. The paleontologist Thomas Huxley, Darwin's most vocal supporter, was active in the search of the record for the corroborating evidence of transitional sequences and in particular fixed on *Archaeopteryx*

as an example of the transitional forms that paleontologists should expect to find. During the late nineteenth and early twentieth centuries the discovery of transitional series in the fossil record to demonstrate Darwin's concept of evolution was a major goal of paleontologists, but few were documented and some of these, as discussed below, have subsequently failed to survive closer examination.

By the 1940s, geneticists had discovered how variations are inherited, and population biologists had modelled the spread of variations through populations. Julian Huxley integrated this new knowledge in a landmark book, *Evolution, The Modern Synthesis*. The model of evolution assembled in this book became known as the modern synthesis, or neodarwinism, and guided evolutionary thought for the next 30 years. The modern synthesis incorporated the following ideas:

1. Variation arises from random changes in the genes and chromosomes.
2. Selection favors the survival and perpetration of particular variants at the expense of others.
3. Changes within evolving lineages result from the gradual accumulation of favorable heritable variations.
4. Interbreeding of individuals within a species maintains a gene pool that gives homogeneity and stability to the population. The morphological and genetic discontinuities that separate contemporaneous species originate by interruption of this gene flow through isolation of small subpopulations and their adaptation to local conditions.

The modern synthesis included the species concept clearly formulated by Ernst Mayr and discussed in Chapter 3—that species are groups of interbreeding populations occupying a territory and genetically isolated from neighboring groups. New species are thought to originate when the gene flow is interrupted by physical or other barriers to interbreeding and local populations diverge genetically and morphologically from the parental group to adapt to local conditions. Most supporters of the modern synthesis perceived evolution as progressing gradually and continuously as favorable mutations spread through large populations at rates depending on their selective advantage. Such change has been called microevolution.

THE SEARCH FOR MICROEVOLUTION IN THE FOSSIL RECORD

Paleontologists in the late nineteenth and early twentieth centuries assumed that the imperfections of the fossil record that had prevented Darwin and his contemporaries from proving the reality of microevolution by means of paleontology would be eliminated by more careful study and the discovery of successions of sedimentary beds that represent continuous deposition. Various successions have been intensively studied and restudied because they appear to show such gradual change.

The Coiling of *Gryphaea*:

A. E. Trueman believed that he could demonstrate slow, progressive, adaptive changes through several species in a lineage of the oyster *Gryphaea* in the Early Jurassic rocks of England (Fig. 10–15B). The major changes involved the coiling of the large left valve until it had reached one and a half turns, the reduction of the area of attachment of the larger valve, the thickening of this valve, its increase in size, the development of a groove in the large valve, and the restriction of the coiling of the valve to its apical portion.

Detailed studies (1) of four species of *Gryphaea* from the Lower Jurassic show that size has increased irregularly (Fig. 15–1). Counts of the growth lines on these bivalves indicate that the increase in size is related to the greater age of the younger species—that is, the specimens higher in the succession are larger because they lived longer. They are more coiled than their ancestors because coiling is correlated by an allometric function with growth. Trueman's conclusion that increased coiling of the valves was an adaptive trend driven by selection can be shown to be incorrect because no change in the basic geometry of the shell took place; they only lived longer. In younger Jurassic rocks decrease in coiling took place (see following discussion). Although a general increase in size is apparent through four species, within individual species the trend is not gradual but episodic as shown in Figure 15–1.

Evolution of the Ammonite *Kosmoceras*:

Ronald Brinckmann collected about 3000 specimens of compressed ammonites from a section of Upper Jurassic Oxford clay in England. Through extensive measurements of seven characters he attempted to establish gradual trends. The 14-meter thick section is thought to represent about 1 million years of evolution. Brinckmann divided his ammonites into four genera each containing several species, but later paleontologists have suggested that only two genera are present and each is dimorphic or that only one genus is represented. The dimorphism explains Brinckmann's observation that parallel adaptive trends occur in the evolution of what he took to be four different genera (Fig. 15–2). He inferred that the parallelism in the trends in different genera was evidence that those trends were adaptations to a changing environment. Although changes occur in the section, these changes

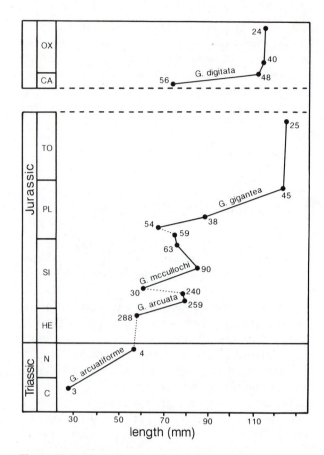

Figure 15-1 Maximum size of the left valve in European species of *Gryphaea* in the Late Triassic and Jurassic. The points represent the means of the length for the number of specimens indicated at each point. Note the general but not regular increase in size with time through the five species. The stages that divide the systems are abbreviated as follows: Carnian (C), Norian (N), Hettangian (HE), Sinemurian (SI), Pleinsbachian (PL), Toarcian (TO), Callovian (CA), Oxfordian (OX). (*Source:* Modified from Anthony Hallam.)

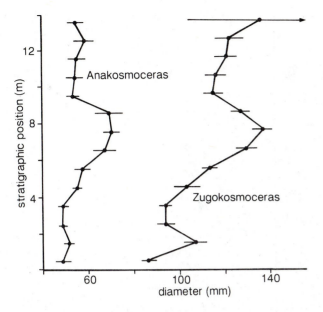

Figure 15-2   Changes in shell diameter of two ammonite genera from the Oxford Clay studied by Brinckmann. The points represent the means for specimens found in 1 m of section and the bars are four standard errors long. The parallelism of the trends suggests that the genera are dimorphic forms of a single taxon. (*Source:* Modified from D. M. Raup and R. Crick, *Paleobiology*, v. 7, pp. 90–100.)

are not obviously directional. As an example (Fig. 15–2), the changes in diameter in two lineages appear to be almost at random. Statistical tests by David Raup and Rex Crick (2) do not eliminate the possibility that the changes in characters in time measured by Brinckmann are random. However, certain segments of the tends can be interpreted as being governed by directional selection, but the direction of this selection must have changed frequently in the course of the evolution of the lineage.

Brinckmann recognized 24 breaks in the succession by sharp boundaries between sets of beds and by the concentration of ammonites on bedding planes (Fig. 15–3). He assumed that these were pauses in deposition and attempted to determine, assuming a constant rate of evolution and deposition, how many centimeters of beds were missing at these breaks (Fig. 15–3). Raup and Crick, by treating his data statistically, could identify 40 horizons in the section at which rapid changes in morphology take place, but only six of these occur at horizons identified by Brinckmann as stratigraphic breaks. They concluded that the rates of evolutionary change vary in the section but the times when the change was

rapid do not necessarily correspond to the presumed stratigraphic breaks and that the magnitude of the breaks cannot be estimated from morphologic trends. These later studies have suggested, but not proved without doubt, that Brinckmann's ammonites do show some microevolutionary trends.

Permian Foraminiferans:

Lepidolina is a fusuline foraminiferan from the Pacific Region. Tomowo Ozawa studied nine characters in the evolution of the species *L. multiseptata*. Some of these characters show little evidence of gradualism but four do show evidence, and the most impressive changes are in the size of the initial chamber (Fig. 15–4). In an interval estimated to be about 20 million years the mean diameter of the chamber changes from 200 to 550 micrometers. This study is based on a large group of samples collected over a wide geographic area. The increase in size of the chamber is believed to be a response to a change of habitat from shallow carbonate to deeper clastic-rich environments.

Ordovician Bivalve *Nuculites*:

Sara and Peter Bretsky (3) studied variation in the bivalve *Nuculites* in Ordovician rocks through a section about 800 m thick in Quebec. The range of variation in their specimens is sufficient for the recognition of four different species of this genus, but they found that the characteristics were gradational through the section and referred them all to one species. Their plot (Fig. 15–5) of the proportion of these shells expressed as the length–height ratio, shows (with an initial reversal) the general, slow change to shells of greater elongation.

Microfossils in Deep-sea Cores:

Sedimentary successions deposited on the borders of continents are often interrupted by gaps produced by erosional episodes when the sea retreated and the site of deposition emerged. The deep parts of the ocean basins are not affected by instability of the margins, and continuous sections that could preserve the record of gradual evolution are more likely to be found there. Only recently have such sections become available as cores from wells drilled in the Deep Sea Drilling Project. Unfortunately, complex life forms are not abundant in the deep

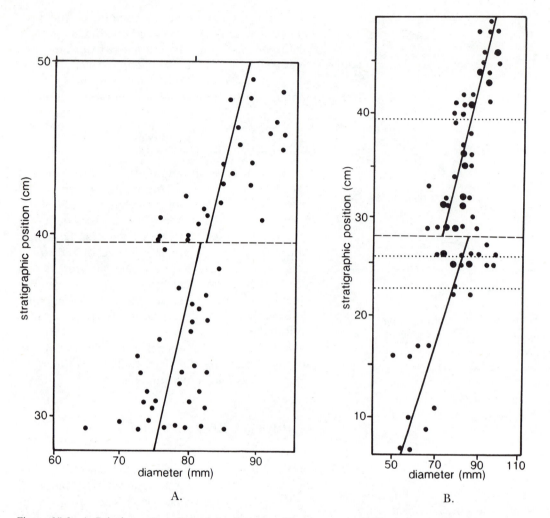

A.   B.

Figure 15-3   A. Brinckmann's analysis of the diameter of *Zugokosmoceras* shells across a horizon on which ammonites are concentrated at 39.5 cm in his section. Each point represents a specimen. The trend lines calculated from the data points on each side of the presumed break show a slight displacement of the trend that Brinckmann used to assess the thickness of sediment apparently missing. B. Raup and Crick's replotting of Brinckmann's data for a longer stratigraphic interval. The plot shows a statistically significant displacement of the trend to increase in size at about 28 cm, but no displacement of the trend at 39.5 cm or at other horizons (shown by dashed lines) where Brinckmann noted stratigraphic breaks. The larger dot is used where two or more specimens have the same coordinates. (*Source:* Modified from D. M. Raup and R. Crick, *Paleobiology,* v. 7, pp. 90–100.)

sea and the major fossils of these cores are of simple planktonic organisms raining down from overlying waters.

Davida Kellogg (4) studied the variation in the radiolarian *Pseudocubus vema* in a deep-sea core representing approximately 2 million years of deposition. Figure 15–6 shows the increase by 50 percent of the width of the thorax through this interval. Although the rate of change of the character is not constant, but stepped, the data can be interpreted as a record of progressive and slow change within

a species as it evolves. These radiolarians illustrate some of the problems of recognizing gradualism in the fossil record. Stephen Gould and Niles Eldredge (5) have objected that the stepped pattern of the trend shows that evolution was episodic rather than gradual. They also pointed out that the changes could have been caused by migration into the area of radiolarians with wider thoraxes rather than by evolution in place. The possibility of migration might be eliminated or established by regional studies of the distribution of the species. However, even the

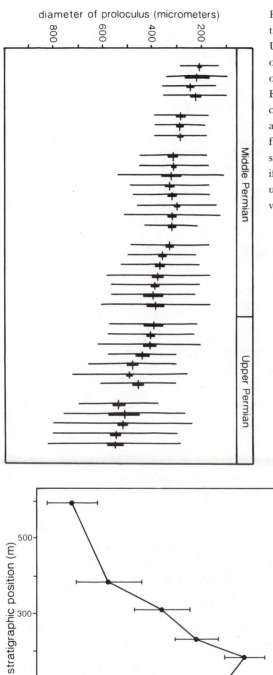

diameter of proloculus (micrometers)

Middle Permian

Upper Permian

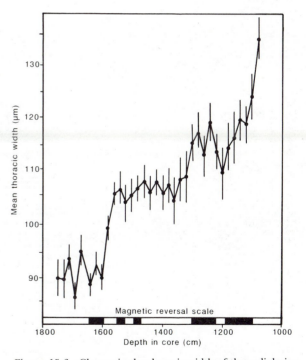

Figure 15-4   Increase in size through the Middle Permian and Upper Permian of the diameter of the initial chamber (proloculus) of the foraminiferan *Lepidolina* in East Asia. Mean values are indicated by vertical lines, black bars are 95 percent confidence limits, finer horizontal lines span two standard deviations. (*Source:* Modified from T. Ozawa, *Memoir*, Faculty Science, Kyushu University, v. D253, pp. 117–164.)

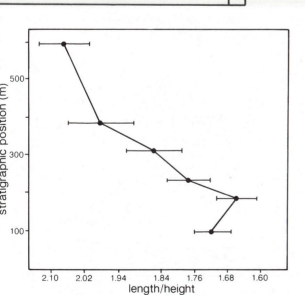

Figure 15-5   Increase in the length-height ratio of the Ordovician bivalve *Nuculites planulatus*. The means and 95 percent confidence intervals are plotted for six collections. (*Source:* Modified from P. & S. Bretsky, *Journal of Paleontology*, v. 51, p. 266.)

Figure 15-6   Change in the thoracic width of the radiolarian *Pseudocubus vema* from a core through Pleistocene and Pliocene beds. The magnetic reversal scale indicates that this deepsea core represents deposition between about 5 and 2.5 million years ago (dark is reversed). The dots are the means of many measurements and the vertical bars are 95 percent confidence levels of the means. A general increase in size is marked by several steps whose significance has been controversial. (*Source:* Modified from D. Kellogg, *Paleobiology*, v. 1, pp. 359–370.)

most thorough study could not eliminate the possibility that change seen in one stratigraphic section was the result of successive migrations rather than gradual evolution in place.

Early Cenozoic Mammals from Wyoming:

Through studies of collections from the Bighorn Basin, Phillip Gingerich (6) has postulated that

gradual change characterizes the history of several genera of mammals. His simplest example, the primitive primate *Pelycodus*, is illustrated in Figure 15–7. The proportions of the first molar, expressed as the logarithm of the product of its length and width, changes almost imperceptibly through the species *P. ralstoni, P. trigonodus,* and *P. jarrovii.* At the top of the succession, *P. jarrovii* appears to give rise to two new species.

Similar studies of the Eocene condylarth *Hyop-*

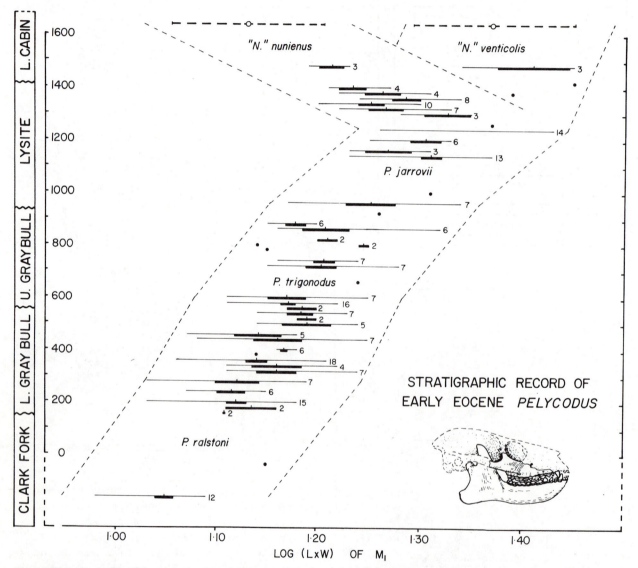

Figure 15-7 Evolution of the early Eocene primate *Pelycodus*. The horizontal scale is a measure of the proportions of the first molar tooth, and the vertical scale is in feet of the stratigraphic section in Wyoming, which is divided into stages. Each horizontal line expresses the variation in the number of specimens noted beside it. The shorter vertical line is the mean, the heavy bar is the standard error, and the light line is the range. The course of the phyletic evolution through this lineage is emphasized by the dotted lines. Several events of anagenesis and one of cladogenesis are suggested. (Courtesy of P. D. Gingerich, *American Journal of Science*, v. 276, p. 16. Reprinted by permission.)

sodus show a more complex pattern (Fig. 15–8) that Gingerich ascribed to gradual change combined with four speciation events.

Phyletic Gradualism:

We can see from these examples that not all cases of gradualism used to support the Darwinian concept of evolution have been shown to be valid by later work. However, enough well-documented examples have now been found to confirm that the ·type of evolutionary change postulated by the modern synthesis was widespread. Some of these gradual changes were within the range of variation included in a single species and no new species were generated. Changes in other lineages were great enough that taxonomists have divided the specimens that define the evolutionary trend into several different species. Evolution by continuous, progressive changes has been called phyletic gradualism.

The appearance of new species may be the result of two kinds of evolutionary change. Anagenesis takes place when one species becomes extinct by gradually evolving into another. The term *speciation* is applied to the process in which one species divides into two reproductively isolated lineages (Fig. 15–9). A division of one lineage into two may also be referred to as cladogenesis. The term clade (the root of the word cladogenesis) refers to all organisms descended from a progenitor species. Only speciation (or cladogensis) accounts for increasing diversity in an evolving lineage; if only anagenesis occurred, the number of species in a lineage would remain constant.

No evidence is found in these examples of initial isolation of small populations at the periphery of species ranges as required by the biogeographic–genetic model of speciation, but in most cases the data is insufficient to eliminate this possibility. Species formation by peripheral isolates has been called peripatric speciation. Supporters of the modern synthesis contend that these processes (anagenesis, cladogenesis, and peripatric speciation) all take place by the continuous accumulation of genetic modifications as organisms adapt to changes in their biological and physical environments.

EPISODIC SPECIATION

Although Darwin was committed to a gradualistic philosophy of the origin of species, he did not insist that evolution took place at a constant slow rate and implied in some of his later writings that species may originate in small populations evolving at faster than normal rates. George Simpson in his seminal book *Tempo and Mode in Evolution* (7) proposed that evolutionary changes giving rise to major groups (such as the initial stage in the radiation of ungulates in South America) took place in small populations evolving rapidly. As one of the architects of the modern synthesis, he held that most change took place in slowly evolving lineages. In the 1930s and 1940s hypotheses that evolution occurred in large steps were proposed by the geneticist Richard Goldschmidt and the paleontologist Otto Schindewolf. Goldschmidt claimed that major changes in the genetic material (chromosomal mutations) could change an individual so drastically that they would lead to a new major taxonomic group, such as a family or order, in one generation. The existence and effectiveness of such massive genetic changes has been doubted by most later workers. Schindewolf pointed out that major changes in the paleontological record of the ammonites took place in large steps rather than slow transitions. These were early proponents of the idea that microevolution alone will not explain the history of life.

Rate of Microevolution:

The rate of production of new species in the examples cited in the last section is slow. Critics of the modern synthesis claim that it is too slow to account for the diversification of life in the 800 million or so years since the origin of the metazoans.

A measure of the rate of evolution of a group is the average time range, or survivorship, of its species. In rapidly evolving lineages species exist for relatively short intervals before giving rise to new species. The average survivorship of all marine invertebrates has been estimated to be several million years. Survivorships of species ranging from .5 to 15 million years have been estimated by specialists of various invertebrate groups. Anthony Hallam (8) has made the following estimates of the average longevity (in millions of years) of invertebrates in Early Jurassic time: bivalves, 5.6; ostracodes, 3; ammonites, 1.5. Thomas Schopf (9) has argued that various biasing factors combine to lengthen these estimates and that a more reasonable figure for average species survivorship is 200,000 years.

Survivorships calculated from examples of phyletic gradualism in the fossil record are also several

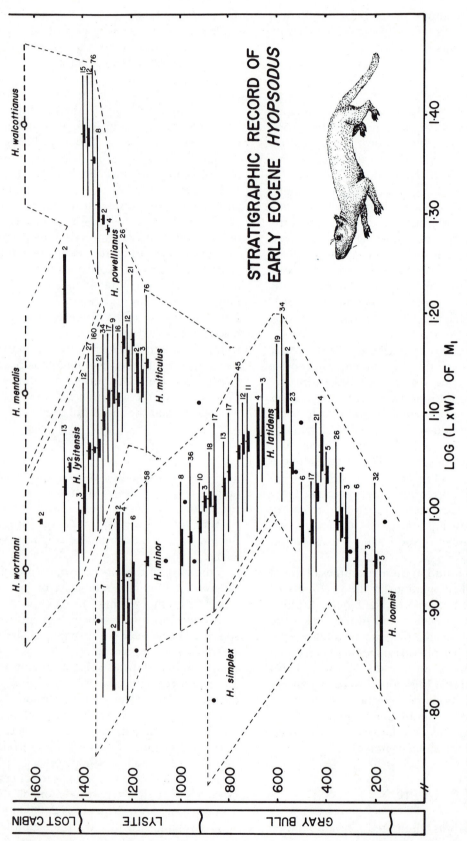

Figure 15-8 Evolution of the early Eocene condylarth *Hyopsodus*. The pattern of anagenesis and cladogenesis in this lineage is interpreted to be more complex. (Courtesy of P. D. Gingerich, *American Journal of Science*, v. 276, p. 13. Reprinted by permission.)

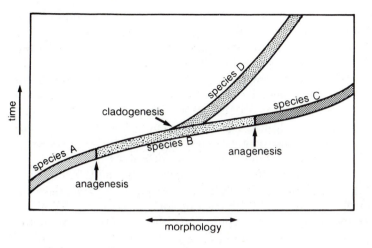

Figure 15-9 Diagrammatic representation of the origin of new species by anagenesis and cladogenesis. The horizontal axis is change in morphology.

million years per species (*Gryphaea,* mammals). Those who believe this rate is too slow to account for the bursts of adaptive radiation documented in the fossil record—for instance, at the start of the Paleozoic when about 135 families of marine invertebrates originated in 25 million years (Fig. 5–15) or at the start of the Cenozoic when 100 new families of mammals appeared in 25 million years—postulate that a faster process is at work and that most evolutionary change takes place during cladogenesis.

Punctuated Equilibrium:

Niles Eldredge and Stephen Gould (10) pointed out that the lack of all but a few examples of phyletic gradualism in the fossil record is what is to be expected if new species arise rapidly as small, peripherally isolated populations. Because emerging species are thought to constitute small populations separated from their ancestral population, we should rarely see the transition from one to the other in a single stratigraphic section. Because the transition takes place rapidly, intermediate forms are unlikely to be preserved anywhere in the fossil record. What we do see is what Eldredge and Gould said we should see: long intervals of no change punctuated by the sudden appearance of a new species as a peripheral isolate differentiated elsewhere spreads into the area of its ancestor. They believed that nearly all evolutionary change takes place at these speciation episodes, and that gradual evolution by anagenesis is an insignificant process in the history of life. The two models can be contrasted in a diagram (Fig. 15–10). Because the period of stability or stasis has

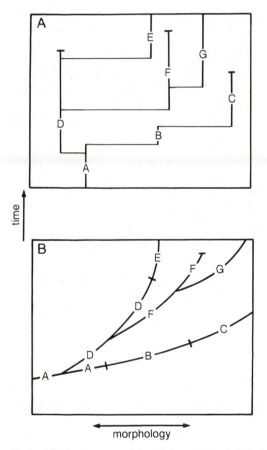

Figure 15-10 Punctuated equilibrium (A) and phyletic gradualism (B) contrasted. A. Species are static in morphology and evolutionary changes are cladogenetic and rapid. B. The origin of some species is anagenetic and some are cladogenetic. Species change gradually along morphologic gradients.

been pictured as a state of equilibrium broken by speciation events, the model has been called punctuated equilibrium. Only detailed sampling of stratigraphic sections of continuous deposition can prove the rapidity of the speciation process and the absence of subsequent change.

Evidence of Punctuated Equilibrium:

During Pliocene time a population of bivalves related to the common cockle, *Cerastoderma*, diversified into 30 genera and many species within a period of 3 million years in the Caspian Sea. During the same time interval the parent population of *Cerastoderma* changed little in the open ocean. Such rapid diversification has been thought to require the concentration of changes at speciation events as an organism adapts to a new environment. The rapid increase in diversity only requires closely spaced cladogenic events; the changes are not required to be episodic.

Niles Eldredge described the changes in the lineage of the trilobite *Phacops rana* in the Devonian of New York (10) using the punctuated equilibrium model. The subspecies and species are distinguished on the basis of the number of vertical rows of lenses in the schizochroal eyes (Fig. 7–26). The oldest species have 18 rows of lenses and the more advanced taxa have 17 or 15 rows. Eldredge believed he could show that the new forms with fewer lens rows first appeared suddenly in the record as peripheral isolates in the eastern part of their geographic range and then spread to the western part. Some doubt has been expressed about this interpretation.

Peter Williamson (11, 12) studied the succession of gastropods and bivalves in the Pliocene and Pleistocene beds of Lake Turkana in northern Kenya. In 19 species lineages he believed he could identify evidence of punctuated equilibrium. At some levels, such as a volcanic ash bed, morphologically distinct subpopulations suddenly appeared without intermediate forms and coexisted with the ancestral population before dying out. Many of the lineages followed the same pattern at the same horizon. Williamson interpreted this sudden appearance as the spread of peripheral isolates into the larger ancestral population. He calculated that the origin and spread could not have taken more than a few thousand years (an instant in geological time). Others have replied that the changes are superficial responses in the growth of the molluscs to changing conditions and not an expression of changes in the genetic makeup of the animals by selection.

The major support for the punctuational equilibrium model according to Gould and Eldredge (10) is the paucity of gradual successions of species despite the search through the record by paleontologists for about 150 years. The most convincing examples of gradualism come from the protozoans whose reliance on asexual reproduction may result in a mode of evolution different from sexually reproducing higher invertebrates and from vertebrates. Although countless paleontologists have approached the paleontological record with hope of finding an evolutionary series, they have found that species generally appear in their sections without obvious ancestors and disappear without obvious descendants.

Conclusion:

Evidence for phyletic gradualism in the fossil record is certainly scarce but it is available. The scarcity of gradualism may be because (1) most stratigraphic successions where paleontologists have searched for evidence of gradualism were deposited on continental borders and are interrupted by many gaps in deposition and on land where even the best-known sequences are too incomplete to show consistent change over less than 5,000 to 10,000 years, or (2), most evolutionary change is concentrated in short periods of speciation (punctuated equilibrium). Several paleontologists have noted that the paleontological record will never be able to demonstrate the nonexistence of either phyletic gradualism or punctuated equilibrium. Times of rapid diversification (adaptive radiation) cannot be accounted for by a background rate of evolution producing species at a rate of about one every few million years. The disagreement among paleontologists is whether this rapid splitting of species and increase of evolutionary pace is just a quickening of the processes of natural selection between individuals that Darwin proposed or whether it is an entirely different process concentrated in speciation.

NATURAL SELECTION

Natural selection is a process that occurs in all populations. It acts upon inheritable variations that result in differences in fitness among individuals—differences that are related either to reproduction of the next generation or to survival of the individual. The distribution of variations within the population will reflect these differences in fitness and will change from the parent population to the offspring population. Evolution can be defined as changes in the relative frequency of alternative characteristics that take place progressively from generation to generation.

The model of evolution proposed by Darwin and incorporated in the modern synthesis was based on competition between individual organisms for resources, mates, safety from predators, and living room. Only recently have extensive studies in the field shown the importance of competition between species and within species (13). One school of ecologists maintains that competitive interactions cannot be demonstrated to be the major driving force of evolution. Some of this school have been implied that Darwin's postulate of the "survival of the fittest" was not based on observation but was an outgrowth of the "laissez faire" economic theories of Victorian England (15). John Endler lists over 125 specific demonstrations of natural selection in natural situations (14). Field studies of competition usually involve the removal or transplantation of one of the competitors and the observation of the reactions of the others. While this may show interspecific or intraspecific competition within or between certain species, the general demonstrations of competition are the adaptive features of organisms (their good fit to the environment, see Chapter 14), convergent evolution, mimicry, and protective coloration.

Convergent evolution was discussed briefly in the last chapter and in Chapter 3. The similarity of organisms of different ancestry living in the same environment suggests that processes of selection for the attributes they share have determined their form.

Mimicry:

Although mimicry exists throughout the plant and animal kingdom, the most striking examples are found in the insects. Insect species that are not poisonous to birds have evolved remarkable similarities to poisonous species. The obvious conclusion is that the nonpoisonous species have achieved relief from predators by the progressive selection of those variants that most resemble poisonous species. Those that were least like poisonous species were preferentially eaten by birds and eliminated from the breeding population.

Protective Coloration:

Many animals have a covering of fur, feathers, scales, or skin that resembles the environment in which they live and makes them difficult to detect by predators. The pepper moth, *Biston betularia*, has been repeatedly cited as an example of selection for protective coloration. A dark phase of this moth differs from a light phase by a single gene. Until the 1840s only the light phase was recorded in the English Midlands. Since this time the dark phase has increased in abundance until it now constitutes about 80 percent of the population in heavily industrialized areas. In these areas the trees on which the moth alights have been blackened by industrial soot in the last 150 years. Against this dark background the light moths are conspicuous and have been selectively eaten by birds. The selective advantage of the dark (or melanic) form has been proved by direct observation of the individual fates of light forms bred in captivity and released into the environment, and by release and recapture experiments with both varieties. In competition with the light phase, the dark phase has selective advantage where trees are darkened. Recent restrictions on the release of industrial pollutants have reduced the darkening of trees and led to a slight comeback of the light phase. This is an example of intraspecific selection—a new species has not been formed by this adaptation but the frequency of occurrence of certain genetically controlled features has been changed by selection over a period of a few tens of years. If numerous genetic changes of this type occurred throughout the range of a species, they could lead eventually to its recognition as a distinct species.

The "Arms Race":

All organisms in a community except the primary producers that derive their energy from inorganic nutrients are, in a sense, predators since they rely on organisms below them in the food chain for sustenance. The evolutionary action-and-response relation between predator and prey has been compared to the arms race between superpowers. An advance in the ability to attack by the predator must be balanced by an advance in the ability to escape by the prey, and vice versa, or one of them will become extinct. The reality of this form of interaction can be illustrated by the change in size of the brains in herbivores (prey) and carnivores (predators) during the Cenozoic Era. Carnivores generally have proportionally larger brains than herbivores because greater intelligence is required to catch prey than to escape from predators (16). Both herbivores and carnivores increase in brain size throughout the Cenozoic but carnivores are always ahead of contemporary herbivores. As the herbivores got smarter, the carnivores got smarter still in order to catch them. This evolutionary trend is difficult to explain unless competition between members of both the predator species (to find more prey) and prey species (to escape more efficiently) is a reality.

These examples leave little doubt that competition within species can be found in nature. Competition between species is more difficult to demonstrate because species generally exclude from their habitat species competing for the same resources.

SPECIES SELECTION

The term *species selection* has been applied to those processes that cause the differential survival of species as opposed to those that affect individual organisms. At present there is controversy over whether these selective processes are distinct.

If phyletic gradualism is rejected as a mechanism of significant evolutionary change and the production of new species is ascribed to major genetic changes that occur at cladogenesis, then selection directed at the individual level loses its importance. The argument has been made by the supporters of punctuated equilibrium that processes which operate only in the production of new species facilitate the rapid evolutionary change necessary to account for periods of rapid diversification in the fossil record. Since gradualistic changes are too slow, the raw material for selection must be produced at these rapid speciation events and the path of evolution must be shaped by selection between the species so produced.

Natural and species selection can be compared in a table which is a modification of one presented by Steven Stanley in his book *Macroevolution*. Stanley, however, identifies the natural selection in this table with microevolution, and species selection with macroevolution.

Although selection among species must take place, disagreement is focused on its importance and whether it is different from, or decoupled from, natural selection. Gould (17) argued that evolutionary processes take place at various levels, or tiers, and that evolution is a hierarchy of processes. Natural selection operates at the level of the first tier and species selection, at the level of the second tier. Biologists have so far discovered few, if any, evolutionary trends that cannot be explained as well by natural selection at the level of the individual as by species selection (18).

MACROEVOLUTION

The term *macroevolution* has been used in many senses. Generally it has been used to refer to evolutionary processes that result in changes giving rise to higher taxonomic categories such as families, orders, and classes. It is in this sense that the term is used here.

The concept of macroevolution comes from the perception that the paleontological record lacks fossils that show the transition between major taxonomic groups. These fossils would be expected if the transition has been bridged by microevolutionary steps. The lack of transitional forms could be due to gaps in the record, to the smallness of the evolving population (and therefore the unlikelihood of preservation), or to the fact that major evolutionary novelties arise by major genetic reorganization or major changes in developmental processes that are too rapid to leave many intermediate individuals. Most of the large evolutionary steps by which the invertebrate phyla differentiated from ancestral metazoans took place before the evolution of skeletons and are therefore not preserved in the fossil record. Whether features of the soft anatomy and physiology that are critical to the origin of many higher taxa of vertebrates appeared abruptly or gradually cannot be determined because these features are not reflected in the bones that are preserved. Examples of such unpreservable features are the amniotic egg that separates all higher vertebrates from fish and amphibians and the mammary glands that characterize mammals.

	Natural Selection	Species Selection
Level of action	Individual	Species
Source of variability	Random individual genetic changes	Major genetic or developmental changes generated randomly associated with speciation
Rewards of success	Survival and reproduction	Speciation
Cost of failure	Death or failure to reproduce competitively	Extinction

Examples of the transitions between major groups that *have* been thought to require macroevolutionary processes include the origin of the invertebrate phyla, the origin of land plants from algae, the origin of the vertebrate body plan from invertebrate ancestors, the transition from fish to amphibians, and that from amphibians to reptiles.

Origin of Higher Taxa:

Ideally, major taxa include all descendants of a single progenitor species. Classes, orders, families, and genera are aggregates of species united by common descent from an ancestor in which the characteristics of the group are first evident.

Methods of classifying organisms make differences between major taxa seem greater than they originally were. Between the end of the Cretaceous Period and the early Paleocene, 16 new orders of early ungulates arose from primitive condylarths in 2 to 3 million years. Many species and genera identifiable in the early part of this period went on to be founders of new families, but had biologists been present to make a classification of the group in Paleocene time, they would have recognized them only as slightly different genera of one family of mammals. The significant characters of founder taxa can be recognized as major steps in evolution only after the successful spread of these characters and after the diversification of the organisms bearing them has taken place.

In well documented transitions that have been described as macroevolutionary events, the initial step that was eventually of great evolutionary importance was not a major reorganization of the organism, as the term macroevolution implies.

Emergence of New Structures:

Three examples from the evolution of the vertebrates will illustrate the nature of some major transitions. Many of the evolutionary changes between major taxa have been associated with changes in habitat, or way of life. The significance of the first small steps in these transitions is that they allowed penetration of a new ecologic zone and led to subsequent diversification into many species that we recognize as constituting a major taxon.

Amphibians evolved from rhipidistian fish in Late Devonian time. The advanced rhipidistians are similar to early amphibians in most anatomical features. The similarities include the capacity to breath air, body shape, and bone structure of the body,

upper limbs, and skull. The only major changes from fish to amphibian were the evolution of feet at the ends of the fins, and the elaboration of ribs to support the lungs and viscera. Forms with limbs intermediate between fins and feet have not been found in the fossil record presumably because there was not a broad adaptive zone to be occupied by animals that were only partially adapted to terrestrial locomotion. However, in a shrinking aquatic environment where fish were being stranded, any limb improvement that would allow the animal to move over the drying surface would have given it a competitive advantage and would have rapidly resulted in the refinement of this method of locomotion. This anatomical change corresponded in time and place with the availability of a new environment. The land had recently been clothed with plants and populated by terrestrial arthropods. Subsequent diversification of vertebrates with limbs capable of walking led to the group we now recognize as tetrapods.

Even the one feature that distinguishes *Archaeopteryx* (the oldest known bird), its feathers, are not new structures. The similarity of their embryological development shows that reptilian scales and avian feathers are homologous. Feathers evolved from scales through size increase and an elaboration of their shape. It is difficult to understand how selection could have acted to favor this modification in the ancestors of birds since an enlarged reptilian scale would have little value as a flight surface. However, feathers also serve for insulation, and even a relatively small increase in size of the reptilian scale may have made it efficient in providing insulation for the small endothermic dinosaurs that gave rise to birds. Once the scales were of larger size, selection may have acted to change their shape to assist in flight. Such a change in function is frequently an important factor in the emergence of major new taxa and their adaptation to different ways of life.

Except for the long forelimb and the presence of feathers, the skeleton of *Archaeopteryx* remains almost totally dinosaurian. The distinctive skeletal features that are typical of modern birds evolved long after the first birds had become committed to flight.

The steps by which the mammals evolved from mammal-like reptiles required over 100 million years through Permian and Triassic time (see Chapter 11). Through countless intermediate forms the anatomical and physiological features that we associate with modern mammals evolved in relation to increased metabolic rate, changes in patterns of feeding, locomotion, and more acute sensory perception. No particular event within this sequence can be recognized as a major macroevolutionary change that in itself produced the higher taxon and yet the final results are as different as those between any of the vertebrate classes.

Conclusion:

Despite more than a hundred years of speculation that mechanisms other than natural selection act on individuals to produce major steps in evolution, no examples from the paleontological record or from studies of living organisms have demonstrated the nature of such processes or proved their existence.

As stated at the beginning of this chapter, this is a time of ferment in paleontology when the record is being searched intensively for examples to corroborate the various models of evolution that have been discussed here. At present a consensus has not emerged from the discussion but when it does, it will probably incorporate parts of both sides of these areas of conflict. The lack of simple answers to many of these questions reflects the complexity and diversity of evolutionary processes. This lack has also stimulated new ways of looking at the paleontological record.

TRENDS AND ONTOGENY

Trends:

Some long-term trends in the fossil record have been discussed in this chapter as examples of phyletic gradualism. By far the most widespread of such trends is increase in size. Size increase among warm-blooded animals has been termed Cope's Law, named after Edward Cope, a vertebrate paleontologist. In the histories of practically all mammalian groups early members are smaller than later ones. Other things being equal, a large animal is more heat efficient than a small one because the rate at which it loses heat is proportional to the area-to-volume ratio. Although size increase has some advantages for invertebrates, such as avoidance of small predators, the trend to size increase in invertebrates is not as widespread. Changes in the coiling of the shell of *Gryphaea* through Early Jurassic time were attributed to allometry and increase in its size. As shells grew bigger, they coiled more strongly. In

later Jurassic beds, although the shells continue to be large, their convexity diminishes; the shells have the size of an adult but coiling of a juvenile *Gryphaea*. The mechanism by which such changes take place can be explained by the study of the development of the individual organism from the fertilized egg to reproducing adult, its ontogeny.

ONTOGENY AND THE FOSSIL RECORD

Embryos and Ancestors:

Since ancient times people have been fascinated by the development of embryos of animals and in particular by the human embryo. The studies in ontogeny made by the ancient Greeks were based for the most part on human fetuses available through spontaneous abortions, and from these they learned that ontogeny is not solely a matter of growth and that embryos are not just miniature adults. But in the nineteenth century the Prussian zoologist Karl Ernst von Baer (1792–1876) established that embryos develop from general characters towards more specific ones, and that the similarity between the embryos of different groups of animals is greatest in their early stages of development.

Comparison of the early stages of embryos led some zoologists of the late eighteenth and early nineteenth centuries to postulate that organisms in ontogeny pass through stages of evolution comparable to those of their ancestors. An example will clarify the type of evidence on which this concept was based. At about four weeks the human embryo develops four internal pouches (pharyngeal pouches) in the throat region that are positioned opposite external folds in the neck region. Between these pouches and folds six blood vessels called aortic arches pass from a ventral artery to a dorsal one. This arrangement is remarkably similar to the same region in the embryos of fish. In fish development the internal pouches and external grooves break through and join each other to form the gill slits. The aortic arches are the vessels by which the blood circulates past the gills to be aerated. In the human embryo only one of the pouches and grooves join to produce a canal. This passage becomes the eustachian tube that joins the ear and throat. Some of the other pouches atrophy; others become glands such as the thyroid and parathyroid. Most of the aortic arches in mammals are lost as the embryo develops. One pair (the sixth) goes to the lungs, and the left half of the fourth becomes the major artery to the head and body.

Ernst Haeckel (1834–1919) formalized and popularized the observation that the anatomy of primitive groups was reflected in the ontogeny of more advanced relatives. He proposed what he called the biogenetic law, usually summarized as: ontogeny recapitulates phylogeny. What he meant was that in its development from embryo to adult the individual passes through (recapitulates) the evolutionary stages of its ancestors. Thus the human embryo progresses from a single cell (protozoan) through a blastula (cnidarian) (Fig. 5–3), through higher invertebrate stages to resemblance to a fish, a reptile, and finally to a primate. This is recapitulation. Evolution consisted of adding on to the end of the ontogenetic sequence new adaptations as they were acquired. Of course, the organism could not repeat the whole of its ancestry, and skipping of stages and acceleration of the sequence was essential if the embryological development were to take a reasonable amount of time. The influence of the concept of recapitulation on paleontological thought of the late nineteenth and early twentieth centuries is hard to overestimate for it seemed to offer zoologists and paleontologists a key to reconstructing the ancestries of all organisms—these ancestries were still preserved in miniature in their ontogenies.

Many kinds of shell-bearing invertebrate animals carry through their lives the record of the development of their skeletons during the embryonic stages. The shells of foraminiferans, most molluscs, and brachiopods are of this type. The horn-shaped skeletons of the rugose corals carry the history of septal insertion in their tips. In other organisms, for example, sponges, arthropods, echinoderms, and vertebrates, the skeleton is either modified during growth or discarded at each growth stage. The early molts of arthropods may be difficult to associate with the adult species into which they eventually grew but this problem has been solved for many trilobites. In the early part of this century much paleontological research was devoted to the search for ancestors in the ontogenetic stages of fossils. Paleontologists ground at the tips of rugose corals to reveal the early septal patterns, pulled ammonites apart to see the patterns made by the sutures in the first-formed chambers, and made thin sections of foraminiferans to study their embryonic chambers.

Yet Von Baer had pointed out in the early 1800s that embryos never resemble the adults of their ancestors, as Haeckel proposed, but only the embryos of related animals. Haeckel admitted that much condensation of ontogenies took place and not infrequently disturbances in the order of embryonic stages led to misleading results in reconstructing phylogenies. In many animals recapitulation did not seem to explain the sequence of developmental changes. In fact, just the opposite took place—early stages of the ontogeny of ancestors became the adult stage in descendants. This process is called paedomorphosis. The process that Haeckel called recapitulation is now referred to as peramorphosis (20) because Haeckel's term had become associated with his evolutionary theories that are now discredited. The basic difference between peramorphosis and paedomorphosis can be illustrated in the following tables.

Peramorphosis

T	↑	A	B	C	D	E	.	F*
I		A	B	C	D	.	E*	
M		A	B	C	.	D*		
E		A	B	.	C*			
				ONTOGENY ⟶				

Paedomorphosis

T	↑	A	B	.	C*			
I		A	B	C	.	D*		
M		A	B	C	D	.	E*	
E		A	B	C	D	E	.	F*
				ONTOGENY ⟶				

*adult stage

A to F are ontogenetic stages of individuals. Ancestors are at the base of the table; descendants are at the top.

When the ontogeny of an ancestor is compared to that of its descendant, peramorphosis has occurred, either if the rate of development of nonreproductive features is accelerated so that new adaptations are added before maturity is reached or if onset of sexual maturity is delayed. Paedomorphosis has occurred if the maturation is accelerated so that the organism can reproduce at a stage when its body resembles that of a juvenile, or if the rate of development of nonreproductive features is retarded. The axolotl, a Central American amphibian, is a classical example of paedomorphosis. Most amphibians pass through a stage in which they breathe in water using gills before they reach maturity, then shed their gills, grow legs, and live as adults on land. The axolotl, however, grows to adult size and matures in the water without ever reaching the adult (land) stage of its ancestors and retains ancestral juvenile features, such as gills, which allow it to remain in the water. The human face is another example of paedomorphosis. Our ancestors, the apes, have sloping foreheads and forward thrust mouths but their young have a facial profile much like that of *Homo sapiens*. The face of the human (the descendant) at maturity is much like that of the ape (the ancestor) at a juvenile stage. Paedomorphosis also describes the decrease in curvature of the *Gryphaea* lineage in Late Jurassic time. Their evolution was marked by the appearance of large, mature individuals with the juvenile proportions (in coiling) of their ancestors, a typical example of this phenomenon.

Although the study of peramorphosis does not solve problems of phylogenetic reconstruction, as it was once hoped it would, it helps the paleontologist to distinguish the more general characteristics of an organism from the more specialized ones and therefore is invaluable in classification and useful in establishing descent and ancestry. Embryos are never like adults; they are always adapted to the embryonic environment and recapitulation, in the sense of Haeckel, does not occur. Major changes in the form of an adult can result from minor changes in the genes that control rates of maturation and growth of the individual.

The relation between the ontogenies of ancestors and descendants is more complex than this discussion has indicated (19). Changes in the rates of (1) growth in size, (2) maturation, and (3) morphological change, between the ontogeny of ancestors and descendants define six different relationships, each of which has a name (Fig. 15–11). Space does not allow a discussion of these relationships, and their recognition in the paleontological record is difficult.

Peramorphosis in Fossils:

Studies of fossil lineages have revealed examples of peramorphosis and paedomorphosis but some structures found in the early stages of ontogenies may be larval adaptations and not directly influenced by the ancestry of the organism.

Peramorphosis is shown by many mammalian lineages in which the size of anatomical features, such as horns or antlers, is allometrically coupled to an

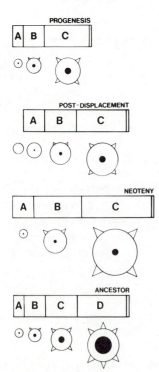

Figure 15-11   Relations of the three paedomorphic processes on the left and three peramorphic processes on the right to the ontogeny of an ancestral form. Stages in the development of a hypothetical organism with a central pore and spines are labeled A, B, C, D, and E. The ontogeny of the ancestor is shown in the lower rows. In neoteny, the rate of development is reduced and in this example, the onset of maturity is retarded, resulting in the descendant adult being larger but having an immature morphology. In postdisplacement, the growth of spines and the pore is delayed; therefore, juvenile characteristics appear in the adult. In progenesis, precocious sexual development results in an adult smaller than the ancestor. In acceleration, increased development rate results in adult characteristics of the ancestor appearing in the descendant juvenile. In predisplacement, the growth of the pore and spines starts at an earlier stage than in the ancestor. In hypermorphosis, the onset of sexual maturity is delayed so that the adult is larger and more developed morphologically than the ancestor. (Courtesy of K. J. McNamara, The Paleontological Society, *Journal of Paleontology*, v. 60, pp. 4–13.)

evolutionary trend for increase in size. For example, the last members of the brontothere family have massive horns which grow from smaller, simpler horns in the ontogeny of the individual (Chapter 12). Early members of the family are smaller in size and have smaller horns. The descendants have gone farther along an evolutionary trend than the ancestors, and the descendant's juveniles resemble the adults of its ancestors. Peramorphosis seems to be a pattern of evolution followed where selective advantage is given (1) to prolonging ontogeny to achieve greater size, and (2) to the advancement into early stages of ontogeny of complex structures that are produced late in growth.

A classic example of a fossil succession ascribed

to peramorphosis was described early in this century by R. K. Caruthers from species of the Carboniferous rugose coral *Amplexizaphrentis*. In what he thought were successive layers, four species of the genus replace each other, but their stratigraphic ranges overlap. The ontogeny of the septal pattern is revealed by cutting through the apical regions of their conical skeletons. The adult pattern of the ancestral form, *Amplexizaphrentis delanoui*, occurs in the early ontogenic stages of its descendant, *A. parallela*. The septal patterns of both adult *A. delanoui* and *A. parallela* appear in successive juvenile stage of the descendant of *A. parallela*, *A. constricta*, and so on (Fig. 15–12). Euan Clarkson (20) has suggested that the pattern of evolution of this coral is

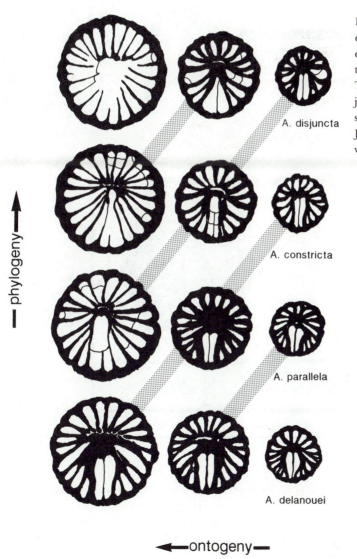

Figure 15-12   Peramorphosis in the Carboniferous rugosan coral *Amplexizaphrentis*. Ontogeny is represented by successive sections across the corallum showing the septal arrangement. The similarity of adult stages of early species to juvenile stages of their successors is emphasized by diagonal bars. (*Source:* Modified from J. A. Caruthers, *Journal of the Geological Society*, v. 66, 1910.)

A. disjuncta

A. constricta

A. parallela

A. delanouei

time

phylogeny

←—ontogeny—→

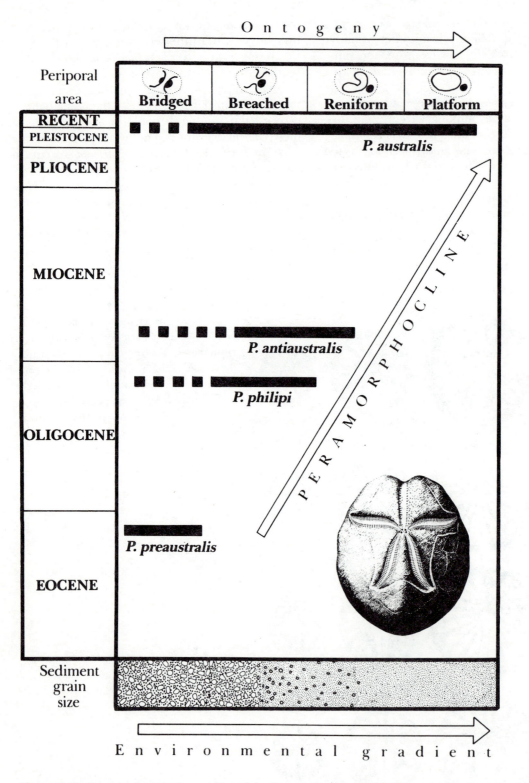

Figure 15-13 Peramorphosis in the evolution of the Cenozoic irregular echinoid *Protenaster* in Australia, shown by changes in the area around the pores illustrated along the top of the diagram. Later species undergo a progressively increased rate of development of this characteristic by acceleration. This change follows the environmental gradient illustrated along the bottom of the figure. The species illustrated in *P. preaustralis*. (Courtesy of K. J. McNamara and *Palaeontology*, v. 28, pp. 311–330.)

complex and not simply explained by Caruther's concept of recapitulation.

Kenneth McNamara (21) has interpreted the evolution of the pores in the plates around the mouth of the burrowing echinoid *Protenaster* as an example of peramorphosis. In the earliest species of the genus of Eocene age, *P. preaustralis*, the pores are all in the bridged condition (Fig. 15–13). In the successively younger species, *P. philipi*, *P. antiaustralis*, and *P. australis*, the pores in ontogeny pass through stages in which they are like the adult pores of their ancestors and progress from a bridged, to a breached, to a reniform condition In the adult stages of the living species the trend is continued to the platform condition. The trend is interpreted as an adaptation to life in finer-grained sediments.

Paedomorphosis in Fossils:

The evolution of the Cenozoic brachiopod *Tegulorhynchia* through several species to the genus *No-*

tosaria is an example of paedomorphosis (Fig. 15–14). In adults of the first species of the lineage, *T. boongeroodaensis*, the pedical hole is small and the apical angle of the pedicle valve is large, but in its juvenile stages the hole is proportionally larger and the angle smaller. As the lineage evolves, adult stages of successively younger species have features that are typical of the juvenile stages of their ancestors. The modifications are probably adaptations to life in more turbulent water where a thicker pedicle was an advantage.

The bivalve species *Cardium plicatum* and its descendant, *C. fittoni*, have been cited as examples of paedomorphosis (Fig. 15–15). The adult of the ancestor has many ribs marked by overlapping growth lamellae. Its juvenile form has fewer, more widely separated ribs with spines on them. The adult of the descendant, although larger than the ancestor, has ribs like its juvenile, few and widely spaced.

Some paleontologists have postulated that major groups arose and major steps in evolution were

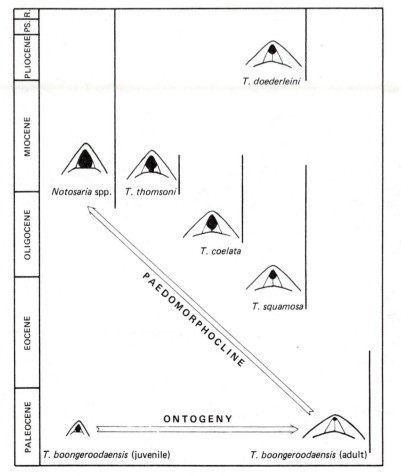

Figure 15-14 Paedomorphosis in four species of the rhynchonellid brachiopod *Tegulorhynchia* and one species of *Notosaria*, all from the Cenozoic of Australia and New Zealand. The ranges in time of the species are shown by vertical lines. The changes occur in the beak region and affect the size of the pedicle opening and the plates closing it, which are illustrated for each species. In the ancestral species *T. boongeroodaensis*, the opening is restricted as the brachiopod develops from a juvenile (ontogenetically). Descendant species progressively retain more juvenile characteristics into adulthood. (Courtesy of K. J. McNamara, The Paleontological Society, and *Journal of Paleontology*, v. 57, p. 465.)

Figure 15-15   Paedomorphosis in two species of late Miocene bivalves of the genus *Cardium*. A. A younger specimen of *C. plicatum* (5 mm). B. Adult of the same species (17 mm). C. The descendant species *C. fittoni* (35 mm) is larger than the adult of the ancestor but has the ribbing of the juvenile ancestor. (*Source:* L. A. Nevasskaya, *Paleontological Journal*, v. 1(4), pp. 1–17.)

taken through paedomorphosis. By small changes in the regulatory genes that control rate of maturation and growth, a paedomorphic adult organism is produced that may be remarkably different from the adult of its ancestor. A highly speculative theory for the origin of scleractinian from rugose corals suggests that the first scleractinians were larval rugosans. The adult sixfold septal symmetry of scleractinians contrasts with the fourfold symmetry of rugosans but the latter pass through a stage early in ontogeny when only six septa are present. Changes in the genes that control rates of development that brought this stage of sixfold symmetry to adulthood might explain the sudden appearance of scleractinians and the lack of intermediates between them and rugosans.

Larval Adaptations:

Larval trilobites have eyes at the front margins in the protaspis stage (Chapter 7). In ontogeny, the eyes move backwards through successive molt stages to their adult position on the upper surface. If this ontogenetic change is interpreted as a result of peramorphosis, then the ancestors of the trilobites must have had forward-looking eyes. An alternative explanation is that the position of the eyes is an adaptation to the free-swimming life-style of the larva to whom forward and downward vision were more important than to the adult living on the bottom.

Ontogenetic changes to greater complexity in the sutures of ammonoids were used by paleontologists of the early part of this century to trace ancestry on the basis of recapitulation theory. Their arguments have now generally been rejected and attention has been focused on the form and ornamentation of the shell. The simpler sutures of larval ammonoids may indicate only that the young lived

in a different habitat from the adults and did not need the shell strength given by complex folding of the septa. Embryos, like adults, are adapted to their environment. If that environment is much different from that of the adult, the paleontologist is faced with the choice of interpreting early ontogenetic stages in terms of larval adaptation or ancestry.

REFERENCES

1. Hallam, A., 1982, Patterns of speciation in Jurassic *Gryphaea:* Paleobiology, v. 8, p. 354–366.

2. Raup, D. M., and Crick, R., 1982, *Kosmoceras:* evolutionary jumps and sedimentary breaks: Paleobiology, v. 8. p. 90–100.

3. Bretsky, S. S., and Bretsky, P. W., 1977, Morphologic variability and change in the paleotaxodont bivalve mollusk, *Nuculites planulatus* (Upper Ordovician of Quebec): Journal of Paleontology, v. 51, 256–271.

4. Kellogg, D. E., 1975, The role of phyletic change in *Pseudocubus vema* (Radiolaria): Paleobiology, v. 1, p. 359–370.

5. Gould, S. J., and Eldredge, N., 1977, Punctuated equilibria: the tempo and mode of evolution reconsidered: Paleobiology, v. 3, p. 115–151.

6. Gingerich, P. D., 1976, Paleontology and phylogeny: patterns of evolution of early Tertiary mammals: American Journal of Science, v. 276, p. 1–28.

7. Simpson, G. G., 1944, Tempo and mode in evolution: New Haven, Conn, Yale Univ. Press, 237 p.

8. Hallam, A., 1987, Radiations and extinctions in relation to environmental change in the marine Lower Jurassic of northwest Europe: Paleobiology, v. 13, p. 152–168.

9. Schopf, T. J. M., 1984, Rates of evolution and the notion of "living fossils": Annual Review of Earth and Planetary Sciences, v. 12, p. 245–292.

10. Eldredge, N., and Gould, S. J., 1972, Punctuated equilibrium: an alternative to phyletic gradualism. In Schopf, T. J. M. (ed.), Models in Paleobiology: San Francisco, Freeman–Cooper, p. 82–115.

11. Williamson, P. G., 1981, Paleontological documentation of

speciation in Cenozoic molluscs from Turkana Basin: Nature v. 293, p. 437–443.

12. Mayr, E., Boucot, A. J., et al., 1982, Punctuationism and Darwinism reconciled? The Lake Turkana Mollusc sequence (replies to P. Williamson): Nature, v. 296, pp. 608–612.

13. Connell, J. H., 1983, On the prevalence and relative importance of interspecific competition: evidence from field experiments: American Naturalist, v. 122, p. 661–696.

14. Gould, S. J., and Lewontin, R. C., 1979, The spandrels of San Marco and the Panglossian paradigm: a critique of adaptationist program: Proceedings of the Royal Society of London, series B., v. 205, p. 581–598.

15. Endler, J. A., 1986, Natural selection in the wild: Princeton, N.J., Princeton Univ. Press.

16. Dawkins, R., and Krebs, J. R., 1979, Arms races between and within species: Proceedings of the Royal Society of London, series B., v. 205, p. 489–511.

17. Gould, S. J., 1985, The paradox of the first tier: Paleobiology, v. 11, p. 2–12.

18. Gilinsky, N. L., 1986, Species selection as a casual process: Evolutionary Biology, v. 20, pp. 249–273.

19. McNamara, K. J., 1986, A guide to the nomenclature of heterochrony: Journal of Paleontology, v. 60 p. 4–13.

20. Clarkson, E. N. K., 1979, Invertebrate Palaeontology and Evolution: London, Allen & Unwin, p. 77.

21. McNamara, K. J., 1985, Taxonomy and evolution of the Cainozoic spatangoid echinoid *Protenaster:* Palaeontology, v. 28, p. 311–330.

SUGGESTED READINGS

Cope, J. C. W. and Skelton, P. W. (eds.), 1985. Evolutionary case histories from the fossil record: Special Papers in Palaeontology 33, Palaeontological Association, London.

Gould, S. J., 1977, Ontogeny and phylogeny: Cambridge, Mass, Harvard Univ. Press, 501 p.

Hoffman, A., 1982, Punctuated versus gradual mode of evolution: Evolutionary Biology, v. 15, p. 411–436.

Pollard, J. D. (ed.), 1984, Evolutionary Theory, paths into the future: New York, Wiley, 271 p.

Stanley, S. M., 1979, Macroevolution, pattern and process: San Francisco, Freeman, 332 p.

Chapter 16

Evolution in Earth History

Paleontological studies may resolve many of the evolutionary problems discussed in the last chapter and only through the study of the history of life can the long-term effects of evolutionary mechanisms be assessed. In this chapter three facets of evolution are examined: rates of evolution, extinctions, and the effects of chance or stochastic processes.

RATES OF EVOLUTION

Reference has been made in preceeding chapters to the measurement of rates of evolution. Questions that the paleontological record may be able to answer about rates of evolution include (1) Can a common pattern of evolutionary rate changes be detected in the history of groups? (2) Have faster evolutionary rates been more widespread at certain times in earth history than at others? (3) Are slowly evolving organisms, so-called living fossils, characterized by a different style of evolutionary change? (4) Are particular environmental conditions responsible for fast and slow evolutionary rates?

Evolutionary rates can be expressed as (1) survivorships (see Chapter 15); (2) the number of taxa that evolved in an interval of time; or (3) the amount of change in a part or a proportion of an organism in a period of time. The third type of rate has been expressed in terms of a unit designated as a darwin. A character of an organism is said to evolve at a rate of 1 darwin when it changes by a factor of "e", the base of natural logarithms (2.718 ...) in 1 million years. For example, the average rate of evolution of bivalve lineages in the late Cenozoic has been estimated to be less than 10 millidarwins.

Because evolutionary studies in paleontology have been focused on the origin of species or higher taxa, the rates usually considered are taxonomic rates, those that refer to the number of species, genera, or orders, originating or becoming extinct in a certain interval of time. Taxonomic rates express two components of evolutionary change: anagenesis, when one species becomes extinct by gradually evolving into another, and cladogenesis, when one species produces a new species and continues to exist with it. Because the passage from one brachiopod species to another does not include morphological changes comparable to the passage from one mammalian species to another, comparison of taxonomic rates between distant groups is of doubtful validity.

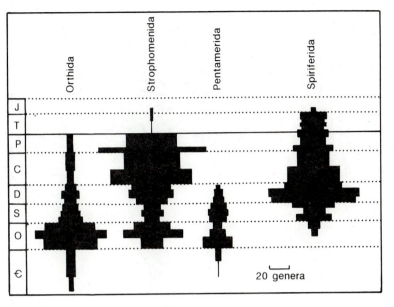

Figure 16-1 Shapes of clades of four major Paleozoic brachiopod orders. Time scale of periods on the left. Note the wide variety of shapes and tendency for a rapid early expansion of diversity. (*Source:* Modified from S. J. Gould and others, *Paleobiology*, v. 3, pp. 23–40.)

Adaptive Radiation:

The history of a clade is usually represented by a diagram in which the varying width of a vertical line represents the variation in the number of taxa through time (Fig. 16–1). Norman Gilinsky and Richard Bambach (1) have shown that on the average the greatest diversity in clades occurs in the early part of their range. Many clades have an initial phase of rapid expansion in diversity followed by a slow decline towards extinction. The shape of the clade in a diagram such as Figure 16–1 is typically a pear or teardrop but a wide variety of clade shapes have been documented. The phase of rapid expansion is the result of a phenomenon known as adaptive radiation—the diversity explosion as the organisms exploit new adaptations or newly available environments.

The potential rate of increase in adaptive radiation may, like the increase in a population by reproduction, be governed by an exponential function:

$$\text{Number of species generated by time } t = e^{Rt},$$

where e is the base of natural logarithms and R is the fractional increase per unit of time. The equation assumes that the clade starts from a single species. The actual number of species present at any time also depends on the rate at which they become extinct. Steven Stanley has calculated the average value of R for families of bivalves is 0.06 per million years and for mammals 0.22 per million years (2). The proportion between these values of R is similar to that between the average time range of species of invertebrates (5 or 6 million years) and mammals (about 1 million years).

In Late Triassic time some scleractinian corals took on symbiotic algae that allowed them to respire and calcify more efficiently (Chapter 10). This adaptation led to a rapid diversification so that from a few Triassic species, 5 orders and 12 families arose in a few million years (Fig. 16–2). The physiological step was not a big one, but the results were far-reaching for the success of the group. Such rapid diversity could not be attained by anagenetic appearance of new species at intervals of several million years that are typical of invertebrates. Rapid cladogenesis, at much shorter intervals than this average figure suggests, is called for. Steven Stanley (2) has used similar examples to justify the concept of macroevolution. If the average time range of invertebrate species is several million years, each step in the adaptive radiation must be a large one in order to bring about the large increase in diversity in the short time available. The diversity could be produced by rapid cladogenesis and progressive, morphological change. The potential of rapid cladogenesis to produce great diversity in a short time can be illustrated by a hypothetical example. If a cladogenetic event occurred every 200,000 years in an adaptive radiation, then from a single species

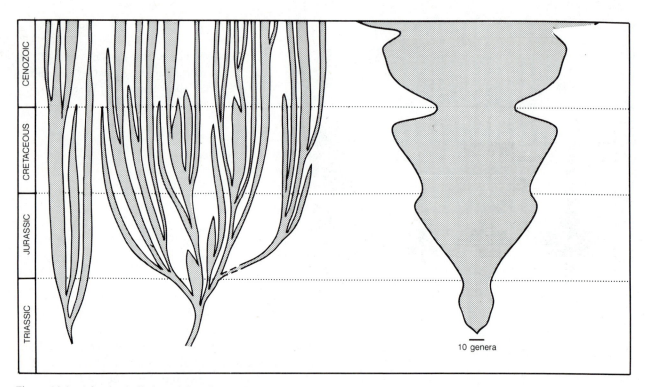

Figure 16-2   Adaptive radiation of the scleractinian corals in Mesozoic and Cenozoic time. At left, the diversification into families is shown after the work of J. W. Wells; at right, the shape of the clade, based on the number of genera as compiled by S. Stanley from information in Wells. (*Source:* J. W. Wells, *Treatise on Invertebrate Paleontology*, v. F, 1956; S. Stanley, *Macroevolution*, 1981.)

1024 species could be generated in 2 million years (2^{10}), and more than 1 million species in 4 million years (2^{20}). This calculation assumes that none become extinct in this time. These figures, though not unrealistic (R must be about 3.5), have no basis in the record but they do illustrate the immense potential of exponential increase. Biologists estimate that only about 10 million species are living. Other examples of adaptive radiation in the history of many groups have been reviewed in preceding chapters. Examples of large-scale radiations are (1) the invertebrate phyla in Late Proterozoic and Cambrian time; (2) the invertebrates with calcareous skeletons in Middle Ordovician time; (3) the land plants in Devonian time; (4) the angiosperms in Cretaceous time; (5) the placental mammals in Late Cretaceous and Paleocene time. Examples of adaptive radiation of intermediate scale include (1) the ammonoids at three times in their history; (2) the dinosaurs in Triassic time; (3) the ungulates of Paleocene time. Adaptive radiations at the grandest scale appear to have resulted from the coincidence of an empty ecologic zone and the evolution of a species with the ability to take advantage of this newly available environment. Such major breakthroughs occur only infrequently in life history.

The availability of ecologic zones may be due to biological or physical events. The accessibility of many land environments to mammals was the result of the extinction of the dinosaurs, but the rapid spread of shelled invertebrates appears to have been made possible by a change in the oxygen content of the atmosphere and oceans.

Cycles in Earth History:

Both directional and cyclical trends have been discovered in the history of the earth. Since Archean time the oxygen content of the atmosphere has increased while the carbon dioxide content has fallen. The luminosity of the sun appears to have increased at about 1 percent per 80 million years since Archean time but the effect on the temperature of the earth of its lower radiance in the Archean was probably compensated for by the greater effectiveness than of the greenhouse effect because carbon dioxide was abundant in the atmosphere until about 500 million years ago. Superimposed on these unidirectional trends in world environments, whose effects on life were discussed in Chapters 5 and 6, are cycles of larger and shorter periodicity that also control evolution. The longest of these cycles is believed to be caused by convective overturn in the mantle of the earth, the layer that lies between the thin crust and the liquid core. The overturn, cycling in a period variously estimated to be 300 to 500 million years, is responsible for the movement of the continents, the expansion and contraction of the ocean basins, and many secondary effects.

Geologists can document only the last two grand cycles in any detail. Late in the Proterozoic Eon the continents were a single mass called Pangaea I. The initial phase of the next to last cycle was the splitting of Pangaea I into continents by rifting, and the widening of the oceans between the continents by the spreading of oceanic crust. During this phase volcanoes were active, carbon dioxide was released by them into the atmosphere (degassing) enhancing the greenhouse effect and warming world climates, the sea level rose expanding the shallow epicontinental seas across the continental platforms, and coastlines were lengthened as the supercontinent split into fragments. Alfred Fischer (3) has called this the greenhouse phase of the cycle. The second phase of the cycle was marked by closure of oceans as the continents again aggregated into Pangaea II, falling sea level, shrinking epicontinental seas and coastlines, decreasing carbon dioxide levels, cooler temperatures, and the icehouse state. In the Paleozoic cycle this phase was marked by extensive glaciation in the southern hemisphere.

These cycles and trends have exerted a strong influence on the history of life. The diversification of the invertebrate phyla and their expansion in numbers occurred in the greenhouse phase of the early Paleozoic when warm epicontinental seas expanded across continental platforms and oxygen content had risen to a level that calcareous shells could readily be secreted. The late Proterozoic icehouse glaciation had passed and the oceans were expanding as Pangaea I broke into continents. The turnover point from the first to second phase occurred in Late Devonian time when oceans began to close and cool and the next icehouse phase came in (Fig. 16–3). This time of extinction was documented in Chapter 7. The most intense biotic crisis in the history of life marks the formation of Pan-

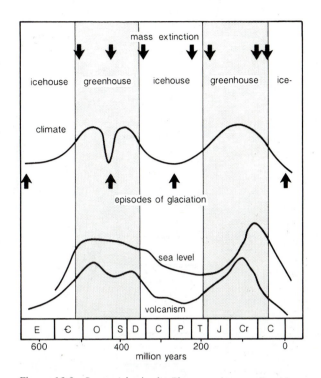

Figure 16-3 Supercycles in the Phanerozoic as outlined by A. Fischer. The letters along the base are initials of the geological periods. (*Source:* Modified from A. Fischer, *Catastrophes in Earth History*, 1984.)

gaea II and the spread of late Paleozoic glaciers across a large part of it. By later Triassic time this supercontinent was breaking up and greenhouse conditions led to the expansion of marine life in what was called in Chapter 10 the second wave of invertebrate diversification. To the present day oceans are expanding but Fischer believes that late Cenozoic glaciations indicate a return to the icehouse condition.

On a broad scale the pace of marine invertebrate evolution can be modelled in terms of this cycle, but did it affect life on the land? Comparison of the effects of the cycles on terrestrial vertebrates and marine invertebrates is possible only for the last 350 million years since vertebrates occupied the land. While restriction and expansion of shallow seas clearly affected marine invertebrates, their effect on land vertebrates is not obvious. The marine biotic crisis of the end of the Paleozoic is not evident in vertebrate communities, or those of marine fish.

Damion Nance and co-workers (4) have traced these supercycles back as far as the Archean Eon and suggested that six rifting episodes took place

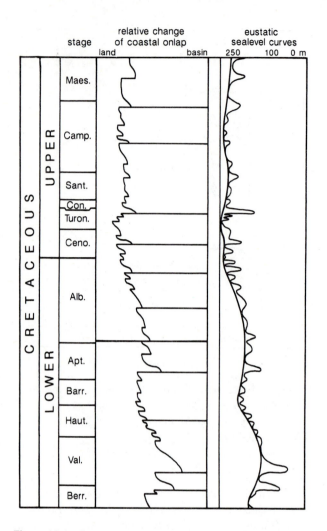

stage | relative change of coastal onlap | eustatic sealevel curves

Figure 16-4 Cretaceous cycles of sea level fluctuation. Short-term changes are represented by thin lines; long-term changes by thick lines. Major cycle boundaries are indicated by horizontal lines. The complete Cretaceous stage names can be found in Figure 13-1. (*Source:* Modified from B. V. Haq, J. Hardenbol, and P. R. Vail, *Science*, v. 235, pp. 1156–1167.)

at 2500, 2000, 1500, 1060, and 200 million years ago. They suggest that pre-Paleozoic cycles controlled the appearance of the eukaryotes and siliceous plankton and the collapse of the stromatolite communities.

Much shorter cycles of transgression and regression of the continental margins also affected the pace of evolution of marine life. Peter Vail (5, 6) and his co-workers have compiled curves that document changes in sea level that caused epicontinental seas to flood and ebb in cycles of various orders from a few tens of thousands of years (6th

order) to tens of millions of years (2nd order) (Fig. 16–4).

Evolution, Transgression, and Regression:

Erle Kauffman (7) studied the effect of transgression and regression on Cretaceous marine faunas in the western interior states. Transgressions provide increasing space for adaptive radiation of fast-breeding organisms that can take advantage of a variety of habitats, but regression places stress on shallow-water organisms, usually as they compete for shrinking space on continental margins. The response in changes of rates of evolution is not simple and varies from organism to organism. Marine invertebrates that can live in a wide variety of environments (those with general ecological requirements) first speciate rapidly as new environments expand during transgression, and secondly when they are forced to compete as their habitats shrink in regression. Species that are specialists, requiring narrowly defined environmental conditions, reach their highest evolutionary rate just past the time of greatest transgression and decrease in rate of speciation during regression.

Anthony Hallam (8) has suggested that the rate of speciation and extinction should increase at times of maximum regression and competition for space. Remnants of the regression crisis would then expand across the shelves with the ensuing transgression.

Some evidence has been found that offshore species inhabiting stable environments have shorter ranges and more specialized requirements, and inshore species responsive to rapid changes in the environment have longer ranges in time. However, no simple relation between rates of evolution and the transgression–regression cycle has yet been formulated.

Although rates of speciation are higher in offshore stable communities, David Jablonski and his co-workers (9) have shown that innovative faunas originate in nearshore waters and force older communities into offshore slope positions (Fig. 16–5). The Cambrian fauna of Sepkoski (see Chapter 7) originally occupied the inner shelf environment but was successively replaced during Late Cambrian to Late Ordovician time by the Paleozoic fauna of articulate brachiopods, gastropods, bryozoans, sponges, and corals, and then by the beginnings of the modern fauna identified by the prominence of bivalves.

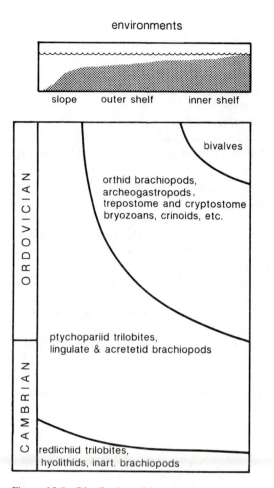

environments

slope   outer shelf   inner shelf

ORDOVICIAN

bivalves

orthid brachiopods,
archeogastropods,
trepostome and cryptostome
bryozoans, crinoids, etc.

ptychopariid trilobites,
lingulate & acretetid brachiopods

CAMBRIAN

redlichiid trilobites,
hyolithids, inart. brachiopods

Figure 16-5 Distribution of four Cambrian–Ordovician fossil communities in time and environment. New communities are introduced nearshore and are forced outward by the appearance of another community inshore. (*Source:* Modified from D. Jablonski and others.)

Living Fossils:

Darwin used this term to apply to living animals and plants that slowly evolved but persisted for long periods of time. Since Darwin's time, the term has been used in a variety of ways. Commonly it refers to living representatives with primitive features in clades that reached their greatest abundance and diversity many millions of years ago. Most so-called living fossil species do not have a paleontological record that allows the history of the particular species to be traced back into the past. *Neopilina galatheae*, a monoplacophoran found in 1957 on a Pacific seamount, is a good example. This genus has no fossil record but it is considered to be a living fossil because it is a representative of a

Cambrian group that is ancestral to the rest of the molluscs (Chapter 7). The living species with the longest record is believed to be the brachiopod *Triops cancriniformis* whose fossil representatives are known from Triassic rocks. The living lingulate brachiopod, *Lingula*, is commonly cited as the oldest living genus because this name has been used for specimens from Ordovician rocks. The ostracode *Bairdia* has a similar long span. *Equisetum*, the horsetail rush, is a good example of a living fossil of the plant kingdom. It is a primitive representative of the sphenophytes that thrived in Carboniferous coal swamps but have since come close to extinction.

The persistence of primitive members of a clade long after more advanced members have become extinct is a common phenomenon in the paleontological record. The dendroid graptolite *Rhabdinopora* is one of the first of its order to appear and the last to become extinct, long after the more specialized graptoloids had died out. The first and most primitive group of the stromatoporoids are the labechiids. This group was also the last to die out at the end of Devonian time after its more advanced relatives had become extinct in the mid-Late Devonian crisis. The genus *Didelphis*, the opossum, is a living remnant of the original radiation of the marsupials in Cretaceous time. The phenomenon may be explained by the tolerance to a range of conditions of early and late members of clades. Early members are generalists capable of living in a wide range of conditions and as they evolve, members divide into smaller parcels the resources in the clade's ecologic zone. Species become specialists with narrow requirements for food and space and therefore are more likely to become extinct when conditions change. When the environment becomes inhospitable, members of the clade that have retained a tolerance for a range of conditions persist the longest.

Paleontologists are not in agreement about what other features living fossils have in common that could account for their longevity. Peter Bretsky and D. M. Lorenz (10) suggested that organisms adapted to highly variable environments develop the ability to survive environmental shifts that eliminate their contemporaries that have adapted to more stable conditions. For the lungfish, adapted to periodically drying lake beds, this hypothesis appears to apply; but for the living fossils of the deep sea, such as the monoplacophoran *Neopilina*, and the coelocanth *Latimeria*, it is inappropriate.

EXTINCTION

Background Extinction:

Living fossils are interesting because they are exceptional; most species that have ever lived are extinct. Extinction is the complement of speciation and, if the diversity of life has remained constant, the rate of speciation and of extinction must balance. However, owing to biases of the record, we cannot be certain whether total species diversity has risen, dropped, or remained essentially constant through geological time. The diversity curves for skeletal marine invertebrates (Fig. 7–60) show an overall Phanerozoic increase but also show periods of static diversity of hundreds of millions of years. Such curves do not reflect the whole range of even marine life (see discussion at end of Chapter 7).

Intuitively, we would expect diversity to have risen as increasingly specialized organisms displaced early generalists carving out more narrowly defined habitats for themselves, dividing the world's resources more finely. Supporters of this concept suggest that the world's resources are like a barrel that can be first filled with melons but space still remains that could be filled will apples, then with walnuts, then blueberries, etc. The fruits of decreasing size represent organisms with increasingly specialized environmental requirements.

Why, or how, do organisms become extinct? The same factors that limit the geographic spread of a species in space are those that limit its continuation in time. Extinction can be considered to be the reduction of the geographic range of a species to zero. These limiting factors can be reduced to three groups:

1. Competition and predation from other organisms.
2. Changes in the physical environment.
3. Chance events.

The response of organisms to factors 1 and 2 is part of the processes included in the concept of natural selection, but processes grouped under factor 3, such as hurricanes, forest fires, and tidal waves, act indiscriminately on all life whatever its adaptation. Because most such chance events are local in their effects, they will lead to the extinction of small populations only. The effect of the three factors acting together has been thought to produce a constant average rate of extinction in individual clades.

Leigh Van Valen (11) plotted survivorship curves (Fig. 16–6) for many taxa and found that the trend of points in all approached a straight line but that the slope of the lines for taxa that differ in their ecological requirements were different. The plots indicate that the probability of extinction is constant through the life of the group but that rates vary from group to group. In its simplest form, Van Valen's law states that extinction in a given group occurs at a constant rate. The survivorship curve is analogous to the curve representing the decay of a radioactive isotope (Fig. 2–1). The probability of decay is the same for all atoms (or the probability of extinction is the same for all species) and constant in time, but the decay of an individual atom is a matter of chance and cannot be predicted. Van Valen's work suggests that processes operating at random determine the survival or extinction of individual taxa, but this conclusion has not been universally accepted.

The causes of extinction of species within historic

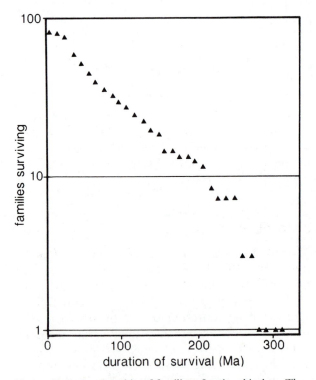

Figure 16-6 Survivorship of families of extinct bivalves. The ordinate plots the number of families having durations at least as long as the value of the abscissa (in millions of years). For example, ten families have survived for 200 million years or more, and only one for more than 300 million years. Note the logarithmic scale of the ordinate. (*Source:* Modified from L. van Valen, *Evolutionary Theory,* v. 1, pp. 1–30.)

times can be assessed accurately. For most of these species, such as the passenger pigeon, the great auk, and the dodo, man has been the immediate cause. He is suspected also of causing the extinction of many species that disappeared at the close of the last ice age. However, for long-extinct species the specific cause of extinction and the nature of their competitors is difficult to determine. Van Valen's law suggests that there is a normal background rate of extinction. However, the rate must have accelerated during certain episodes in earth history when many taxa became extinct at the same time.

Mass Extinctions:

At the end of the Paleozoic and the Mesozoic eras, so many organisms became extinct that the contrasts between life before and after these times were recognized as the basis for the division of the history of life into eras as early as 1821. Since then four additional major episodes of mass extinction and many less important ones have been recognized. The six events in which much of the world biota became extinct occurred at or near the ends of the Eocene Epoch, the Cretaceous, Triassic, Permian, and Ordovician periods, and between the Frasnian and Famennian stages of the Late Devonian Epoch. The extent of these mass extinctions has been sum-

marized by Steven Stanley (12) and has been referred to in previous chapters. On diversity graphs (Fig. 7–60) these extinctions appear as sharp, downward points in the curve. In a relatively short time many clades became extinct. In the crisis just before the end of Ordovician time nearly a third of the brachiopod families disappeared along with major groups of the conodonts, trilobites, bryozoans, graptolites, corals, and stromatoporoids. In the Late Devonian crisis, 30 percent of invertebrate families, and in the Permian crisis 50 percent, became extinct. By far the most research has been focused on the event that closed Cretaceous time when extinctions decimated the dinosaurs, marine reptiles, ammonites, rudist bivalves, and much of the marine plankton. Analyses of curves showing percent extinction of marine families since the Permian (Fig. 16–7) has suggested to David Raup and John Sepkoski that extinctions reach peaks above the background extinction rate in a regular cycle with a period of 26 million years. However, the reality of this cycle has been doubted by other paleontologists (13, 14, 15).

The Impact Hypothesis:

Although paleontologists have speculated about the causes of mass extinctions since the general outlines

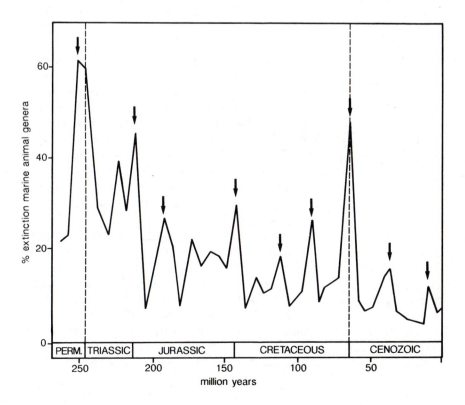

Figure 16-7 Percentage extinction of marine animal genera since Permian time. Arrows indicate times when extinction rate exceeded the background rate, according to Raup and Sepkoski. The line connects means of estimates at each inflection point. Error bars have been omitted to simplify the diagram from the original. Note that the highest points correspond to the Late Permian, Late Triassic, and Late Cretaceous biotic crises. (*Source:* Modified from D. M. Raup and J. J. Sepkoski, *Science*, v. 231, pp. 883–886.)

of the history of life became known, the discussion has intensified during the past decade, stimulated by the hypothesis that the event at the end of the Cretaceous was caused by the impact of a meteorite or asteroid, and that other biotic crises had similar causes. The hypothesis was formulated by Walter and Luis Alvarez and their co-workers (16) to explain the high concentrations of the rare element iridium in the clay layer that marks the boundary between the Mesozoic and Cenozoic sedimentary rocks at Gubbio, Italy. The concentration of this element, and several others of the platinum group found in this layer, is 10,000 times greater in meteorites than in the earth's crust. Subsequently, the so-called iridium spike was found to be widely distributed at the boundary both in cores of deep-sea drill holes and in stratigraphic sections on land. The Alvarezes proposed that a meteorite about 10 km in diameter struck the earth and the impact raised enough debris into the atmosphere to plunge the world into darkness for several months killing most of the plants, cooling the surface by shutting off the sun's heat, disrupting food chains, and causing the extinction of a large part of the world biota. The search for similar trace element anomalies at the horizons of other extinction events produced some successes and some failures. Anomalies were found at the Late Devonian and Eocene events but (apart from a single doubtful occurrence) not at the Permian–Triassic, nor late Ordovician events. However, stratigraphic sections in which deposition across the Permian–Triassic boundary is continuous are rare. Astronomers who accepted the meteorite hypothesis and the hypothesis of cyclic extinctions have suggested that the earth periodically encounters conditions favorable to impact by meteorites or comets. Statistical analysis of the orbits of asteroids that cross the orbit of the earth have shown that during Phanerozoic time the earth is likely to have been struck by an asteroid of the size required by the Alvarezes on average about once every 30 million years but the ages of meteorite craters does not reflect such frequency. Space in this book is not sufficient to include an extensive discussion of the meteorite hypothesis. It has become one of the most controversial theories of the last decade and has stimulated great debate over the importance of catastrophes in shaping the history of life.

If survival of an organism depends not on its fitness tested in competition with other organisms, but on the chance it will escape from a sudden unpredictable catastrophe, then the role of natural selection over long periods of life history differs significantly from that envisaged by Darwin. The mechanism of long-term adaptation cannot be applied to sudden but extremely rare events. No member of a biota can adapt to the chance of a meteorite impact. Stephen Gould (ref. 16, Chapter 15) termed these catastrophic processes the third tier of selection (the first is natural selection acting at the individual level, the second is species selection). Differential survival of catastrophes as a major control on life history is in sharp contrast to selection by competition between adaptations of different utility. Because acceptance of the impact theory radically changes the interpretation of the history of life, it has come under much scrutiny, and many paleontologists react to it with skepticism.

Alternative Theories of Extinction:

If extinction at these crises was sudden and unrelated to the earth's environment, then it should have overtaken the groups affected in an instant of geological time. Evidence has now been accumulated which shows that mass extinctions in many groups occurred over a considerable period of time, rather than at a single stratigraphic horizon. Figure 16–8 shows that ammonite diversity declined steadily through the Late Cretaceous as if responding to declining environmental conditions, and relatively few genera persisted to the end of the Maastrichtian stage when the last genus became extinct. A similar history is shown by the the stromatoporoids before their near extinction in the Late Devonian crisis. The most diverse fauna of dinosaurs is found at the start of the Maastrichtian stage about 10 million years before the close of the Cretaceous Period, and from this time onward their diversity decreased to extinction. Decline in diversity before the extinction event suggests that extinction was due to accumulating effects of environmental deterioration and not to sudden disaster.

Theories of mass extinction must account for the selectivity of the operative process. Many mechanisms for destroying life indiscriminately can be imagined, but in the major biotic crises, the organisms that became extinct had relatives that survived. When dinosaurs and marine and flying reptiles became extinct, other reptiles, the turtles, crocodiles, and lizards, survived. When the ammonites died out, the nautiloids lived on to expand

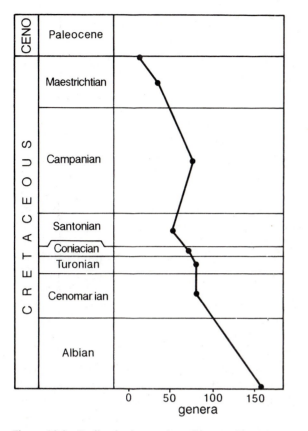

Figure 16-8   Decline in the number of ammonite genera through Late Cretaceous time to extinction at the close of the period. (*Source:* Plotted from information of Jost Wiedmann, *Biological Reviews*, 1969, p. 594.)

teorite impact are various worldwide environmental changes. Steven Stanley (12) summarized the evidence that extinctions are climatically controlled. He cited faunal and isotopic evidence that cooling of world temperatures took place at the major extinctions and several of them were accompanied by glaciation in parts of the world. Another cause that has been proposed is worldwide fall of sea level restricting the space available for marine biotas. Both the Cretaceous–Cenozoic and the Permian–Triassic crises correspond to exceptionally low stands of sea level. Other causes proposed include the rise of deep, oxygen-impoverished water onto the continental shelves destroying marine life, violent volcanism that could inject sufficient dust into the atmosphere to produce effects like those of meteorite impact, and spillover of cold Arctic water into the warmer southern oceans, causing mass mortality of microorganisms and the collapse of food chains. None of these mechanisms is as successful as decrease in temperature in explaining either the decline of diversity before extinction or the geographic distribution of extinctions.

The hypothesis of global cooling is supported by the distribution of victims and survivors. If victims are largely equatorial species and survivors are high-latitude species, this would be good evidence of a cooling event. This pattern of change can be detected in the graptolites at the Late Ordovician crisis, and in the stromatoporoids at the Late Devonian one. Through Permian time reef-forming organisms became progressively restricted to equatorial latitudes indicating cooling. Leigh Van Valen and Robert Sloan (17, 18) have concluded from detailed study of the Cretaceous–Cenozoic transition beds in Montana that subtropical forests in which the dinosaurs lived were progressively displaced southward by temperate forests with a fauna of warm-blooded mammals as winters became colder. They also raised the possibility that dinosaurs died out earlier in high latitudes than at low ones as world climate deteriorated. Unfortunately, well-dated, dinosaur-bearing deposits in the tropics have not been found to substantiate this hypothesis.

In summary, mass extinction events can be explained by the occasional intensification and expansion of the processes of normal background extinction. Hypotheses of sudden catastrophe, terrestrial or extraterrestrial, have stimulated reexamination of the evidence at critical stratigraphic sections, but most paleontological work does not support a sudden, indiscriminant event.

into the abandoned habitats in Cenozoic time. Although only 13 percent of planktonic foraminiferans survived the Cretaceous crisis, 75 to 85 percent of benthonic genera survived. A key to understanding mass extinctions must lie in the differences between the victims and the survivors. Catastrophic mechanisms to explain extinction offer an indiscriminant instrument like a bomb when what is needed is a bullet.

Competing hypotheses explain the various iridium anomalies as the result of the eruption of volcanoes, its concentration by dissolution of seafloor sediments, or its accumulation preferentially in organic matter. Evidence for meteorite impact at biotic crises other than that at the Cretaceous–Cenozoic boundary, is meager and, even if the iridium anomaly at that level is accepted as proof of impact, such impacts may have been extremely rare events. The agents of extinction that have been postulated as alternatives to catastrophes by me-

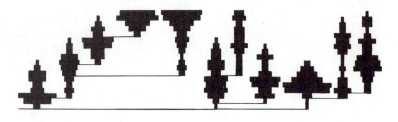

Figure 16-9   Random clade shapes constructed by a computer program. Compare with the real clades of Figure 16-1. (*Source:* Modified from S. J. Gould and others, *Paleobiology*, v. 3, p. 24.)

STOCHASTIC PROCESSES

The impact of a meteorite on earth is a chance event. Although the probability that a 10 km meteorite will strike the earth can be calculated statistically from trajectories and abundance, the year, century, or millenium of its impact cannot be predicted. Events such as this, or the tossing of a coin, are probabilistic in nature and are said to be controlled by stochastic processes. We can predict how many times in a hundred tosses, a coin is likely to yield the result heads but we cannot predict for any one toss what the result will be. Events that predictably follow from a series of causes are said to be deterministic. To say that the results of coin tossing are regulated by stochastic processes is not to imply that the result of a single coin toss is not determined by a series of forces such as the way the coin is propelled, the air resistance, the wind, the surface on which it lands, etc. These interacting forces are so many and so difficult to assess that at present their result can only be treated statistically.

We have already considered hypotheses that mass extinctions are stochastic events whose individual occurrence cannot be determined. The survival of individuals may also be stochastically controlled. The rabbit killed by the fall of a tree is not less well adapted than the rabbit that the tree missed. What are the contributions to the evolution of life of such processes; are they trivial, like the death of the rabbit, or far reaching, like the extinction of the dinosaurs?

The course of evolution is determined by natural selection and by random changes in the gene pool of the population. The second of these alternatives is called genetic drift. It is a stochastic process that may be responsible for the course of evolution taken by small populations where selection pressure is low. Many geneticists take a neutralist view of the importance of natural selection. They note that many genetic differences seem to be neutral from the standpoint of survival, and therefore propose that natural selection is of minor importance in controlling their frequency. Thomas Schopf (19) compares species to atoms in a gas in constant, random motion, whose individual positions cannot be predicted but whose overall effect on the behavior of the gas can be described statistically. From this viewpoint the course of evolution is a statistical summary of the histories of species whose origination and individual fates are determined randomly (compare Van Valen's law) and, if the game of evolution could be replayed, the result would certainly be different. The contrasting deterministic viewpoint is that, given the same physical environmental changes in time, the replay would show life evolving in the same pattern.

David Raup and his co-workers (20, 21) constructed computer programs to draw artificial phylogenies based on arbitrary assumptions about rates of speciation and extinction. At any step in the program, branching, continuation, or extinction of the lineage is determined by a set of random numbers. The shapes of the artificial clades are therefore largely determined randomly but under rules limiting their size and rates of expansion established by the programmers. They found the similarity of the artificial clades to those found in nature to be remarkable (Fig. 16–9).

Nearly all of the factors discussed in this chapter are controversial in terms of their importance in the history of life. The scientific debates between supporters of deterministic and stochastic processes in shaping evolution, between supporters of catastrophic and environmental causes of extinction, and between gradualists and punctuationists, continues in the scientific literature. Meanwhile scientists are driven to unearth more paleontological data to support their viewpoints and our knowledge of nature advances to a point where some of these hypotheses may be rejected, but in most of these controversies, there is some truth on both sides.

REFERENCES

1. Gilinsky, N. L., and Bambach, R. K., 1986, The evolutionary bootstrap: new approach to the study of taxonomic diversity: Paleobiology, v. 12, p. 251–268.

2. Stanley, S. M., 1979, Macroevolution, pattern and process: San Francisco, Freeman, 332 p.

3. Fischer, A. G., 1984, The two Phanerozoic supercycles. In Berggren, W. H., and van Couvering, J. A. (eds.), Catastrophes in earth history: the new uniformitarianism: Princeton, NJ., Princeton Univ. Press, p. 129–150.

4. Nance, R. D., Worsley, T. R., and Moody, J. B. 1986, Post-Archean biogeochemical cycles and long term episodicity in tectonic processes: Geology, v. 14, p. 514–518.

5. Vail, P. R., Mitchum, R. M., and Thompson, S., 1977, Seismic stratigraphy and global sea level, part 4. In Peyton, C. E. (ed.), Seismic Stratigraphy: American Association of Petroleum Geologists Memoir 26, p. 83–97.

6. Haq, B. V., Hardenbol, J., and Vail, P. R., 1987, Chronology of fluctuating sea levels since the Triassic: Science, v. 235. p. 1156–1167.

7. Kauffman, E. G., 1977, Evolutionary rates and biostratigraphy. In Kauffman, E. G., and Hazel, J. E. (eds.), Concepts and methods of biostratigraphy: Stroudsburg, Pa., Dowden, Hutchison and Ross, p. 109–141.

8. Hallam, A., 1978, How rare is phyletic gradualism and what is its evolutionary significance? Evidence from Jurassic bivalves: Paleobiology, v. 4, p. 16–25.

9. Jablonski, D., Sepkoski, J. J., Bottjer, D. J., and Sheehan, P. M., 1983, Onshore–offshore patterns in the evolution of Phanerozoic shelf communities: Science, v. 222, p. 1123–1124.

10. Bretsky, P. W., and Lorenz, D. M., 1972, Adaptive response to environmental stability: Proceedings North American Paleontological Convention, v. 1, p. 522–550.

11. Raup, D. M., 1975, Taxonomic survivorship curves and Van Valen's law: Paleobiology, v. 5, pp. 337–352.

12. Stanley, S. M., 1987, Extinctions: New York, Scientific American Books, 242 p.

13. Raup, D. M., and Sepkoski, J., 1986, Biological extinction in earth history: Science, v. 231, p. 1528–1533.

14. Raup, D. M., 1987, Mass extinction: a commentary: Palaeontology, v. 30, p. 1–13.

15. Hoffman, A., 1985, Patterns of family extinction depend on definition and geological time scale: Nature, v. 315, p. 659–662.

16. Alvarez, W., Alvarez, L. W., Asaro, F., and Michel, H. V., 1980, Extraterrestrial cause for the Cretaceous–Tertiary extinction: Science, v. 208, p. 1095–1108.

17. Sloan, R. E., et al., 1986, Gradual dinosaur extinction and simultaneous ungulate radiation in Hell Creek Formation: Science, v. 232, p., 629–633.

18. Van Valen, L., and Sloan, R. E., 1977, Ecology and the extinction of the dinosaurs: Evolutionary Theory, v. 2, p. 37–82.

19. Schopf, T. J. M., 1979, Evolving paleontological views on deterministic and stochastic approaches: Paleobiology, v. 5, p. 337–352.

20. Raup, D. M., 1977, Probabalistic models in evolutionary biology: American Scientist, v. 65, p. 50–57.

21. Gould, S. J., Raup, D. M., Sepkoski, J. R., Schopf, T. J. M., and Simberloff, D. S., 1977, The shape of evolution: a comparison of real and random clades: Paleobiology, v. 3, p. 23–40.

SUGGESTED READINGS

Alvarez, W., Kauffman, E. G., Surlyk, F., Alvarez, L., Asaro, F., and Michel, H., 1984. Impact theory of mass extinction and the invertebrate fossil record. Science, v. 223, p. 1135–1140.

Berggren, W. A., and Van Couvering, J. A. (eds.), 1984, Catastrophes and earth history: the new uniformitarianism: Princeton, N.J., Princeton Univ. Press, 464 p.

Campbell, K. S. W., and Day, M. F. (eds.), 1987, Rates of Evolution: Winchester, Mass., Allen & Unwin, 336 p.

Eldredge, N., and Stanley, S. M. (eds.), 1984, Living Fossils: New York/Berlin, Springer-Verlag, 291 p.

Elliott, D. K. (ed.), 1986, Dynamics of Extinction: New York, Wiley, 294 p.

Gingerich, P. D., 1983, Rates of evolution: effects of time and temporal scaling: Science, v. 222, p. 159–161.

Larwood, G. P. (ed), 1988, Extinction and Survival in the Fossil Record: Systematics Association Special volume 34, Oxford, Clarendon Press.

Nitecki, M. H. (ed.), 1984, Extinctions: Chicago, Univ. of Chicago Press, 354 p.

Officer, C. B., and Drake, C. L., 1983, The Cretaceous–Tertiary transition: Science, v. 219, p. 1383–1390.

Silver, L. T., and Schultz, P. H. (eds.), 1982, Geological implications of the impact of large asteroids and comets on earth: Geological Society of America Special Paper 190.

Stanley, S. M., 1984, Mass extinction in the ocean: Scientific American, v. 250, no. 6, p. 64–67.

Valentine, J. W. (ed.), 1985, Phanerozoic diversity patterns: profiles in macroevolution: Princeton, N.J., Princeton Univ. Press, 441 p.

Walliser, O. H. (ed.), 1986, Global bioevents: a critical approach. Lecture Notes in Earth Sciences: Berlin/New York, Springer-Verlag, 442 p.

Chapter 17

Paleoecology

INTRODUCTION

Ecology is the study of the relations between modern organisms and the environments in which they live; paleoecology is the study of these relations between species represented in the fossil record and the environments they inhabited. Ecologists can study the relations between organisms in a community by observing their behavior and can measure the characteristics of modern environments precisely, but how can paleontologists reconstruct the complex interrelationships of organisms and their environments when all the organisms are dead and the environments in which they lived have all changed? Paleontologists can only deduce behavior from the functional morphology of fossils and from the effects of the organisms on the sediments. The mineralogy, structures, and textures of these sediments can be used to reconstruct the environment of deposition, and studies of continental movements can add information on geographic position and climate of the habitat. Despite the limitations of dealing with dead and imperfectly preserved organisms, and the lack of direct information on the characteristics of the ancient environments, paleontologists can reconstruct ancient communities in considerable detail.

Paleoecological Inference:

This is largely, but by no means completely, uniformitarian in nature. In living communities ecologists have found that organisms can be grouped according to the roles they play in the flow of energy or food through the unit of interdependent organisms. These roles are called trophic roles (from the Greek *trophe* = nourishment). Plants take nutrients and light from the environment, converting them into organic compounds. They are commonly referred to as primary producers (Fig. 17–1). Plants are eaten by herbivorous animals, which are in turn eaten by carnivores, when alive, and scavengers, when dead. In death, plants and animals are decomposed by bacteria and their organic compounds are broken down into nutrients that are returned to the environment. The organisms united by this flow of energy are referred to as an ecosystem.

Paleoecologists infer that similar relationships existed in the past between organisms and their environment. Obviously many groups in the reconstructed ecosystem will be poorly represented or missing entirely. The primary producers of marine ecosystems are microscopic planktonic plants that are rarely preserved, but are presumed to have existed since the Archean because they are essential

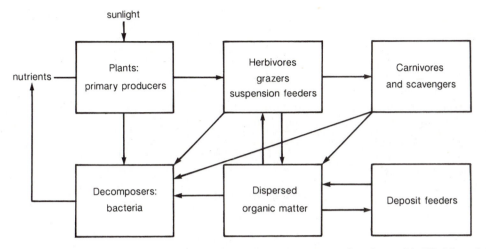

Figure 17-1   Major trophic groups and energy pathways in a community. (*Source:* Modified from J. W. Valentine.)

to life in the marine environment. Within the major trophic groups many subdivisions based on feeding behavior can be defined, and paleoecologists can compare the forms of modern animals of known trophic roles with those of fossils to infer their positions in ancient ecosystems.

This uniformitarian approach does not imply that conditions in the past were exactly the same as those of today—in fact, we know that they were not. It does imply that general laws governing the functioning of organisms have not changed in time, although environments may have. Different kinds of organisms were adapted to the unique atmospheres of the pre-Paleozoic eras, but some must have been primary producers and others decomposers, although herbivores and carnivores were probably absent at first. As life has become more diverse, the complexity of ecosystems must have increased, but the laws governing the flow of energy through them are assumed not to have changed.

Some paleoecological inference is independent of comparisons with modern communities. Some fossils preserve the evidence of interactions between the organisms they represent. Predator—prey relationships are shown by the ammonite shell bearing the tooth marks of the mosasaur that killed it (Fig. 17–2), or the starfish caught in the act of feeding on a bed of bivalves. The presence of scavenger rodents is shown by tooth marks on fossil mammal bones. Parasitism is evidenced by pathologically developed specimens. Interdependence can be deduced from the constant occurrence together of two organisms, such as the gastropod *Platyceras* on the anal openings of Paleozoic crinoids. Fossil

lungfish found in burrows give clear evidence of their ability to aestivate. To reconstruct the past environment successfully the paleoecologist must combine comparisons with modern communities with evidence available within the fossil deposit itself.

Figure 17-2   Jurassic ammonite *Cardioceras* with the shell broken into by a predator, probably a mosasaur. (C. W. Stearn, Redpath Museum.)

Niches, Communities, and Ecosystems:

Differences in physical and biological features of its environment limit the range of an organism. Physical features include such parameters as temperature, light, oxygen availability, and nutrient supply. Biological features include competition for resources with adjacent species of similar ecological requirements. The "space" bounded by values of these parameters acceptable to the organism constitute its niche. Niches in medieval cathedrals are indentations in the wall made to display the image of a saint. The "fences" to this "space", both physical and biological, restrict the expansion of the species just as the stone walls of the cathedral contain the statue. The word *space* is placed in quotation marks to indicate that the niche of an organism may be bounded by other than geographical changes. Its expansion may be limited as much by internal factors, such as its cycle of reproduction, or the number of offspring produced. A more theoretical definition of a niche is "an abstract hypervolume situated in a space whose axes correspond to the biological and environmental variables affecting the organism in question" (1).

Some organisms are adapted to a narrow range of conditions and are said to be specialized or to have narrow niches. Many parasites are of this nature and can live only in a particular part of their host organism. The narrowness of the niche of many animals is determined by their restriction to a specific food. The koala spends most of its life in eucalyptus trees because it only eats the shoots of this plant. Animals with broad niches include the common rat (*Ratus norvegicus*) and the house fly (*Musca domestica*). Species that live in rapidly changing environments must adapt to a wide range of conditions and have broad niches. These environments are common at middle latitudes and at shorelines. The modern bivalve *Hiatella arctica*, which is found from Greenland to Hawaii, is such a species. Species that live in stable environments, such as tropical seas, tend to have narrow niches and more specialized life-styles.

Ecologists use a hierarchy of terms to refer to groupings of organisms. The members of a species living together are commonly referred to as a population. The different species that live together in a particular environment may be referred to as a community. The deciduous forest community or the intertidal marine community are examples. Communities are commonly named after the dominant or diagnostic species. Ecosystems are larger units consisting of all the organisms that interact in an area. A small ecosystem might be an isolated island; the largest constitutes the whole planet. Paleoecologists have concentrated on the reconstruction of ancient communities as a first step in formulating a history of the world ecosystem.

PALEOECOLOGICAL EVIDENCE

Taphonomy:

In Chapter 1, the variety of pathways by which organisms enter the fossil record was traced. The probability that a pathway will open is much less for some organisms than for others and therefore the fossil record is neither a complete, nor even a random, sample of the life of the past but is biased. The path to preservation is almost always barred to soft-bodied organisms and is rarely open to organisms inhabiting environments where sediment is not being deposited rapidly. Even hard-shelled organisms must run the gauntlet of predation, abrasion, bioerosion, and dissolution in order to reach a permanent place in the fossil record. The branch of paleontology that is concerned with the way organisms enter the fossil record is called taphonomy. The effects of the preservational processes must be taken into account before a paleontologist can start the interpretation of an ancient community from its fossil remnant. Distinctive modes of preservation may aid in reconstructing ancient environments.

The extent of information about an ancient community lost to the paleontologist by the selectivity of the preservational processes can be estimated for specific communities but results are not applicable to other communities. Comparison of living communities with fossil assemblages derived from similar communities has given interesting but discouraging results. Only Cenozoic fossil assemblages contain species whose ecological requirements we can be confident were like those of living species. In one of the first studies, David Lawrence (2) compared the macroscopic organisms of a modern oyster bank community on the southwest coast of the United States with a fossil assemblage from an Oligocene oyster bank in the same area. He found that almost 50 percent of the species of the modern oyster community had no chance of preservation because they did not have hard shells. Representatives of only 25 percent of all the macroscopic organisms of the oyster bank community were found in the equivalent fossil assemblage—a loss of 75 percent of the species.

Taphonomic studies by Anna Behrensmeyer and her co-workers (3) of bones in the Amboseli basin in east Africa showed that only 74 percent of the vertebrate fauna was represented by bones lying on the plain, and these were rapidly falling apart under the influence of scavengers, bacterial decay, and weathering during the ten-year observance period. Most of the species that are not represented in the bone deposits are of animals under 1 kg in weight and are believed to have been eaten by predators and scavengers.

The most valuable communities for paleoecological analysis are those that have been overwhelmed by a catastrophic storm or a flow of sediment that by burial "freezes" the assemblage on the bedding plane. Such assemblages of fossils have been called census assemblages or, less correctly, life assemblages. In a true life assemblage all the species of the living community are present in the same proportions as they were in life. However, no fossil assemblage could be preserved in this way. All bear the impression of preservational processes and are therefore death assemblages, not life assemblages. The effect of these processes must be assessed by the paleoecologist in reconstructing ancient environments.

Comparisons can also be made between the living community and the potential fossils, that are accumulating in the sediment from its skeletal parts. Charles Peterson (4) studied the shelled animals in lagoons along the California coast, sampling the living and dead mollusc shells in the sediment. He found that more species were represented in the deposits than in the living community on its surface. From season to season different species occupy the sample site and leave their shells behind so that what accumulates is a sample, not merely of the group of species that happens to be at the site when the survey was taken, but of a much wider community spread over a larger area. He found that the difference between the species composition of live and dead communities at a site was greatest where the communities are most variable in time.

Time Averaging:

Most marine, fossil-bearing, sedimentary beds contain an accumulation of shells representing the fauna that occupied the site over a period of many years. If sedimentation is slow and conditions at the site change, different communities may get mixed in the fossil deposit, that is, a succession of communities are condensed into the bed. If deposition is rapid, the effects of this time averaging are not as great because the record of the ecologic events is distributed through a greater thickness of sediment (Fig. 17–3).

Most fossil collections consist of specimens collected from a single bed of sedimentary rock. The relative proportion of the various species living in an environment is rarely constant over the years required to deposit a bed and may change as a response to the seasons (Fig. 17–4), catastrophic environmental events such as hurricanes, diseases, or the invasion of a new predator. After a hurricane in Barbados the area covered by the most abundant coral, *Porites porites*, dropped from 11 to 1 percent. Following a disease that decimated the common Caribbean sea urchin *Diadema antillarum* in the early

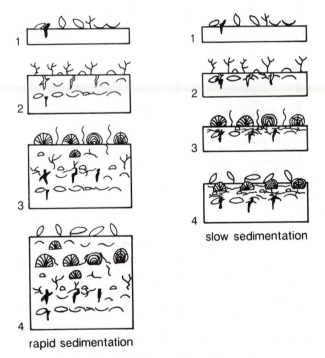

rapid sedimentation

slow sedimentation

Figure 17-3   The effect of rate of deposition on the condensation of a succession of faunas at a depositional site into a single bed. If sedimentation is rapid, the four faunas (1, 2, 3, 4) are separated in the accumulating sediment; if it is slow, they are difficult to distinguish from one another and the infaunal organisms in one are mixed with epifaunal organisms in earlier communities.

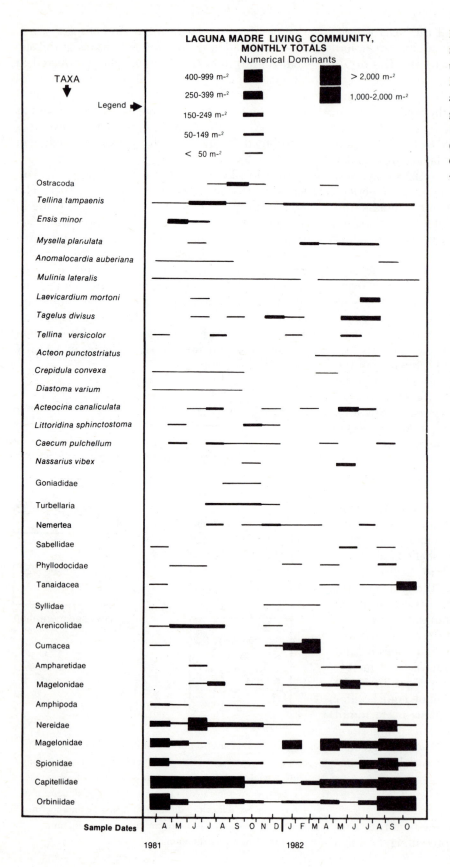

LAGUNA MADRE LIVING COMMUNITY, MONTHLY TOTALS
Numerical Dominants

Figure 17-4   Variation in time of the numerical abundance of the dominant taxa in a living community in the Laguna Madre, Texas Gulf coast. The abbreviations along the base of the diagram are the months of 1981 and 1982, when the samples were taken. (Courtesy of George Staff and others, Geological Society of America Bulletin, v. 97, p. 433.)

1980s the density of this species on the same reef dropped within a few months from 23 per m^2 to 0.3 per m^2. The plague of the coral predator starfish, *Acanthaster*, on Indo−Pacific reefs in the 1970s caused sudden radical changes in the coral biotas that will affect the composition of the communities for many tens of years. These biotic changes may not be caused by changes in the physical parameters of the environment and may have little long-term effect on the community. Swings in species composition are evened out by the condensation of successive faunas into a single bed and the paleoecologist may get a better picture of the average community composition from the fossil record than the ecologist who measures the distribution of species over only a few years. However, these short-term variations in communities should make the paleoecologist wary of using proportions of species to conclude that adjacent fossil assemblages came from different environments and of taking census communities as typical of past environments.

George Staff and his co-workers (5) assessed the effects of taphonomy and time averaging on death assemblages in two Texas bays (Fig. 17−4). They concluded that because fossil assemblages are described using time-averaged data, they cannot be accurate representations of the living community. The fossil assemblages will be less similar to the living community the greater the temporal variability of the living community and the longer the time averaged in the fossil assemblage. They found that the order of numerical abundance of species is rarely the same in life and death assemblages.

Most studies of marine taphonomic processes have been made on coastal molluscan communities in areas of soft sediment deposition. These studies suggest that new principles may have to be formulated to interpret fossil assemblages because ecological principles formulated to interpret life assemblages may have limited application. Closer resemblance between fossil and living assemblages may be found in reef communities whose members are largely cemented in place and in deeper water communities in areas of constant sedimentation.

Dissolution of shells:

The destruction of hard parts does not end with their entombment. During millions of years of burial many shells are dissolved by water seeping through the sediment. Peterson (4) estimated the rate of dissolution of shells in the sediment of the

Californian lagoons and determined by extrapolating short-term rates that different molluscs had residence times of from 5 to nearly 900 years in the sediment.

Of the two calcium carbonate minerals common in invertebrate shells, aragonite and calcite, aragonite is more soluble and therefore shells composed of it soon dissolve and such shells are rare in older sediments. If the aragonite shell is dissolved after the sediment is cemented, a mold of the shell may be left behind but if the sediment is poorly consolidated and collapses into the cavity, no trace may remain. C. F. Koch and Norman Sohl (6) used the extent of preservation of aragonite in many bulk collections in a particular Cretaceous clayey sandstone of the Gulf Coastal Plain to assess the effects of postdepositional dissolution on species composition. Collections in which aragonite shells are well preserved (and are deduced to have been little affected by dissolution) contained 75 different species; those from which aragonite was dissolved contained only 18 species. The authors concluded that a great many species of the community will not be recognized in the most altered samples, and therefore only those collections in similar states of preservation should be compared in paleoecological analyses to minimize postdepositional modification of species distribution.

Transported Assemblages:

The hard parts of organisms, like pebbles and sand grains, can be carried by streams, currents, and waves from one place to another. If the remains of organisms are transported from the environments in which the organism lived, they will give false information about the environment in which they are eventually deposited. Tidal currents flush the shells of animals living in brackish water of estuaries onto the continental shelves and wash shelf shells into estuaries. Shells of shallow-water invertebrates are carried to abyssal depths down continental slopes by slumping sediment and turbidity currents. Many Paleozoic deep-water successions consist of shales containing fossils of planktonic animals, such as graptolites, and no bottom fauna, interbedded with limestones bearing shallow-water fossils, such as brachiopods and crinoids. These deposits are not a record of two basically different environments alternating as the site of deposition went repeatedly from shallow to abyssal depths; the deposit was formed as muds accumulated in deep water at the base of the continental slope and shallow-water sediments and organisms were introduced periodically by the slumping of sediments from a shallow shelf into the abyss. Because the wind carries pollen and spores from plants and drops them in all environments of deposition, these tiny fossils are particularly valuable for correlating diverse facies but not necessarily diagnostic of the environment of the sediment in which they are found. The bones of dryland species of animals may be carried by streams in flood into swamp and lake deposits far from where they lived.

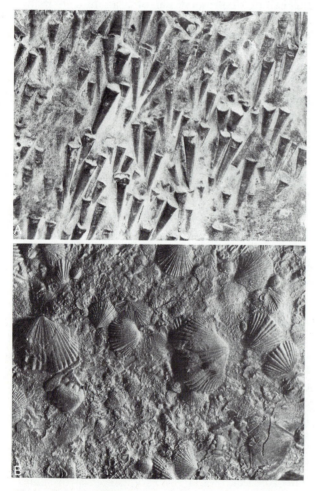

Figure 17-5   Orientation of fossil shells. A. The enigmatic conical Cambrian fossil *Salterella* aligned by currents or waves on a bedding plane. (Courtesy of William Fritz, Geological Survey of Canada (#85898); *Canadian Journal of Earth Sciences*, v. 25, pp. 403–416.) B. Brachiopods (*Hebertella*) and trilobite gabellas (*Ceraurus* and *Flexicalymene*) in the hydrodynamically stable position (convex upward) on the bedding surface of an Ordovician limestone. (C. W. Stearn.)

Evidence that most invertebrate fossil assemblages have been affected by waves and currents is everywhere. On bedding planes most saucer-shaped shells of brachiopods and bivalves are preserved with the convex side upward—the most stable position in moving water (Fig. 17–5B). Shells in a fossil assemblage are usually similar in size because currents have carried away the small, immature shells, sorting the deposit just as a stream sorts sand, clay, and cobbles. Bivalved shells should be represented in an undisturbed deposit in equal numbers, but if the valves have different sizes and shapes and are deposited in moving water, they will be sorted like any other sedimentary particles and deposited in different areas. Log-shaped shells are commonly oriented by waves or currents on a bedding plane (Fig. 17–5A). The paleoecologist must watch for such signs of transported assemblages and take into account that not all the fossils in the collection represent animals that lived in the area where they were collected. However, most studies of recent biotas suggest that extensive transportation of marine shells is uncommon.

Eliminating Collection Biases:

The collection from an outcrop should be representative of the fossils in the rock, yet casual collectors do not make such collections. They are naturally attracted to large specimens, unbroken specimens, and rare specimens. Once they have collected a few of the common species, they will tend to ignore them and seek rarer members of the assemblage. Such collections are not representative of what was preserved in the rock.

These collection biases can be avoided by bulk collecting. A certain volume or weight of rock (e.g., 0.5 m^2 or 2 kg) is collected at each site. Many Cenozoic sediments are poorly consolidated and can easily be broken down into their grains and fossils in the lab. All the fossils in the rock are identified and counted. This method is difficult for well-cemented rocks, such as those of Paleozoic age, and the paleoecologist who uses it must patiently break the rock into small pieces with hammer and chisel to identify all the fossils and their fragments.

An estimate of the proportion of fossil species on a rock surface may be made by quadrats (a square sample area) or line transects. Typically a bedding plane is marked out in a grid pattern whose dimensions depend on the area to be surveyed. All the fossils in each square may be identified and counted if the area is small, or the squares can be numbered and those in which the fossils are to be counted are identified from a table of random numbers. The line transect method is used by both paleontologists and ecologists. Paleoecologists draw lines or stretch strings across a bedding surface at regular intervals or at random. Each fossil or fragment intersected by the line is identified and counted. The result may be expressed simply as a count of the number of individuals of each species, or the distance along the line occupied by the fossil may be measured to estimate the area of the bedding plane occupied by each type of organism. Point count methods are simpler. The line is marked at regular intervals, for example every 2 cm, and a count is made of the species (or sedimentary rock) at each mark. The counts for each species will be proportional to the area occupied on the surface.

If the proportions of organisms preserved in a deposit are so dependent on temporal changes in the living community, transportation of some of the constituents from the burial site, and loss of many community members during preservation, why should the paleoecologist be concerned with the precise composition of the fossil assemblage? Paleoecologists can hope to make some adjustments for the previously described biases but the first step in any analysis must be the determination of exactly what has been preserved.

LIMITING ENVIRONMENTAL FACTORS

Niches are bounded by an organism's tolerance for environmental change. Important physical parameters of the environment that limit niches are temperature, light, salinity, nature of the soil or bottom sediment, and rainfall. Study of the distribution of living animals controlled by these environmental factors may allow the paleontologist to assess the environmental requirements of fossil assemblages and determine the conditions under which the organisms lived.

Temperature:

Decrease in temperature is correlated with increase in latitude, depth in the sea, and height on the land. These gradients are major controls on the worldwide distribution of organisms at present, and are assumed to have been in the past. Cooling climates result in the migration of organisms toward the equator and towards sea level from mountain heights and ocean depths. For marine invertebrates control by temperature has its greatest effect on larval stages. Unless the temperature of the water is within the range of their tolerance, the larvae will not develop into adults.

The fossils of living species whose temperature tolerance is known are obviously of value for determining temperatures of ancient seas. However, most fossil species are extinct and we do not know their tolerances. Morphologies of living species that can be correlated with environmental temperature may also be shown by fossils. For example, the plants of tropical rain forests have large leaves, large breathing pores (stomates), and tips of leaves shaped for dripping moisture. These features in extinct fossils plants, such as those of the coal forests, are interpreted as indicative of warm, moist climates. Among warm-blooded animals the largest members of many groups live at the highest latitudes and have smaller appendages (such as ears and fingers) than their warm climate relatives. Both are clearly adaptations for greater efficiency in retaining heat. The distribution of fossil mammals would be expected to show the same pattern. At present reef-building organisms thrive only in shallow water in the tropics, although some corals do build substantial mounds in deep, cold water. Geologists reason that most ancient reefs must also have been confined to tropical latitudes and construct paleogeography on this basis. However, the assumption that ancient reef builders (rugosans, tabulates, and stromatoporoids) had ecological requirements comparable to modern ones (scleractinians and coralline algae) is not firmly based.

Because no other major environmental feature varies as regularly with latitude as temperature, latitudinal biotic changes are often ascribed to temperature effects even though the action of temperature on the organisms may be obscure. The diversity of animals and plants is correlated with latitude— a greater variety of organisms inhabit equatorial environments than polar environments. When the number of species of a group is plotted against latitude, a gradient of diversity is defined from equator to pole (Fig. 17–6). Although the general conclusion that fossil biotas of high diversity are likely to have lived in low latitudes and those of low diversity in high latitudes can be used in reconstructing paleogeography, sufficiently detailed information on the diversity of fossil groups on a worldwide basis is not available to allow diversity

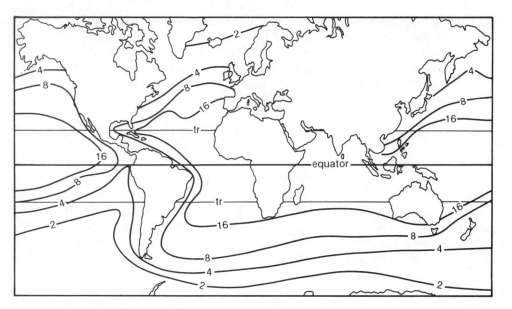

Figure 17-6 Contoured pattern of species diversity in recent planktonic foraminiferans. (*Source:* Modified from F. G. Stehli and C. E. Helsey, *Science*, v. 142, pp. 1057–1059.)

studies to be useful in determining the location of ancient equators and poles.

Decrease in diversity with latitude may be controlled by the greater variability of temperatures in higher latitudes rather than by the latitudinal gradient in average temperature. In low latitudes the temperature changes little from season to season but in temperate and polar latitudes, organisms must be able to withstand wide seasonal and daily variations in temperature. Many ecologists believe that highly diverse biotas can only develop under stable environmental conditions and that environmental instability of any type results in low diversity biotas.

Although the ability of the land to support plants and animals decreases poleward, the productivity of the sea does not and the decrease in diversity of marine biotas is not accompanied by decrease in the total mass of plants and organisms (biomass); that is, the few kinds of organisms that live at high latitudes have large populations. The "cake" representing available marine resources is sliced thinly into many morsels in the tropics and into large portions in subpolar and polar seas.

Determination of temperatures from fossils has been of particular interest to paleontologists examining drill cores of ocean floor sediments. During the recent Ice Age the temperature of the oceans changed cyclically, and continental ice sheets expanded and shrank. The shifting of the climatic zones caused changes in the distribution of planktonic foraminifera whose tests are major constituents of deep-sea sediments. The tests are composed of a series of chambers in a trochoid spiral and may be secreted in a dextral or sinistral pattern (see Chapter 7 for a discussion of these terms for gastropod shells). The proportion of the two types of test secreted by the forams depends on the temperature of the water in which they live (Fig. 17–7). For *Neogloboquadrina pachyderma* the water temperature dividing populations in which most of the tests are dextral from those in which most are sinistral is about 9°C. For *Globigerina bulloides* 80 percent are sinistral in water at 4°C; but only 60 percent at 15°C. Other test features that vary with temperature include shape, number of chambers in the final whorl, porosity of the wall, and overall size. In addition, the specific temperature tolerances of many living foram species are known and, if these species are found in ancient deposits, an estimate of the temperature at which they were laid down can be made.

The proportion of the isotopes of oxygen incorporated in the shells of marine invertebrates varies with the temperature of the ocean water in which they live. The ratio of the isotopes ^{18}O and ^{16}O in fossil shells is used to estimate the temperatures of ancient seas. As invertebrates extract oxygen in the form of carbon dioxide from seawater to form the

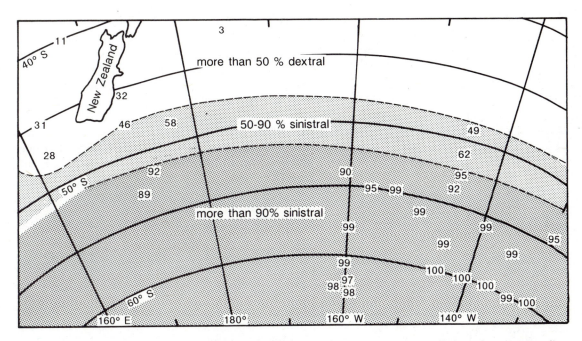

Figure 17-7   Relative proportion of sinistral and dextral specimens of *Neogloboquadrina pachyderma* from recent sediment samples in the South Pacific Ocean. The figures plotted represent the percentage of sinistral specimens. (*Source:* Modified from J. P. Kennett, *Micropaleontology*, v. 14, pp. 305–318.)

calcium carbonate of their shells, fractionation takes place, that is, the proportions of the oxygen isotopes are changed by incorporation in the shell.

$$\frac{^{18}O \,/\, ^{16}O \text{ of shell}}{^{18}O \,/\, ^{16}O \text{ of water}} = 1.021 \text{ at } 25° \text{ C}$$

This relationship indicates that ^{18}O isotope is concentrated slightly in the shell relative to the water. The value of the fractionation factor (1.021 in the equation) is dependent on temperature, and its value at various temperatures is known (Fig. 17–8). If the isotope content of the shell and the water are known, the temperature is easily calculated. The change in ^{18}O in fractionation ($\delta^{18}O$) is usually expressed in parts per thousand with respect to a standard which may be a belemnite guard from the Cretaceous Peedee Formation (PDB) or standard mean ocean water (SMOW). The isotopic ratios in fossil shells can be measured by means of a mass spectrometer but the isotope ratio of the ocean water in which the shells lived must be calculated on the basis of several assumptions. A major control on the sea water ratio is the amount of water removed from continuous circulation in precipitation and evaporation in the weather cycle into continental ice caps, and most paleotemperature measure-

ments involve an assumption as to the extent of glaciation at the time the shell was secreted. The seawater isotope ratio also varies with salinity and for shells from estuaries, or other areas where salinity is abnormal, a correction factor must be applied. Unfortunately, processes of preservation that

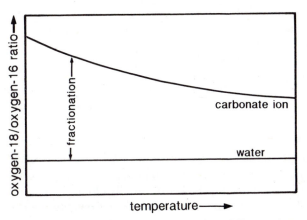

Figure 17-8   Schematic representation of the difference in the $^{18}O/^{16}O$ ratio in water and carbonate precipitated in equilibrium with that of water at different temperatures. The separation of the lines is a function of the temperature-sensitive fractionation. (Courtesy of J. R. Dodd, *Paleoecology, concepts and applications*, Wiley Interscience, 1981.)

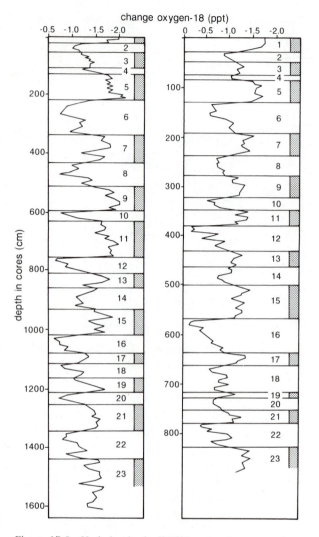

change oxygen-18 (ppt)

Figure 17-9 Variation in the $^{18}O/^{16}O$ ratio of two cores from the Pacific Ocean, showing the use of the ratio in correlation. The numbers and the shaded bars identify the temperature-based time divisions of this interval, as proposed by C. Emiliani. (*Source:* Modified from N. J. Shackelton and N. D. Opdike, *Geological Society of America Memoir 145,* pp. 449–464.)

ratios have been correlated from hole to hole and interpreted as variations in temperature although they may also reflect changes in the volume of continental ice caps (also temperature dependent) (Fig. 17–9). General temperature curves for Late Cretaceous and Cenozoic time (Fig. 17–10) show high early Cenozoic ocean temperatures and decrease toward the late Cenozoic glacial period in which we now live. Separate measurements from planktonic and benthonic foraminifera in the same sample indicate the temperature difference between surface

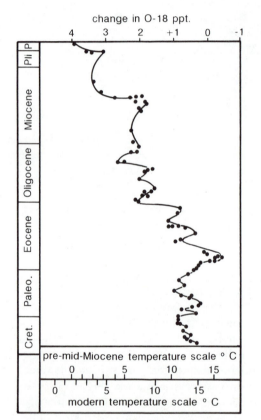

Figure 17-10 Oxygen isotope composition of multispecies assemblages of benthonic foraminiferans from several deepsea cores. The figure shows how they can be used to interpret temperature trends. The pre-Miocene temperature scale is calibrated on the assumption that there were no glaciers at this time. The modern temperature scale is based on the present distribution of glaciers. Temperatures based on oxygen isotopes can only be interpreted on the basis of assumptions about the extent of glaciation. The oxygen isotope scale is in parts per thousand. (*Source:* Modified from S. M. Savin, R. G. Douglas, and F. Stehli, *Geological Society of America Bulletin,* v. 86, pp. 1499–1510.)

alter the shell in any way usually change the isotopic ratios and make accurate paleotemperature determinations impossible. Most shells older than late Mesozoic have been so changed that paleotemperatures cannot be measured from them.

The greatest success of the oxygen isotope method has been in the study of foraminiferal tests of Cenozoic age, particularly those recovered from the Deep Sea Drilling Project. Variations in the isotope

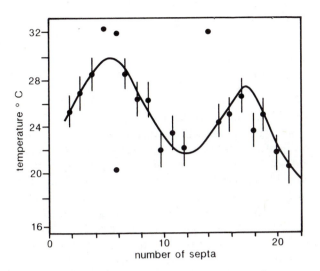

Figure 17-11 Temperature of formation of septa of a Jurassic ammonite, based on oxygen isotope values. If the variation with age is interpreted as seasonal, the ammonite must have been born in the spring and died in the winter of its second year. The values far off the curve are questionable. (*Source:* Modified from W. Stahl and R. Jordan, *Earth & Planetary Sciences Letters*, v. 6, pp. 172–178, Elsevier Scientific Publishing Co.)

and bottom waters for that location. Studies of variation of oxygen ratios within a single shell, such as an ammonite (Fig. 17–11), may reveal not only the temperature of the ocean in which it lived, but also its seasonal range, the age of the animal, and the time of year that it died.

Salinity:

Salinity is an important control on the distribution of aquatic animals and plants. Organisms living in salinities greater than that of their internal fluids tend to lose water through their cell walls and must have mechanisms to retain or replace it. Those in salinities less than that of their fluids must have mechanisms to get rid of the water that enters through their cell walls. The laws of osmosis apply to these relationships. Through such mechanisms organisms are adapted to a particular salinity environment, and only a few can tolerate large changes, or ranges, of salinity. Marine fish that spawn in fresh water are an example of such animals. However, echinoderms, foraminiferans, and brachiopods are now confined to water of close to normal marine salinity (about 35 parts per thousand) and

no evidence has been found to suggest that they lived differently in the past. Their fossils therefore indicate marine conditions of deposition. Bivalves and gastropods have adapted to freshwater environments but other molluscan classes are confined to marine water. Most sponges, cnidarians, brachiopods, bryozoans, cephalopods, and hemichordates are exclusively marine. The shells of freshwater invertebrates may be distinguished by the thickness of the organic covering formed to prevent dissolution of the calcium carbonate in the undersaturated water, or by signs of post-mortem dissolution where the covering has broken down.

Biotas in hypersaline environments, such as coastal lagoons in areas of high evaporation, are of low diversity and are rich in ostracodes and stromatolites. Hypersaline environments can also be identified by the association of the sediments with evaporite minerals such as halite.

Estuaries are characterized by salinity gradients that may be identified in the fossil record. As salinity lessens upstream in the estuary, the amount of calcium carbonate available for shell secretion decreases and the shells of animals of wide salinity tolerances become thinner. The living mussel, *Mytilus edulis*, can live in a wide range of salinities from 5 parts per thousand to normal seawater and penetrates far into estuaries, but its shell in lower salinities is thinner and smaller than in normal sea water. Gradients in shell thickness in fossil species may point to gradients in salinity. Since most marine species have limited tolerance for salinity changes, the number of species in estuaries decreases as the water freshens. The Baltic Sea is not a simple estuary but the salinity drops from normal at its North Sea entrance to about 5 parts per thousand at the Gulf of Bothnia (near Finland). Figure 17–12 shows a diversity gradient in molluscs that accompanies this salinity gradient. A similar gradient dependent on salinity can be traced in the fossils of the sediments of the Champlain Sea that flooded into the St. Lawrence estuary at the close of the last glaciations, about 10,000 years ago (Fig. 17–13).

The proportion of carbon isotopes ^{12}C and ^{13}C has been used to estimate the salinity in which fossil shells secreted calcium carbonate. The method is based on the deficiency of the ^{13}C isotope in land plants and humus that supply carbon to estuarine brackish waters. The ^{18}O /^{16}O ratio also varies with salinity as fresh water derived from rainfall and

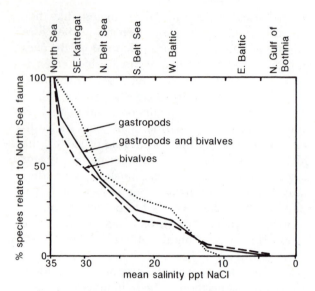

Figure 17-12 Decrease in the number of molluscan species from the normal salinity environment of the North Sea, along the salinity gradient of the Baltic Sea. The mean salinity is expressed in parts per thousand (ppt). (*Source:* Modified from D. V. Ager and T. Sorgenfrei; courtesy of Geological Survey of Denmark, 1958.)

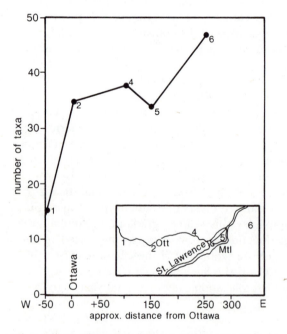

Figure 17-13 Diversity gradient believed to be related to decreasing salinity westward in the late Pleistocene Champlain Sea, in the region of the Ottawa and St. Lawrence valleys. The position of the six collections with relation to Ottawa (Ott) and Montreal (Mtl) is shown on the inset map. (*Source:* Plotted from information of F. Wagner, *Geological Survey of Canada*, 1970.)

ultimately from evaporation of seawater is enriched in the light isotope ^{16}O.

Illumination:

Light is essential to the process of photosynthesis and the life of all green plants. As all animals are dependent on plants, directly or indirectly, most life forms are limited in their range by the availability of light. At the land surface light is generally available everywhere but is locally restricted by a thick overhead canopy of vegetation that limits the type of plant that can grow at ground level. In the sea, sunlight penetrating the surface is absorbed in passing downward through the water. Light of red wavelengths is the first to be absorbed so that to a diver deeper than 10 m most objects appear to be shades of blue. The depth at which all light has been absorbed is about 200 meters but varies with the clarity of the water and the amount of sunlight penetrating the surface. This level is the base of the photic zone and the top of the aphotic zone. The amount of sunlight refracted into the water depends on the angle at which the sun strikes the water, and hence the latitude and season, and also on the turbulence of the water surface. Planktonic plants, the primary food producers of the sea, are confined to the photic zone and are concentrated near the surface. As a result animal life is most abundant in surface waters where food is plentiful because the light is strong. Animals living below the photic zone must rely for food on dead organisms that "rain" down from above. The abundance of fossil algae is direct evidence of a shallow-water environment. Any marine deposit in which invertebrate marine fossils are abundant is likely to have been formed in shallow water in the photic zone.

Several groups of invertebrates rely on algae within their cells to supply their oxygen needs through photosynthesis while the algae use the cell's waste carbon dioxide for their own metabolism. In modern reef corals the symbionts are a form of dinoflagellate called zooxanthellae. In terms of light requirements the corals act as if they were plants because they have become completely dependent on their symbionts. Reef corals only thrive within a few meters of the sea surface and are sparsely distributed and slow growing at depths below 50 meters. The lower limit for survival of corals dependent on zooxanthellae is about 70 meters; below this level the light is not strong enough for the photosynthesis of the zooxanthellae to satisfy the oxygen requirements of the coral. The rapid secretion of calcium carbonate which is characteristic of scleractinians is also believed to depend on the zooxanthellae because corals secrete skeleton faster in sunlight. Several mechanisms have been proposed to explain how they help in secretion but no consensus has been reached. Some paleoecologists have argued that the massive carbonate secretion evidenced by Paleozoic corals points to a symbiotic relationship between the corals and zooxanthellae, but as the microorganisms are not preserved, no direct evidence supports this suggestion. The strong domal form of many Paleozoic corals has suggested to some the opposite conclusion: that they had to grow wave-resistant colonies because their relatively slow growth rate in the absence of zooxanthellae made the repair of broken skeleton difficult. They reason that one difference between the scleractinians and Paleozoic corals is that the former much more commonly grow to branching fragile forms because their relatively rapid growth rate, helped by zooxanthellae, allows them to repair wave damage without difficulty. George Stanley (7) presented evidence that the scleractinians did not become diverse and abundant until latest Triassic time yet they first appeared in the Middle Triassic. He suggested that they did not achieve their present dominant position in reefs until they teamed up with zooxanthellae at the end of the Triassic Period.

The larger foraminiferans of today also have a variety of algal symbionts including zooxanthellae, red and green algae and diatoms, and those of the past, such as the nummulites and fusulines, probably also had them. The giant clam *Tridacna* supports its extravagant deposition of calcium carbonate through zooxanthellae in the mantle and paleontologists have speculated that the rudists of the Cretaceous satisfied their massive skeletal requirements through a similar symbiosis. All these animals, being light dependent, must have been confined to shallow water.

Dissolved Gases:

The distribution of organisms in the sea is also controlled in part by the concentrations of the two gases involved in metabolism, carbon dioxide and oxygen. As these gases dissolve in the sea from the atmosphere, their concentration might be expected to decrease with depth but the situation is more complex. The flow of ocean water along the deep-sea floor from pole to equator brings dissolved gases to great depths and allows animals to live in abyssal environments.

A larger amount of carbon dioxide is dissolved in the oceans than is present in the atmosphere. The availability of the gas does not limit the range of marine plants that use it, but its distribution in the ocean controls the solubility of calcium carbonate and therefore the occurrence of modern and fossil shells. The solubility varies inversely with the concentration of carbon dioxide: the more CO_2 present in the water, the more soluble is $CaCO_3$. Carbon dioxide, like all gases, is more soluble in cold water than hot. The warm surface waters of the ocean are therefore relatively deficient in dissolved CO_2 and are supersaturated in $CaCO_3$. Shells secreted in this environment have no tendency to dissolve and extraction of the carbonate from seawater is easy. The colder water deep in the ocean can dissolve more CO_2 and is undersaturated with $CaCO_3$. In this water $CaCO_3$ dissolves and the shells of planktonic organisms, such as foraminiferans, that fall from warm, saturated surface water begin to disappear like snowflakes melting as they reach warmer air. This dissolution occurs at depths of 4000 to 5000 meters in the sea and the zone is known as the lysocline. Sediment deposited below the lysocline contains no carbonate fossils despite the steady supply from above of calcareous tests. The lack of carbonate fossils in some post-Jurassic sedimentary rocks is therefore an indication of deposition below the lysocline. Planktonic foraminiferans did not evolve until Late Jurassic time and the absence of their tests in older sediments cannot be used as an indication of deposition in deep water.

The oxygen content of the sea is between 5 and 6 milliliters per liter at the surface and decreases to values of less than 1 ml/l at depths of about 800 meters before gradually increasing again into abyssal water. As abyssal sediments amount to only a small part of the stratigraphic record, the paleontologist is most concerned with environments in the upper 800 meters of the ocean in which O_2 content declines. The water in which oxygen content ranges from surface values to about 1 ml/l is the aerobic zone. That in which oxygen content is 1 to 0.1 ml/l is the dysaerobic zone, and that with less oxygen is the anaerobic zone. The last of these cannot support any significant amount of metazoan life. The effect of the oxygen gradient on life has been studied in the Black Sea and off the California coast (8). Faunal changes that take place along the gradient and might be recognized in the paleontological

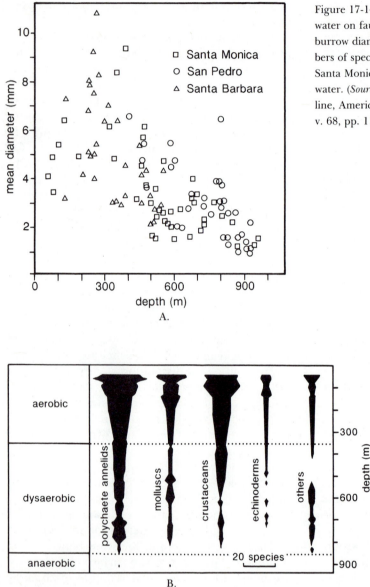

Figure 17-14   Influence of the decreasing oxygen content of the water on faunas off the California coast. A. Decrease of average burrow diameter with depth in three basins. B. Decrease in numbers of species of the major organism groups with depth in the Santa Monica basin and the zones of oxygen content in the water. (*Source:* Modified from C. Savrda, D. Bottjer, and D. Gorsline, American Association of Petroleum Geologists Bulletin, v. 68, pp. 1179–1192.)

record include (1) decrease in number of species; (2) decrease in those that secrete calcareous shells; (3) decrease in the size of organisms and burrows they make in sediment; and (4) relative increase in the number of animals living in burrows. Studies on the California continental slope showed a surprising number of brittle star echinoderms living in the dysaerobic environment. Charles Savrda, David Bottjer, and Donn Gorsline (9) have suggested that the size of burrows made by organisms in ancient sediments could be used as an index of the degree of oxygenation of the water (Fig. 17-14). The biofacies models for recent continental margins have been applied to Cretaceous and Devonian formations (10,11). Anaerobic environments are important to the petroleum geologist for many of the source beds of petroleum accumulated in this zone where lack of oxygen excluded scavengers and retarded bacterial decay so that organic matter was retained in the sediments as it accumulated. At times of transgression in the past the waters of the anaerobic zone spilled onto continental shelves, depositing black, lifeless shales and causing widespread extinction of shelf organisms. Such events (called anoxic events) in Late Cretaceous and Late Devonian time were important not only because they resulted in the spread of potential oil source beds, but also for their deleterious effect on shelf biotas.

Turbulence and Substrate Stability:

Marine organisms live in an environment that is rarely still. The surface waters of the sea are almost always in motion, shaped into waves by the wind acting on the surface. The motion is transmitted downward in the water decreasing in its intensity to stillness at a depth that is dependent on the wavelength. The depth to which wave motion penetrates is defined as wavebase. During intense storms wave motion may reach to depths of about 70 meters, but under average conditions it rarely extends to more than 10 meters.

Shallow-water and shoreline organisms are most affected by turbulence in their environment. At the shoreline of a rocky coast animals and plants must cling tenaciously to the substrate to maintain their position. Some bore into the rock for protection from the surf. On shorelines with beaches invertebrates must have heavy shells to maintain stability and strong muscles to control the shell. Some invertebrates in this zone also burrow in the sediment for protection when the tide is out but permanent homes there are difficult to maintain as the sediment is constantly being moved about by the waves.

The organisms that live below the sediment surface constitute the infauna, or are said to be infaunal. Those that live on the surface of sediment or attach themselves to rocks or to the shells of other animals constitute the epifauna. Most invertebrate phyla include both types of animals but in some, such as Cnidaria and Bryozoa, infaunal members are rare.

The nature of the seafloor sediment influences the type of organisms that live in an environment and its nature, in turn, is influenced by the organisms. The size of sedimentary particles influences the organic matter content and degree of oxygenation of the sediment. In turbulent environments fine sediment is always kept in suspension—even coarse sand is moved about in storms. Such environments are the deserts of the sea, difficult for infauna and almost impossible for epifauna. Silts and clays are laid down in tranquil water and accumulate organic matter because the sediment is not continuously stirred up by oxygen-rich water favoring bacterial decay. Such environments are rich in infaunal organisms that live on the interstitial

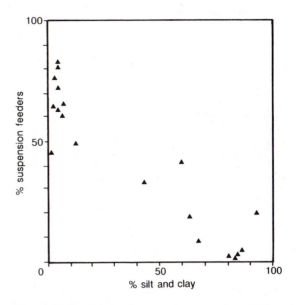

Figure 17-15   Relation between the proportion of infaunal filter feeders and fineness of the sediment in Buzzards Bay, Massachusetts. Filter feeders prefer coarse-grained sediments. (*Source:* From the work of H. L. Saunders.)

organic matter but are not favored by suspension feeders whose apparatus is clogged by the fine sediment in the water (Figure 17–15). The mixing of sediment by the action of animals moving through it will be most intense in fine-grained, organic-rich sediments and evidence of this is common in the fossil record. The association between organisms and the sediments in which they prefer to live is established in the larval stage. The free-swimming larvae of many invertebrates are capable of testing potential settlement sites chemically and physically to see whether they are suitable for their needs and can move on if the first settlement point is not suitable.

Although suspended mud must be detrimental to filter feeders, some of the most fossiliferous sedimentary rocks rich in this trophic group are alternations of clay-rich limestones and calcareous shales.

Attached invertebrates living in nearshore waters must be capable of withstanding storm waves if they are to survive. Obviously fragile branching animals will not be able to live in surf and the animals there, in fact, grow to low encrusting and domal forms. Only encrusting specimens of the species of *Millepora*, a reef-building hydrozoan, live in the nearshore zone of Caribbean reefs. The less robust branching and bladed specimens live in less turbulent water offshore. Individuals of a species that

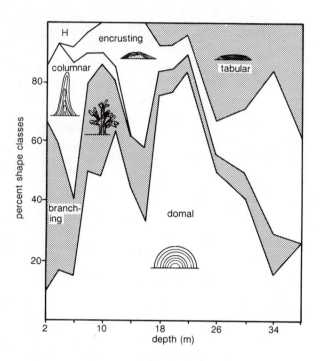

Figure 17-16   Variation across the reef at Aldabra (Indian Ocean) in the shape of scleractinian corals. The percentage of cover by each shape is plotted against depth in meters. *H* is the abbreviation for the shape class honeycomb. Note that at all depths, the reef has a mix of various shapes. (*Source:* Modified from C. W. Stearn, *Paleobiology*, v. 8, p. 231.)

differ in form as a result of the direct action of an environmental factor, such as turbulence, are called ecophenotypes. Although fragile forms of corals are found mostly in less turbulent water than encrusting forms, the distribution of shapes of corals, and other fixed invertebrates, on a reef is not controlled simply by turbulence. Each zone of the reef is characterized by a range of growth forms from branching to domal to encrusting (Fig. 17–16). The shape to which corals grow is controlled by a growth program that is part of their genetic makeup; it is not solely a response to the turbulence of the environment (see Chapter 14). The application of the paradigm approach to the form of corals leads to gross simplification.

With the exception of reef builders, plants and animals attached to hard surfaces are not generally preserved in place because rocky bottoms are not areas of deposition. In reefs the attached biota is continually being overgrown and incorporated into the reef limestone. When the accumulation of loose sediment is interrupted and cementation takes place on the seafloor, a hard bottom surface, or hard-

ground, may be formed. Infauna may bore into this hardground and epifauna become attached to it. If sedimentation resumes with the sudden intrusion of mud or sand, encrusters and borers may be preserved in place attached to and within the hardground (Fig. 17–17).

In reefs and hardgrounds, cavities may be excavated by organisms or formed by the overlapping growth of framebuilders. These shaded or dark areas support a distinctive community that may include some types of organisms that are unable to compete for space with the fast-growing and light-dependent framebuilders of the reef. At present such faunas include sclerosponges, bryozoans, and brachiopods that may be remnants of the Paleozoic order Strophemenida; all animals that were more successful in the reef environments of the past. These cavity-dwelling organisms living under low light conditions are called coelobites. The community has been identified in rocks as old as Early Cambrian age.

COMMUNITY PALEOECOLOGY

Communities and Assemblages:

Most of us have observed that certain organisms are commonly found together in recurring groups that ecologists call communities. The simplest definition of a community is that it is the animals and plants that live together. Erle Kauffman and Robert Scott (12) would add that they should be linked by their effects on one another and on the environment they share. Those that emphasize the biological dependency of organisms in the community view it as a superorganism with properties such as trophic structure, diversity, size distribution, and potential for generating new species. Those who object to the superorganism concept suggest that communities have no reality in nature but we construct them in our minds from the randomly overlapping geographic ranges of organisms. The truth probably lies somewhere between these two extreme viewpoints.

Paleoecologists do not observe ancient communities (paleocommunities) but assemblages of fossils. Through taphonomic analysis they try to reconstruct the missing components of the paleocommunity but complete reconstructions are beyond their grasp. Unfortunately, some write as if fossil assemblages and paleocommunities were the same.

Paleoecologists may recognize assemblages on a

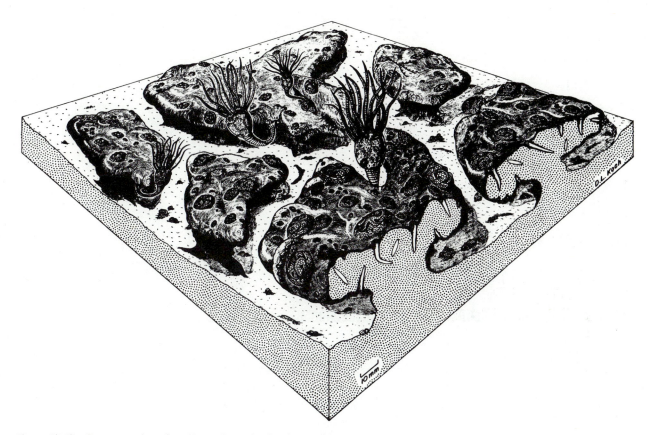

Figure 17-17   Reconstruction of an Upper Devonian hardground in Iowa. The surface was given relief by dissolution. Edioasteroids and cystoids were attached to it and preserved on it. Boring organisms also penetrated the cemented surface. (Courtesy of D. L. Koch and H. L. Strimple, *Geological Survey of Iowa, Report of Investigations No. 5.*)

purely qualitative basis by observing the association of certain fossils or they may use statistical analysis to quantify the similar species compositions of different collections. Cluster analysis, in which taxonomic compositions of a series of samples are compared, is commonly used to establish the degree of similarity among them. Programs for computers are readily available to make these calculations (Q-mode cluster analysis). For example, Figure 17–18 is a dendrogram that is the output of such a program and illustrates the percentage similarity with respect to species composition of a series of collections. Samples from the same assemblage should have larger similarity coefficients than those from different assemblages but the paleoecologist has the choice of the level of dissimilarity required to separate assemblages. The fossils that repeatedly occur together in samples are then used to define the assemblages. For instance, in the assemblage discussed in the following section, the occurrence together of *Cyrtodonta, Loxoplocus, Stictopora,* and *Bathyurus* are sufficient for its recognition, and the name

of one is commonly used to designate it (the *Cyrtodonta* assemblage).

Some communities have abrupt boundaries (the edge of a coral reef) and others grade into neighboring communities. The first types of boundaries are said to be ecotonal; the second, ecoclinal.

In the following sections the properties that have been ascribed to marine invertebrate communities and their application to fossil assemblages are discussed.

Trophic Structure:

At the beginning of this chapter the flow of food energy in an ecosystem through the major groups—producers, herbivores, predators, decomposers—was outlined. The animals can be further divided on the basis of how they obtain their food. The organic matter required by all organisms in the sea that are not primary producers is most abundant at the sediment–water interface. Dead organisms in the water fall to the bottom, or organisms be-

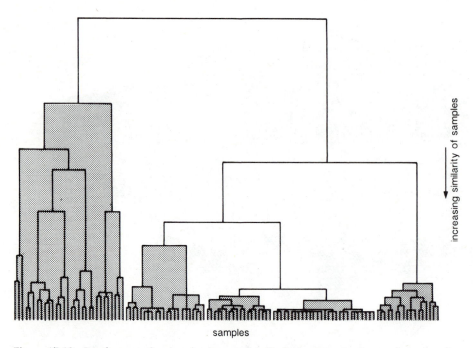

increasing similarity of samples

samples

Figure 17-18 Dendrogram showing the degree of similarity between the fauna of a series of samples from Silurian reefs in northern Ontario. Each vertical line at the base represents a sample containing a suite of species. The computer program arranges these so that those with the greatest number of common species are together and joins the samples and groups of samples at levels of similarity that decrease upward. The major groups distinguished as separate communities in this study are shaded for emphasis. (Modified from Andre Chow.)

neath the surface die, and their organic matter is incorporated in the upper few centimeters of sediment where bacteria swarm to break it down. Bacteria are generally so effective that little survives to become more deeply buried in the sediment. Most invertebrates live near this surface, either in the upper layers of the sediment or on the surface itself.

By applying methods of functional morphology the paleontologist can determine how extinct species fed. Most epifaunal invertebrates are filter or suspension feeders and obtain their food from microscopic organisms and organic particles suspended in the water. Some suspension feeders, such as crinoids, sample the water high above the sediment surface and others, such as brachiopods and bivalves, reach only the lowest layers. This vertical distribution of filterers which divides the resources in the community, is known as tiering. Suspension feeders may be epifaunal or infaunal. Many bivalves, although living in the sediment, feed by passing water through their siphonal system. Deposit feeders are animals that pass sediment through their gut in order to feed on the interstitial organic matter and the organic molecules adsorbed on the surfaces of the grains. Most deposit feeders are in-

faunal but some are epifaunal. A great many infaunal deposit feeders are soft bodied and leave behind as fossils only burrows that are further described in Chapter 20. Other groups that may be recognized in paleoecological trophic analysis are grazers (that eat algal films and seaweed), predators (that eat live animals), scavengers (that eat dead organisms), and parasites (that live on or within a host). The ratio of deposit to suspension feeders has been used as an indication of the wave energy of ancient environments, for only in relatively calm environments does much organic matter come to rest on the bottom and enter the accumulating sediments.

Species are assigned to their various feeding groups in trophic analysis of fossil assemblages. In living communities and fossil assemblages one species is usually found to be numerically dominant and less abundant species do not belong to the dominant's feeding group. For example, the following species were listed by Kenneth Walker (13) from an Ordovician assemblage in New York State:

Cyrtodonta   Low-level filterer   60 percent
  (bivalve)

Loxoplocus (gastropod)	Grazer	20 percent
Stictopora (bryozoan)	High-level filterer	10 percent
Bathyurus (trilobite)	Scavenger, grazer	3 percent
Strophomena (brachiopod)	Low-level filterer	2 percent

The assignment of various species in the assemblage to their trophic roles helps the paleoecologist to reconstruct the environment in which they lived.

The roles of the organisms in the community can be summarized in a cartoon (Fig. 17–19) that illustrates their relationship to the sediment accumulating there.

Diversity:

A community of high diversity is one including many different organisms; one of low diversity includes few. The simplest form of diversity measurement is a count of the number of species present. All estimates of species diversity are based on samples

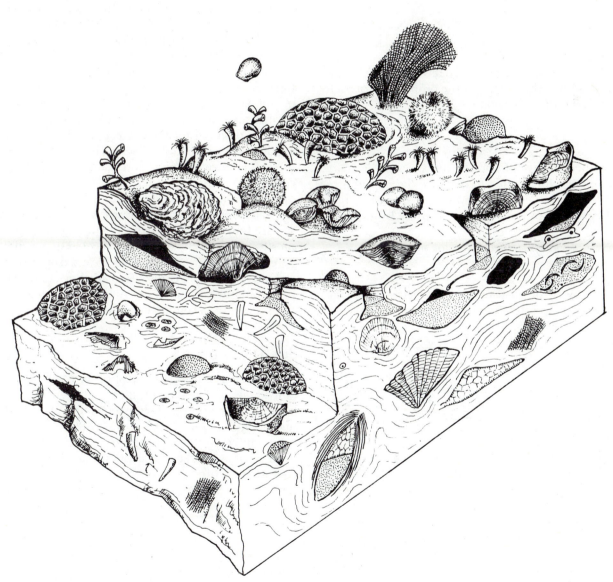

Figure 17-19   Reconstruction of the community on a Carboniferous algal reef. The top surface is the reconstruction; the step represents a bedding plane on which the fossils have been preserved. The community is dominated by the large tabulate coral *Michelinia*, the small rugosan corals *Cyathaxonia*, the bryozoan *Fenestella*, and brachiopods *Streptorhynchus* and *Leptagonia*. (Courtesy of W. S. McKerrow, *The Ecology of Fossils*, 1978, by permission of MIT Press and Duckworth.)

of the community because counting every member of even a small living community would be impossible. Most communities consist of both rare and abundant species. Small samples are likely to contain only the abundant species, but by chance some of the rare ones may be included. As the size of the sample taken increases, more of the less common species will be taken and the species count will rise toward the total number actually present (Fig. 17–20). One simple method of accounting for the difference in number of specimens when comparing two samples is to divide the species count by the size of the sample:

$$\text{Diversity} = \frac{\text{number of species}}{\text{number of specimens}}.$$

George Simpson suggested a diversity index which is the sum of the squares of the proportions of each species in the collection:

$$q = (p_1^2 + p_2^2 \ldots p_n^2).$$

Other more sophisticated measures of diversity have been formulated to decrease the importance of the chance encounter with rare species. Ecologists now favor an index proposed by Brouillon.

$$H = \frac{\log N! - \log n_1! - \log n_2! - \ldots \log n_i}{N},$$

where N is the total number of specimens in the collection, n_i is the number of specimens in the ith

species, and the log of factorial n can be read from tables in statistics texts. A widely used index of diversity is referred to as the Shannon–Weaver function,

$$H' = -\sum_{i=1}^{N} p_i \log_e p_i,$$

where p_i is the proportion in the sample of the ith species. This index has been criticized because, along with the concept of species richness, it includes a measure of equitability, that is, how evenly the species numbers are distributed in the sample. These indexes are pure numbers useful mostly for comparing samples. For instance, the H' coral diversity of Caribbean reefs at present is commonly less than 2, whereas Indo–Pacific reefs have diversities of over 3. Diversity indexes can be calculated from fossil assemblages but will be much less than those of comparable living communities.

Diversity gradients in fossil assemblages may be useful in defining ancient environments. Reference has already been made to diversity gradients related to temperature, salinity, and stability. In marine communities diversity increases steadily offshore onto the continental shelves before decreasing into abyssal waters. The communities of highest diversity are found in stable environments and these are mostly tropical ones. However, ecologists are not agreed that high tropical diversity is caused by greater stability only (14). As discussed further later, mature communities are more diverse than those in the process of becoming established on a new site.

Size Distribution:

Some fossil assemblages consist of unusually small individuals, individuals of species that grow to a larger size elsewhere; such assemblages are called micromorph faunas. If this size distribution has not been caused by the sorting action of waves and currents, it may indicate exceptional environmental conditions. Unfortunately, identification of sorted assemblages in the fossil record has not been easy and unless indications of fossil orientation or wear are found, the distinction between a fauna whose average size is below normal owing to ecological factors and a sorted one is difficult. Micromorph faunas are usually found in iron-rich, pyritic shales. Ernest Mancini (15) proposed three causes of micromorph faunas (apart from sorting): stunting, juvenile mortality, and paedomorphosis. Stunted

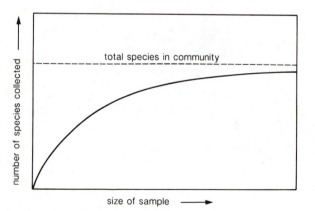

Figure 17-20   Relation between the sample size collected of an assemblage and the number of species recorded. As the sample size is increased, rarer species are collected, but the rate of addition of species slows as the total approaches that preserved.

faunas are the result of the direct action on the organisms of an unfavorable environment reducing their growth rate. Stunted faunas have been ascribed to the action of environments of low or high salinities, low oxygen content, and high iron content. The small size of a group of Cretaceous molluscs from Maryland has been attributed to the low salinity of their environment and this conclusion is reinforced by their association with species that are known to have favored brackish waters. A decrease in size of ostracodes through a Late Jurassic stratigraphic sequence in central England has been interpreted as a reflection of decreasing salinity of the marine environment in which they lived. The inverse relation between the size of the specimens and the amount of pyrite in fine-grained sedimentary rocks containing them has been assumed to indicate that these faunas were stunted by the deficiency of oxygen in their environment and the consequent production of hydrogen sulfide by anaerobic sulfate-reducing bacteria which combined with the iron to make pyrite. A juvenile fauna results from mass mortality of a fauna that is composed mostly of juvenile specimens and should be easy to distinguish if the immaturity of the fossil specimens can be established. Mancini has interpreted a dwarf fauna of bivalves, gastropods, cephalopods, and echinoids from the Cretaceous of Texas as an adaptation of animals to a soft, unsupporting sediment. They adapted by paedomorphosis (see Chapter 15), accelerating sexual maturity (adulthood) into early growth stages because their larger adult forms sank into the mud and died. Such micromorph faunas should be characterized by specimens with a combination of juvenile and adult features.

Succession:

Most nonevolutionary changes in assemblages of fossils collected through several layers of a stratigraphic section are responses to changes in the physical environment. The changes that result from modification of the environment by the organisms themselves are what ecologists call succession. The original studies of succession were made on land cleared of vegetation by forest fires or for farming. As this land again reverts to forest, communities of plants appear and give way in orderly succession to other communities until the stable community of the forest is reestablished. For instance, in temperate latitudes grasses and weeds are replaced by

low bushes. These in turn are replaced by medium-sized hawthorns and small hardwood trees, and these by hardwoods forming a canopy excluding light for the growth of smaller trees, and finally all are replaced by conifers that are the stable, or so-called climax community in this environment. The physical parameters of the environment have not changed but the community has. Early stages appear to prepare the environment for later ones by their biological activity, and are displaced by species of later stages with which they cannot compete. Three stages are commonly distinguished in such a succession: (1) a pioneer stage of low diversity in which species that grow rapidly, expend much of their energy in producing multitudinous offspring, are relatively small, and have short life spans occupy the open area; (2) a mature stage in which diversity rises to a peak; and (3) a climax stage in which slower growing, larger, long-lived species that have fewer offspring displace early stage species, and with their domination diversity declines. The successions described by ecologists have time spans in the tens to hundreds of years. Whether such geologically short-term changes can be discerned in the pre-Cenozoic fossil record is a matter of controversy.

Basic to the succession concept is control by biological changes within the community. Such control is said to be autogenic (originating internally) and contrasted with control by the physical environment which is said to be allogenic (originating externally). Temporal and nonevolutionary changes in fossil assemblages certainly can be found in the paleontological record—separating changes controlled autogenically from those caused allogenically is a problem. The term succession should be applied only to sequences of faunas that are controlled autogenically. Allogenic successions are better termed community replacement.

The succession concept has been most successfully applied in paleontology to the study of vertical changes in reefs. In an analysis of several Paleozoic reefs, Kenneth Walker and Leonard Alberstadt (16) distinguished four stages. In the stabilization stage, soft sediment is populated by organisms that produce a platform on which epifauna can establish itself (Fig. 17–21). In Paleozoic reefs these stabilizers were usually crinoids and other stalked echinoderms. During the following colonization stage the solid debris from the stabilizers is colonized by the first framebuilding organisms, such as bryozoans, tabulates, rugosans, and stromatoporoids. The greatest diversity is reached in the diversifi-

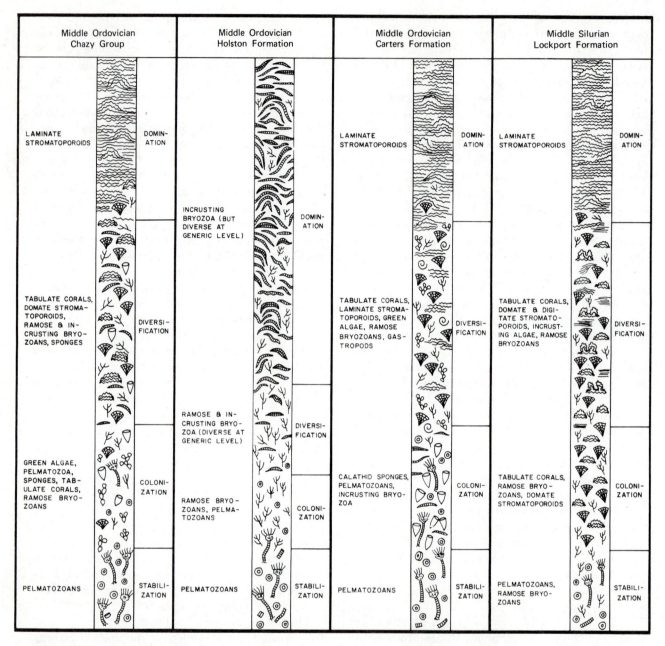

Figure 17-21 Comparison of the developmental stages in four ancient reefs, showing the vertical succession of the reef biota. (Courtesy of K. R. Walker, L. P. Alberstadt, and The Paleontological Society, *Paleobiology*, v. 1, pp. 238–257.)

cation stage when all the frame builders thrive. In the domination stage one highly successful organism excludes most of the rest and the diversity falls. Laminar stromatoporoids usually play this role in mid-Paleozoic reefs. The last of these assemblages appears to be adapted to life in the surf zone and its appearance may be allogenically controlled by the change to an environment of high turbulence. The other stages are believed to follow each other autogenically. Since these changes usually take place through meters or tens of meters of sedimentary rock, some have objected that they must have occurred over an interval of tens of thousands of years because sediments accumulate very slowly in shallow shelf seas where reefs grow. If so, the changes took place at an entirely different rate from that of modern successions. However, reef limestones may build up at rates of 1 m in 100 years much more rapidly than sediments on level sea-floors and therefore these changes may be comparable to those observed by ecologists today.

In nonreefal sections where accumulation is slower, succession takes place within beds a few centimeters thick (17). The impetus for succession appears to be modification of the sediment surface by colonists that allows epifauna requiring a hard base to succeed them. In the Ordovician of Tennessee (Fig. 17–22) the flat brachiopod *Strophomena basilica* forms a pavement at the base of limestone layers underlain by shales. On this pavement branching trepostome bryozoans were established. The bryozoans were used as attachment sites by the rhynchonellid brachiopod *Rostricellula rostrata* which dominates the final stage of the succession. These three assemblages form a true succession, as the first prepared a base for the growth of the second, and the second for the third.

To make a convincing case for succession under autogenic control the paleoecologist must demonstrate that early stages modify the environment in favor of later ones. Most temporal changes in the paleontologic record are likely to be allogenically controlled; true autogenic succession is likely to be found only on reefal assemblages or within a single, thin bed of a level-bottom assemblage.

Communities in the Stratigraphic Record:

Paleocommunities have been named from the environment they are inferred to have inhabited; for example, the beach community, or the reef community, or the level-bottom community. They have also been named from the fossil taxa that characterize them; for example, the *Clorinda* community. The first are not true communities but biofacies that recur over the whole of the Phanerozoic eras. The second only recur when the environment favoring their particular mix of organisms returns to the area during the time range of its constituent taxa. Because organisms evolve, such taxonomically based communities can only be recognized for parts of a geological period but their successors in a particular environment can usually be identified.

A primary basis for relating fossil marine assemblages to each other is their distribution. Most of the environmental factors that limit the ranges of organisms, such as light, turbulence, temperature, gas content, and sediment nature, change with depth. Contemporaneous communities of bottom-living organisms are, therefore, arranged along depth gradients in zones parallel to the shore. Arthur Boucot recognized six benthic zones into which fossil assemblages could be divided (18). Communities in zones 1 and 2 are intertidal. Those in zone 3 live below the tide level in well-lit shelf seas. Zones 4 and 5 comprise the area of the continental shelves below the photic zone, and zone 6 comprises the top of the bathyal zone on the continental slope. Within these depth-related zones different communities, represented by fossil assemblages, occupy level and rocky bottoms, and turbulent and quiet water (Figure 17–23).

Alfred Zeigler's study of brachiopods of the Lower Silurian of Wales was among the first analyses of depth-related assemblages (Fig. 13–4). A reconstruction of the third assemblage is shown in Figure 17–24. A book edited by W. S. McKerrow (referenced at the end of the chapter) is the most extensive compilation of community descriptions. Although paleocommunities should comprise all the organisms living together at one site, they have usually been established by specialists dealing with one group, such as brachiopods. The same suite of specimens divided into communities by a specialist in corals might have different boundaries.

Summary:

Although the community concept is widely used in paleoecological analysis, it is not precisely defined. Assemblages themselves may have such vaguely defined boundaries that they are difficult to recognize outside of their area of definition, so that each paleoecologist is tempted to define assemblages to fit a particular area of study. They are a convenient way of discussing commonly recurring sets of species but whether they are discrete ecological or evolutionary units with properties that amount to more than the sum of their constituent species is a matter of controversy. If communities have an individuality of their own, their descriptors can be compared to those of an individual species as follows:

Community	Species
Diversity	Variation
Trophic mix	Trophic role
New species production	Rate of reproduction
Rate of species extinction	Rate of mortality
Succession	Evolution

Whether the biological integration of communities is great enough for them to act as evolution-

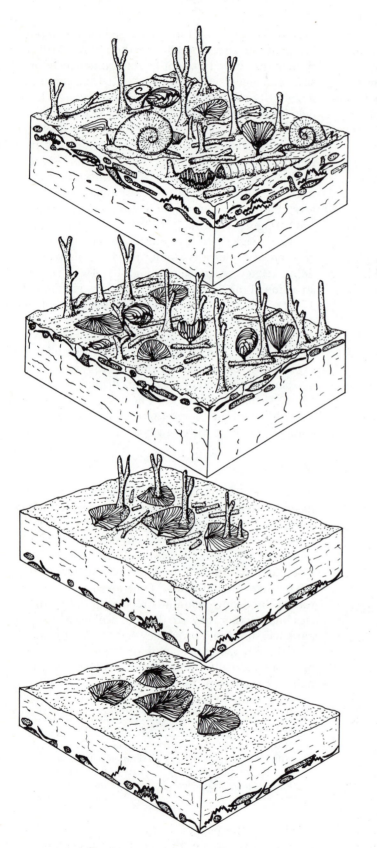

Figure 17-22 Reconstruction of the *Strophomena–Rostricellula* succession. The *Strophomena* provide a solid base for the colonization of a muddy sea floor and allow an increasingly diverse community to populate it. The final community is dominated by the rhynchonellid brachiopod *Rostricellula* but also includes the trepostome bryozoan *Stictopora*, the gastropod *Maclurites*, and the bivalve *Cyrtodonta*. (*Source:* From information of K. R. Walker and W. W. Diehl, *Palaios*, v. 1, p. 70.)

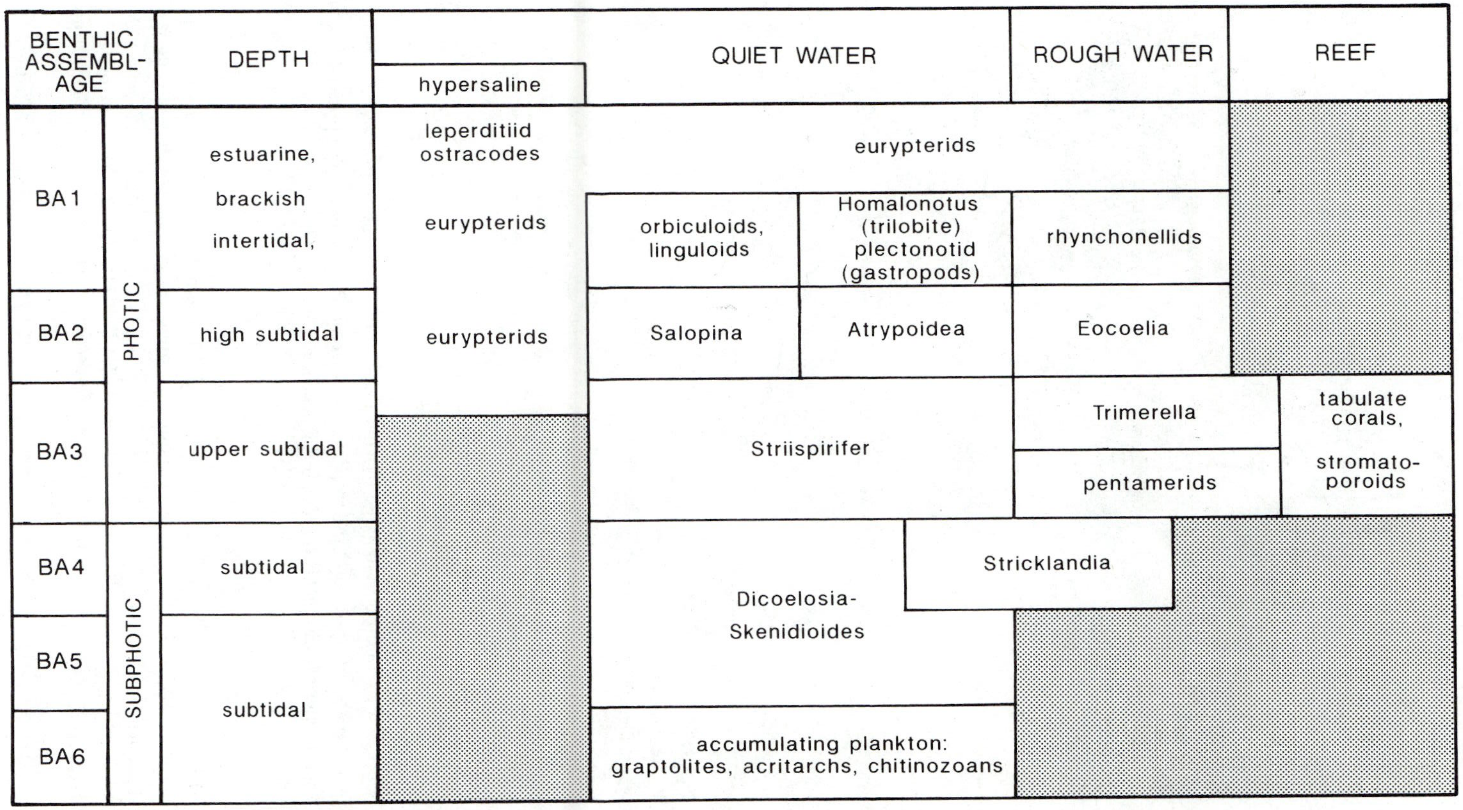

Figure 17-23 Benthic assemblages recognized by Arthur Boucot and their inferred relationship to water depth and conditions based on brachiopod communities (unless otherwise indicated) in Upper Silurian rocks of the North Atlantic area. The names of the fossils in the table are the names of the communities of organisms inhabiting the environments. (*Source:* From information of A. J. Boucot.)

Figure 17-24   Reconstruction of the Silurian *Pentamerus* community. The large brachiopods are the pentamerid *Pentamerus*. The smaller brachiopods are *Eocoelia* and *Atrypa*. The chain coral *Halysites* and the trepostome bryozoan *Hallopora* are also present. (Courtesy of W. S. McKerrow, *The Ecology of Fossils*, by permission of MIT Press and Duckworth, 1978.)

ary units, that is, to be subject to some form of selection independent of that which effects their constituent species, has not been satisfactorily answered. As biological integration and interdependency within fossil assemblages can be assessed only crudely, the study of paleoecology is unlikely to supply an answer.

REFERENCES

1. May, R. M., and Seger, J., 1986, Ideas in Ecology: American Scientist, v. 17, p. 256–267.

2. Lawrence, D., 1968, Taphonomy and information loss in fossil communities: Bulletin Geological Society of America, v. 79, p. 1315–1330.

3. Behrensmeyer, A. K., Western, D., and Dechant Boaz, D. E., 1979, New perspectives in vertebrate paleoecology from a recent bone assemblage: Paleobiology v. 5, p. 12–21.

4. Peterson, C. H., 1976, The relative abundance of living and dead molluscs in two California lagoons: Lethaia, v. 9, p. 137–148.

5. Staff, G. M., Stanton, R. J., Powell, E. N., and Cummins, R., 1986, Time averaging, taphonomy and their impact on paleocommunity reconstruction: death assemblages of Texas Bays: Bulletin Geological Society of America, v. 97, p. 428–443.

6. Koch, C. F., and Sohl, N. F., 1983, Preservational effects in paleoecological studies: Cretaceous mollusc examples: Paleobiology, v. 9, p. 26–34.

7. Stanley, G. D., 1981, Early history of scleractinian corals and its geological consequences. Geology, v. 9, p. 507–511.

8. Thompson, J. B., Mullins, H. T., Newton, C. R., and Vecoutere, T. L., 1985, Alternative biofacies models for dysaerobic communities: Lethaia v. 18, p. 167–179.

9. Savrda, C. E., Bottjer, D. J., and Gorsline, D. S., 1984, Development of a comprehensive oxygen-deficient marine biofacies model: evidence from Santa Monica, San Pedro, and Santa Barbara Basins, California: American Association of Petroleum Geologists Bulletin, v. 68, p. 1179–1192.

10. Byers, C. W., 1977, Biofacies patterns in euxinic basins; a general model: Society of Economic Paleontologists and Mineralogists, Special Publication 25, p. 5–77.

11. Byers, C. W., and Larson, D. W., 1979, Paleoenvironment of Mowry Shale (Lower Cretaceous) western and central Wyoming: American Association of Petroleum Geologists Bulletin, v. 63, p. 354–375.

12. Kaufman, E., and Scott, R. W., 1976, Basic concepts of community ecology and paleoecology. In Scott, R. W., and West, R. R. (eds.), Structure and Classification of Paleocommunities: Stroudsburg, Pa., Dowden, Hutchison and Ross.

13. Walker, K. R., 1972, Trophic analysis, a method of studying the function of ancient communities: Journal of Paleontology, v. 46, p. 82–93.

14. Rosen, B. R., 1981, The tropical high diversity enigma. In Forty, P. L. (ed.), Chance, change, and challenge: the evolving biosphere: Cambridge, England British Museum (Natural History) and Cambridge Univ. Press. p. 103–129.

15. Mancini, E. A., 1978, Origin of micromorph faunas. Journal of Paleontology v. 52, p. 311–322.

16. Walker, K. R., and Alberstadt, L. P., 1975, Ecological succession as an aspect of structure in fossil communities: Paleobiology v. 1, p. 238–257.

17. Walker, K. R., and Diehl, W. W., 1986, The effect of synsedimentary substrate modification on the composition of paleocommunities: paleoecological succession revisited. Palaios v. 1, p. 65–74.

18. Boucot, A. J., 1975, Evolution and Extinction Rate Controls. Elsevier, Amsterdam, 427 p.

SUGGESTIONS FOR FURTHER READING

Behrensmeyer, A. K., 1984, Taphonomy and the fossil record. American Scientist, v. 72, p. 558–566.

Boucot, A. J., 1981, Principles of Benthic Marine Paleoecology. Academic Press, New York. 463 p.

Dodd, J. R., and Stanton, R. J., 1981, Paleoecology, Concepts and Applications. John Wiley and Sons, New York, 559 p.

Kennett, J. P., 1976, Phenotypic variation in some Recent and late Cenozoic plankton foraminifera. In Hedley, R. H. and Adams, C. G. (eds.) Foraminifera v. 2. Academic Press, New York, p. 111–170.

McKerrow, W. S. (Ed.), 1978, The Ecology of Fossils. MIT Press, Cambridge, 384 p.

Schopf, T. J. M., 1980, Paleoceanography. Harvard University Press. Cambridge, 341 p.

Whittington, H. B. and Conway Morris, S., 1987, Extraordinary fossil biotas, their ecological and evolutionary significance. Philosophical Transactions of the Royal Society of London, v. B311, 194 p.

Chapter 18

Paleobiogeography

INTRODUCTION

Modern Faunal Realms:

The terrestrial faunal realms were first outlined by Alfred Wallace, Darwin's co-author of the first paper on natural selection. Wallace established biogeography, the study of the geographic distribution of animals, as a science and divided the continents into five faunal realms: the Holarctic, comprising North America, Africa north of the Sahara, Europe and northern Asia; the Neotropical, comprising South America; the Ethiopian, comprising most of Africa; the Oriental, comprising India and southeast Asia; and the Australasian, comprising Australia, eastern Indonesia, and New Zealand (Fig. 18–1). The differences between the realms are not basically environmental. Tropical rain forests exist in South America, Africa, and Indonesia but, although environments are duplicated, the fauna and flora differ for reasons that cannot be strictly environmental. The contrast of the larger animals of the Neotropical and Ethopian realms illustrates the order of magnitude of biogeographic differences. The monkeys of Brazilian jungles are a completely

different suborder from those of the Congo. There are no proboscidians, giraffes, or hippos in South America and no sloths or armadillos in Africa. These differences are accounted for by the barrier, the South Atlantic Ocean, that separates the realms and prevents mixing, and by the history of the separation of the two continents in mid-Mesozoic time so that their faunas, then similar, have evolved along different paths in the intervening 150 million years. In the previous chapter we considered differences in contemporary biotas caused by environmental differences; in this one we consider differences arising from barriers and history.

Because the oceans flow into each other and are not completely separated by land barriers as the continents are by the sea, modern marine realms are less distinct. Pacific and Atlantic realms are usually distinguished and the Indian Ocean fauna is recognized as a subdivision of the Pacific realm. The barriers separating these realms are the continental masses of the Americas and Africa–Asia. They are effective in restricting the mingling of the warm-water faunas of the two realms because they extend into cold, or polar, waters at each end and warm-water animals cannot migrate through these inhospitable temperatures. The differences be-

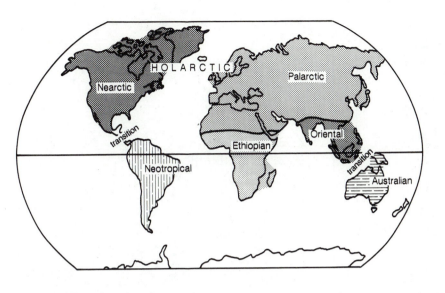

Figure 18-1 Modern terrestrial faunal realms.

tween Pacific and Atlantic faunas are illustrated by the contrast on opposite sides of Panama where they are separated by a narrow land barrier. Similar environments are available, but no species of coral is common to the two faunas although typical reefs on the Caribbean side include 50 species and those on the Pacific side, 15 species. Many years ago William Dall studied the shelled invertebrates on either side of the isthmus of Panama and found that of the 517 species on the Caribbean side and 805 species on the Pacific side, only 24 were common to the two oceans. Although the isthmus is cut by a canal, the canal is filled with fresh water and is as much a barrier to the migration of marine species as land.

Ancient Faunal Provinces:

Comparable faunal differences are found in the fossil record. William Oliver (1) has described the distribution of rugose corals in the Devonian of North America in terms of two faunal realms. The western part of North America was a part of the Old World realm which included much of the rest of the world. It was separated during Early Devonian time from the Eastern Americas realm by a low ridge extending through the central part of the continent (Fig. 18–2). The number of species that occur exclusively in this realm increases in Early Devonian time from 57 to 91 percent. Species that are confined to one biogeographic unit are called endemics. Barriers to mixing with the Old World fauna were effective; otherwise this degree of en-

demism could not have developed in the Early Devonian sea. During Middle Devonian time the percentage of endemics decreased from 67 percent in the early part of the epoch to 45 percent in the late part. By Late Devonian time none of the species in the eastern part of the continent was endemic. The barrier had begun to leak Old World corals in Middle Devonian time and by Late Devonian time it had become ineffective and probably inundated. The distinction between the two provinces disappeared. The study gives useful information about the placement of land and sea in the Devonian Period.

In mid-Mesozoic time a long equatorial ocean separated the united continents of North America, Europe, and Asia from India, Africa, and South America (Fig. 18–3). From this seaway, called Tethys, arose the Alpine, middle eastern, and Himalayan mountains as Africa and India collided with Europe and Asia in Cenozoic time. The Mediterranean, Black, Caspian, and Aral seas are remnants of the Tethys. A distinctive fauna is found in Jurassic rocks deposited in this seaway. Almost all of the rudist bivalves, brachiopods, corals, and some families of ammonites were confined to the Tethyan faunal realm. New groups evolved in the Tethyan realm and spread northward onto the continental shelf seas of Europe. The uniqueness of the Tethyan faunal realm may have been the result of contrasts between its tropical waters and the cooler water towards the poles or to the greater depth of water in it. It was not isolated from neighboring oceans by physical barriers yet maintained a distinct faunal character for tens of millions of years.

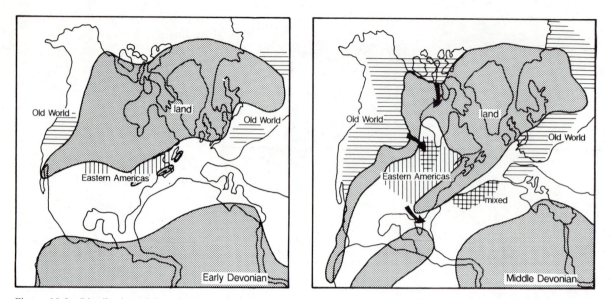

Figure 18-2 Distribution of faunal provinces based on rugosan corals in Early and Middle Devonian time. The areas of the Old World province are indicated by horizontal ruling; those of the Eastern Americas province, by vertical ruling. Where the two patterns intersect, mixed faunas are found. Possible migration routes are shown by arrows. (*Source:* Modified from W. A. Oliver, *Journal of Paleontology*, v. 50, pp. 365–373.)

BOUNDARIES AND DEFINITIONS

Paleoecologists have difficulties distinguishing between dissimilar contemporaneous faunas separated by barriers and those separated by differences in environment. Barriers between some biogeographic units are environmental changes such as latitudinal changes in temperature. The faunal contrasts between modern realms are partly the result of environment, not barriers or history. For instance, the holarctic realm does not include tropical rain forest environments as the other four do, and therefore organisms adapted to this environment are missing from the holarctic realm. If differences between fossil biotas cannot be explained by differences in environment, paleontologists conclude that they were produced by barriers even though such barriers may be difficult to identify in distant geological periods.

Realms are the largest biogeographic units. Some lesser units in order of their decreasing size are: region, province, and subprovince. Quantitative definitions of this hierarchy of units on the proportion of endemic species have been proposed. One proposal requires a province to have between 50 and 25 percent endemic species. If the area has over 75 percent, it is a realm, and under 25 percent is a subprovince. Most paleontological studies do not use such strict definitions and are more qualitative in nature.

Statistical procedures may be used to assess quantitatively the similarity of biotas of various regions. The comparison is made by using one of the similarity coefficients described in the previous chapter and by cluster analysis, which suggests which regions have sufficient similarity to be placed in the same realm, province, or subprovince.

Division of faunas into provinces and subprovinces is rarely based on an analysis of the whole fauna or flora, but usually on a single group. James Valentine (2) divided the continental shelves of the world into many provinces on the basis of molluscs, but a biologist studying corals would distinguish only two or three major provinces. Biogeographic boundaries are different for different groups because the effectiveness of barriers separating provinces depends on the way in which an animal disperses itself.

Obviously, land is a barrier to marine migration and seaways are barriers separating continental biogeographic units, but other barriers may be equally effective. Mountain chains separate the modern holarctic and oriental realms. The Sahara desert forms the boundary between the holarctic and Ethiopian realms. Rainfall and temperature control the distribution of plants as well as the distribution of

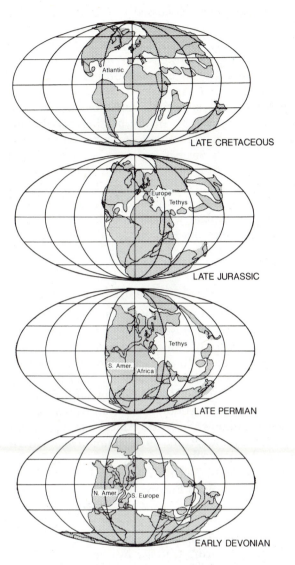

LATE CRETACEOUS

LATE JURASSIC

LATE PERMIAN

EARLY DEVONIAN

Figure 18-3   Paleogeographic reconstructions of the positions of the continents in Early Devonian, Late Permian, Late Jurassic, and Late Cretaceous time. The shading emphasizes the outlines of the continents and is not a representation of the land and sea distribution. The precise positions of the continents in the Paleozoic is still open to question. (*Source:* Compiled from various sources, particularly the maps of C. R. Scotese.)

the animals that depend on them. For benthonic animals of shallow seas, the deep ocean may be an effective barrier because their free-swimming larvae mature in a relatively short time and must then find a place to settle to the bottom. They will die if they are swept out into water of abyssal depth.

CONTINENTAL MOVEMENT AND FOSSIL DISTRIBUTION

Since the early 1960s geologists have established that the continents move laterally in a 350 million year cycle. Early cycles of Proterozoic time are obscure. The first cycle that can be clearly discerned starts with the continents collected into one supercontinent (Pangaea I) in late Proterozoic time. From that time until the Ordovician Period, they continued to separate as oceans widened. From Ordovician to Carboniferous time they moved together, coalescing to form the supercontinent of late Paleozoic and early Mesozoic time called Pangaea II. Since mid-Mesozoic time the earth's crust has returned to the rifting phase of the cycle that continues to the present day as Pangaea splits apart (Fig. 18–3).

The foundations of biogeography were laid in the nineteenth century when the positions of the continents and oceans were regarded as fixed. Each group of plants and animals was considered to have first appeared in a location where it is now most diverse, and to have extended its range outwards as it was successful in competition with similar species. Where the range of a group of related taxa is now separated into segments by barriers, those barriers were postulated to have been passable when the organism established its range. The shallow-water aquatic reptile *Mesosaurus* is found in late Paleozoic sediments of Africa and South America that are now separated by the barrier of the South Atlantic Ocean. In the last century a hypothetical late Paleozoic land bridge, like the isthmus of Panama, spanning the Atlantic was required to explain how *Mesosaurus* could have extended its range from one continent to the other. Many other animals and plants (such as the seed fern *Glossopteris*) that are common to the southern continents at that time required the bridge. We now know that the two continents were part of Pangaea in late Paleozoic time and that *Mesosaurus* would have had no problem in crossing from one to the other. The geographic range of *Mesosaurus* was split into parts in the Mesozoic Era when South America and Africa separated, long after the animal was extinct. Distributions of terrestrial animals that were puzzling before the acceptance of continental drift, have become much clearer in the last 20 years. However, geologists are not in agreement about the positions of the continents before their union in late Paleozoic time and much has yet to be learned about the relationship of continental disposition and marine organism distribution in early and middle Paleozoic time.

Closing Oceans:

When the faunas of two regions become increasingly similar in time, the change may indicate the decreasing effectiveness of a barrier separating them as we have seen for the Devonian rugose corals. In the early Paleozoic, North America and Europe were separated by an ocean called Iapetus which was closing as the continents united into Pangaea. The approach of the continents is reflected in the fossil faunas on either side. In the Ordovician only animals that were planktonic, such as graptolites, were able to cross the deep ocean barrier and establish themselves on both sides. In the Silurian benthonic organisms that had swimming larvae, such as brachiopods, stromatoporoids, and corals, were able to spread across the gap. By Middle Devonian time the northern part of Iapetus had closed and species that could not cross a marine barrier, such as freshwater fish, were common to the two continents.

During the late Mesozoic South America separated first from North America and then from Africa and was then surrounded by water. In the early Cenozoic episode of isolation a distinctive mammalian fauna rich in marsupials developed. In mid-Cenozoic time no families of mammals were common to the two Americas; 29 families lived in South America, and 27 families lived in North America. In late Pliocene time, only a few million years ago, the isthmus of Panama was raised between the continents, breaking the ocean barrier between the lands and making a land barrier between the oceans. Mammals migrated in both directions but mostly into South America, and 22 families are now common to the two continents. North America received as immigrants the opossum (a marsupial), the armadillo, and two mammals that became extinct in the Pleistocene Epoch, the glyptodont and ground sloth.

Opening Oceans:

During the past 150 million years the continents around the Atlantic Ocean have drifted apart. Australia has separated from Antarctica and Antarctica from Africa (Fig. 18–3). Only the Tethys Ocean has been closing. Those groups of terrestrial animals that evolved and spread before the breakup of Pangaea have had their ranges fragmented (ex-

amples have been discussed in Chapter 12). The marsupials are a good example. During the Late Cretaceous the marsupials arose from primitive mammalian stock in the New World and dispersed across an island chain linking North to South America and thence to Australia by way of Antarctica, which had not yet entirely separated from South America (Fig. 18–3). In the northern continents the marsupials had limited success in competition with the rapidly diversifying placental mammals but in the southern continents, which were becoming increasingly isolated, they thrived. The similarity of the marsupials in South America and Australia is difficult to explain without knowledge that these continents were once connected through Antarctica.

Faunas of Accreted Terranes:

Split geographic ranges may lead to important conclusions about the history of the continents. The Cambrian rocks of the eastern seaboard of North America from Newfoundland to Florida contain trilobite faunas unlike those of the rest of the continent but similar to those of northwestern Europe. For instance, the Middle Cambrian trilobite *Paradoxides* is unknown in North America except in this so-called Avalon terrane but is common in northern Europe. Yet in the early Paleozoic, fossil faunas from other parts of North America indicate that it was separated from Europe by the Iapetus Ocean. The trilobite faunas tell us that during the early Paleozoic the Avalon terrane must have been part of, or close to, Europe and therefore separated from the rest of North America by the Iapetus (Fig. 18–4). When the Iapetus closed in mid-Paleozoic time, the Avalon terrane was attached to North America. In Jurassic time when the continents again separated and the Atlantic Ocean formed, the rift occurred east of the old Iapetus suture, leaving the Avalon terrane, a bit of the opposite shore of Iapetus, attached to North America. Geologists have discovered that many mountain chains are composed of masses containing exotic fossil faunas.

Many out-of-place terranes can now be recognized in the North American Cordillera. They are sections of the crust brought eastward from the Pacific Ocean basin. As the oceanic crust spread and was subducted beneath the Cordillera, these terranes collided with and were attached to the west side of the continent and are, therefore, called accreted terranes. Each contains an exotic fossil fauna brought by the movement of the crustal plates into

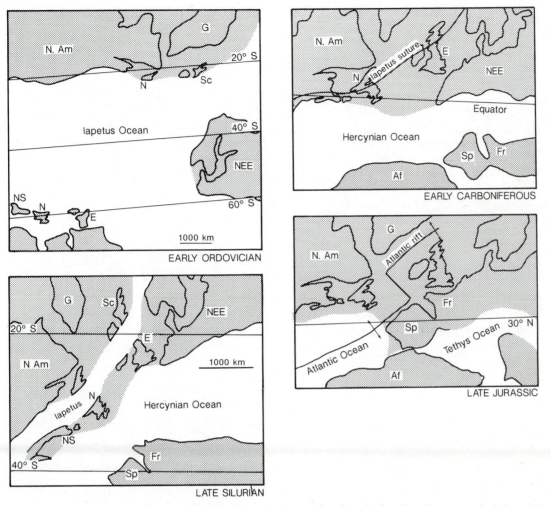

Figure 18-4 Development of the North Atlantic region from early Paleozoic time. Land areas are shaded, and the coastlines of the present land masses identify the movement of the land masses through time. Note that the maps are not all drawn to the same scale. N. Amer. = North America, N = Newfoundland, N.S. = Nova Scotia, Sc = Scotland, G = Greenland, E = England, NEE = northeast Europe, Fr = France, Sp = Spain, Af = Africa. (*Source:* Compiled from maps of L. R. Cocks and R. A. Fortey, A. G. Smith and others, and C. R. Scotese and others.)

contact with contemporary faunas that lived in North America many thousands of kilometers away (3). They are now identified, and their pathways across the Pacific reconstructed, by a combination of paleomagnetic and paleontological studies. Figure 18–5 shows various exotic terranes of the Cordillera spread out across the Pacific as they might have appeared in Carboniferous time.

Biogeographic Theories:

Biogeographers who believe in dispersal from centers of origin require special mechanisms to explain the occurrence of a taxon—order, family, or genus—in two or more areas now separated. For them each newly evolved species spreads from a center as it competes with less well adapted organisms. The ranges of these latter are fragmented as they find refuge in the less desirable corners of the world. Many biogeographers ascribe a major influence to repeated advances and retreats of ice sheets in the recent Ice Age which split formerly continuous ranges into segments and caused latitudinal migration of biotas. In Europe the southward advance of the continental ice sheet from Scandinavia and the northward advance of mountain glaciers from the Alps formed a frigid zone across central Germany, splitting the European fauna into segments that

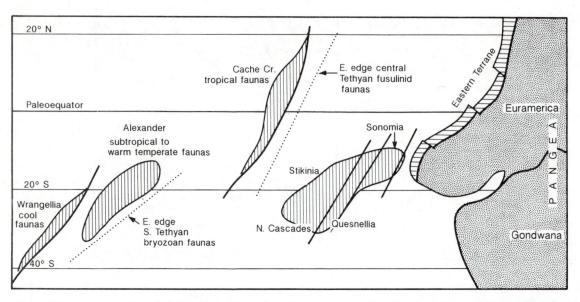

Figure 18-5   Reconstruction of the position of the accreted terranes of the Cordillera in their positions in the Pacific Ocean in Late Carboniferous and Early Permian time. The terranes carry faunas typical of various Tethyan realms and foreign to the Eurasian continent (which was then becoming part of Pangaea). In Mesozoic time, these terranes docked with North America and became part of the Cordillera. (*Source:* Modified from C. A. and J. R. P. Ross, with their assistance, SEPM, Pacific Section, 1983.)

found refuge in the relatively warmer climates of the Balkans and Spain (Fig. 18–6). As a result of the genetic isolation, eastern and western species of many animal groups developed and have remained distinct even though the barrier separating them disappeared with warming climates.

Leon Croizat (4) formulated concepts that have come to be called vicariance biogeography. Vicariance biogeographers explain modern distributions largely on the basis of separation of ranges of organisms as the components of Pangaea drifted apart and ascribe a major role in allopatric speciation to the breakup. Primary dispersal of organisms is postulated to have taken place at times of continental union, and speciation and "vicariating" or fragmenting of the ranges, to have occurred during breakup phases.

The union and separation of continents was an important, but not exclusive, control on the distribution of organisms throughout geological time. Many processes are active in dispersal and fragmentation of organisms that together determine their geographic range at any one time.

Continents and Evolution:

James Valentine and Eldredge Moores (5) have drawn attention to the relation between episodes

Figure 18-6   Extent of ice sheets in Europe during their maximum advance in Pleistocene time. Although Alpine and Scandinavian glaciers did not meet, central Europe became a frigid corridor that restricted the mixing of animals between Spain and the Balkans. Coastlines are shown as extended by the drawdown of sea level. (*Source:* Modified from the work of R. F. Flint.)

of expansion and extinction of marine life and the rifting and suturing of the continents. The breakup of the continents leads to extension of coast lines, expansion of shallow-water environments of continental shelves, and diversification of marine life. The first rifting episode took place at the end of Proterozoic time and influenced the expansion of the metazoa at the beginning of the Paleozoic Era. The second, in early Mesozoic, was accompanied by the rise of modern invertebrate faunas in Late Triassic and Jurassic time (Fig. 7–60).

REFERENCES

1. Oliver, W. A., 1976, Biogeography of Devonian rugose corals: Journal of Paleontology, v. 50, p. 365–373.
2. Valentine, J., 1973, Evolutionary paleoecology of the marine biosphere: Englewood Cliffs, N.J., Prentice–Hall, 511 p.
3. Ross, C. A., and Ross, J. R. P., 1985, Carboniferous and early Permian biogeography: Geology, v. 13, p. 27–30.
4. Nelson, G., and Rosen, D. E. (eds.), 1981, Vicariance biogeography: New York, Columbia Univ. Press, 593 p.
5. Valentine, J. W., and Moores, E. M., 1972, Global tectonics and the fossil record: Journal of Geology, v. 80, p. 167–184.

SUGGESTED READINGS

Hallam, A., 1973, Atlas of paleobiogeography: New York, Elsevier.

Marshall, L. G., 1988 Land Mammals and The Great American Interchange: American Scientist, v. 76, p 380–388.

McKenna, M. C., 1973, Sweepstakes, filters, corridors, Noah's arks, and beached Viking funeral ships in palaeogeography. In Tarling, D. H., and Runcorn, S. K. (eds.), Implications of Continental Drift to the Earth Sciences, v. 1: Academic Press, London, p. 295–308.

McKenna, M. C., 1983, Holarctic landmass rearrangement, cosmic events, and Cenozoic terrestrial organisms: Annales Missouri Botanical Gardens, v. 70, p. 459–489.

Ross, C. A. (ed.), 1974, Paleogeographic provinces and provinciality: Society of Economic Paleontologists and Mineralogists Publication, No. 21, 233 p.

Robison, R. A., and Teichert, C. (eds.), 1979, Biogeography and biostratigraphy. Treatise on invertebrate paleontology, part A: p. 79–569 Boulder/Lawrence. Geological Society of America and Univ. of Kansas.

Vermeij, G. J., 1978, Biogeography and adaptation: Cambridge, Mass., Harvard Univ. Press, 332 p.

Chapter 19

Fossils as Builders
of Sedimentary Rocks

INTRODUCTION

Fossils make significant contributions to the volume of sedimentary rocks that cover most of the earth's surface. Many sedimentary rocks, such as limestone and coal, may be composed almost entirely of the remains of animals and plants. Many sandstones and shales also contain a high proportion of fossil particles. When the organisms that formed them die, shells, bones, or leaves become sedimentary particles carried by streams or ocean currents and deposited in sediments just as are grains of sand or pebbles. The growth of organisms also controls sedimentary environments such as oyster banks, tropical reefs, and coastal and deltaic swamps. In order to understand the role of fossil organisms in forming rocks and controlling depositional environments, we must examine the minerals they use to secrete tests, shells, and skeletons and the microscopic structure of these hard parts. With this information the geologist may be able to identify the organisms that secreted the shell fragments that form sedimentary rocks.

Limestone is by far the most important of the sedimentary rocks formed by fossils. Many invertebrates in the sea and in lakes secrete some form of calcium carbonate by extracting the ions dissolved in water. In the tropics where the shallow seawater is supersaturated with calcium carbonate, the tests and shells of invertebrates do not dissolve in the water when they die but accumulate on the seafloor to form limestone. A high proportion of limestone is composed of fragments of the skeletal tissue secreted by animals and plants. Some constituents are precipitated by physical processes, such as evaporation, from the sea and others are precipitated by modifications of the carbon dioxide content of seawater by the metabolic processes of plants. At present about 7 percent of the sediments on the deep-sea floor are carbonate deposits in the process of becoming limestones. About 5 percent of all the sedimentary rocks in the earth's crust, or 3500×10^{14} metric tons, are limestone. Seventy-five percent of this vast quantity is of Phanerozoic age, for only since the beginning of the Paleozoic has carbonate been extracted in quantity from the sea and stored in the crust (1).

Limestones are uncommon in Archean sedimentary sequences but their rarity is probably related to the absence of shallow-water sediments in general. Remember that limestone does not accumulate below the lysocline (Chapter 17) and most Archean seas appear to have been deeper. The abundance of carbon dioxide in the Archean atmosphere and ocean would also have raised the level of the lysocline to near the surface. In some Proterozoic shallow-water deposits carbonates are relatively common rocks. Dolomite, the double carbonate of calcium and magnesium, is much more abundant in these older sediments than in post-early Paleozoic sedimentary sequences. Pre-Paleozoic limestones and dolomites could not have been produced by the accumulation of the skeletal carbonates, for calcified skeletons did not appear until the Cambrian Period. They must be precipitates from saturated marine water produced either physically by the evaporation of the water, or biochemically by the extraction of carbon dioxide from the water by algae.

Even Phanerozoic limestones do not consist entirely of fossil fragments but include carbonate particles a few micrometers across that were originally a fine lime mud. Some of this mud is derived from the breakdown of skeletal parts of animals and from carbonates in the tissues of marine plants, but other mud is precipitated physically or biochemically from marine water. After calcium carbonate in crystals a few micrometers across is consolidated into limestone, it is called micrite. Micritic limestones of Phanerozoic age commonly contain or are entirely made up of ellipsoids less than a millimeter in diameter of fine grained carbonate. Many of these particles, technically called peloids, are produced as pellets when deposit-feeding organisms pass the lime mud through their guts; there is some evidence that others may be produced by bacterial or inorganic processes after the deposition of the limestone. Filter-feeding organisms also form waste products into pellets that are ejected in the effluent stream. Aggregation of waste into pellets insures that it will be too heavy to stay in suspension and reenter the organism in the incoming current. Pellet-forming organisms may be soft-bodied and may leave behind only the peloidal limestone as evidence of their former presence on the seafloor.

MINERALOGY OF SHELLS AND TESTS

Carbonate Minerals:

Invertebrates secrete calcium carbonate in three forms: high-magnesium calcite, low-magnesium calcite, and aragonite. The difference between calcite and aragonite is not in the proportion of the calcium, carbon, and oxygen that constitute the minerals but in the way the ions are arranged in the crystal lattice. The packing of the ions in calcite results in crystals of rhombohedral shape and symmetry; packing in aragonite results in needlelike crystals of the class known as orthorhombic. Because the ions are packed differently, the space available for the calcium ion is different in calcite and aragonite. All the sites in the crystal lattice available for the calcium ion are not occupied by it in naturally occurring calcium carbonates; other ions may substitute for calcium if they fit into its space. In calcite magnesium ions fit into the space available with little disturbance of the crystal structure. Organically secreted calcites have been found to contain up to 20 percent magnesium. Calcites with more than 4 percent magnesium are classed as high-magnesium calcites, those with less than 4 percent as low-magnesium calcites. Iron and manganese ions are also of appropriate size to fit into the calcite lattice. Larger ions, particularly strontium, prefer the aragonite lattice but do not occupy more than 3.3 percent of the positions.

At surface temperatures and pressures calcite is the stable form of calcium carbonate, and aragonite is unstable. Despite its theoretical instability, aragonite is secreted by many organisms and converts to calcite only very slowly. It is said to be metastable. The oldest aragonite fossils are of Carboniferous age but in most deposits older than a few million years aragonite shells have been dissolved away or replaced by calcite. Even in replaced fossils the higher strontium content characteristic of aragonite may persist and allow the paleontologist to identify the original shell mineralogy.

Calcite can be distinguished from aragonite most effectively by X-ray diffraction analysis where the spacing of the ions in the crystal is measured. In thin sections a stain called Feigel's solution can be applied to the polished surface, depositing a dark coating on the aragonite, but leaving the calcite clear. The crystallographic differences of the two minerals give them different optical properties, but these can rarely be distinguished under the light microscope because aragonite is secreted by organisms in extremely fine microcrystals whose properties are obscured by their aggregation into compound crystals.

The ratio of marine organisms secreting calcite to those secreting aragonite appears to change in earth history. Calcite is secreted by more fossil groups in post-Cambrian Paleozoic time but aragonite became more abundant in skeletons early in the Mesozoic and has continued to be favored to the present. Does this change indicate that the geochemistry of Paleozoic seas was different from that of Mesozoic and Cenozoic seas? Philip Sandberg (2) has studied the changing mineralogy of calcium carbonate precipitated inorganically from the ocean and postulated that it was controlled by the carbon dioxide level in the atmosphere and hydrosphere. Times of high concentration of carbon dioxide and high sea levels favor calcite precipitation; times of low carbon dioxide concentration favor aragonite. These intervals correspond with the greenhouse and icehouse phases described in Chapter 16. From late Proterozoic to Cambrian time aragonite was favored. From then until the beginning of the Carboniferous, oceanic conditions favored the precipitation of calcite. The aragonite phase returned from Carboniferous to Cretaceous time, was briefly replaced by the calcite phase in Cretaceous and early Cenozoic time, and resumed in early Cenozoic time until the present. The trend in skeletal carbonates does not show such frequent changes but probably is in some way related to the icehouse–greenhouse cycle.

Opal:

Opal is hydrous silica of amorphous structure, which means it is not crystalline. It is relatively soluble in seawater and waters that circulate in sedimentary rocks. Therefore organisms with shells of this material are usually poorly preserved or have been replaced by less soluble minerals soon after burial. Much of the opal secreted by microorganisms near the surface of the sea is dissolved before it reaches the bottom.

Chitin:

Chitin is a complex polysaccharide, a polymer that forms long chains which unite into fibers. It is secreted in the ectoderm and commonly forms the exoskeleton of invertebrates. It is a material much like collagen, which forms the connective tissue of many advanced animals, but collagen is secreted in the mesoderm. In the exoskeletons of invertebrates chitin is one component of a protein–chitin complex. Chitin may be decomposed by bacteria and digested by enzymes but, if buried, may persist for hundreds of millions of years and is known from

rocks as old as Cambrian in age. When its volatile constituents are lost during preservation, it is carbonized to an opaque film on rock surfaces.

Dahlite:

Dahlite is the calcium phosphate mineral in bone. It is basically a variety of the mineral apatite, a calcium hydroxy phosphate $[Ca_5(OH)(PO_4)_3]$. Some carbonate generally enters the structure and its presence is expressed in the following formula for dahlite: $Ca_{10}(PO_4)_6CO_3.H_2O$. Phosphates are secreted largely by the chordates but are present in a few invertebrate groups where the mineral has been identified as a similar variety of apatite called francolite $[Ca_5(PO_4,CO_3OH)_3(F,OH)]$.

MINERALOGY AND MICROSTRUCTURE OF MARINE INVERTEBRATES

If each group of organisms secreted a particular mineral in a particular pattern, even fragments of their hard parts would be easy to identify in sediments. Unfortunately for the paleontologist, few groups of animals consistently secreted a single carbonate mineral or secreted it in a single crystalline form. In addition the mineralogy and microstructure originally secreted may have been changed by the time the limestone is examined. The arrangement of the crystals in the skeletal tissue of an animal can be investigated microscopically by means of slices of the material ground to a few micrometers in thickness so that light is transmitted through them, or by means of the scanning electron microscope which forms an image of a broken or etched surface of the object.

The table on p. 413 shows in summary form the distribution of these minerals in animals and plants.

Algae:

The red algae secrete skeletal carbonate of high-magnesium calcite in crystals that are fractions of a micrometer across and difficult to resolve with the light microscope. The magnesium content ranges from about 2 to 18 percent and appears to be greater when the seawater in which the plant lived was warmer. The green algae secrete aragonite needles. Coccoliths are composed of tiny plates that are single crystals of low magnesium calcite.

The cyanobacteria that form stromatolites trap

MINERALOGY OF THE SKELETAL TISSUE OF ANIMALS AND PLANTS
(only the minerals secreted by a major section of each group are shown)

ORGANISM	CALCITE LO–MG	CALCITE HI–MG	ARAGONITE	OPAL	PHOSPHATES
ALGAE coccoliths	x				
diatoms				x	
green			x		
red		x	x		
FORAMS planktonic	x				
benthonic		x	x		
RADIOLARIANS				x	
PORIFERANS Calcarea		x			
hexactinellids				x	
demosponges				x	
sclerosponges	x		x		
stromatoporoids			?		
CNIDARIANS			x		
scleractinians					
BRYOZOANS		x	x		
BRACHIOPODS					
articulates	x				
lingulates					x
MOLLUSCA gastropods			x		
bivalves	x		x		
cephalopods			x		
ECHINODERMS		x			
ARTHROPODS		x			
PROTOCHORDATES			x		
VERTEBRATES					x

(Table modified from Chave and Erben, Encyclopedia of Paleontology)

whatever carbonate grains are suspended in the water and do not have a characteristic mineralogy. There is some evidence that precipitation of carbonates also takes place in the cyanobacterial film at the stromatolite surface.

Foraminifera:

Porcellanous tests of foraminifera are composed of high-magnesium calcite in which the crystals are very small and randomly oriented. Fusulines and endothyrids have similar microstructure but their tests are now low-magnesium calcite. The hyaline or glassy tests, such as those of the Rotalliina, are composed of sheets of calcite that may be of the high or low variety, or mixtures of the two. Only a small group of forams secrete aragonite tests.

Sponges:

The spicular skeleton of demosponges and hexactinellids is composed of opal. When preserved as fragments in sedimentary rocks, the spicules can be distinguished from glassy volcanic fragments, which they resemble optically, by the canal which runs through the center of most of them. Sponges of the class Calcarea secrete spicules of high-magnesium calcite.

The nonspicular skeletons secreted by some demosponges, sphinctozoans, and sclerosponges are of a variety of compositions and microstructures. The microstructure of many is duplicated in corals. Some are composed of spherical aragonite masses in which the needlelike crystals radiate from the center (Fig. 19–1B). This microstructure is said to be spherulitic. In others the aragonite fibers are inclined outward radially from a linear axis in a bundle that is called a trabecula (Fig. 19–1A). The fibrous and spherulitic microstructures are not confined to sclerosponges composed of aragonite but also develop in those with calcite mineralogy. Fibrous carbonates in which the fibers spread outward from the axis of the plates and rods of the skeleton either perpendicularly or obliquely to their surface are also found in Mesozoic stromatoporoids. The Paleozoic stromatoporoids generally do not show this fibrous microstructure. Both types are generally preserved in calcite but rare Mesozoic stromatoporoids are aragonite and the Paleozoic ones may

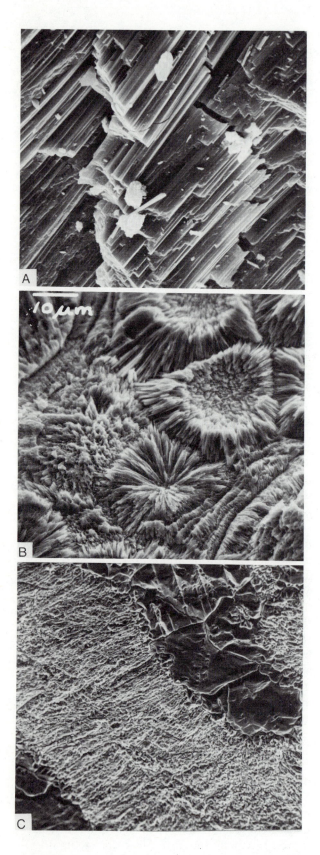

Figure 19-1   Scanning electron micrographs showing the microstructure of nonspicular carbonate skeletons secreted by sponges. A. Fibrous aragonite crystals of the modern sclerosponge *Ceratoporella* ($\times$2500). B. Spherulitic aragonite of the modern sclerosponge *Astrosclera* ($\times$1200). C. Fibrous calcite of the Devonian stromatoporoid *Hammatostroma*, probably derived from an originally aragonitic skeleton ($\times$220). (C. W. Stearn.)

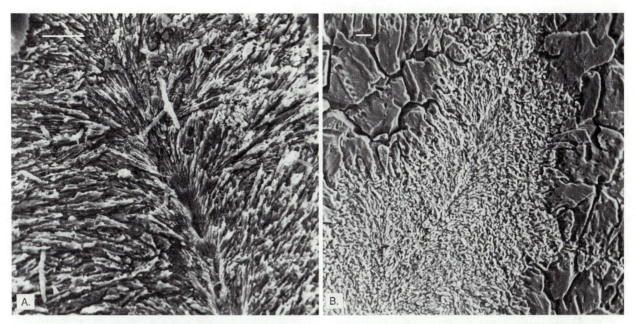

Figure 19-2   Scanning electron micrographs of coral microstructure. The scale bars are 20 micrometers. A. Trabecular microstructure of the aragonite of the septum of the Pleistocene scleractinian *Siderastrea*. B. Fibrous calcite of the Devonian colonial rugosan coral *Pachyphyllum*. (Courtesy of James Sorauf, and The Paleontological Society, *Journal of Paleontology*, v. 45, pp. 23–32.)

have also secreted aragonite and been completely replaced in preservation (Fig. 19–1C).

Corals:

All modern scleractinian corals secrete aragonite. The septa consist of upward and outward-radiating bundles of aragonite microcrystals, called trabeculae, set side by side. In cross section the structure appears to be spherulitic, but in longitudinal section the relationship of the crystal fibers to the axis of the trabeculae is clear (Fig. 19–2A).

Rugose corals are now all composed of calcite but some paleontologists have suggested that this is a replacement of an original aragonite skeleton. However, the faithfulness with which the fine fibrous carbonate microstructure is preserved has convinced most that it could not have survived the replacement of aragonite by calcite, and therefore the rugose corals must have secreted calcite originally (Fig. 19–2B). The same reasoning can be applied to similar microstructures of the tabulate corals. In rugose corals individual trabeculae are uncommon and the fibers grow outward from the axial plane of septa. The lamellar calcite found in some rugose corals is believed by some paleontologists to be primary and by others to be diagenetic.

Brachiopods:

Articulate brachiopods have consistently formed their shells of low-magnesium calcite since their first appearance. The shell is composed of a thin, outer, primary layer of fibers perpendicular to the surface and a thicker, inner, secondary layer of much coarser fibers lying oblique to the surface (Fig. 19–3B,C). Some thick-shelled brachiopods, pentamerids for example, add an inner prismatic layer. Lingulate brachiopod shells combine chitin and francolite.

Bryozoans:

Bryozoan microstructure is also characterized by laminar and fibrous units (Fig. 19–3A). Fossil bryozoa are all calcite, but modern ones use calcite, aragonite, and mixtures of the two.

Molluscs:

Molluscs secrete both aragonite and calcite in a bewildering array of complex microstructures. These can be simplified into four basic patterns. In prismatic microstructures the carbonate prisms are parallel-sided and generally elongated at right angles to the surface. Laminar structures are formed of

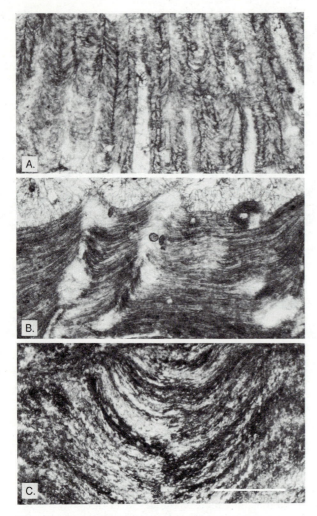

Figure 19-3   Laminar calcite microstructures of bryozoans and brachiopods in thin section. The scale bar is 0.5 mm. A. Ordovician trepostome bryozoan. B. Strophomenid brachiopod, showing the rodlike pseudopunctae characteristic of this order forming spines on the shell surface. C. Rhynchonellid brachiopod. (Photomicrographs C. W. Stearn.)

oblique to the surface, and the third order is oblique to the second order (Fig. 19–4). In homogeneous microstructures, the equant crystals are unoriented in a mosaic. Although nacreous and cross lamellar structures are generally aragonite and laminar structures are generally calcite, the microstructures of molluscs are largely independent of shell mineralogy.

Cephalopod shells are mostly aragonitic, and therefore pre–Mesozoic cephalopods tend to be poorly preserved. Belemnite guards are now calcite but some paleontologists believe they were originally aragonite. Gastropods also have shells composed mostly of aragonite but bivalves use both

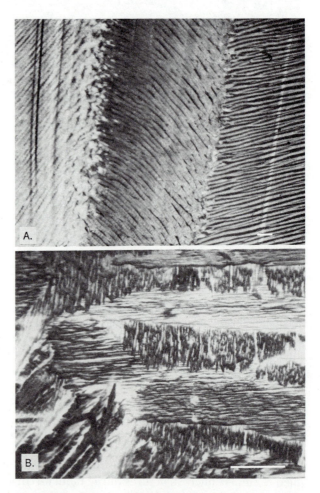

Figure 19-4   Cross lamellar microstructure in molluscs. The scale bars are 0.1 mm. A. Thin section of the modern gastropod *Strombus*. B. Scanning electron micrograph of molluscan shell showing the lamels at various angles. (C. W. Stearn.)

rods, laths, and blades parallel to the growth surface producing a foliated appearance. A special type of laminar structure composed of stacked polygonal tablets of aragonite forms mother-of-pearl or nacre on the inside of many mollusc shells. Nacreous layers are deposited by certain oysters around irritating sand grains in their shells to produce pearls. Cross lamellar structures consist of three orders of carbonate units called lamels. The first order plates or lenses are elongated parallel to the surface. The second order cross them perpendicularly and are

aragonite and calcite in their shells in various proportions in the various orders.

Echinoderms:

Echinoderms secrete high-magnesium calcite exclusively in a distinctive pattern that was described in Chapter 7. Each plate is a single calcite crystal formed as an open network and capable of being changed in shape and size during the growth of the organism by the organic matter that permeates it (Fig. 7–36).

Arthropods:

Arthropods form an exoskeleton that is based on chitin, but in many marine arthropods is reinforced by high-magnesium calcite. The carapaces of most trilobites are usually dark owing to the carbonized organic matter, whose original nature has not been determined, but as preserved they are almost pure calcite. The thin outer layer consists of very small calcite needles perpendicular to the surface and the thick inner layer is finely crystalline calcite marked by laminae parallel to the surface.

Vertebrates:

Vertebrate bone is a combination of the phosphatic mineral dahlite (70 percent) and organic matter that is for the most part the fibrous connective tissue collagen. Like echinoderm calcite it is capable of growth, healing of broken parts, and thickening in stressed areas because it is permeated by living organic matter. Within the vertebrate body bone takes on various microstructures whose description is beyond the scope of this chapter.

FOSSIL DIAGENESIS

Individual fossil particles become rock through processes that are collectively known as diagenesis. Not only are the grains cemented together and voids within and between them filled, but mineralogical and textural changes take place within the grains. These changes often degrade and, if carried far enough, destroy the microstructures of fossils and inhibit the analysis of the paleontologist. When paleontologists refer to fossils as being poorly preserved, they mean that the processes of diagenesis

have advanced to a stage of obliterating much of the original structures.

Cementation:

Fossil fragments on the seafloor are bound together in the first stage of diagenesis by very small crystals of aragonite or high-magnesium calcite apparently derived from the ions dissolved in the seawater. Usually the nature of the grain on which this cement is deposited determines which mineral precipitates (Fig. 19–5). For instance, rims of needle crystals of calcite may grow on echinoderm fragments and of aragonite on adjacent scleractinian fragments. The cement crystals usually grow in the same orientation as the crystals in the grain forming what is known as a syntaxial rim. Similar crystals begin the filling of the internal cavities in the fragments reducing the porosity. If cementation takes place in a freshwater environment, such as exists on emerging coral islands, the calcite that initially binds the grains is low-magnesium calcite.

The initial stage in the diagenesis of bone is the infilling of the canals and pores from which the collagen and other organic matter has decayed. This infilling by calcite or silica results in the permineralized condition described in Chapter 1.

Figure 19-5 Scanning electron micrograph of acicular aragonite cement at the corner of a dissepiment and septum in a recent scleractinian coral. The crystal mass is about 130 micrometers across. (Courtesy of D. Boucher, *Bulletin of Marine Science*, v. 30, p. 500.)

Conversion of Unstable Carbonates:

Both aragonite and high-magnesium calcite are unstable phases of calcium carbonate and in time and in a suitable aqueous environment they convert to the stable phase, low-magnesium calcite. The loss of magnesium from calcite occurs early in diagenesis and releases large quantities of magnesium ions into solutions permeating the nascent limestone. Some of these solutions may be influential in forming dolomites in zones where fresh and salt waters intermix beneath the shorelines of tropical landmasses. Although the conversion is thought to take place by dissolution of one phase and reprecipitaton of the other, the loss of magnesium does not cause obvious textural changes in fossil fragments.

The replacement of aragonite by low-magnesium calcite usually does cause major textural changes in fossils. The replacement is postulated to take place in solution in a thin film that separated the aragonite and calcite. Examination at high magnifications of the scanning electron microscope of the front separating the advancing calcite phase from the original aragonite in partially replaced fossils shows no space between the phases for this film (Fig. 19–6).

Carbonate petrologists disagree on the question of whether such textures of aragonite as spherulitic and trabecular can survive the replacement process. Some claim that all organisms whose calcite fossils show such textures must have secreted them as calcite, and others claim that the textures have survived the calcitization of an originally aragonite skeleton. Possibly the preservation of fossil micro-

Figure 19-6   Calcite (Ca) replacing aragonite (Ar) in a Pleistocene specimen of the gastropod *Strombus*. Note the abrupt contact between the two minerals. A. Scanning electron micrograph. B. Thin section. (Courtesy of Norman Wardlaw, *Canadian Journal of Earth Sciences*, v. 15, pp. 1861–1866.)

structures only takes place when replacement occurs above the water table (3). The original mineralogy of several Paleozoic groups, such as the stromatoporoids and rugose corals, therefore remains uncertain (4).

Changes in Grain Size:

Fossil particles in limestones usually have narrow rims that are opaque in transmitted light under the microscope because they are composed of very small grains of carbonate. These opaque borders of grains are termed micrite envelopes. Carbonate grains now being deposited in shallow water show similar envelopes. The rims are formed when algae that bore minute tubes a few micrometers across penetrate the carbonate grain (Fig. 20–8) and are subsequently filled by fine micrite. Some skeletal grains may lose all microstructure in this micritization process.

The processes of carbonate diagenesis that involve substitution of one mineral for another, such as calcite for aragonite, generally result in increase in size of the crystals. The size change is explained by the second law of thermodynamics. Processes of replacement in carbonate diagenesis are collectively termed neomorphism and when they result in crystal growth, they are distinguished as aggrading neomorphism. Larger crystals of low-magnesium calcite, the stable phase, replace smaller ones of less stable phases and eventually the fossil becomes a coarse mosaic of calcite. Many older fossils may be in this state of preservation; the form of the shell is still clear but the original microstructure and mineralogy have been replaced by a coarse calcite mosaic.

Replacement of Carbonates by Other Minerals:

How one mineral replaces another when rocks are buried remains a geological mystery. That it occurs is shown by the common occurrence of fossils, originally formed as carbonates, replaced by silica, pyrite, phosphates, iron oxides, and more rarely by many other minerals. The faithfulness of the replacement is so great that controversy over the original composition of the fossils may arise. For example, in earliest Cambrian rocks fossils are now composed of phosphate (Chapter 5). Whether these minute shells were secreted as calcite or aragonite and replaced during a unique chemical phase of the sea when such replacement was favored, or were secreted as francolite or a similar phosphate, is a question of current discussion. There is no doubt that pyrite or hematite fossils were not secreted as these minerals, but there is discussion of the time of the replacement and the conditions under which it occurred. Many of these replaced fossils are from micromorph faunas whose small size is indicative of exceptional environments at the time of their entombment (Chapter 17). If exceptional marine conditions were responsible for the size distribution of the fauna, they may also have caused early replacement of the shells before they were buried.

POROSITY

Petroleum geologists study limestones to find oil and gas. About 30 percent of the world's petroleum resources are produced from carbonate rocks. The nature of many limestone reservoirs is determined by the organisms that formed the limestone. Geologists and paleontologists concentrate their attention on grains and cements but to the reservoir engineer the important part of the reservoir rock is the part that is not there—the pore space. The oil in a reservoir is held in the pore spaces within and between grains, and our ability to recover it is related both to the number and size of the pores (the percent porosity) and the degree to which the pores are connected so that oil may flow through the reservoir towards the oil well (the permeability). The average porosity of productive carbonate reservoirs is about 15 percent, but the range of porosity values is great.

We are concerned here only with the porosity produced by, or dependent on, fossils. However, it should be noted that most carbonate reservoirs have had their porosity and permeability enhanced by postdepositional solution, faulting, and fracturing.

Intraparticle Porosity:

Some fossils have internal cavities that are not filled with sediment when they are buried and may later be filled with oil, water, or gas. In some reef reservoirs a significant part of the porosity is made up of cavities within corals, stromatoporoids, sponges, or rudists (Fig. 19–7A). Cavities may be made by boring organisms in skeletal grains before they are buried in sediment.

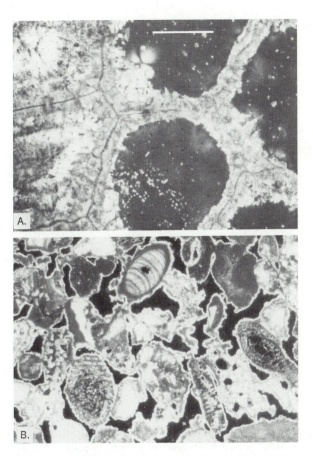

Figure 19-7 Thin sections showing porosity in limestones related to fossils. Scale bar is 1 mm. A. Interfossil porosity. Silurian tabulate coral *Favosites*. The corallites on the left are filled with cement, but those on the right are empty.
B. Interparticle porosity. A Pleistocene grainstone from Barbados, with grains of coralline algae and foraminiferans, cemented by a thin carbonate rim (light tone) leaving large pore spaces which are black in the petrographic microscope with crossed polars.

Interparticle Porosity:

This occurs in limestones composed of coarse, sand-sized grains (Fig. 19–7B). Either the pore space between the grains never filled with cement, or the cement that filled the space early in diagenesis was leached at a subsequent stage of the diagenetic history when unsaturated solutions flowed through the rock. Shelter porosity is a type of interparticle porosity that is formed when large fragments, commonly fossils, prevent cavities below them from being filled with sediment.

Growth-Frame Porosity:

This occurs in reef limestones. Modern corals and their partners in reef building, the coralline algae, tend to overgrow cavities and form overhangs. By shading their neighbors, these light-dependent organisms compete for space on the reef surface. This overgrowing habit leaves large cavities in reefs that may not fill with sediment or cement. When some reef reservoirs have been entered by drilling, the drill has dropped into cavities several meters across. These may not be produced by the overgrowing phenomena alone but may have been enlarged by later solution. Paleozoic reef builders apparently did not have comparable ability to construct frameworks and overgrow cavities (possibly because they were not light dependent), but growth frame porosity may also be important in Paleozoic reefs.

Solution Porosity:

Pores are produced when constituents of a limestone are preferentially dissolved. If the dissolved parts are fossils, molds of fossils are left in the rock and the porosity is said to be moldic. Such moldic porosity is particularly characteristic of limestones that have been replaced by dolomite. The porosity of many reef reservoirs has been greatly enhanced by dolomitization. Solution may open out cavities of all sizes that are potential sites for petroleum accumulation. Smaller solution cavities up to a few tens of centimeters across are called vugs and this type of porosity is referred to as vuggy porosity (Fig. 19–8).

FOSSIL FUELS

Most of the organic matter of the animals and plants of the past is decomposed by bacteria and recycled into new life. If it escapes bacterial decay it may be stored in sediments as diffuse organic matter or converted to petroleum and coal. Although coal, oil, gas, and oil shale are all fossils in the sense that they are remains of organisms of the past, these remains have been so changed after burial that the form and even the composition of the organisms that accumulated to make them are now obscure. The source of oil and gas is fossil organic matter diffused in buried sediment. When the buried, organic-rich sediment is heated over a sufficient period of time, chemical changes take place that re-

Figure 19-8   Vuggy solution porosity with the cavities stained with oil. An oil well core from the Fenn Oilfield, Alberta.

lease hydrocarbons that constitute oil. Under pressure of the overlying sediments the minute droplets are flushed out of the source rocks as they are compacted and into the pores of reservoir rocks. Deeper burial results in higher temperatures during the maturation of the organic matter and in the generation of gas rather than oil.

Oil shale is formed when algae increase greatly in numbers producing an ooze rich in waxy hydrocarbons on the bottom of a lake or seaway.

Coal is formed of plants that have not completely

decayed but have been sufficiently changed that little trace of the original structure can be seen. Most coals appear to be amorphous, brown, organic matter when examined in thin section under the microscope. Only when the accumulating plant matter has been permeated with calcite to form a concretion or coal ball is the fine structure of the plants well preserved (Fig. 8–7).

REFERENCES

1. Garrels, R. M., and Mackenzie, F. T., 1971, Evolution of Sedimentary Rocks: New York, Norton, 379 p.

2. Sandberg, P., 1983, An oscillating trend in Phanerozoic non-skeletal carbonate mineralogy: Nature, v. 305, p. 19–22.

3. James, N. P., and Choquette, P. W., 1984, Diagenesis 9–Limestones–the Meteoric Diagenetic Environment: Geoscience Canada, v. 11, p. 161–194.

4. Sandberg, P. A., 1984, Recognition criteria for calcitized skeletal and non-skeletal aragonites: Paleontographica Americana, v. 54, p. 272–281.

SUGGESTED READINGS

Bathurst, R. C., 1975, Carbonate sediments and their diagenesis: Amsterdam/New York, Elsevier, 620 p.

Horowitz, A. S., and Potter, P. E., 1971, Introductory petrography of fossils: New York/Berlin, Springer–Verlag, 302 p.

Morrison, T. A. M., and Brandt, U., 1987, Geochemistry of recent marine invertebrates: Geoscience Canada, v. 13, p. 237–254.

Flügel, E., 1982, Microfacies analysis of limestones: Berlin/New York, Springer–Verlag; 633 p.

Milliman, J. D., 1974, Marine Carbonates: New York/Berlin, Springer–Verlag, 375 p.

Scholle, P. A., 1978, A color illustrated guide to carbonate rock constituents. Textures, cements and porosities: American Association of Petroleum Geologists Memoir 27, 241 p.

Scoffin, T. P., 1987, An introduction to carbonate sediments and rocks: New York, Chapman & Hall, 274 p.

Chapter 20

Paleoichnology: The Study of Trace Fossils

DEFINITIONS AND DESCRIPTIONS

What Are Trace Fossils?:

Most of the fossils described in this book are the preserved parts of the bodies of organisms and are called body fossils to distinguish them from tracks, trails, burrows, and borings which are referred to as trace fossils. The study of traces left by living animals is called ichnology; the study of fossil traces, paleoichnology. Because a track or burrow is a partial impression or mold of the exterior of the organism making it, the line between trace fossils and body fossils is not easy to draw. Usually the surface of the animal is not preserved in sufficient detail for identification in a trace fossil because the animal moves as it makes the trace or only a small part of the animal is represented in the impression (typically a foot). Some paleontologists also include in paleoichnology the study of the excreta of animals. The fossils of the droppings of larger animals are coprolites; smaller sandlike particles in which many marine invertebrates package their waste products are fecal pellets.

Most trace fossils or ichnofossils tell us little about

the form of the animal or plant that made them but much about its behavior in feeding, moving, escaping, making its home, and resting. The behavior of trace makers can be determined by observing the effect of living animals on sediment in or on which they live, for most trace fossils are the product of organism–sediment interaction. The approach to the interpretation of ichnofossils is uniformitarian, and many of the first ichnological observations were made on animals that inhabit the tidally exposed mudflats of the North Sea on the German coast (1). Intertidal traces are usually removed by the waves at the next high tide and as a result the more easily studied organisms are those whose traces are not apt to be preserved to become ichnofossils. The traces produced by subtidal and deep-water animals can be investigated by SCUBA, coring, and photography but the amount of information available from these sources is relatively limited.

Types of Ichnofossils:

Tracks are the individual impressions of the feet of organisms, and trackways are the assemblage of tracks produced when an individual crosses soft

Figure 20-1   Aerial view of dinosaur trackways on the banks of the Purgatoire River, Colorado. (Courtesy of Martin Lockley, *Geological Society of America Bulletin*, v. 98, p. 1165.)

sediment that will take impressions. The most familiar trackways were made by dinosaurs. Notable North American dinosaur trackways are preserved in the Lower Jurassic sandstones of the Connecticut River Valley in New England, in Cretaceous rocks at Paluxy Creek, Texas, and on the Purgatoire River in southern Colorado (Fig. 20–1). The Texas tracks reveal that the sauropods walked on land on vertical legs, rather than laterally splayed legs and were able to support their weight without the buoyancy of water. The traces formed by epifaunal marine invertebrates, such as trilobites and crabs, are called tracks and trackways when their individual appendages impressed the sediment (Fig. 20–2).

Trails are continuous grooves and ridges formed by the passing of an animal living on the sediment surface. The trace fossil *Cruziana* (Fig. 20–3) was formed by the passage of a trilobite along the surface of muddy sediment. *Aulichnites* is a trail that does not have the transverse ridges that characterize *Cruziana*.

Burrows are made by many infaunal animals either for protection or to find their food in interstitial organic matter in sediment. Burrows may be simple excavations in the sediment or they may be lined by the burrower with grains coated in mucus or with calcareous cement making a tube. *Ophiomorpha* is a burrow lined with pellets that give it a granular

Figure 20-2   Trackways of two large animals, probably arthropods, from the Upper Cambrian to Lower Ordovician Potsdam Sandstone. The track that resembles that of a tire is *Climactichnites*. The track that appears to be a tail impression bordered by foot prints is *Protichnites*. The slab is about 1.5 m in width. (C. W. Stearn, Redpath Museum.)

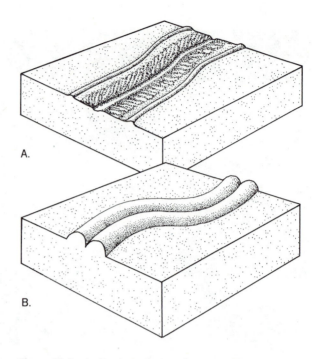

Figure 20-3   A. *Cruziana*, the crawling trace of a trilobite. B. *Aulichnites*, a crawling trace. Both are illustrated as they would appear on the bottom of a bedding plane that had been overturned (reversed relief).

in U-tubes each live at a particular depth below the surface and respond to reestablish this depth when disturbed by sediment burying them or the erosion of sediment bringing them nearer to the surface. If buried, the animal brings the tube upward by

surface (Fig. 20–4A). Similar lined burrows are being formed today by the shrimp *Callianassa* in many areas of the tropics. Many invertebrates make U-shaped burrows so that they can have the protection afforded by living in the sediment yet can filter feed from the water above. At one end of the tube the animal feeds and at the other it passes out excrement. Simple, vertical U-tubes in sandstone are given the name *Arenicolites* (Fig. 20–4B). When the organism periodically moves sideways or vertically and backfills the burrow behind it, the remains of the wall left in the sediment form concentric or parallel layers that are called by the German term spreite (plural = spreiten). The trace fossil genus *Rhizocorallium* consists of a U-tube lying horizontally whose arms enclose a set of spreiten and are bent at their ends to open on the surface (Fig. 20–4C). Species

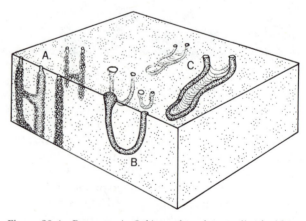

Figure 20-4   Burrows. A. *Ophiomorpha*, a burrow lined with cemented sand grains. B. *Arenicolites*, a simple U-shaped tube. C. *Rhizocorallium*, a U-shaped tube with spreiten. The sediment in which they are embedded is shown here as transparent.

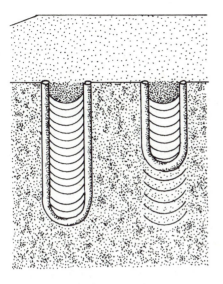

Figure 20-5   *Diplocraterion*. Formation of spreiten by the upward (right) or downward (left) movement of an animal in a U-shaped tube.

taking sediment from the ceiling of the base of the U and plastering it on the floor, and if threatened with exposure by erosion, it does the opposite (Fig. 20–5). In either case, spreiten are left behind as the outlines of the former positions of the base of the burrow. Spreiten may also be formed as an animal progressively samples the sediment for interstitial food in a broad arc and fills the burrow behind with digested sediment (Fig. 20–6).

On rocky coastlines marine animals bore into hard rocks making cavities that are a form of ichnofossil. Some invertebrates penetrate hard substrates by mechanical rasping with claws or shell, others by chemical action, and many by a combination of both. As rocky shorelines are areas of erosion, such cavities are rarely preserved in the fossil record. However, where a hard ground is produced by rapid cementation during a pause in deposition on the ocean floor, borings in that surface may be pre-

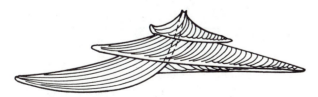

Figure 20-6   *Zoophycus*, a complex feeding trace with spreiten formed in sweeping arcs.

Figure 20-7   X-ray photograph of a slice of the recent scleractinian coral *Montastrea annularis*, showing seasonal growth bands and voids caused by the activity of the boring sponges *Siphonodictyon* (1) and *Cliona* (2). (Courtesy of Kirk MacGeachy, *Bulletin of Marine Science*, v. 30, p. 487.)

served when sedimentation resumes. Disconformities may also preserve hard rock borers (Fig. 17–17). The cavities of boring organisms are common in the hard tissue of fossils that they attacked. Corals are extensively bored by sponges and bivalves at present (Fig. 20–7) and fossil corals and stromatoporoids also show evidence of hard tissue borers in ancient seas. Predatory gastropods bore neat holes in bivalves, and similar holes drilled by unknown organisms are known from brachiopods as old as Ordovician in age. Algal and fungal filaments a few microns across currently bore through many carbonate skeletons, and did so in the past (Fig. 20–8).

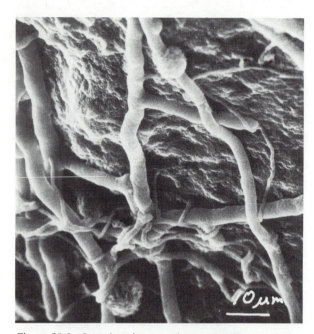

Figure 20-8   Scanning electron micrograph of epoxy molds of borings of microorganisms in the septum of the scleractinian coral, *Montastrea*. The tubes have been injected with the plastic and the carbonate has been dissolved from around them. The large tubes are ascribed to algae and the smaller to fungi. Note the swellings that may be reproductive structures, and the fact that the borings avoid other borings. (C. W. Stearn.)

PRESERVATION OF TRACE FOSSILS

Most traces when originally formed are depressions in the sediment surface or cavities beneath it. Their likelihood of being preserved depends on the firmness of the sediment and the availability of sediment of contrasting color or texture to fill the cavity before it collapses. Many trace makers whose fossils are found in alternating sandstone and shale successions lived in the soft muds represented by the shales and were only preserved when the muds were covered by coarser sands which filled the depressions. When such a deposit is exposed to weathering, the shales are less resistant and are washed away, leaving the bedding planes of the resistant sandstone exposed (Figs. 20–3, 20–9). The sand molds of the grooves of the traces appear in convex relief on the bottom of the sandstone bed. Burrows completely within the shale may be filled with sand, and weather from their matrix as cylindrical, wormlike, or branching bodies. Grooves on the top of the sandstone bed appear in normal relief

on its surface. Whether the trace is preserved in true or reversed relief (whether it is a groove or a ridge) may not be obvious unless the orientation of the sample is marked when it is collected.

In carbonate rocks the obvious burrows are filled with coarser grained sediment, usually pelleted either by the trace maker or by other organisms. In the absence of some difference in texture between burrow and matrix, ichnofossils are difficult to detect in carbonates. In many limestones the contrast between matrix and burrow is increased by the introduction of dolomite into the fillings. The dolomite brought in by magnesium-rich solutions percolating along permeable pathways in the burrows is coarser in texture than the limestone matrix and browner in color owing to a higher iron content. Gray limestones mottled with rust-colored burrows of the trace fossil *Thalassinoides* are common in the

Figure 20-9   Under surface of a bed of Lower Silurian sandstone with the trace fossil *Arthrophycus*. (C. W. Stearn, Redpath Museum.)

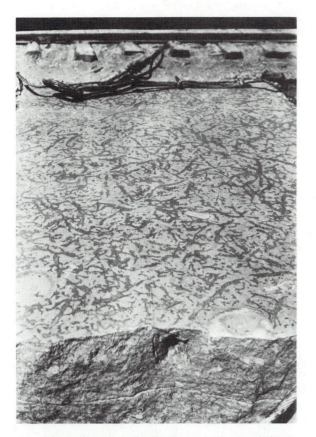

Figure 20-10 Burrows of *Thalassinoides* in lime muds have been partially dolomitized so that they stand out in Late Ordovician limestone. (Courtesy of the Geological Survey of Canada.)

Upper Ordovician of western and arctic North America and are popular as building stones (Fig. 20–10).

NAMES OF ICHNOFOSSILS

Trace fossils are given binomial Latin names just as are body fossils, but since 1930 the names have not been protected by the priority rules of the International Code of Zoological Nomenclature because they are not of the same nature as the names of body fossils. A trace fossil does not represent a unique organism. For example, U-shaped burrows identified as species of the genus *Arenicolites* can be shown by observations of living organisms to be constructed by echiurid worms (annelids or a phylum of their own), enteropneusts (protochordates), holothurians (echinoderms), polychaetes (annelids), amphipods (arthropods), and bivalves (molluscs). Although ichnogenera and ichnospecies are described formally, higher categories, such as families and orders, are not recognized, and in systematic descriptions of taxa, genera are commonly listed alphabetically.

Many trace fossils have names with the suffixes *ichnus* (Greek *ichnos* = trace) or *phycus* (Greek *phucos* = seaweed). The *phycus* names were proposed in the nineteenth and early twentieth centuries when many trace fossils, then called fucoids, were believed to be remnants of soft seaweeds.

We do not know which animals formed most ichnofossils because animals are rarely preserved in tracks and burrows. Most trace makers of the past were soft-bodied and constitute a fauna entirely different from the body fossil fauna. Although trace fossils are commonly found in beds that contain body fossils, they also occur in beds that contain no other fossils. From the first appearance of burrows until the first Ediacara body fossils appear, trace fossils are the only evidence of metazoan life.

INTERPRETATION OF TRACE FOSSILS

Adolf Seilacher (2) proposed a genetic classification into five groups, but these have been expanded by later workers. Fossils illustrative of these divisions are shown in Figures 20–11 and 20–12.

1. *Resting Traces* are made when animals settle to the bottom, leave impressions, or dig shallow depressions, and move off. *Rusophycus* and *Asteriacites* are resting traces ascribed to the impressions of arthropods and asteroids respectively.
2. *Crawling Traces* are linear or sinuous grooves or trackways formed as animals move across sediment surfaces. Some of them end or begin in resting traces. *Climactichnites* is a large crawling trace that resembles a tire track in Cambrian sandstone (Fig. 20–2). It was probably formed by a large trilobite, but no trilobite of this size has been found as a body fossil in these beds. Many other trails, such as *Cruziana* and *Aulichnites* (Fig. 20–3), have been ascribed to trilobites. *Scolicia* is a complex trail apparently made by the passage of a many-legged organism such as an arthropod.
3. *Grazing Traces* are grooves or pits made by mobile deposit feeders or algal grazers at or

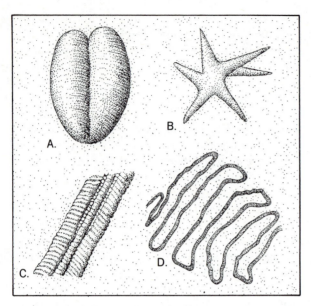

Figure 20-11 Trace fossils on a bedding surface.
A. *Rusophycus* (resting trace. B. *Asteriacites* (resting trace).
C. *Scolicia* (crawling trace). D. *Helminthoida* (grazing trace).

just beneath the sediment surface. Generally they are unbranched and nonoverlapping grooves made in one plane as the animal efficiently crops algal turf. The meandering grooves of *Helminthoida* are typical grazing traces.

4. *Feeding Structures* are burrows dug by deposit feeders mining sediment for nutrients. They are usually branching galleries with vertical connections that may form intricate networks.

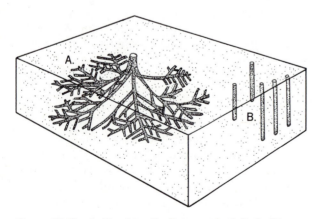

Figure 20-12 A. *Chondrites* (feeding structure). B. *Scolithos* (dwelling structure). (*Source:* Modified from work of T. P. Crimes; courtesy of G. M. Pemberton, *Short Course Notes*, Society of Economic Paleontologists and Mineralogists, 1984.)

Chondrites is a typical genus. Feeding traces may be difficult to distinguish from both grazing traces and dwelling structures.

5. *Dwelling Structures* provide more-or-less permanent homes for animals many of which are infaunal filter feeders. Simple or complex vertical tubes, such as those of *Skolithos* and *Diplocraterion*, are typical of this group. The burrow walls are usually lined because they are primarily used for protection, but the walls of feeding structures are not because the burrows are used to gain access to food in the sediment.

6. *Escape Structures* are produced when animals in dwelling structures move vertically to keep their optimum depth in response to sedimentation or erosion at the surface above them. Escape structures show vertically stacked spreiten (Fig. 20–5).

Classification of an ichnofossil into these groups defined on the basis of behavior requires an interpretation of the intent of the fossil animal in making the ichnofossil. The paleontologist must also remember that the animal may have had more than one purpose in constructing the trace. For example, it may have served as both a feeding and dwelling structure.

USE OF TRACE FOSSILS

Studies of recent trace makers show that not only can a wide range of animals make the same trace but a single animal can make a variety of traces. Because trace fossils in marine and lacustrine rocks are not the products of specific, evolving organisms but merely reflect the behavior of a whole range of aquatic animals, they do not change much in time and are of little value in biostratigraphic and evolution studies. However, detailed study has shown that they may have value in defining the Proterozoic–Cambrian boundary. Many ichnogenera have ranges extending throughout the Phanerozoic Eras. Improvement in efficiency of grazing appears to have taken place over hundreds of millions of years, but such a change is not comparable in scale or nature to the vast changes in body fossils over the same period.

Trace fossils reveal something about the conditions of life in sedimentary deposits that are otherwise without fossils. In many nonmarine deposits the footprints of terrestrial animals and the impressions of the roots of plants may be the only fossils. The footprints of the dinosaurs from the Connecticut Valley Jurassic sandstones are not accompanied by body fossils and most of the traces cannot be connected to skeletal remains. The footprints of *Cheirotherium* are abundant in Triassic rock of Europe, but skeletal remains of the reptile that made them were not found until recently, 130 years after the footprints were first described.

The tracks of dinosaurs give information, not only on how they walked and ran, but also on their social organization. On a single bedding plane at Mt. Tom, Massachusetts, 19 trackways of the carnivorous dinosaur *Eubrontes* show movement in the same direction. Consistent orientation is also evident in 25 trackways of a large ornithopod dinosaur at Lake Innes, Texas. Both carnivorous and herbivorous dinosaurs must have travelled in herds for otherwise the trackways would trend back and forth across a well-traveled trail. Trackways indicate that, contrary to many reconstructions, sauropods did not drag their tails and some theropods could attain speeds of 40 kilometers per hour. They can also give information on the diversity and local abundance of dinosaur populations, the distribution of age groups, and the geographic ranges of dinosaur types. Most tracks are made on the shores or in the shallow water of lakes and rivers and can be used to recognize shoreline trends, and to estimate water depths, the water content of sediments, and current directions.

For the interpretation of ancient environments trace fossils have two advantages over body fossils. First, they are always found in place, unmoved by currents or wave action. Secondly, the occurrence of a particular genus or species of ichnofossil is almost totally controlled by the suitability of the environment, not by the age of the beds in which they are found. Because trace fossil species and genera have such long ranges, assemblages of ichnofossils recur throughout the Phanerozoic and have been designated as ichnofacies.

ICHNOFACIES

Seven recurrent assemblages of trace fossils have been generally recognized as useful in the interpretation of ancient environments. Their occurrence, like that of other paleocommunities, is dependent on environmental factors such as turbulence, substrate, oxygen availability, and salinity. The assemblages were originally considered to be

diagnostic of various depths because these factors change systematically with depth in most marine environments. Further observations have shown that the four fundamental environmental variables previously listed are the primary controls, and that wherever appropriate combinations occur ichnofacies will be present. The facies are illustrated in Figure 20–13.

Scoyenia Ichnofacies:

All nonmarine environments and their trace fossils, from dinosaur footprints, to insect trackways, to burrows of lake-dwelling invertebrates, are included in this ichnofacies. Ichnofossils formed in lake sediments may be similar to their marine counterparts and many genera, such as *Arenicolites* and *Skolithos*, are recognized in both this and marine ichnofacies.

Trypanites Ichnofacies:

Organisms of this facies bore into fully consolidated substrates on rocky coasts, in beach rock, reefs, and hardgrounds. The structures produced by borings cut across the consolidated grains of the host rock.

Glossofungites Ichnofacies:

Members of this facies occur in substrates that are firm but not consolidated into rock. Such substrates are generally intertidal or supratidal. The trace fossils are typically vertical burrows, without linings but with spreiten in the U-tubes. Pouch-shaped burrows called *Gastrochaeonolites*, that are probably formed by bivalves, are typical.

Skolithos Ichnofacies:

This is characteristic of clean, well-sorted sands in nearshore, highly turbulent environments. The animals are infaunal filter feeders inhabiting vertical burrows like the simple vertical tubes of *Skolithos*. Alternating times of deposition and erosion in this rapidly changing environment cause the formation of spreiten.

Cruziana Ichnofacies:

This is a subtidal facies of muddy sediments in offshore areas of low turbulence. Vertically and horizontally branching burrows such as *Thalassinoides*

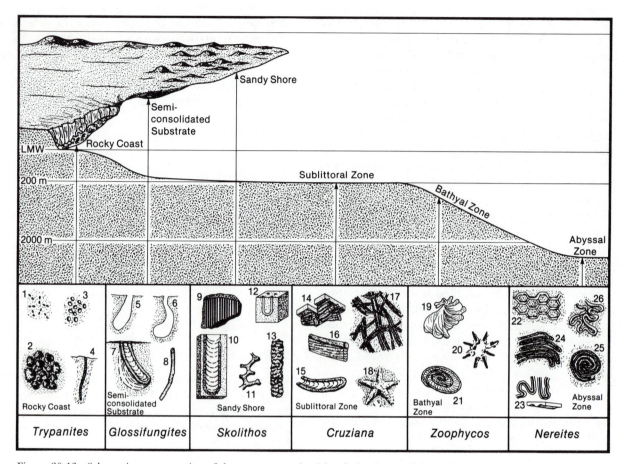

Figure 20-13   Schematic representation of the common marine ichnofacies. Some of the fossils common in these facies are illustrated as follows: 1. *Polydora*. 2. *Entobia*. 4. *Trypanites*. 7, 10. *Diplocraterion*. 9. *Scolithos*. 11. *Thalassinoides*. 12. *Arenicolites*. 13. *Ophiomorpha*. 14. *Phycodes*. 15. *Rhizocorallium*. 16. *Teichichnus*. 17. *Crossopodia*. 18. *Asteriacites*. 19, 21. *Zoophycus*. 20. *Lorenzinia*. 22. *Paleodictyon*. 23. *Taphrhelminthopsis*. 24. *Helminthoida*. 25. *Spiroraphe*. 26. *Cosmoraphe*. (*Source:* Modified from T. P. Crimes; courtesy of G. M. Pemberton, *Short Course Notes*, Society of Economic Paleontologists and Mineralogists, 1984.)

and *Ophiomorpha* are formed by infaunal deposit feeders. Both trails and burrows reach their greatest abundance in this zone which today is developed on the continental shelves.

Zoophycus Ichnofacies:

This occurs in fine-grained, muddy sediments with abundant organic matter in oxygen-deficient environments of low turbulence. These conditions may occur in relatively deep water on continental slopes or in nearshore environments protected from turbulence, where water circulation is restricted. The animals are surface grazers and deposit feeders working a short distance below the surface and making mostly horizontal traces. The facies is of

low diversity but some ichnospecies may be abundant.

Nereites Ichnofacies:

Animals of this facies lived in deep water below wavebase, usually in sediments that were deposited from turbidity flows. Planar, horizontal, complex grazing traces are characteristic of the facies. The meandering and spiral grazing traces of *Helminthoida* and *Spiroraphe* are typical. The curious hexagonal grid of *Paleodictyon* is most commonly found in this facies but has also been found in shallow water deposits. Photographs of the modern deep-sea floor have recently shown that an animal that made *Paleodictyon* still lives.

As study of the distribution of ichnofossils continues, these broad facies divisions will be subdivided and new ichnofacies will be introduced. The identification of an ichnofacies and deductions about its ecological significance can be made only after a study of the fossils, the nature of the sediment, and the relationships to adjacent ichnofacies. Identification of a single trace fossil, or even several, is insufficient to place a bed in a particular ichnofacies.

REFERENCES

1. Schafer, W., 1972, Ecology and paleoecology of marine environments: Chicago, Oliver Boyd and Univ. of Chicago Press, p. 1–568.
2. Seilacher, A., 1967, Bathymetry of trace fossils: Marine Geology, v. 5, p. 413–429.

SUGGESTED READINGS

Ekdale, A. A., 1985, Paleoecology of the marine endobenthos: Palaeogeography, Palaeoclimatology, and Palaeoecology, v. 60, p. 63–83.

Ekdale, A. A., Bromley, R. G., and Pemberton, S. G., 1984, Ichnology: trace fossils in sedimentation and stratigraphy: Society of Economic Paleontologists and Mineralogists, Short Course Notes No. 15, 317 p.

Frey, R. (ed.), 1975, The study of trace fossils: New York/Berlin, Springer–Verlag, 562 p.

Frey, R., and Pemberton, S. G., 1985, Biogenic structures in outcrops and cores, 1. Approaches to ichnology: Bulletin of Canadian Petroleum Geology, v. 33, p. 72–115.

Miller, M. F., Ekdale, A. A., and Picard, M. D. (eds.), 1984, Trace fossils and paleoenvironments. Marine carbonate, marginal marine terrigenous and continental terrigenous settings: Journal of Paleontology, v. 58, p. 283–597.

Lockley, M. G., 1986, The paleobiological and paleoenvironmental importance of dinosaur footprints: Palaios, v. 1, p. 37–47.

Index

Page numbers in *italics* refer to pages on which the item is illustrated. Page numbers in **bold face** indicate pages on which the feature is defined.

Aardvark, 305, 306
Acanthaster, 379
Acanthocephala, 69
Acanthochaetetes, 90
Acanthodii, 198, *199,* 200
Acanthostyle, **104**
Accleration, **356**
Accreted terranes, 406–*408*
Acipenseriformes, 199, 200
Acoelomate, **55,** 69, 75–101
Acritarch, 218, *219*
Acropora, 222
Acrotretida, 109
Actinoceratoidea, 123
Actinopteria, 121
Actinopterygii, 198–201
Actinostroma, 89, 90
Actinostromatida, 88
Adapidae, 290, *291*
Adapiformes, 289
Adaptation, 327–329
Adaptive radiation, 363–*364*
Adductor muscle:
  bivalve, *117,*120, 223, 224
  brachiopod, 109, 111, 112
  Aegyptopithecus, 291
Aeluroidea, 295, 296
Aerobes, 42, 43

Aerobic zone (in sea), 387–*388*
Aestivatation, 204
Aetosauria, 249
Agaricia, 170, 222
Agelas, 82
Agelocrinites, 156
Agglutinated test, 77
Agnatha, 29, 186–192
Agnostida, 138, *139,* 140
Ahermatypic corals, 220, 221
Aigialosaurs, 248
Aistopoda, 205, 207, *208*
Albatrosses, 263
Alberstadt, L., 395
Alcohols, 42
Aldanella, 126, 128
Algae, 54, 169
  endolithic borers, 425, *426*
  green, 171, 172
  phylloid, 171
  red, 169, *170, 171*
  reef, *393*
  skeletons, 412, 413
  Alievium, *219*
Allogenic succession, 395
Allogromia, 76
Allogromiina, 77, 78
Allometry, 330–332

Gryphaea, 342
Allopatric speciation, 408
Allosaurus, 253
Allotheria, 270
Alphadon, 278
Alternation of generations:
  forams, 76, *78*
  plants, 173
  Alveolinidae, 215
Ambonychia, 121
Ambulacral area, **146,** 147, *154, 236, 237*
Amiiformes, 199, 200
Amino acid, 42
  sequences, 29, 30, 43
Ammobaculites, 214
Ammodiscus, 77, 213, *214*
Ammonia-atmospheric, 46
Ammonite, *8, 34*
  habitat, 231–232
  heteromorph, 231, *232*
  history, 232–233
  shell, *375*
  shell form, 333–*334,*
  temperature, *385*
  zone, 318, 319, 322
Ammonitic suture, 229
Ammonitida, 232, 233

Ammonoidea, 119, 123, **125**, 126, 232
   Mesozoic, 229–233
   Paleozoic, 125–*127*
   shell form, 335–336
   suture, 126–*127*, 229, 360
Amniota, **27**, 29, 205, *209*
Amoeboid cell, 80, *81,*
Amorphognathus, 162
Amphiaspidae, 187, 189
Amphiaerobes, 42, 43, 49
Amphibia, 205–209, 277
   evolution from fish, 353
Amphicyonidae, 296
Amphioxus, 164, 184–186, 192
Amphizona, 141
Amplexizaphrentis evolution, *357*
Ampulla, 146, *236,* 239
Anabarites, 65
Anaerobes, 42, 43
Anaerobic zone (in sea), 387–*388*
Anagale, 298
Anagalida, 284, 289–294, 297
Anagenesis, 346, **347**, *348, 349,* 362–363
Anakosmoceras, 343
Anal plate, 148, *149*
Analogous structures, **23**, 335
Anapsida, 187, 189, *190,* 191, **211**
Anaptychus, 231
Anarcestida, 123, 232, 233
Ancestor identification, 25, 26
Anchiceratops, 258
Anchitherium, 311
Ancyloceras, 232
Ancyloceratina, 231, 233
Ancylopoda, 310
Ancyrochitina, 220
Andrewsarchus, 312
Anemone, 220
Aneurophyton, 175, 180
Angiosperm, 173, 179–183
Animalia, 54
Anisoceras, 232
Anisograptid, 160, 161
Ankylosauria, 251, 253, 256
Annelida, 56, 57, 61, 68, *143*
Annularia, 175
Anomalodesmata, 223
Anoxic events, 388
Antarctodolops, 281

Antelopes, 306, 307–309
Antennule, *140*
Anthophyta, 173, 179–183
Anthozoa, 95–101
Anthracobundidae, 304
Anthracosauria, 205, 207, 210
Anthracotheriidae, 307
Anthropoidea, 289, 290
Antiarchi, 195
Antilocapridae, 308, 309
Antler, 310
Antorbital fenestra, 249, *251*
Ants, 285
Anura, 205, 207
Aortic arches, 25, 354
Apatemyidae, 283
Apatosaurus (Brontosaurus), 255
Aperture, gastropod, 129, *131*
Apes, 290
Aplacophora, 116, 165
Aplodontoidea, 297
Apomorphy, **24**
Appendages
   arthropod, 131
   trilobite, 136, *137*
Apternodontidae, 287
Aptychus, 231
Arachnida, 132, 133
Araeoscelida, 245
Aragonite, 75, 411–*418*
   solubility, 379
Arandaspis, 187, 188
Araucarioxylon, 178, *181*
Arca, 223, 225
Archaebacteria, 43, 54
Archaeocyathida, 65, 68, *70,* 78, 82, *83, 84, 85*
Archaeogastropoda, 128–*130,* 228
Archaeohippus, 311
Archaeopteris, 175, 180
Archaeopteryx, 263, 341, 353
Archaic *Homo sapiens,* 293
Archanodon, 121
Archean:
   divisions, 39, 40
   environment, 40, 41
   fossils, 43
   sediments, 410–411
Archeoceti, 313
Archimedes, 106

Archosauria, 249–*259*
Archosauromorpha, 245, 246, 249–259
Arctinurus, 139
Arctocyonidae, 301
Arctoidea, 295, 296
Arctostylopidae, 302
Arenicolites, 424, 427, *430*
Argon, 15
Argyrolagidae, *281*
Argyrolagus, 281
Aristotle, 5
Aristotle's lantern, *237,* 238
Arkansas novaculite, 220
Arkarua, 61
Armadillo, 285, 406
Armenoceras, 125
Arms race, **351**
Arthrodira, 195
Arthrophycus, 426
Arthropoda, 56, 57, 61
   biology, 130–133
   classification, 132–133
   microstructure, 417
Arthrostylus, 106
Articulata (crinoid), 148, 151
Articulata (brachiopod), 108, 109
Artiodactyla, 284, 306–310
Ascocerida, 123, *125*
Asioryctes, 282
Assemblage:
   fossil, 390
   transported, 379–*388*
Asteriacites, 427, *428, 430*
Asteroidea, 145, *146,* 147
Asterophyllites, 175, 179
Asteroxylon, 174
Asterozoa, 145
Astraeospongium, 85, *86*
Astragalus, 306
Astrapotheria, 284, 301, 302
Astraspis, 187, 188
Astrohippus, 311
Astroites stellis, 5, *6*
Astrorhizae, 88, *89, 90*
   function, 332–333
Astrosclera, 414
Astylospongia, 87
Astylospongiids, 85, *87*
Atdabanian Stage, 65
Atelostomata, 235

Atlantic ocean development, *407*
Atmosphere development, 40, 73, 464–466
Atrium, 186
Atrypa, 113, 400
Atrypella, 32–34
Atrypida, 109, 112, *113*
Auditory bulla, 280, 290, 295, 296, 302
Auks, 263
Aulacophore, 156,
Aulacophyllum, 100
Aulechinus, 239, *240*
Aulichnites, 423, *424,* 427
Auloporid, 97
Auricle, bivalve, 225, *226*
Australasian realm, 402, *403*
Australopithecus afarensis, 291, *292, 294*
    africanus, 292, *294*
    boisei, 292
    robustus, 292, *294*
Autapomorphy, **24**
Autogenic succession, 395
Autozooecium, **104**
Avalon terrane, 406
Aves, 29, 209, 262–265
Ax, P., 26
Axolotl, 355–366
Aysheaia, 132, *143*

Bactritina, 123, 233
Baculites, 231, *232*
Bairdia, 141, 367
Bakker, R., 337
Balancing shells, 123–125
Baleen whales, 313
Baltic salinity, 385–*386*
Baluchitherium, 310
Bambach, R., 363
Bandicoots, 280, 281
Barbados reef, *93*
Barbourofelis, 296
Barghoorn, E. S., 45
Barnacle, 133
Barriers to migration, *323,* 404
Barunlestes, 282
Basal plate, *148–149,* 153, *154, 155*
Basilosauridae, 313
Basilosaurus, 313
Bathyurus, 393

Batoidea, 187, 195
Batostoma, 105
Bats, 287
Beak, bivalve, 117
Bear-dogs, 296
Bears, 293, 296, 297
Beavers, 297, 298
Behrensmeyer, A., 377
Belcher Is. 46
Belemnitella, 234
Belemnitida, 232, 233–235
Belemnoidea, 119
Belemnopsis, 234
Bell, A., 334
Bellerophontina, 128, *129*
Benthic assemblage, 397–398
Beresford L., 51
Bering land bridge, 278
Bermuda, *92*
Beyrichia, 141
Biarmosuchidae, 266–268
Biarmosuchus, 268, 269
Bias, fossil record, 11–12, *167,* 371
Billingsaria, 97
Bindstone, 93
Bioerosion, **7**, 92, 237, 425
Biofacies, **319**
Biogenetic law, 355
Biogeographic theories, 407, 408
Bioherm, **92**
Biostratigraphic unit, **326**
Biostratigraphy, 4–5, 60, 317–326
    ammonite, 232
Biozone, **319**, 326
Bird, 27, 246, 262–265
    ancestry, 353
    of prey, 264
Bison, 309, *310*
Biston, melanism, 351
Bitter Springs chert, *45,* 46
Bivalvia:
    biology, 116–118
    borer, 227–228
    byssate, 224–225
    cemented, 225
    classification, 222–223
    freshwater, 122
    Mesozoic, 221–228
    microstructure, 416
    Paleozoic, 119–122

rudist, 225–*228*
shell form, *334*
survivorship, *368*
swimming, 225
Bivariate plot, *33*
Black chert fossils, 43–46
Blastoidea, 153, *154, 155*
Blastozoa, 145
Blood circulation, amphioxus, 186
Blue-green algae, *45*
Bolandiceras, 126
Bone, origin and function, 188
Bony fish, 277
Borer, *425, 426*
    bivalve, 222, 227, 228
Borhyaenidae, 279, 281
Botanical Code, 22
Bothriocidaris, 239
Bothriocidaroidea, 235
Bothrieolepis, 194
Bottjer, D., *388*
Boucot, A., 167, 397, 399
Bovidae, 309
Bovoidea, 307
Brachial:
    plate, *148,* 149
    valve, 107, *108*
Brachiole, *151,* 153,
Brachiopoda, 56, 107–114, 222
    biology, 107–109
    Cambrian, 109
    microstructure, 415, *416*
    Ordovician, 110–112
    shell form, *334*
    shell function, 332
    silicified, 8
    Silurian/Devonian, 112–113, 324
Brachyceratops, 258
Brain size, 269, 270, 273, 292, 293, 300, 352
Braincase, 269
Branchiopoda, 132, 133
Branchiostoma, 164, 184
Brasier, M. D., 73
Bretsky, S. & P., 343, 367
Brinckmann, R., 342
Brongniart, A., 17, 317
Brontosaurus, 255
Brontotheria, 357

Brontotherioidea, 310, 311
Brood pouches, 102, 103, 240, *241*
Brooksella, 60
Brouillon index. 394
Brown body, 102
Bryophyta, 169, 173
Bryozoa, 56, 101–107, 240
    biology, 101–104
        Mesozoic/Cenozoic, 240
        mineralogy, 415, *416*
        Paleozoic, 104–107
Bufo-subspecies, 31
Buildup, **92**
Bulk sampling, 380
Bullfrog, subspecies, 31
Bunodont, **308**
Buoyancy, nautiloids, *124–125*
Bur, 310
Burgess Shale, 137, 142–144, 150, 338
Burgessochaeta, 143
Burne, R. 46
Burrow, **423**, *424*
Burrower, bivalve 222, 223, 224
Busycon, 229
Byssal notch, 225, *226*
Byssus, 121, 222, 224, *225, 226*

Cache Creek Group, 220
Caecilia, 207–209
Caenolestoidea, 280
Cairns-Smith, A. G., 42
Calamites, 175, 178, 179, 181
Calcarea, 82, 85, 86, *87*
Calcichordata, 154–157
Calcification, zooxanthellae, 387
Calcite, 75, 411–419
    carbonate, solubility, 387
    varieties, 411, 418
        Calice, *98*
Calippus, 311
Callianassa, 424
Calyx (crinoid), 146, *148*
Cambrian fauna, 65–67
Camelidae, 308
Cameloidea, 307
Camenella, 66
Camerate crinoids, 151
Canadia, 143
Cancrinella, 114
Canidae, 296

Caprina, 226, *228*
Captorhinidae, 211
Carapace of turtle shell, 243
Carbon isotopes, 49
    and salinity, 385
Carbon dioxide:
    atmospheric, 42, 365
    oceanic, 387, 410, 412
Carbonate minerals, 411–419
    in diagenesis, 418
Carbonate production, 221
Carbonization, 9, 11
Carcharodon, 196
Cardinal teeth, *223*
Cardinal process, *109–112*
Cardiocephalus, 206
Cardioceras, 231, 375
Cardium, 359, 360
Cardonia, 302, 303
Carnassial teeth, 295
Carnivora, 284, 293, 295–297, 313
Carnivores, 282
Carnosaurs, 251
Carpoidea, 154–157
Carroll, R., 208
Cartilaginous fish, 195, 277
Caruthers, R. K., 357
Caryocrinites, 151
Caseidae, 265, 266
Caspian sea cockles, 349
Cassian Formation, 85
Cassiduloidea, 235
Cassowaries, 265
Castorimorpha, 297
Castoroides, 298
Catarrhini, 289, 290
Cattle, 306–309
Caviomorpha, 297, 299
Cellularia, 82
Cellulose digestion, 288, 308
Cement aragonite, *417*
Cementation, 417
Cenozoic, name, 17
Census assemblage, **377**, 379
Centipede, 132
Cephalaspidomorpha, 187, 189, 191
Cephalaspids, 189, 191
Cephalochordata, 155, 157, 164, 184
Cephalon, **134**, *135*

Cephalophytarion, 45
Cephalopoda, 116,
    biology, 119, 122–123
    early Paleozoic, 123–125
    late Paleozoic, 125, 126
    microstructure, 416
    shell form, *334*
Cephalothorax, *131, 133*
Cerastoderma, 223, 349
Ceratites, 230
Ceratitic suture, 229
Ceratitida, 232, 233, 242
Ceratocystis, 155, 156
Ceratomorpha, 310
Ceratoporella, 414
Ceratopsia, 251, 253, 256
Ceraurus, 139, 380
Cercopithecoidea, 289, 291
Cervidae, 308, 331
Cervoidea, 307
Cetacea, 284, 312, 313
Chaetetes, 89
Chaetetida, 82, *89, 90, 91*
Chaetognatha, 164
Chalicotheres, 310, *311*
Chalicotherium, 311
Chalk, 216, *217,* 225
Chama, 226, 227
Champlain sea salinity, 385–*386*
Chancelloria, 66
Character state, 24
Charadriiformes, 263
Charniodiscus, 61, 63
Chasmatosaurus, 249, 251, 252
Chasmosaurus, 258
Cheilostomata, *102,* 104, 240, *241*
Cheirotherium, 429
Chelicerata, 132–133
Cheliped, *131*
Chelonia, 2, 243–246
Chelonioidea, 244
Chimaeroids, 195, 196
Chimpanzee, 289, 291, 292
Chiniquodontidae, 266, 270, *271*
Chiroptera, 284
Chitin, 130, 140, 412, 417
Chitinozoa, *220*
Chiton, 116
Chlamys, 226
Chlorophyta, 169, *171*-172
Choanocyte, *81*

Chondrichthyes, 29, 187, 195
Chondrites, 428
Chondrostei, 199, 200
Chonetida, 114
Chonostrophia, 114
Chordata, 55, 57, 184
    origin, 157
Chromosome, plant, 173, 181
Chron, **19**
Chronostratigraphic unit, **326**
Chrysochloroidea, 287
Chrysophyta, 169, 217
Cidaris, 238
Cidaroidea, *235*, 240
Cirripedia, 132, 133
Cirrus, *147, 148,* 149
Civets, 296
Clade-shape, *363, 364, 372*
Cladistics, 23–29
Cladodus, 196
Cladogenesis, 346, **347**, *348, 349,*
    352, 362, 363
Cladogram, *28, 29*
Cladoselache, 196, 197
Clark, R., 68, 71
Clarkson, E., 357
Classification, 22, 26–31
Clathrochitina, 220
Clathrodictyida, 88
Clathrodictyon, 89
Climacograptus, 160
Climactichnites, 424, 427
Climatic change, 289
Climax community, **395**
Cliona, 425
Clonal organisms, 94, 102
Clonograptus, 159
Clorinda-assemblage, *324*
Cloud, P., 52, 60, 73
Cloudina, 62
Cluster analysis, 391–*392*
Clydagnathus?, 164
Clymeniina, 123, 233
Clypeaster, 239
Clypeasteroidea, *235,* 239
Cnidaria, 56, 68, 69, *70*
    biology, 94–96
Coal, 420
    ball, 174, *175*
    swamps, 174–*178*
Coal forest environment, 381

Coccolith, 169, *217*
Coccosphere, *217*
Coccosteus, 194
Codaster, 153, *155*
Coelacanthini, 199, 200, 204,
    205
Coelenterata, 94
Coelobites, 390
Coelolepida, 191, *192*
Coelom, **55**, 56
    development, 57–59, 69–70
Coelomate, **55,** 57
Coelophysis, 252, 254
Coelurosauravidae, 245, *246*
Coelurosauravus, 247
Coelurosaurs, 252
Coenostele, 88
Coenostrom, 88
Coenozone, **319**
Coleoidea, 119, 123, 232, 233
Collagen, 29, 417
Collar cell, *81*
Collection bias, 12, 380–381
Collins, D., 142
Colonial organism, 94
Colonization stage, 395, *396*
Color alteration index, 161
Columella, 100
Columnal plates, 149, *150*
Comatulid crinoid, *147,* 149
Community, **376**
Community:
    depth controlled, 323–324, 397,
        399
    descriptors, 397
    paleoecology, 390–400
    replacement, 395
Compensating sack, *241*
Compsognathus, 263
Computer simulation of form,
    333–335
Conception Group, 62
Condensed assemblages, *377*
Condylarthra, 284, 300, 301, 306
    evolution, 347–*348,* 353
Conies, 305, 306
Coniferophyta, 173, 178
Coniform element, 161, 163
Conodonta, 161–165
    affinity, 164

biostratigraphy, 165
color, 161
zones, 321
Constellaria, 104, *105*
Continental drift, 259, *405,* 406
Contortothrix, 45
Conus, 228, *229*
Convergence, 279, 290, 304
Convergent evolution, 23,
    328–*329,* 351
Conway Morris, S., 142
Cooksonia, 173, *174*
Cope's law, **354**
Coprolites, 422
Corallinacae, *170*
Corallite, **95,** *96*-100
Corallium, 96
Corallum, **98**
Corals, 94–101, *see* Scleractinia,
    Rugosa, Tabulata.
Cordaites, 178, *181*
Cormohipparion, 311
Correlation, 319, 321–324
    lithostratigraphic, 321
    oxygen isotope, *384*
    quantitative, 322–323
    time, 321
Cortex (graptolite), *158*
Corthurnocystis, 156, 157
Corynexochida, 138, *139,* 140
Coryphodon, 283
Cosmoraphe, 430
Cotylorhynchus, 266
Cowrie shell, 229
Crab, 133
Craniida, 109, 111
Cranioceras, 310
Craniopsida, 109
Crassostrea, 227
Crawfordsville, IN, 149
Crawling traces, 427
Crayfish, *131*
Creodonta, 284, 295
Cribilina, 241
Cribrilaria, 241
Crick, R., 343
Cricoconarids, 67, *68*
Crinoidea, 68, 145, 146, *148-*
    151
Crinozoa, 145
Crises, extinction, 166–168

Crocodylia, 29, 211, 246, 249, 250
  metabolism, 336, 338
Croizat, L., 408
Crompton, A., 274
Cross lamellar microstructure, *416*
Crossopodia, 430
Crossopterygii, 200, 202, 204, 205
Crustacea, 132, 133
Cruziana, 423, *424*, 427
Cruziana ichnofacies, 429, *430*
Cryptodira, 243–246
Cryptolithus, 135, *139*
Cryptonella, 113
Cryptostomata, 104, *106*
Cryptozoon, 48
Ctenacanthiformes, 195
Ctenacanthoidea, 187, 195
Ctenidodinium, 219
Ctenophora, 94
Ctenostomata, 104, *107*
Cuckoos, 264
Currie, P., 208
Cuttlefish, 119
Cuvier, G., 17, 317
Cyanobacteria, *45*, *46*, 47, 49, 51, 52, 171, 412
Cyathaxonia, 393
Cycadeoidophyta, 173, 179, *182*
Cycadophyta, 173, 178
Cyclammina, 77
Cycles:
  eustatic, 366
  in earth history, 364–366
Cyclocystoidea, 154
Cyclomedusa, 61
Cyclonema, 130
Cyclostome bryozoan, 104
Cyclotella, 218
Cylindromphyma, 87
Cylindroteuthis, 234
Cymbospondylus, 261
Cynodontia, 265, 266, 267
Cypraeidae, 229
Cypridina, 140
Cyrtodonta, 392, *398*
Cyrtonella, 128
Cystiphragm, **104**
Cystiphyllida, 100, 101
Cystiphyllum, 100
Cystoidea, 68, 145, *151–152*, 155, *391*

Cystoporata, 104, *105*
Cytochrome-c, 30

d'Orbigny, A., 317, 325
Dahlite, 412, 417
Dall, W., 403
Darwin, C., 18, 23, 26, 183, 318, 327, 328, 347, 367
Darwin unit, 362
Dasypoidea, 285
Dasyuridae, 281
Dasyuroidea, 280, 281
Dawson, W., 209
Death assemblage, **377**, 379
Deep sea cores, *384*
Deer, 306–308
Degassing, 41
Deinitheroidea, 304
Deinonychus, 253, *254*
Deltoblastus, 155
Deltoid plate, 153, *154*, *155*
Demiplate, *236*
Demospongia, *82*, *85*, 87
Dendrogram, 391–*392*
Dendrograptus, *159*, 160
Dendroidea, 157, *159*, 160
Dentalina, 77
Dentition, mammals, 277, 278
Derived character, **24**, 28
Dermoptera, 284, 288, 289
Desmostylia, 284, 301, 304, *305*
Deuterostome, 56, 58, 59, 71
Devonian System, founding, 325
Dextral, gastropod, *118*
Diabolichthys, 203
Diacodexis, *306*, 307
Diadema, 236, 237, 377
Diadematacea, 235
Diadiaphorus, 302
Diagenesis of fossils, 417–419
Diaphragm, **104**
Diapsida, 244–265
Diastema, 274
Diatom, 169, 217, *218*
Diatomite, 217
Diatremiformes, 264
Diatryma, 264
Dice coefficient, 322
Dice diagram, **32**, 33
Diceras, 228
Dichobunidae, 307

Dichrometra, 147
Dickinsonia, *61*, *63*, 68, 71
Dicranograptus, 160
Dictyonema, 160
Dicynodon, 268
Dicynodontia, 265, 266–268
Didductor muscles, *109*, *111*, *112*
Didelphidae, 279
Didelphis, 367
Didelphoidea, 280–282
Didolodontidae, 301
Didymograptus, 160
Diexallophasis, 219
Digitigrade posture, **295**
Dimeropyge, 138
Dimetrodon, *181*, 266, *267*, *268*
Dimorphism:
  ammonite, *34*, 230, 343
  Foraminiferida, 34, 76–78
  sexual, *34*
Dineley, D., 189
Dinocerata, 283, 303
Dinoflagellate, 169, 217, 218, *219*, 221
  symbionts, 221, 386, 387
Dinohippus, 311
Dinophyllum, 100
Dinosaur, 250–258
  extinction, 338
  metabolism, 336–338
  movement, 337
  posture, 336
  size and temperature, 337
  trackway, *423*, 429
Dinosauria, 246
Diplocraterion, *425*, 428, *430*
Diplodus, 196
Diplograptid fauna, 161
Diplograptus, 160
Diplopore, *152*
Diploporita, 145, 151
Diplorhina, 187
Diploria, 221, *222*
Dipnoi, 199, 200
Diprotodonta, 280
Diprotodontidae, 282
Diprotodontoidea, 281
Dipterus, 202
Discocyclina, 215
Disconformity, 320
Discovery Bay, Jamaica, 221

Dispersal of fossil organisms, 322, *323*, 407
Dispersal theories, 407
Dissepiment, 88, *96, 98*
Dissolution of shells, 379
Dissorophidae, 208
Diversification:
  causes, 72–74
  invertebrate, *72*
  metazoan, 68–73
  stage, *396*
Diversity, 216, 368
  ammonite, 233, 370–*371*
  anthophytes, *182*
  archaeocyathid, *85*
  Cambrian, 68, *69*, 75
  community, 393–394
  curves, *166–167*
  index, 394
  latitudinal variation, 381, *382*
  Ordovician, *91*
  Paleozoic, 165, *166–167*
  and salinity, 385–*386*
  and stability, 382
  stromatoporoid, 370, 371
DNA and phylogeny, 29
Dobb's Linn, Scotland, 326
Docodonta, 270, 274
Dogs, 293, 296
Dolomite, 411
Dolomitization, 420
Dolphin, 328, *329*
Domestic fowl, 264
Domination stage, *396*
Dormice, 297
Dorsal hollow nerve cord, 184
Doublure, *135*, 136
Draco, 245, *247*
Drepanoistodus, 163
Dromomerycidae, 308, 310
Dryopithecus, 291
Dsungaripterus, 259
Ducks, 263
Dusisiren, 305
Dwelling structures, 428
Dysaerobic zone (in sea), 387–*388*
Dysodont dentition, *223*

Earth, origin, 40
Earthworm, *71*
Eastern Americas realm, 403, *404*

Echidna, 271
Echinacea, 235
Echinocardium, 238
Echinochimaera, 198
Echinocorys, 238
Echinocystitoidea, 235, 239
Echinodermata, 56, 57, 145–157, *see also* its classes
  biology, 145–148
  microstructure, 417
  skeleton, *145*
Echinoidea, 145
  evolution, *358*
  irregular, 237–239
  Mesozoic/Cenozoic, 235–240
  Paleozoic, 239–*240*
  regular, 235–237
Echinozoa, 145, 147
Echinus, 236
Echolocation, 287
Ecoclinal, **391**
Ecophenotype, **390**
Ecosystem, 374, 376
Ecotonal, **391**
Ecphora, 228, *229*
Ectoconus, 300
Ectoderm, 55–*58*, 69, 70, 94
Ectoprocta, 101
Edaphosauridae, 265, 266
Edaphosaurus,181, 266, *267*
Edentates, 285–287
Ediacaran fauna, 60–*63*, 95
  interval, **60**, 64
Edmontosaurus, 256
Edrioasteroidea, 61, 145, *153*, 154, *156, 391*
Edrioblastoidea, 154
Eggerella, 77
Eglonaspis, 189
Ektopodon, 281, 282
Ektopodontidae, *281*, 282
Elasmobranchii, 195
Elasmosauridae, 260
Eldredge, N., 344, 349
Electrosensory structures, 189, 201
Eleganticeras, 34
Elephant birds, 264, 265
Elephant shrews, 287, *288*, 289
Elephantidae, 304
Elephants, 301, 304, 305
Elk, Irish, 331, *332*

Elliot L. Mines, 50
Embrithopoda, 301
Embryology:
  and ancestors, 354–360
  human, 354, 355
  and polarity, 25
  relationships, 59,
Emiliana, 217
Enaliarctos, 297
Endemic fauna, **404**
Endler, J., 351
Endocasts, 258
Endoceras, 125
Endoceratoidea, 123–125
Endoderm, 55–*58*, 69, 94
Endolithic algae, 425, *426*
Endothyra, 79
Endothyrida, 213, 413
Enopleura, 156
Entalophora, 104
Entelodont, *308*
Entelodontidae, 307
Entelodontoidea, 307
Enterocoel, 108
Enterocoelomate, *55*, 58, 59, 145–165
Enteropneusta, 157
Entobia, 430
Eoastrion, 44, 45
Eocene Epoch, name, 17
Eocoelia, 400
Eocoelia assemblage, *324, 325*
Eocrinoidea, 154
Eofletcheria, 97
Eoharpes, 139
Eospermatopteris, 180
Eosphaera, 44, 45
Eosuchia, 245, 246, 250
Eosynechococcus, 46
Eotitanosuchia, 266
Epifauna, **389**, 390, 392
Epihippus, 311
Epiphyses, **246**
Epiphyton, 171
Epipterygoid, 269
Equidae, 311
Equisetum, 175, 367
Equitability, **394**
Equoidea, 310
Equus, 311, *312*
Erinaceomorpha, 287

Ernietta, 62
Estuarian environments, 385–386
Ethiopian realm, 402, *403*
Eubacteria, 54
Eubrontes trackway, 429
Eucalyptocrinites, 149
Eucidaris, 236
Eudimorphodon, 259
Euechinoidea, 235
Euhoplites, 231
Eukaryotes, 42, *43*, 50–53, 54, 72, 169
Euphemites, 130
Euplectella, 86
Eurotamandua, 286
Eurymylus, 298
Eurypterida, 132, 141–*142*
Eurypterus, 142
Euselachii, 195
Eustatic cycles, *366*
Eusthenopteron, 204
Eusuchia, 250
Eutheria, 270, 282–313
Event, magnetic, 19
Evolution:
    convergent, 328–*329*
    definition, **351**
    explosive, 73
    fossil evidence, 4
    mechanisms, 340–361
    rates, 362–367
Evolutionary systematics, 23, 24
Excurrent pore, 80, *81*
Exogyra, 225, *227*
Exoskeleton, 130–132
Extinction:
    ammonite, 232–233
    background, *368*-369
    causes, 368–371
    climatic, 371
    Cretaceous, 240, 242
    cyclic, *369*
    impact hypothesis, 369–370
    mass, 346–372
    Paleozoic, 166–168
Extraembryonic membranes, 209
Eye, trilobite, *135*

Fabrosaurus, 255
Facial suture, *135*, 138
Facies, **319**

Facies fossil, **319**
Falites, 10
Fallotapsis, 65
Faunal realm, barriers, 404
Favistina, 99, 101
Favosites, 97
Favosites, porosity, *419*
Favositida, 96
Feathers, evolution, 353
Fecal pellets, 422
Feeding structures, 428
Feigel's solution, 411
Felidae, 296
Fenestella, 106, 393
Fenestrata, *102,* 104, *106*
Fermentation, 43, 51
Ferns, 173, 175
Figured stones, *5–6*
Filter feeding, 185, 187, *389*
Fischer, A., 365
Fistulipora, 105
Fistuliporid, 104
Fitness, **327,** 328
Flamingos, 263
Flatfish, 195
Flatworms, 69, 117
Flexicalymene, 102, 136, 380
Flinders Ra., 60
Flood, Biblical, 6
Flying foxes, 287
Flying lemurs, 287, *288,* 289
Flying reptiles, 257–259
Food grooves, 153
Foraminiferida:
    benthonic, 213–215
    biology, 76–78
    diversity, *382*
    evolution, 343–345
    Mesozoic, 213–216
    mineralogy, 413
    Paleozoic, 78–80
    planktonic, 215–*216*, 387
    reproduction, 76, *78*
    test, *76, 77-80*
    zones, 318, 322
Fordilla, 121
Formaldehyde, 42
Formation, 321
Formic acid, 42
Fortey, R., 334
Fossil fuels, 420–421

Fossil Gardens, N.Y., 48
Fossil record, bias, 11, 12, 167, 371
    inadequacy, 11, 12, 341
Fossil:
    chemical, 49
    definition, **3**
    living, 175, 367
    mineralogy, 411–417
    practical value, 4–5
    preservation, 6–11
    range, 319–321
    relative value, 321–322
Fossula, **100,** 101
Framebuilders, 93
Framestone, 93
Franciscan Formation, 220
Francolite, 419
Frequency distribution, *33*
Freshwater turtles, 244
Frigilaria, 218
Frogs, 207, 209, 266
Fucoid, 427
Functional morphology, 329–332
Fungi, 54
    boring, *425*
Funnel (cephalopod), *120*
Fusulinina, 77–*80, 79,* 213, 387, 322
    mineralogy, 413

Galeaspida, 187, 189, *190*
Galeoid, 187, 195, 196
Galesauridae, 266
Galton, P., 255
Gametophyte, 173, 176
Gas exchange, 186
Gases:
    limiting, 387–389
    sea, 387
Gastrochaeonolites, 429
Gastropoda, 68, 116
    biology, *118*-119
    drills, 119, 229
    Mesozoic, 228–229
    microstructure, 416
    Paleozoic, 126–130
    shell form, *333–334*
Gene pool, 31, 341
Gene flow, 31, 341
Gene, 31, 52
Genetic drift, 372

Genital plate, *236*, 237, *238*
Genus name, 21
Geochemistry, 4
Geochronology, 14–16
Geolabididae, 287
Gibbons, 291
Gilinsky, N., 363
Gill:
  bivalve, *117*
  echinoid, *237*, 239
Gill slits, 155, *156*, 184–186, 354
  embryology, 354
Gingerich, P., 346
Gingkophyta, 173
Giraffe, 308, 310
Giraffidae, 309
Girtycoelia, 87
Glabella, *135*
Glaciation:
  cycles, 365
  late Proterozoic, 60
  Varangian, 65
Glaessner, M. F., 60
Glauconite, 16
Glenobotrydion, 45
Gliding possum, 281
Gliroidea, 297
Globigerina, *214*, 216, 382
Globigerina-ooze, 216
Globigerinidae, 215
Globogerinoides, 216
Globorotaliidae, 215
Glossofungites ichnofacies, 429, *430*
Glossopteris, 405
Glycogen, 337
Glyptodont, 406
Glyptodontoidea, 285
Glyptorthis, *110*
Gnathostomata, 29, 186, *193*, 235
Goat, 309
Gobiconodontidae, *274*
Golden mole, 287
Golden Lane, Mexico, 226
Goldschmidt, R., 347
Gomphonema, *218*
Gomphotheriidae, 304
Goniatitic suture, 229
Goniatitida, 123, *126–127*, 232, 233
Goniograptus, *160*
Gorgonopsia, 266–268
Gorilla, 289, 291

Gorsline, D., 388
Gould, S., 328, 330, 340, 344, 349, 370
Gracilosuchus, 249
Granuloreticulosa, 78
Graphite, 48
Graptolithina, 57, 157, *158*-161
  alteration, 161
  Dendroidea, *11*, *159*-160
  Graptoloidea, 157, *159*, *160*-161
  polyphyletic, 161
  zones, 321
Grass, 183, 311
Grazers, 183, 392
Grazing traces, 427
Great Barrier Reef, 92
Greenhouse effect, 364, 365
Greenhouse phase, 412
Greenops, *135*
Greererpeton, *206*
Growth in ostracoderms, 188
Gryphaea, 225, *227*
  evolution, *342*, 349, 354, 355
Guard (belemnite), *234*
Gubbio, Italy, 370
Gulls, 263
Gunflint chert, *44*-45
Gymnolaemata, 103, 104
Gymnophonia (caecilians), 205, 207
Gymnosperm, 176, 179, 180
Gypsina, 215

Hadean:
  environment, 40, 41
  time, 39, 40
Hadrosauridae, 255
Haeckel, E., 355
Hagfish, 186, 187, 190–*192*
Half-life, *15*
Halimeda, *171*
Hallam, A., 347, 366
Hallopora, 104, *105*, *400*
Hallucigenia, *338*
Halysites, *97*, *400*
Halysitida, 97
Hammatostroma, *414*
Hantkenina, *214*
Haplohippus, 311
Haramiyidae, 274
Hard parts, 11, 12

Hardground community, 390–*391*
Hearing in vertebrates, 208
Hebertella, *380*
Hedgehogs, 287
Hegetotheroidea, 301, 302
Helicoplacoidea, 154
Heliolites, *97*
Heliolitida, 98
Heliopora, 96
Helminthoida, *428*, *430*
Helodontidae, *262*
Hemichordata, 157–161, 184
  biology, 157
  Graptolites, 158–161
Hemiphragma, *105*
Hemoglobin, and phylogeny, 30
Hennig, W., 24, 25, 26
Henodus, *262*
Heomys, *298*
Hercynian ocean, *407*
Hermatypic corals, 220, 221
Heroditus, 215
Hesperornis, 263, *264*
Hesperornithiformes, 263, *264*
Heteractinida, 82, *83*, 85, *87*
Heterocercal caudal fin, **199**, 202
Heteroconchia, 223, 227
Heterodont dentition, *223*
Heterodontosauridae, 253
Heteromorph ammonite, 231, *232*
Heterostraci, 187, 188, 189
Hexactinellida, 82, 84–*86*, *87*
Hexagonaria, 101
Hibbardella, *163*
Hildoceras, *231*
Hindiids, 85
Hipparion, 311
Hippomorpha, 310
Hippopotamidae, 307
Hippopotamoidea, 307
Hippos, 306
Hippurites, 226, *228*
Histogram, *33*
History of paleontology, 5–6
Hog, 307
Holarctic realm, 402, *403*
Holaspid, *138*
Holasteroidea, 235
Holectypoidea, 235
Holmes, R., 208
Holmesina, *286*

Holocephali, 187, 195, 196, *198*
Holochroal eye, 135
Holosteans, 200
Holotheca, 101
Holothuroidea, 145, 147
Holotype, 35
Holzmaden Shale, Germany, 233
Homagnostus, 139
Homalozoa, 154–157
Homeoplastic structure, **23**
Hominidae, 289, 291–294
Hominoidea, 289, 291
Homo, 289, 293
Homo erectus, 293, *294*
Homo habilis, 293
Homo sapiens, 292, *294*
Homologous structure, **23**, 24, 335
Homotrema, 215
Homotryblium, 219
Hoofed mammals, 300–313
Hopson, J., 337
Hormatoma, 129, *130*
Horn core, 310
Horner, J., 255
Horses, 309, 310
Horseshoe crab, 132, *133*
Huayangosaurus, 257
Human paleontology, 30, *see*
    Primates
Hunsrück Slate, 8, *9*
Huroniospora, 44
Huxley, T. H., 341
Huxley, J., 341
Hyaenas, 293
Hyaenidae, 296
Hyaenodon, 295, *296*
Hyaenodontidae, 295
Hyaenodontinae, 295
Hybodontoidea, 187, *196*
Hybodus, 197
Hydnoceras, 87
Hydrospire, *152,* 153, *154, 155*
Hydrozoa, 95, 96
Hylobatidae, 289, 291
Hylonomus, 210, 211, 245
Hyolithellus, 65, 67
Hyolithelminthes, 67
Hyolithes, 66, 116
Hyolithida, 65, *67*
Hyopsodus, 347–348
Hypermorphosis, **356**

Hypersaline environments, 385
Hyperstrophic gastropods, 127
Hypohippus, 311
Hypopetraliella, 102
Hypostome, *135,* 136
Hypsilophodon, 252
Hypsodontidae, 301
Hyracoidea, 284, 305
Hyracotherium, 311, *312*
Hyraxes, 305, 306
Hystricognathi, 297
Hystricognathous, **298**

Iapetus ocean, 406, *407*
Icaronycteris, 288
Ice Age distributions, 407, *408*
Icehouse phase, *365,* 412
Ichnofacies, 429–*430*
Ichnofossils, 10, 11, 60, 422–431
Ichnogenera, 427
Ichnospecies, 427
Ichthyornis, 263
Ichthyornithiformes, 263
Ichthyosaur, 328, *329*
Ichthyosauria, 246, 259–261
Ichthyostega, 204
Ichthyostegalia, 205, 207
Idioprionoidus, 163
Idmonea, 104
Igneous rocks, 4, 16
Iguanodontidae, 255
Illumination, limiting, 386–387
Imitoceras, 127
Impact hypothesis, 369–370
Impedance matching middle ear,
    246
Inarticulata, 108, 109
Incurrent pore, 80, *81*
Incus, 269, *270,* 271
Infauna, **389,** 390
Information loss, 376
Infrabasal plate, 148, *149*
Iniopterygiformes, 195–*197*
Innes L., TX, 429
Inoceramus, 225, *227*
Insecta, 132
Insectivora, 282, 284, 287
Interambulacral area, **146,** *236,*
    237
Interatherium, 303
Interbrachial plate, 148, *149*

Intercentrum, 207
Intraspecific combat, 255
Invertebrates, **55,** 56
Iridium anomaly, 370, 371
Irish Elk, 331, *332*
Iron formation, 50, *51,* 52
Iron ores, 4
Iron-rich soil, 50
Ischadites, 171, *172*
Ischyromyoidea, 297
Isocrinoid, *147,* 148
Isofilibranchia, 223
Isometric growth, 330, *331*
Isotope, **15**
    carbon, 49, 385
    corals, 221
    dating, 14–16
    fractionation, 382–385
    oxygen, 382–385
    stable, 49
Isua Group, 41, 49, 50

Jablonski, D., 366
Jaccard coefficient, 322
Jasper, 50
Jaw muscles:
    Adductor mandibulae externus,
        209
    Deep masseter, 269
    Masseter, 269
    Superficial masseter, 269
    rodents, *299*
Jaws, origin in vertebrates, 193
Jefferies, R., 156, 157
Jellyfish, 11, 60
Jenkins, F., 274
Johnson, M., 324
Juvenile faunas, 395

Kakabekia, 44, 45, 46
Kangaroo, 281
Kannemeyeria, 268
Kauffman, E., 60, 366, 390
Kayentachelys, 244
Kellogg, D., 344
Kennalestes, 282, *283*
Kenyapithecus, 291
Keratosa, 82
Kerogen, 48, 49
Kielan-Jaworowska, Z., 274
Kingdom, *54*

Kingfishers, 264
Kirk, N., 161
Kiwi, 265
Klonk, Czechoslovakia, 325
Koala, 280, 281
Koch, C., 379
Komodo dragon, 248
Kosmoceras evolution, 342, *343, 344*
Krause, D., 274
Kronosaurus, 260
Kuehneotheriidae, 274
Kuehneotherium, 272
Kyptoceras, 309

LaBarbera, M., 72
Labechia, 88, 89
Labechiida, 88, 367
Labyrinthine enfolding, 207
Labyrinthodontia, 205, 207
Lacertilia, 245, 246
Lagena, 77
Lagenina, 78
Lagomorpha, 284, 297
Lagosuchidae, 249, *252,* 253
Lagosuchus, 252
Laguna Madre, TX, 378
Lambeophyllum, 101
Lambeosaurus, 256
Lamellibranchia, 117 *see* Bivalvia
Lamels, *416*
Lamina, 88
Laminar microstructure, 415, *416*
Lamprey, 185–187, 190–*193*
Land bridges, 405
Laoporus, 12
Lapis stellaris, 5, *6*
Lappets, 230, *231*
Lapworthella, 64–*66*
Larva:
    adaptation, 360
    dispersal, 322, *323*
Larval stage, amphibians, 209
Lateral line canals, 189
Lateral teeth, *223*
Lateral temporal openings, 265
Latimeria, 201, *202,* 204, 205, 367
Lawn mower ecology, 275
Lawrence, D., 376
Lead, 15
Lecanospira, 127, *130*
LeGrand, Iowa, 150

Lemuriformes, 289
Lemurs, 289, 290
Lenticulina, 214
Leperditiids, 140
Lepidocyclus, 113
Lepidocystis, 153
Lepidocystoidea, 154
Lepidodendron, 174, 176, 178, 181
Lepidolina, 343, *345*
Lepidophloios, 174
Lepidosauria, 29, 245
Lepidosauromorpha, 245, 246
Lepidosiren, 204
Lepidotes, 201
Lepisosteiformes, 199, 200
Lepospondyli, 205, 207, 208
Leptagonia, 393
Leptoceratops, 258
Leptolepids, 200
Leucetia, 82
Lewontin, R., 328
Lichen, 172
Lichenaria, 96, *97*
Lichiida, 138, *139,* 140
Life:
    assemblage, **377**, 379
    definition, 6
    origin, 42, 51
Light, limiting, 386–387
Lima, 225
Limestone, 410
    fossil fragment, 91
    reef, 92
    reservoirs, 419
Limiting factors, 381–390
Limnoscelis, 206
Limnosceloids, 210
Limpet, 12
Limulus, 133
Line transect sampling, 381
Lingula, 367
Lingula assemblage, *324, 325*
Lingulata, 108, 109, 415
Lingulida, 110, 109
Linnaeus, 21
Linné, Carl von, 21, 22
Linsley, R., 129
Lion, 293
Liopleurodon, 262
Liquidambar, 182
Lissamphibia, 207, 208

Lithistida, 82, *83,* 85,
Lithofacies, **319**, 321
Lithophaga, 227, *228*
Lithophyllum, 170
Lithostrotion, 99, 101
Litopterna, 284, 301
Living fossil, 26, 110, 175, 367
Lizards, 211, 245
Lobe (suture), 229
Lobe-finned fish, 198, 201
Lobster, 133
Loligo, 120
Lonsdaleia, 99, 101
Loon, 263, 264
Lophophorate, 59, 71, 101–115
Lophophore, 56
    brachiopod, 107, *108,* 113
    bryozoan, 102, *103*
    graptolite, 157, *158,* 161
Lophospira, 130
Lorenz, D., 367
Lorenzinia, 430
Loricata, 285
Lorisids, 290
Loxonema, 131
Loxoplocus, 393
Lucy, 291
Lung, 199
Lungfish, 201–204, 375
Lycaenops, 267, *268*
Lycophyta, 173–175
Lycopod, 209
Lysocline, 387–388
    Archean, 410, 411
Lystrosaurus, 267
Lytoceratida, 232, 233

Machaeroidinae, 295
Maclurites, 127, *130, 398*
Macluritina, 128–*130*
Macraucheniidae, 301, *302*
Macroevolution, **352**-354, 363
Macroscelidae, 287, 289
Madreporite, *146, 236,* 237, *238*
Magellania, 108
Maglio, V., 304
Magnetic reversals, 18, *19*
Magnetism, remanent, 19
Magnetopolarity unit, **326**
Malacostraca, 132
Malleus, 269, *270,* 271

444 INDEX

Mammalia, 27, 29, 209, 265, 270,
    277–313
  diversification, 349
  evolution, 346–347
  origin, 353, 354
Mammal-like reptiles, 265–271, 278
Mammary glands, 271
Mammoth, preservation, 7
Mammutidae, 304
Manatee, 305
Mancini, E., 394
Manis, 286
Manitounuk I., *46*
Mantle overturn, 365
Mantle, brachiopod, 107, *108*
Manton, S., 132
Marlin, 259
Marmot, *298*
Marrella, 143
Mars, 39, 42
Marsupial mammals, 265, 274,
    277, 278, 406
Marsupialia, 279
Mastondontosaurus, 206
Mature community, **395**
Mayr, E., 23
McGhee, G., 167
McKenna, M., 278, 285
McKerrow, W., 397
McNamara, K., 359
Medullosa, 175
Medusa, *95*
Medusoid, 61, 62, 68, 144
Megaceros, 310
Megachiroptera, 287
Megahippus, 311
Megaloceros, 332
Megalodon, 122
Megalodont bivalve, 225
Megalonychioidea, 285
Megalosauridae, 253
Megalospheric test, 34, 77, *78*
Megaspore, 176
Megatherioidea, 285
Melanism, industrial, 351
Mellita, 239
Melonechinus, 236, 240
Meniscotheriidae, 301
Meraspid, *138*
Mercenaria, 117, 223
Meryceros, 310

Merychippus, 311, 312
Merycoidodontidae, 307
Merycoidodontoidea, 307
Mesaxonic, **310**
Mesentery, 95, *96*
Mesoderm, *55-58*, 69, *70*
Mesogastropoda, 128, 129, *131*,
    228
Mesohippus, 311, 312
Mesonychia, 284
Mesonychidae, 301, 313
Mesosauria, 259, 260
Mesosaurus, 260, 405
Mesosuchia, 250
Mesozoic mammals, 270–275
Mesozoic, name, 17
Mesozooecium, **104**
Messel fauna, 286
Metabolic rate, 266, 268, 271, 273
Metabolism:
  bird, *337*
  dinosaur, 336–338
  mammal, *337*
  unicellular, 43
Metacheiromys, 286
Metacone, *272*, 278
Metamerism, 56, 58, 70, *71*, 185
Metamorphic rocks, 4, 16
Metatheria (marsupials), 270,
    279–282
Metazoa, 52
  early, 60–68
Meteorite impact hypothesis, 370,
    372
Methanogens, 43, 45, 49, 51, 54
Methylacetylene, 42
Methylamine, 42
Miacidae, 295
Miacoidea, 295
Mice, 297
Michelinia, 393
Micraster, 238, 239
Micrite, 411
Micrite envelope, 418
Micritization, 418
Microbiolite, 46
Microbiotheriidae, 280
Microchiroptera, 287
Microevolution, 342–349, 352
  rate, 347, 349
Micromorph fauna, 394–395, 419

Microsauria, 205, 207
Microspheric test, 34, 77, *78*
Microspore, 176
Microstructure of fossils, 412–417
Middle ear, 269, 271
Migration and correlation, 322,
    323
Miliolina, 77, 78, 389
Millepora, 95, 389
Milleratia, 141
Mimia, 200
Mimicry, 351
Mimolagus, 298
Mineralogy of fossils, 411–417
Miocene Epoch, name, 17
Miocidaris, 240
Miohippus, 311, 312
Mistassinia, 40
Mistiaen, B. 90
Mixopterus, 142
Mixosaurus, 261
Moas, 264, 265
Modern Synthesis, 341
Modiolus, 224, 225
Moeritherioidea, 304
Moeritherium, 304, 305
Molds and casts, 9, *10*
Molecular clock, 30
Moles, 287
Mollusca, 56, 116–130
  biology, 116–119
  mineralogy, 415, *416*
Monaxon, *83*
Monera, 42, 54
Mongoose, 296
Monkey, 290
Monoclonius, 258
Monograptid fauna, 161
Monograptus, 160, 161
Monophyletic groups, **23**, 26
Monophyletic (plants), 183
Monoplacophora, 68. 116, 117,
    124, 127, *128, 129*
Monorhini, 187, 189
Monotremata, 270
Monotremes, 265, 271, 273, 282
Montastrea, 222, 425
Monterey Formation, 217
Moon, 40, *41*
Moore, L., 46
Moores, E., 73, 408

Morganucodon, 272, *273*, 274
Morosovella, *214*
Mosasaur, 375
Mosasauridae, 248, 249
Mother-of-pearl, 416
Mount Tom, MA, 429
Mountain beaver, 297, 298
Mountain goat, 309
Mountain sheep, 309
Mourlonia, *130*
Mouth, embryonic, 58
Murchison, R., 325
Multimembrate assemblage, 163
Multituberculata, 265, 270, 272, 280
Murchisoniina, 128–*130*
Murex, 228, *229*
Muroidea, 297, 299, 300
Muscles:
    invertebrate, 70, *71*
    reptilian, 337
Musk ox, 309
Mustelids, 296, 297
Mutation, chromosomal, 347
Mya, 223, *224*
Mylagaulidae, 298, *299*
Mylodontoidea, 285
Myomorpha, 297
Mysticeti, 313
Mytilus, 223, *224*, 225
Mytilus, salinity control, 385
Myxiniformes, 187

Nacreous microstructure, 416
Nama Group, 62
Names:
    fossils, 21, 22
    ichnofossils, 427
Nance, D., 365
Nannippus, 311
Nannofossil zone, 318
Nasohypophyseal opening, 191
Natural selection, 329, 350–352, 368, 370, 372
Nautiloidea, 119, 123, 124. *125*
Nautilus, 119, *120*, *122*, 123, 231, 334, *335*-336
Neanderthals, 5, 293, 294
Nearshore/offshore communities, *367*
Necrolemur, *291*

Nectridea, 205, 207
Nemakit-Daldym Fm., 65
Nematocyst, 94
Nematode, 57, 69
Neoceratodus, 204
Neodarwinism, 341
Neodymium, 15
Neoflabellina, *214*
Neogastropoda, 128, 228, *229*
Neogene System, name, **17**
Neogloboquadrina, *382*–*383*
Neohipparion, 311
Neomorphism, 418
Neopilina, 367
Neopterygii, 200
Neoselachians, 196
Neostromatoporoid, 89
Neoteny, **356**
Neotropical realm, 402, *403*
Neritopsina, 128
Neural arch, 207
Neutralist position, 372
New World monkeys, 290, 299
Niche, **376**, 381
Nimravidae, 296
Nobel, A., 217
Nodosinella, *79*
North Pole, Australia, 44
Notharctus, *291*
Nothosauria, 245, 246, 260, 262
Notochord, 155, 184
Notoprongonia, 301, 302
Notosaria, *359*
Notoungulata, 284, 301, 302
Novacek, M., 282, 285
Nuculites, 343, *345*
Nuculoidea, *121*
Numees tillite, 62
Numerical taxonomy, 23, 24
Nummulites, *215*, 387
Nummulitic Epoch, 215
Nyctitheriidae, 287
Nyctosaurus, 259

Obturator process, 252
Ocean, history, 4
Oceanic Formation, 220
Octocoralla, 96
Octopus, 119
Ocular plate, *236*, 237, *238*
Odontella, *218*

Odontoceti, 313
Odontopleura, *139*
Odontopleurida, 138, *139*, 140
Oil glands, 271
Oil shale, 420
Oil well zonation, 320
Old World:
    monkeys, 291
    realm, 403, *404*
Oldest rocks, 40, 41
Olenellus, *134*, *139*, 140
Olenoides, 136, *137*
Oligomerism, 58, 71
Oliver, W., 403
Omomyidae, 290, *291*
Ontogeny, 354–360
    coral, *100*
    trilobite, 137–*138*
Onychophora, 132
Opal, 412
Operculum bryozoan, 240
Operculum, gastropod, *118*, 119
Ophiacodon, 267
Ophiacodontidae, 265, 266
Ophiocystoidea, 154
Ophiomorpha, 423, *424*, 430
Ophiomorphus, 248
Ophiuroidea, 145, 147
Opisthobranchia, 119, 128
Opossum, 279, 406
Oppel, A., 318
Orangutan, 291
Orbitolites, *215*
Oreodont, 307
Organic matter (dispersed), 48, 49
Organs, 55, 56
Oriental realm, 402, *403*
Orientation of fossils, *380*
Origin of Species, book, 18, 341
Ornithischia, 249, 251, 253, 254–257
Ornithomimidae, 253
Ornithopoda, 251, 252, 253
Ornithosuchia, 249
Orodus, *196*
Orohippus, 311, *312*
Orthida, 109, 110
Orthoceratoidea, 124
Orthograptus, *160*
Orthosphaeridium, 219
Osculum, 80, *81*, 86

Osmosis, 385
Ossicones, 310
Osteichthyes, 195, 198–201
Osteolepiformes, 205
Osteolepis, 202, 206
Osteostraci, 187, 189–*190*
Ostracoda, 132, 133, *140–141*
   replacement, *10,*
   stunted, 395
Ostracoderms, 187
Ostrich dinosaurs, 253
Ostriches, 265
Ostrom, J., 253
Otarioidea, 295
Otoites, 231
Otsuka coefficient, 322
Ovary, 180
Owls, 264
Oxyaenidae, 295
Oxygen:
   atmospheric, 42, 49–52, 73, 364
   isotope (temperature), 382–385
   isotope (salinity), 385
   in sea water, 387–388
Oyster, 117
   community, 376
Ozawa, T., 343
Ozone shield, 51

Pachycephalosauria, 251, 253, 255, *256*
Pachyphyllum, 415
Pachypleurosaurus, 262
Pachyrhachis, 248
Pachyrukhos, 303
Pacing, **308**
Paddle fish, 200
Padian, K., 257
Paedomorphocline, *359*
Paedomorphosis, 354–360, 395
Paired fins, origin of, 190
Pakicetus, 312
Palaeanodonta, 285, *286,* 287
Palaeodonta, 307
Palaeomerycidae, 308
Palaeomyrmedon, 286
Palaeonisciformes, 199, 200
Palaeoniscoidea, 199, *200,* 202, 203
Palaeoparadoxia, 305
Palaeotheriidae, 310, 311
Paleobiogeography, 93, 402–408

Paleocommunities, **390**
Paleodictyon, 430
Paleoecology, 5, 374–401
Paleofavosites, 97
Paleofelids, 296
Paleogene System, name, **17**
Paleogeography, 5
Paleoherpeton, 206
Paleoichnology, 422–431
Paleophonus, 142
Paleophyllum, 99, 101
Paleotaxodonta, 223
Paleozoic, history, 165–168
   name, 17
Paliguana, 248
Pallial:
   line, 117
   sinus, 223
Palynomorph zone, 318
Panama isthmus, 403
   interchange, 406
Panda's thumb, *330*
Panderodus, 164
Pangaea, 213, 278, 365, 405, *408*
Pantodonta, 283
Pantolestidae, 283, 285
Pantotheria, 265, 270, 272, 275
Paracone, *272,* 278
Paracrinoidea, 154
Paradigm method, 332
Parahippus, 311, 312
Paranyctoides, 287
Paraphyletic groups, **27**, 28
Parapuzosia, 230
Parasaurolophus, 256
Parasitism, 375, 392
Paratoceras, 309
Paraxonic, **310**
Parrots, 264
Parsons, T., 208
Parvancorina, 71
Passerines, 264
Pasteur Point, 50
Patellina, 128
Paterina, 66
Patterson, B., 275
Peccaries, 307
Pecopteris, 11
Pecora, 307–309
Pecten, 225, 227

Pectinacea, 225
Pectiniform element, 161, *163*
Pectinirhomb, *152*
Pedder, A., 167
Pedicellate teeth, **208**
Pedicle:
   brachiopod, *108*
   valve, 107, *108*
Pelagiella, 128
Pelecypoda, 117, *see* Bivalvia
Pelicans, 264
Pelleted sediment, 426
Peloidal limestone, 411, 426
Peloids, 411, 422
Pelycodus, 346
Pelycosauria, 265–267
Pen (squid), *120*
Penguin, 264
Penicillus, 171
Pentaceratops, 258
Pentacrinites, 150
Pentamerida, 109, 111, *112,* 415
Pentamerus, 112, 400
Pentamerus assemblage, *324, 325*
Pentremites, 153, 154
Perameloidea, 280, 281
Peramorphocline, *358*
Peramorphosis, 355–359
Perignathic girdle, *237*
Period, **17**
Peripatric speciation, **347**
Periptychidae, 301
Perischoechinoidea, 235
Perissodactyla, 284, 309–312
Permeability, 419
Permian System, founding, 325
Permineralization, **7,** 8
Peronidella, 85, 86
Petalodont, *196*
Petalonamae, 62
Peterson, C., 377, 379
Petrel, 263
Petrified wood, 7, *8*
Petrified forest, Ariz., 7, 178, *181*
Petrolacosaurus, 245, 246
Petroleum, 93, 420
Petromyzontiformes, 187
Pflug, H. D., 62
Phacopida, 138, *139,* 140
Phacops, 136
   eye evolution, 349

Phaenopora, 106
Phaeophyta, 169
Phalangerids, 281
Phalangeroidea, 280
Pharetronida, 82, 85, *87*
Pharyngeal pouch, 354
Pharynx, 186
Phascolarctoidea, 280
Phenacodontidae, 301
Phenacolophidae, 301, 304
Phenetics, 23
Phillipsastrea, 101
Phillipsia, 139
Pholas, 227, *228*
Pholidophoroids, 200
Pholidota, 284–286
Phoronida, 71, 109
Phoronis, 71
Phorusrhacidae, 264
Phorusrhacus, 264
Phosphate:
    Cambrian, 65
    replacement, 8, *10*, 418, 419
    skeletal, 412, 417
Photic zone, **386**
Photodissociation, 51
Photosystem-I, 42, 52
Photosystem-II, 42, 52
Phragmoceras, 126
Phycodes, 430
Phylactolaemata, 103, 104
Phyletic gradualism, **347**, *350*, 352
Phylloceras, 230
Phylloceratida, 232, 233
Phylloid algae, 171
Phylloporina, 106
Phylogenetic systematics, 23–25
Phylogeny, *28*
    adaptation, 329
    molecular, 29, *30*
Phylozone, **320**
Phytoplankton, 217
Phytosauria, 249
Pigeon, 264
Pig, 306, 307
Pilina, 128
Pillar, 88
Pilosa, 285
Pinna, 224
Pinnipedia, 297
Pinnularia, 218

Pinnule, *148, 150*
Pioneer community, **395**
Pistosaurus, 260, 262
Placental mammals, 265, 274, 277, 278
Placodermi, 29, 187, 194, 195
Placodontia, 246, 262
Plagiolophus, 306
Plagiomenidae, 288, 289
Planktonic forams, 213, 215–*216*
Planktonic microorganism, 216–220
Plantae, 54
Plantigrade posture, **295**
Plants, 169–183
    early Paleozoic, 173
    flowering, 179–183
    late Paleozoic, 174–178
    Mesozoic, 178–183
Planula larva, 69, *70*
Plasmopora, 97, 98
Plastron of turtle shell, 243
Plate tectonics, 278
Plateosauria, 251
Platyceras, 129, 130, 375
Platycrinites, 149
Platyhelminthes, *59, 70,* 117
Platypus, 271
Platyrrhini, 289, 290
Plectodina, 162
Plesiadapiformes, 289, 290
Plesiadapis, 290
Plesiomorphy, **24**
Plesiosauria, 245, 246, 260, 262
Plesiosauroidea, 245, 260
Plethodon subspecies, 32
Pleurocentra, 207
Pleurocystites, 151
Pleurodira, 243–245
Pleurosaurs, 247
Pleurotomariina, 128–*130*
Pliocene Epoch, name, 17
Pliohippus, 311, *312*
Pliosauroidea, 245, 260
Pneumatic foramina, 258
Point count sampling, 381
Pojeta, J., 119
Polacanthus, 257
Polarity:
    of characters, **24**, 25
    commonality, 25

outgroup comparison, 25
Pollen, 180
Polychaete annelid, 144
Polydolopoidea, 280, 281
Polydora, 430
Polygnathus, 163
Polyp, *95*
Polyphyletic groups, 26
    arthropods, 132
    graptolites, 161
Polyplacophora, 116
Polypteriformes, 199, 200
Polypterus, 200
Polyzoa, 101
Pongidae, 289
Pongo, 291
Porcupine, *298*
Pore-rhomb, *152*
Porifera, 56, 69
    biology, 80–83
    mineralogy, 413
    Paleozoic, 83–91
    skeleton, 81–*83*
Porites, 377
Porolepiformes, 205
Porosity, 419–420
    growth frame, 420
    interparticle, *419*
    intraparticle, *419*
    moldic, 420
    shelter, 419
    solution, 420
    vuggy, *420*
Portugese man-of-war, 94
Post-displacement , **356**
Potamotherium, 297
Potassium-40, 15
Potsdam sandstone, *424*
Pound quartzite, 60
Prasopora, 105
Pre-displacement, **356**
Pre-Paleozoic life, 39–53
Precambrian life, 39–53
Predation, 351–352, *375*
Predator, 392
Predentary bone, 254
Prepubic process, 252
Preservation, 6–11, 209
    potential, 12
    trace fossils, 426
Priapulid annelid, 144

Primary producer, 374, 375
Primary rocks, 16
Primates, 282, 284, 289
    evolution, *346*
Primitive character, **24**
Prioniodus, 163
Prismatic microstructure, 415
Prismatophyllum, 99
Proboscidea, 284, 304, 305
Procellariiformes, 263
Procerberus, 278
Procheneosaurus, 256
Proconsul, 291
Procynosuchidae, 266
Procyonids, 296
Productida, 114
Proetida, 138, *139*, 140
Profusulinella, 80
Proganochelydia, 244, 246
Proganochelys, 244, 245
Progenesis, **356**
Progress in evolution, 340
Progymnospermophyta, 173, 175,
    180
Prokaryotes, 42, *43*, 52
Prolecanitida, 232, 233
Pronghorn antelope, 310
Prorichtofenia, 114
Prosauropoda (Plateosauria), 251,
    253
Prosimii, 289, 290
Prosobranchia, 119, 128
Prosoma, 141, *142*
Protaspid, *138*
Protective coloration, 351
Proteins, 29, 30
Protenaster, 358, 359
Proterosuchia, 249, 250
Proterotheriidae, 301, *302*
Proterozoic:
    divisions, 39, 40
    fossils, *45*-48
Proteutheria, 283
Protichnites, 424
Protista, 54, 75, 78
Protoceras, 306, 309
Protoceratidae, 308, *309*
Protoceratops, 258
Protocetidae, 313
Protochordate, 56, 57, 184
Protocone, *272, 275, 278*

Protoconodonts, 165
Protohertzina, 65
Protohippus, 311
Protopterus, 204
Protorosauria, 246, 249, *250*
Protostome, 56, 58, 59
Protosuchia, 249
Prototheria, 270
Protozoa, 55, 75, 78
Protrogomorpha, 297
Protungulatum, 301
Provinces faunal, 403, 404
Proviverinae, 295
Pseudocoelomate, 55, 57, 58, 69,
    70
Pseudocubus, 344, 345
Pseudohipparion, 311
Pseudometamerism, 58
Pseudonuclei, *46*
Pseudopods, *76*
Pseudopunctate microstructure,
    416
Psilophyton, 173, 174
Pteranodon, 259
Pteraspidae, 187, 189
Pteraspidomorphi, 187, 188
Pteraspis, 189
Pteria, 226
Pteriacea, 225
Pteridinium, 62
Pteridophyta, 173, 175
Pteridospermophyta, 173, 176, 177
Pteriomorphia, 223, 227
Pterobranchia, 157, *158*
Pterocanium, 219
Pterodactyloidea, 246, 249
Pteropoda, 119
Pterosauria, 249, 257–259
Pterygoid muscle, 210
Pterygoideus, 269
Pterygotus, 141, 142
Ptychopariida, 138, *139*, 140
Ptycnodontida, 195
Ptyonius, 206
Pulchrilamina, 93, 94
Pulmonata, 119, 128, 130
Punctuated equilibrium, **349**-*350*,
    352
Purgatoire R. CO, *423*
Pygidium, **134**, *135*
Pygmy possums, 281

Pyramids, Eygpt, 215
Pyrite replacement, *8, 9*, 419
Pyrotheria, 284, 301, 302
Pyrrhophyta, 169

Quadrat sampling, 380
Quahog, 223
Quaternary rocks, 16
Quetzalcoatlus, 258, 259
Quinqueloculina, 77

Rabbits, 297, 282
Raccoons, 296
Radial plate, *148, 149*, 153, *154,
    155*
Radial water canal, *236, 237*
Radiation, adaptive, 363–*364*
Radioactivity, 14, 15
Radiolaria, 76, 219–220
Radiolites, 226, 228
Radula, ammonite, 231
Ramapithecus, 291
Ramiform element, 161, 163
Range, fossil, 319–321
Rangea, 62
Rates of evolution, 362–367
Ratfish, 195, 196
Rats, 297
Raup, D., 222,333, 334, 343, 369,
    372
Ray-finned fish, 198
Rays, 195
Realms, faunal, 402–409
Recapitulation, 355
Receptaculites, 171, 172
Recrystallization, 7
Recycling of life, 6, 7
Red Queen hypothesis, 328
Redbeds, 50, 52
Redlichiida, *134*, 138, *139*, 140
Reef, 91–94
    archaeocyathid, *84*
    builders, 381
    community, Paleozoic, 91–94
    definition, **92, 93**
    limestones, 420
    organism shapes, *88*, 94, 387
    patch, *92*
    succession, *396*-397
Reflected lamina of the angular,
    266, *268*, 269, *260*

Regression:
  and correlation, 324
  and evolution, 366
Relationship of organisms, 22–26
Relief, reversed, 426
Renalcis, 83, 171
Replacement, 7, 8
  in carbonates, 418
  phosphate, 8, 418
  pyrite, 419
Reproduction:
  amphibians, 209
  early mammals, 271, 273
  ichthyosaurs, 260
  placental mammals, 282
  reptiles, 209
Reptiles (paraphyletic), 27
Reptilia, 209, 277
Respiratory canal, 238
Resting trace, 427
Revolution in earth history, 317
Rhabdinopora, 11, 159, 160, 367
Rhabdopleura, 157, *158, 159*
Rhamphorhynchoidea, 246, 249,
  258
Rheas, 265
Rhenanida, 195
Rhenops, 9
Rhinocerotoidea, 310, 311
Rhinoceros, 7, 309
Rhipidistia, 122, 199, 200, 203,
  205, 207
  evolution, 353
Rhizocorallium, 424, 430
Rhodophyta, 169–171
Rhombifera, 145, 151
Rhopalonaria, 107
Rhynchonellida, 109, 112, *113,*
  114, 332, *416*
Rhynchosauria, 246, 249, *250*
Rhynia, 174
Rhynie, Scotland, 173
Rhyniophyta, 173
Richtofeniidae, 114
Ring canal, *146, 148,* 236, *237*
Riojasuchus, 251
Rodentia, 282, 284, 297–300
Romer, A., 205
Rose, K., 285
Rostricellula, 397, *398*
Rostroconchia, 116, 120, *129*

Rostrum, 233
Rotaliina, 78, 213–216, 413
Rotifer, 57, 69
Rubidium, 15
Rudist, 225–*227,* 387
Rudwick, M., 72
Ruga, *98*
Rugosa, 98–101
  biogeography, 403, *404*
  microstructure, *415*
  ontogeny, *100*
Ruminantia, 307–309
Runnegar, B., 73, 117, 119
Rusophycus, 427, *428*
Russell, D., 242

Saddle (suture), 229
Saetograptus, 158
Sagenocrinites, 149
Salamander, 32, 207–209
Salinity controls, 385–*386*
Salmon, 195
Salterella, 380
Samarium, 15
Sampling method, 380–381
Sand dollar, 239
Sandberg, P., 412
Sanitary canal, *238*
Sansabella, 141
Santa Monica basin, CA, *388*
Saratoga Springs, N.Y., 48
Sarcinulida, 98
Sarcodina, 78
Sarcomastigophora, 78
Sarcopterygii, 198, 199, 201–205
Sarkastodon, 295
Saurischia, 249, 251–254
Sauropoda, 251, 253
Sauropodomorpha, 251, 255
Sauropsida, 29
Sauropterygia, 245, 260, 262
Saurornithoididae, 253
Savrda, C., 388
Scallop, 117, 225, *226,* 330
Scaly pangolins, 286
Scandentia, 284
Scaphites, 231, *232*
Scaphopoda, 116
Scatter diagram, *33*
Schaeffer, B., 192
Schindewolf, O., 347

Schizochroal eye, 135, *136*
Schizocoel, 108
Schizocoelomate, *55,* 58, *59,*
  116–142
Schopf, J., 44, 50
Schopf, T., 347, 372
Sciuroidea, 297
Sciurognathi, **297**
Sciurognathous, 297,*298*
Sciuromorpha, 297
Scleractinia, 220–222, 387, *425*
  cement, *417*
  microstructure, *415*
  origin, 360
  radiation, *364*
  symbiosis, 220, 221, 363
Scleromochlus, 257
Sclerospongea, 82, *414*
Scolecodont, 57
Scolicia, 427, *428*
Scolithos, 428, 429, *430*
Scorpionida, 132, 141
Scott, R., 390
Scoyenia ichnofacies, 429, *430*
Scrutton, C., 97
Scutellosaurus, 255, *257*
Scutellum, 139
Scyphozoa, 95, 96
Sea horse, 195
Sea lion, 296, 297
Sea otter, 283
Sea turtle, 244
Sea-fan, 61, 96
Sea-pen, 61, 96
Seal, 296, 297
Secondary palate, 267, *268*
Secondary rock, 16
Sedimentary rock, 4, 15, 410–421
Seed-fern, *175,* 176, 178
Seed, origin, 176, 178
Segmentation, 58
Segmented invertebrate, 184
Seilacher, A., 329, 427
Selection:
  intraspecific, 351
  natural, 329, 341, 350–352,
    354
  sexual, 329, 341
  species, 352
Selenodont, **307,** *308*
Semicircular canals, 188, 189

Semiorbis, 218
Sensory input, 273
Sepkoski, J. J., 68, 91, 165, 366, 369
Septal insertion, 220
Septal neck, *122*
Septum:
  ammonite, 229
  archaeocyathid, *83*
  cephalopod, 119, *120, 122,*
    125–126
  coral, *96, 98*
  rugosan, 100
Serial sections, *100*
Series, definition, 11
Serpentes. 245, 246
Sewellel, 298
Sexual dimorphism, 267
Sexual reproduction, 52
Shannon-Weaver index, 394
Shark Bay, Australia, 46, *47*
Shark, *193,* 195
  convergence, *329*
Sheehan, P., 91
Sheep, 306, 307, 309
Shonisaurus, 261
Shrew, 287
Sicula, *158,* 160
Side-necked turtle, 244
Siderastrea, 415
Sigillaria, 174, *176, 177,* 209
Signor, P., 166
Silicification, *8*
Similarity coefficient, 322
Simons, E., 291
Simpson, G., 23, 347
Simpson coefficient, 322
Simpson index, 394
Sinistral gastropod, 118
Sinohippus, 311
Sinuites, 130
Siphon:
  bivalve, *117,* 222, 223
  gastropod, *118,* 127–129
Siphonodictyon, 425
Siphuncle, *120, 122–125*
  ammonite, 229, 231
  belemnite, *234*
  function, *335,* 336
Sirenia, 284, 305
Sister group, 24
Sivapithecus, 291

Sivatherium, 310
Size increase trend, 354
Size distribution, 394–395
Skates, 195
Skeletal tissue:
  carbonate, 73,75, 84
  metazoan, 73
  mineralogy, 412–417
  phosphatic, 73
Skolithos ichnofacies, 429, *430*
Skunks, 296
Sloan, R., 371
Sloth, 285, 406
Smilodon, 296
Smith, W., 17, 317
Snakes, 211, 245, 248, 249, 277
Sneath, P., 23, 24
Soft-shelled turtles, 244
Sohl, N., 379
Sokal, R., 23, 24
Solenodontidae, 287
Solenopora, *170*
Solenoporacae, 170
Somasteroidea, 154
Sonar, 313
Songbirds, 264
Sorauf, J., 167
Soricidae, 287
Soricoidea, 287
Soricomorpha, 287
South American anteaters, 285
South American ungulates,
  301–304
Spatangoidea, 235
Speciation, **347**
Species:
  biogeographic, 31, 32
  concept, 341
  definition, 31–35
  Linnean, 32–35
  name, 21
  new, 22
  paleontological, 35
  selection, 352
  variation, 33–35
Sphaeractinoid, 89
Sphenacodontia, 265
Sphenacodontidae, 266
Sphenodon 210, 211, 244–247
Sphenodontida, 245–247
Sphenophyta, 173, 175

Sphenosuchia, 250
Spherulitic microstructure, 413,
  414
Sphinctozoa, 82, 85, 86, *87*
Spicule, *81–83,* 88
Spider, 132, *133*
Spine, echinoid, *235, 236, 238*
Spinocyrtia, 113
Spiracle, 153, *154*
Spiriferida, 109, *113*
Spiroraphe, 430
Spondylium, *112*
Sponge, 56, 68, *see also* Porifera
  biology, 80–83,
  boring, *425*
Spongocoel, 80, *81, 83*
Spore (plant), 173, 176
Sporophyte, 173, 176
Spotila, J. R., 337
Spreite, *424, 425,* 428
Spriggina, 61
Squalomorpha, 187, 195
Squamata, 245, 247–249
Squid, 119
Squirrel, 297, 298
Stabilization stage, *396*
Stabilizing shells, 123–125
Staff, G., 379
Stage, *317,* 318, 325
Standard deviation, 32
Stanley, G., 221, 387
Stanley, S., 352, 369
Stapes, 208, 246
Staurikosauria, 251, 253
Staurikosauridae, 253
Staurikosaurus, 250, 252
Stegodontidae, 304
Stegosauria, 251, 253, 255
Stegosaurus, 255
Steidtmann, J., 60
Stellar's sea cow, 305
Stellispongia, 85, 87
Stenacanthus, 197
Stenolaemata, 103, 104
Stephanozyga, 131
Stephen Formation, 142
Stictopora, 106, 393, 398, *398*
Stigmaria, 174
Stigmatella, 105
Stolon (graptolite), 157, *158*
Stomatopora, 104

Stone canal, *146, 237*
Stratigraphic Code, 326
Stratigraphic Guide, 326
Stratotype, 324–325
Strebulus, 77
Streptelasma, 100, 101
Streptorhynchus, 393
Streptostyly, **248**
Stricklandinia-assemblage, *324*
Stromatolite, 40, 46–*48,* 51, 52, 93,
    366, 412
Stromatopora, 89
Stromatoporella, 89
Stromatoporellida, 88
Stromatoporida, 88
Stromatoporoidea, 82, 86, 88–90,
    93, *94,* 332–333, *414*
    mineralogy, 88
Strombus, 416, 418
Strongylocentrotus, 145, 236, 237
Strontium, 15, 16
Strophodonta, 111
Strophomena, 393, 397, 398
Strophomenida, 109, *111, 112,*
    114, 416
Stunted faunas, 395
Sturgeon, 200
Stylar shelf, 278
Styliolina, 67, 68
Stylohipparion, 311
Styracosaurus, 258
Subcephalic muscle, 203
Subspecies, 31
Subulites, 131
Succession, 395–397
    level bottom, 397, *398*
    reefs, *396*
Suidae, 307
Suina, 307
Suiodea, 307
Sunnaginia, 66
Supraorbital bone, 254
Survival of the fittest, 327, 351
Survivorship, **347,** 349, 362, 368
Suspension feeder, 73, *389,* 392
Suture:
    ammonite, 229, *230*
    ammonoid, 336
    goniatite, *126–127*
Sweat glands, 271
Swifts, 264

Swimming invertebrate, 71
Swordfish, 195, 259, 329
Symbiosis:
    algal, 386
    bivalve, 226
    zooxanthellae, 220, 221, 386
Symmetrodonta, 265, 270, 274
Symplasma, 82
Synapomorphy, 21, 28
Synapsida, 211 265–271
Syndactyly, **281**
Syndyoceras, 309
Synplesiomorphy, 24, 28
Syntaxial rim, 417
Synthetoceras, 309
Syringopora, 97
System definition, **17,** 319
Systematics:
    evolutionary, 23, 24
    phylogenetic, 23, 24–29

Tabula, 95, 96, 98
Tabulata, 96, *97,* 98
Taeniodontia, 283
Talonid basin, *272*
Talonid, *272,* 275
Talpidae, 287
Tanystropheus, 250
Taphonomy, 6, 7, 376–381
Taphrhelminthopsis, 430
Tapiroidea, 310, 311
Tapir, 309
Tarsiiformes, 289
Tarsoidea, 290
Tasmanian devil, 279, 281
Tasmanian wolf, 279, 281
Taxodont dentition, 223
Taxon, **22**
Taxonomy, 22, 21–22, 31–35
    numerical 23, 26
Tayassuidae, 307
Tegmen, 148
Tegulorhynchia, 359
Teichichnus, 430
Teleostei, 199, 200, *201,* 277
Teleostomi, 29
Telson, *131, 133, 142*
Temnospondyli, 205, 207, 208
Temperature, limiting factor,
    381–385
Tempo and Mode in Evolution, 347

Temporal openings, 2l0, *211*
Temporalis, 269
Tenrec, 287
Tenrecoidea, 287
Tentacullites, 67, 68, 116
Terebratulida, 109, *113,* 114
Teredo, 227
Termite, 285
Terrestrisuchus, 252
Tertiary rocks, 16
Testudinata, 243–245
Testudinoidea, 244
Tethyan faunal realms, 403, *408*
Tethys ocean, 278, 403, *405, 407*
Tetraclaenodon, 306
Tetracoral, 98–101
Tetradium, 97, 98
Tetragraptus, 160
Tetrapoda, 29
Tetraxon, *83*
Teuthida, 232, 233
Textularia, 77
Textulariina, 77–79, 213, *214*
Thalassinoides, 426, 427, 429, *430*
Thecideacea, 114
Thecodontia, 246, 249
Thelodontida, 187, 191, *192*
Theocampe, 219
Theologian, natural, 328
Theosodon, 302
Therapsida, 265, 266
Theria, 270
Therian mammals, 274
Theropoda, 251, 253
Thomson, K., 192
Thorium, 15
Thrinaxodon, 268, 269
Thylacinidae, 281
Thylacoleo, 281
Thylacoleonidae, 281
Thylacosmilidae, 281
Thylacosmilus, 296
Ticinella, 214
Tiering, in communities, **392**
Tiers, in evolution, 370
Tiger, 293
Tillodontia, 283
Time scale:
    relative, 16–18
    standard, 19–20
Time averaging, 377–379

Time measurement, 14–20
Tissue rejection, 282
Tissues, 55, 56
Tommotian Stage, 64–67
Tommotid, 65, *66*
Tooth:
    formula, **278**
    occlusion, 268, 270, 272
    replacement, 268, 277, 278
Toothed whale, 313
Torosaurus, 258
Torsion, gastropod, 119
Tortoises, 244
Toxodontia, 301, 302
Trabecular microstructure, 413, *415*
Trace fossils, 422–431
Tracheophyta, 169
Tracks, 9, 10, 12, 60, **422**
Trackway, 12, **422,** *423, 424*
    dinosaur, *423,* 429
Tragulidae, 308
Traguloidea, 307
Trail, productid, *112,* 114
Trail, **423-***424*
Transgression:
    and correlation, 324
    and evolution, 366
Transition rocks, 17
Translation rate, *333*
Transported assemblages, 379–*380*
Transverse flange of pterygoid, 210
Tree shrews, 287, *288*
Trematosaurs, 259
Trepostomata, *102, 103,* 104, *105, 416*
Triadobatrachus, 209
Triarthrus, 139
Tribosphenic tooth, *272,* 275
Tribrachidium, 61, 63, 72
Triceratops, 258
Triconodonta, 265, 270
Tridacna, 226, 387
Trigonia, 223, *224*
Trilete spores, 173
Trilobita, 65, 68, 132, *133–140*
    appendages, 136, *137*
    classification, 138–139
    history, 139–140
    onotgeny, *137–138,* 360
    replacement, *9*

zones, 321
Trilobitoidea, 133
Trilobitomorpha, 132, 138
Trimerella, 111
Trimerellida, 109, 111
Trimerophytophyta, 173
Trionychoidea, 244
Triops, 367
Trithelodontidae, 266, 270
Triticites, 80
Trituberculata, 270
Tritylodontidae, 266, 269–271
Trivial name, 21
Trochiina, 128, *130*
Trophic role, 374, 375
Trophic structure, 391–393
Trophoblast, 282
Trout, 195
Trueman, A., 342
Trypanites ichnofacies, 429, *430*
Trypostega, 241
Tube foot, *146, 238,* 239
Tubipora, 96
Tubulidentata, 284
Tubuliporata, *104,* 240
Tuna, 259
Turbulence, limiting, 389–390
Turgai Strait, 278
Turkana L. molluscs, 350
Turrilites, 232
Turtle, 211, 243–245
Tyler, S. A., 45
Tylopoda, 307
Tympanic ring, 269, *270*
Tympanum, 246
Type section, 17, 324
Type specimen, 35
Typotheroidea, 301, 302
Tyrannosauridae, 251
Tyrannosaurus, 251, 253, *254*

Ubaghs, G., 156
Uintathere, 283
Ulrichia, 141
Ultradextral gastropod, 127
Umbo, bivalve, *117,* 223
Umbo, brachiopod, *108*
Ungulates, 282, 300–313, 353
Unicellular life, 42–52
Uniformitarianism, 375
Unimembrate assemblage, 163

Unionoida, 224
Uniramia, 132
Uranium, 15, 16
    conglomerate, 50
    ores, 4
Urochordata, 157, 184
Urodela (salamanders), 205, 207
Uscia, 241
Utatsusaurus, 261

Vaceletia, 86
Vail curve, *366*
Valentine, J., 73, 166, 404, 408
Van Valen's law, 368, 372
Van Valen, L., 328, 368, 371
Varangian glaciation, 65
Varanus, 248
Vascular tissue, 172–173
Vascular plants, 169, 172–174
Vendian System, 62
Venus, 39, 42
Vermeij, G., 222, 228
Vermilingua, 285
Vertebrae, 207
Vertebrata, 29
Vertebrates, origin of, 186
Vicariance biogeography, 408
Vieraella, 209
Vinella, 107
Virgella, *158,* 160
Vis plastica, 5
Viverravidae, 295
Viverrids, 296
Vombatoidea, 280
Von Baer, E., 59, 354, 355

Walchia, 181
Walcott, C., 142
Walker, K., 392, 395
Wallabies, 281
Wallace, A., 402
Walrus, 296, 297
Walter, M. 44
Warrawoona Group, 44, 47, 48
Water vascular system, *146*
Wavebase, 389
Weasel, 296
Wedges metaphore, 328
Westermann, G., 336
Whale bone, 313
Whale, 301

Whittington, H., 142
Whorl expansion, *333*
Williams, E., 208
Williamson, P., 349
Witwatersrand Mines, 50
Wombat, 281
Woodpecker, 264
Wynyardiidae, 281

Xanthos of Sardis, *5*
Xenacanthida, 195, 196

Xenarthra, 284, 285
Xiphosura, 132

Zalambdalestes, 282, *283*, 297
Zalambdodonta, 287
Zangerl, R., 195
Zeigler, A., 324, 397
Zoantharia, 96–101
Zone, **317**, 318
    assemblage, **319**
    concurrent range, **319**

    interval, **319**
    phylozone, **320**
    range, **319**
Zooecium, **101**, *103*
Zoogeography, 278–280
Zoological Code, 22
Zoophycus, *425*, *430*
Zooplankton, 216
Zooxanthellae, 221, 386–387
Zosterophyllophyta, 173, 175
Zugokosmoceras, *343*, *344*